The Cell Biology of Cyanobacteria

Edited by

Enrique Flores and Antonia Herrero

Instituto de Bioquímica Vegetal y Fotosíntesis
CSIC and Universidad de Sevilla
Seville
Spain

Caister Academic Press

Caister Academic Press
Norfolk, UK

www.caister.com

British Library Cataloguing-in-Publication Data
A catalogue record for this book is available from the British Library

ISBN: 978-1-908230-38-6 (hardback)
ISBN: 978-1-908230-92-8 (ebook)

Cover design adapted from various images. Background: scanning electron microscopy (SEM) micrograph of a microcolony of *Anabaena* sp. (courtesy of Mareike Rudolf and Mireia Burnat). Front, left: transmission electron microscopy (TEM) micrograph of a cell of *Synechocystis* sp. in the process of division (courtesy of Claudia Menzel and Iris Maldener). Front, right: TEM micrograph of part of a filament of *Anabaena* sp. (courtesy of Victoria Merino-Puerto and Iris Maldener).

Contents

Contributors

Eva-Mari Aro
Laboratory of Molecular Plant Biology
Department of Biochemistry
University of Turku
Turku
Finland

evaaro@utu.fi

Natalia Battchikova
Laboratory of Molecular Plant Biology
Department of Biochemistry
University of Turku
Turku
Finland

natbat@utu.fi

Devaki Bhaya
Department of Plant Biology
Carnegie Institution for Science
Stanford, CA
USA

dbhaya@stanford.edu

Bianca Brahamsha
Marine Biology Research Division
Scripps Institution of Oceanography
University of California,
San Diego, CA
USA

bbrahamsha@ucsd.edu

Donald A. Bryant
Department of Biochemistry and Molecular Biology
The Pennsylvania State University
University Park, PA
USA

dab14@psu.edu

Mireia Burnat
Instituto de Bioquímica Vegetal y Fotosíntesis
CSIC and Universidad de Sevilla
Seville
Spain

mireia.burnat@ibvf.csic.es

Adam N. Bussell
Department of Biology
Indiana University
Bloomington, IN
USA

bussella@indiana.edu

Jeffrey C. Cameron
Department of Plant and Microbial Biology
University of California
Berkeley, CA
USA

jcameron@lbl.gov

Corinne Cassier-Chauvat
Laboratoire de Biologie et Biotechnologie des Cyanobactéries
Gif sur Yvette Cedex
France

corinne.cassier-chauvat@cea.fr

Franck Chauvat
Laboratoire de Biologie et Biotechnologie des Cyanobactéries
Gif sur Yvette Cedex
France

franck.chauvat@cea.fr

Enrique Flores
Instituto de Bioquímica Vegetal y Fotosíntesis
CSIC and Universidad de Sevilla
Seville
Spain

eflores@ibvf.csic.es

Alexander Hahn
Institute of Molecular Biology
Goethe University
Frankfurt am Main
Germany

Hahn@bio.uni-frankfurt.de

Antonia Herrero
Instituto de Bioquímica Vegetal y Fotosíntesis
CSIC and Universidad de Sevilla
Seville
Spain

herrero@ibvf.csic.es

Judith Herzfeld
Department of Chemistry
Brandeis University
Waltham, MA
USA

herzfeld@brandeis.edu

David M. Kehoe
Department of Biology
Indiana University
Bloomington, IN
USA

dkehoe@indiana.edu

Cheryl A. Kerfeld
Department of Plant and Microbial Biology
University of California
Berkeley, CA
USA

ckerfeld@lbl.gov

Iris Maldener
Interfaculty Institute of Microbiology and Infection Medicine/Organismic Interactions
University of Tübingen
Tübingen
Germany

iris.maldener@mikrobio.uni-tuebingen.de

Vicente Mariscal
Instituto de Bioquímica Vegetal y Fotosíntesis
CSIC and Universidad de Sevilla
Seville
Spain

vicente.mariscal@ibvf.csic.es

Enrico Schleiff
Center of Membrane Proteomics
Cluster of Excellence 'Macromolecular Complexes' and the Institute of Molecular Biology
Goethe University
Frankfurt am Main
Germany

schleiff@bio.uni-frankfurt.de

Dirk Schneider
Johannes Gutenberg-University Mainz
Department of Pharmacy and Biochemistry
Johann-Joachim-Becher-Weg
Mainz
Germany

Dirk.Schneider@uni-mainz.de

Assaf Sukenik
Israel Oceanographic and Limnological Research
Yigal Allon Kinneret Limnological Laboratory
Migdal
Israel

assaf@ocean.org.il

Michael L. Summers
Department of Biology
California State University
Northridge, CA
USA

michael.l.summers@csun.edu

Markus Sutter
Department of Energy Joint Genome Institute
Walnut Creek, CA
USA

msutter@lbl.gov

Yvonne Zilliges
Humboldt Universität zu Berlin
Institute of Biology/Biochemistry
Chausseestrasse 117
Berlin
Germany

yvonne.zilliges@cms.hu-berlin.de

Current books of interest

Microarrays: Current Technology, Innovations and Applications	2014
Metagenomics of the Microbial Nitrogen Cycle: Theory, Methods and Applications	2014
Proteomics: Targeted Technology, Innovations and Applications	2014
Biofuels: From Microbes to Molecules	2014
Human Pathogenic Fungi: Molecular Biology and Pathogenic Mechanisms	2014
Applied RNAi: From Fundamental Research to Therapeutic Applications	2014
Halophiles: Genetics and Genomes	2014
Phage Therapy: Current Research and Applications	2014
Bioinformatics and Data Analysis in Microbiology	2014
Pathogenic *Escherichia coli*: Molecular and Cellular Microbiology	2014
Campylobacter Ecology and Evolution	2014
Burkholderia: From Genomes to Function	2014
Myxobacteria: Genomics, Cellular and Molecular Biology	2014
Next-generation Sequencing: Current Technologies and Applications	2014
Omics in Soil Science	2014
Applications of Molecular Microbiological Methods	2014
Mollicutes: Molecular Biology and Pathogenesis	2014
Genome Analysis: Current Procedures and Applications	2014
Bacterial Membranes: Structural and Molecular Biology	2014
Bacterial Toxins: Genetics, Cellular Biology and Practical Applications	2013
Cold-Adapted Microorganisms	2013
Fusarium: Genomics, Molecular and Cellular Biology	2013
Prions: Current Progress in Advanced Research	2013
RNA Editing: Current Research and Future Trends	2013
Real-Time PCR: Advanced Technologies and Applications	2013
Microbial Efflux Pumps: Current Research	2013
Cytomegaloviruses: From Molecular Pathogenesis to Intervention	2013
Oral Microbial Ecology: Current Research and New Perspectives	2013
Bionanotechnology: Biological Self-assembly and its Applications	2013
Real-Time PCR in Food Science: Current Technology and Applications	2013
Bacterial Gene Regulation and Transcriptional Networks	2013
Bioremediation of Mercury: Current Research and Industrial Applications	2013
Neurospora: Genomics and Molecular Biology	2013
Rhabdoviruses	2012

Full details at www.caister.com

Preface

Cyanobacteria form a phylogenetically coherent group of organisms whose appearance on Earth can be traced back to more than 2.4 billion years ago, when oxygen had accumulated at significant levels (albeit much lower than present day levels) in the atmosphere. This estimate of the age of cyanobacteria is based on the notions that oxygenic photosynthesis likely evolved in predecessors of cyanobacteria and that this biological process is the responsible for the presence of oxygen in the Earth's atmosphere. Notwithstanding this ancient origin of cyanobacteria, they are a sister group of other bacterial groups, implying a diversification of bacteria since life appeared on Earth, about 3.4 billion years ago. In global phylogenies of living organisms, cyanobacteria lie close to Gram-positive bacteria such as the actinobacteria. Nonetheless, cyanobacteria are didermic like Gram-negative bacteria, bearing an outer membrane outside of the cytoplasmic membrane and the murein (peptidoglycan) sacculus. Interestingly, the latter is thicker in cyanobacteria than in bacteria such as enterobacteria.

In most cyanobacteria, oxygenic photosynthesis is carried out in internal membranes that form a closed compartment within the cytoplasm, the thylakoid. Thus, the subcellular structure of these organisms is more sophisticated than that of most other bacteria, which lack intracellular membranes. Additionally, many cyanobacteria grow as chains of cells that form filaments (also known as trichomes), and some filamentous cyanobacteria present more than one type of cell with specialized functions, making the supracellular structure of the filament a topic of further interest. Collectively, these characteristics make the cell biology of cyanobacteria particularly appealing in a time when the discipline of bacterial cell biology has significantly developed.

In this book, Donald A. Bryant enlightens us with a personal proposal for looking at the history of cyanobacterial research (Chapter 1). We then have a contribution by Corinne Cassier-Chauvat and Franck Chauvat that covers a key aspect of cell biology, cell division (Chapter 2). The three membrane systems present in cyanobacteria are addressed, from different perspectives, in the three next contributions. Alexander Hahn and Enrico Schleiff extensively cover the cell envelope (including the plasma membrane, the peptidoglycan layer and the outer membrane; Chapter 3), Natalia Battchikova and Eva-Mari Aro cover the thylakoids as seen from proteomic studies (Chapter 4), and Dirk Schneider addresses the mechanisms of protein targeting, transport and translocation to and through biological membranes (Chapter 5). These three chapters together provide an updated discussion of the interesting and important, but difficult, topic of protein targeting to the three membrane systems of cyanobacteria.

Cyanobacteria are visible to the naked eye in natural sites thanks to their coloration, which is in part due to the presence of particular photosynthetic pigments, the phycobiliproteins. Notably, the complement of these pigments changes in some cyanobacteria in response to light quality, a phenomenon known as chromatic acclimation that is covered by Adam Bussell and David Kehoe (Chapter 6). Central to photosynthesis after light absorption and energy conservation is carbon dioxide fixation that in cyanobacteria takes place in specialized cellular structures called carboxysomes, which are covered in this book by Jeff Cameron, Markus Sutter and Cheryl Kerfeld (Chapter 7). When in excess, fixed carbon is accumulated in a reservoir, glycogen. Similarly, when in excess, nitrogen is accumulated in a reservoir called cyanophycin. These reservoirs constitute cellular deposits that are covered in this book by Yvonne Zilliges in Chapter 8 and by Antonia Herrero and Mireia Burnat in Chapter 9, respectively.

An important aspect of the biology of cyanobacteria is the ability of some of their representatives to move. There are cyanobacterial strains that show buoyancy, a property that is due to the presence of gas vesicles, which can be regulated producing vertical movement in water columns. Research on gas vesicles is reviewed by Judith Herzfeld in Chapter 10. There are also many cyanobacteria that exhibit motility on a surface. Motility, including phototactic responses, is covered in this book by Bianca Brahamsha and Devaki Bhaya (Chapter 11).

An outstanding property of some filamentous cyanobacteria is their capability to experience different developmental possibilities, which are covered in this book by Iris Maldener, Michael L. Summers and Assaf Sukenik (Chapter 12). Related to multicellularity, Vicente Mariscal describes in Chapter 13 recent advances in the knowledge of cell–cell joining proteins, which can be involved in intercellular communication, in filamentous cyanobacteria. Interestingly, the study of these two topics, cellular differentiation and cell–cell communication, has particularly benefited from the application of modern cell biology techniques.

We are deeply indebted to all the contributors to this book, who have done an excellent work and with whom it has been a pleasure to work. We hope that researchers in this field will find this book on *The Cell Biology of Cyanobacteria* to adequately complement the topics covered by the book *The Cyanobacteria: Molecular Biology, Genomics and Evolution* that we previously edited for this publisher, and we thank Annette Griffin from Horizon Press for actively demanding this new book. Editing a book requires considerable attention, and therefore we thank the members of our laboratory for their patience and interest.

Work in our laboratory is supported by Plan Nacional de Investigación, Desarrollo e Investigación from the Spanish government, co-financed by the European Regional Development Fund (grants BFU2010-17980 and BFU2011-22762), and by Plan Andaluz de Investigación, Regional Government of Andalucía (grants P08-CVI-03838 and P10-CVI-6665).

Enrique Flores and Antonia Herrero

A Brief History of Cyanobacterial Research: Past, Present, and Future Prospects

1

Donald A. Bryant

When Antonia and Enrique invited me to write an introductory chapter for their latest book summarizing recent research on cyanobacteria, I thought that it would be a simple task. However, as I thought about it more and more, and as is typical for me procrastinated longer and longer, I became ever more blocked as I looked backward and as time marched forward. Approaches to this chapter seemed either to require too much detail or too little, or so it seemed to me. One would think that after 40 years, one might actually have found something worthwhile to say. Yikes!! Forty years! It has now been 40 years since I started my long affair with cyanobacteria and research during my first rotation project as a graduate student at UCLA so many years ago in September 1972. Clearly, I believe that it has been a long but very exciting journey. No doubt about it, though: 40 years is a long time to do any one thing, and still love it! I finally decided to write just a few brief overview comments concerning my perceptions about this field of science and those, including myself, who have tried in recent years to summarize the collective progress in monographs. So, especially for you newcomers to this field, this will be my limited attempt to describe where this field started, where I started, where the research focus has been recently, and where it might be going for the next 10 years. If one should be so inclined, it also provides a pathway to obtain a historical perspective on the field, especially over the last 60 years.

I believe that the study of cyanobacteria can be divided into five 'eras,' at least one of which, thankfully, predates my entry into the field. With no disrespect intended to the many legions of scientists who have conducted studies of these wonderful organisms, I will refer to these five eras as 'The Dark Ages' (1800 to 1950), 'The Middle Ages' (1950 to 1979), 'The Renaissance' (1980–1995), 'The Age of Enlightenment' (1996 to present) and 'The Post-Modern Era' (i.e. the future). For me, these five eras are largely defined by the prevailing methodological approaches used to the study of microorganisms during those time periods and by how our perceptions of cyanobacteria, and even microbiology, have changed over time. *Schizophyceae* (i.e. then blue-green algae, now cyanobacteria) were first described in the early nineteenth century and were the first photosynthetic prokaryotes to be discovered. For the first 150 years or so during 'The Dark Ages', field observations, light microscopic observations, enrichment methods and physiological studies were employed, and many important properties of these organisms were discovered. Most of the important treatises concerning cyanobacteria during this time were attempts by various individuals (e.g. Ferdinand Cohn, Carl Agardh, Camille Sauvageau, Friedrich Kützing, Lothar Geitler, Pierre Bourrelly, etc.) to classify these morphologically variable and metabolically complex microorganisms on the basis of light microscopic and simple physiological observations. Although the classification of cyanobacteria as either algae (subject to the rules of the Botanical Code) or bacteria (subject to the rules of the Bacteriological Code) has been hotly disputed for more than 100 years, and still is by a few individuals, molecular taxonomic methods and whole genome sequences have effectively ended this debate. There is no longer any justifiable reason to deny that cyanobacteria are bona fide members of the domain Bacteria.

Cyanobacterial research began to develop as a specific discipline within microbiology during 'The Middle Ages', when the first pure cultures became available thanks to the efforts of Mary Belle Allen, Chase van Baalen, and others in the 1950s and 1960s (e.g. Allen, 1952; van Baalen, 1962). By the mid-1970s, the Pasteur Culture Collection had obtained more than 300 axenic cyanobacterial cultures, which were categorized into five sections (Rippka *et al.*, 1979). With the availability of a large and diverse collection of pure cultures, the prevailing approaches at the time could be broadly applied to these organisms for the first time. From about 1950 until the late 1970s, most studies concerning cyanobacteria relied on physiological and biochemical approaches. Importantly, electron microscopy revealed that these organisms lacked nuclei and had a cellular architecture like that of other bacteria. I began to study cyanobacteria during 'The Late Middle Ages' and, like many others during this period, I learned to apply biochemical approaches to isolate and characterize phycobilisomes and phycobiliproteins; electron microscopy to study the organization of these light-harvesting complexes (Bryant *et al.*, 1979); and other physiological and biochemical methods to study other photosynthetic processes, including complementary chromatic acclimation, which was called complementary chromatic adaptation at that time (see Cohen-Bazire and Bryant, 1982).

During 'The Late Middle Ages', I was a PhD student with Alexander N. Glazer and Frederick A. Eiserling at UCLA and subsequently a post-doctoral fellow with Roger Y. Stanier and Germaine Cohen-Bazire at the Institut Pasteur in Paris. While in Paris, I met Noel Carr, who was visiting while on sabbatical leave. Noel, from whom I learned so much, would become a dear friend and mentor within the world of cyanobacteria. His two books edited with Brian Whitton, *The Biology of the Blue-Green Algae* (1973) and *The Biology of the Cyanobacteria* (1982) summarized the state of knowledge of cyanobacterial structure, physiology, and metabolism through 'The Middle Ages'. As noted by Roger Stanier in the Foreword to the latter treatise, the change from blue-green algae to cyanobacteria was a notable break with tradition, because cyanobacteria had been classified as algae for some 180 years by that time, but it had become clear that these organisms were indeed bacteria – and as noted above, importantly lacked a nucleus and all the other attributes of eukaryotic cells.

Thanks to the solid biochemical and physiological framework produced during 'The Middle Ages', cyanobacterial research flourished when powerful molecular biological methods were finally applied to these organisms. Led by Bob Haselkorn and many others, *'The Renaissance'* began around 1980, and the application of molecular biological and genetic methods in cyanobacterial research continues to the present. I was lucky enough to start my independent academic career at Penn State in 1981, and I have now spent more than 30 years applying molecular biological and genetic methods to study a wide range of physiological and metabolic questions in the naturally transformable, marine cyanobacterium *Synechococcus* sp. PCC 7002. Molecular methods for taxonomy, including DNA–DNA hybridization and 16S rRNA sequence analysis, became commonplace. During this period, many details about the structural and functional features of the photosynthetic apparatus were established, and many details of cyanobacterial physiology, development, and metabolism were elucidated. Much was discovered about gene regulation, cellular differentiation and patterning processes important for the formation of heterocysts, hormogonia, and akintes. With strong encouragement from Noel Carr, I took on the challenge of editing the next logical volume in the series of books summarizing knowledge of cyanobacteria. *The Molecular Biology of Cyanobacteria*, Volume 1 in the series *Advances in Photosynthesis and Respiration* [Kluwer (now Springer), Dordrecht, The Netherlands] was published in 1994 (Bryant, 1994). I believe it is fair to say that this book served as a major resource for many aspects of cyanobacterial physiology and metabolism for a generation of students who have studied, or still study, these organisms. Amazingly (at least to me), Springer is still selling several copies of this book per year after nearly 20 years. For me, this was truly a labour of love, and although I have been asked many times to produce a second edition, it is something I have not had the time – or perhaps the energy and willpower – to do. Perhaps that is still something in my future, but as you already know if you are reading this, others have stepped forward and have accepted the challenge of helping to organize and deliver information to students and researchers in this field.

I now realize that there was a very important omission in the topics covered in that treatise published in 1994, although considering the size and content of that book, I do not see how more material could have been added to it. Until 2005, I was strictly a 'model organism' scientist – the perfect product of the pure-culture school of microbiology. My students and I worked with genetically tractable, axenic model organisms, and I did not really concern myself too much with the messy world of natural microbial communities and microbial ecology. Looking back, I now realize this was a major shortcoming in my research career. Fortunately, Brian Whitton and Malcolm Potts did not ignore these subjects, and they produced an excellent, highly complementary monograph to my own entitled *The Ecology of Cyanobacteria: Their Diversity in Time and Space*, in 2000 (Whitton and Potts, 2000). After spending a summer with David Ward at Montana State in 2005, I too have finally realized the advantages – no, the necessity – of studying cyanobacteria and other chlorophototrophs in the context of their native environments and with their natural community partners. I have really enjoyed getting down in the muck and the mess of natural communities of phototrophs!

In my view, 1995 marked a critical turning point for microbiology. Craig Venter, Hamilton Smith, Clare Fraser, and their colleagues at The Institute for Genomic Research published two landmark papers in that year, which described the shotgun sequencing of the genomes of *Haemophilus influenzae* (Fleischmann *et al.*, 1995) and *Mycoplasma genitalium* (Fraser *et al.*, 1995). These publications divide modern microbiology into two other time periods, BG (before genomes) and AG (after genomes). In 1996, Satoshi Tabata and his collaborators at Kazusa DNA Research Institute in Japan described the third bacterial genome, that of the cyanobacterium *Synechocystis* sp. PCC 6803 (Kaneko *et al.*, 1996). Cyanobacterial research moved into AG period. I have chosen to call the era that ensued to present *The Age of Enlightenment*, because I believe that few things impact global knowledge of a microorganism in the same manner that sequencing its genome does. In the nearly two decades that have followed, close to 200 cyanobacterial genomes have been sequenced. The resulting database represents information that has forever changed how cyanobacteria are studied. After bacterial genomes started to become available, I used to think that there were really only two types of microbiologists – those who worked on organisms for which the genomes had been sequenced, and those that did not. At least for a while, one certainly did not want to be in the latter category when grant proposals were being considered. However, almost anyone can now study an organism with a sequenced genome, because the cost of sequencing is small and automated pipelines for genome analysis are readily available to all. I do not believe that we have seen the end of genome sequencing, but I think there is likely to be a shift towards using sequencing in more ways, for example to identify and characterize organisms within their natural communities – something that had been almost entirely lacking in our field until recently. Axenic strains with little connection to the natural world still predominate, but it is often difficult or even impossible to connect our favourite model organisms to real-world microbial communities and environments.

The availability of genomes for genetically tractable model cyanobacteria, including *Synechocystis* sp. PCC 6803, *Synechococcus* sp. PCC 7942, *Nostoc* (also known as *Anabaena*) sp. PCC 7120, *Nostoc punctiforme*, *Thermosynechococcus elongatus* BP-1, and *Synechococcus* sp. PCC 7002, and others, has stimulated research on cyanobacteria and enabled studies that could not have been imagined only 20 years ago. In addition to speeding up and enabling more sophisticated analyses of individual genes and their functions, system-wide approaches (i.e. all of the –omics methodologies), network analysis, and metabolic modelling have now become possible with cyanobacteria. Furthermore, interest in cyanobacteria as platforms for the production of high-value commodity chemicals and biofuels has greatly expanded in the past decade, in large part driven by genomic information and improvements in tools to perform genetic analyses in these organisms. This positive feedback loop will probably only accelerate research in the future. It has now become a nearly impossible task to summarize the vast amounts of information on cyanobacteria, but some excellent volumes covering a wide variety of subjects have recently appeared, including *The Cyanobacteria: Molecular Biology, Genomics and Evolution*, (Herrero and Flores, 2008), and *Genomics of Cyanobacteria* (Chauvat and Cassier-Chauvat, 2013). A new edition of Whitton's treatise on cyanobacterial ecology, *Ecology of Cyanobacteria II* (Whitton, 2012), has also very recently appeared. These volumes will help to serve as the platform on which new students in this field will be able to build and frame their own research contributions.

Another important development in microbiology over the past 20 years has been the application of cell biology methodologies, especially fluorescence-based methods, to the study of diverse processes in microorganisms. The ability to use fluorescence (or luminescence) to visualize proteins, to study their localization, and to characterize cellular dynamics has changed our views of many fundamental processes that occur in microbes, including cyanobacteria. Susan Golden, Takao Kondo, Carl Johnson, and their colleagues provided an important early application of such methods by demonstrating and characterizing circadian rhythms in cyanobacteria (for a summary, see Golden *et al.*, 1997). Protein visualization methods have recently been very important in studies of heterocyst differentiation. Genome sequencing, comparative genomics and improved genetic methods have certainly contributed greatly to the successful application of these methods in cyanobacteria. This volume, *The Cell Biology of Cyanobacteria*, summarizes recent progress on these organisms with emphasis on how cell structures are built and function.

So, what comes next in 'The Modern/Post-Modern Era', what will students be reading about 10 years from now, and importantly, who will be organizing all of this knowledge? The answer, in part, is more of the same, as there remains so much to learn about the fundamental physiology and metabolism of cyanobacteria. However, because of the desperate need to address the interrelated problems of rising CO_2 levels in the atmosphere, climate change, and the replacement of fossil fuels with carbon-neutral solutions, cyanobacteria will increasingly become organisms of choice for various applications in industry and solar energy. For this reason and others, I think we can expect research in the areas of systems biology and metabolic modelling to increase substantially over time for cyanobacterial platforms. These studies are likely to include research in which cyanobacteria will be co-cultured (intentionally or unintentionally) with various microorganisms. Understanding the relationships between and/or among these partners might provide solutions to serious problems that are already being encountered in commercialization trials. An important point is that one can let nature provide the guiding light here – cyanobacteria often occur with other phototrophs in natural microbial communities.

Comparative physiology and comparative genomics will continue to be important approaches in the immediate years ahead. However, the current genomic database also emphasizes how diverse and complex cyanobacteria really are. Unfortunately, based on my personal experience, one is likely to learn quickly that the capacity to use comparative genomics will be severely limited because of deficiencies in biochemical and metabolite catalogues for cyanobacteria. For example, even the carotenoid contents of the sequenced organisms will have to be re-determined if one wishes to use comparative genomics to identify novel genes involved in the biosynthesis of a particular carotenoid. Genomic data is great for creating hypotheses, but those hypotheses can only be validated using genetics and biochemistry. Biochemists are not likely to run out of things to do any time soon!

Although aspects of core metabolism are of course similar in most cyanobacteria – after all, nearly all perform oxygenic photosynthesis, fix CO_2 by the Calvin–Benson–Bassham cycle, and synthesize biomass, chlorophylls, haems, carotenoids, and phycobilins – it is equally important to accept how strikingly diverse these organisms are from a genomic perspective. This should hardly be a surprise, considering their morphological and developmental complexity, the many different environmental niches in which they occur, and their more than 2.5-billion-year evolutionary history on Earth. There is an enormous pool of genes encoding 'proteins of unknown function' in cyanobacteria (i.e. hypothetical and conserved hypothetical open reading frames). This genetic 'dark matter,' as some have called it, may well remain so if we continue to study cyanobacteria in isolation from their 'friends and neighbours' in natural communities. I believe that it is important that more and more cyanobacteriologists adopt a holistic view and attempt to study these organisms by starting from the communities in which they naturally occur. Understanding the relationships between and among microorganisms is one of the next big challenges in microbiology and will take us beyond the 'pure culture concept' and hopefully to a more natural and complete view. Molecular tools to study natural communities, including meta-omics approaches, stable isotope labelling methodologies, mass spectrometry (including nano-SIMS) and advanced microscopy methods, are revolutionizing microbial ecology and ecophysiology. We will not truly understand the post-genomic biology of cyanobacteria until we understand their complex interactions with other organisms in natural communities. For me, this is one of the major challenges for the future.

Acknowledgements

The author would like to thank his mentors (Alexander N. Glazer, Frederick A. Eiserling, Roger Y. Stanier, Germaine Cohen-Bazire, and Roderick K. Clayton), his many friends, colleagues, collaborators (with special thanks to Noel G. Carr, John H. Golbeck and David M. Ward) and competitors, as well his many graduate students and postdoctoral associates, all of whom helped to introduce (and re-introduce) him to cyanobacteria and photosynthesis over the years. If I forgot to mention you, I probably just forgot, period! The author would also like to thank the National Science Foundation, US Department of Energy, Air Force Office of Scientific Research, US National Aeronautics and Space Administration, National Institutes of Health, and US Department of Agriculture for their generous support of his cyanobacterial studies during the past 32 years.

References

Allen, M.B. (1952). The cultivation of *Myxophyceae*. Arch. Mikrobiol. *17*, 34–53.

van Baalen, C. (1962). Studies on marine blue-green algae. Bot. Mar. *4*, 129–139.

Bryant, D.A. (1994). The Molecular Biology of Cyanobacteria, Advances in Photosynthesis and Respiration, Vol. 1 (Dordrecht, The Netherlands: Kluwer Academic Publishers).

Bryant, D.A., Guglielmi, G., Tandeau de Marsac, N., Castets, A.M., and Cohen-Bazire, G. (1979). The structure of cyanobacterial phycobilisomes: a model. Arch. Microbiol. *123*, 113–127.

Cohen-Bazire, G., and Bryant, D.A. (1982). Phycobilisomes: composition and structure. In The Biology of the Cyanobacteria, Carr, N.G., and Whitton, B.A., eds. Botanical Monographs, Vol. 19 (Blackwell Scientific Publications, London), pp. 143–190.

Carr, N.G., and Whitton, B.A. (1973). The Biology of the Blue-Green Algae (Blackwell Scientific Publications, London).

Carr, N.G., and Whitton, B.A. (1982). The Biology of Cyanobacteria. Botanical Monographs, Vol.19. (Blackwell Scientific Publications, London).

Chauvat, F., and Cassier-Chauvat, C. (2013). Genomics of Cyanobacteria. Advances in Botanical Research, Volume 65, (Academic Press, New York).

Fleischmann, R.D., Adams, M.D., White, O., Clayton, R.A., Kirkness, E.F., Kerlavage, A.R., Bult, C.J., Tomb, J.F., Dougherty, B.A., Merrick, J.M., *et al.* (1995). Whole-genome random sequencing and assembly of *Haemophilus influenzae*. Science *269*, 496–512.

Fraser, C.M., Gocayne, J.D., White, O., Adams, M.D., Clayton, R.A., Fleischmann, R.D., Bult, C.J., Kerlavage, A.R., Sutton, G., Kelley, J.M., *et al.* (1995). The minimal gene complement of *Mycoplasma genitalium*. Science *270*, 397–403.

Golden, S.S., Ishiura, M., Johnson, C.H., and Kondo, T. (1997). Cyanobacterial circadian rhythms. Annu. Rev. Plant Physiol. Plant Mol. Biol. *48*, 327–354.

Herrero, A., and Flores, E. (2008). The Cyanobacteria: Molecular Biology, Genomics and Evolution (Caister Academic Press, Norfolk, UK).

Kaneko, T., Sato, S., Kotani, H., Tanaka, A., Asamizu, E., Nakamura, Y., Miyajima, N., Hisosawa, M., Sugiura, M., Sasamoto, S., *et al.* (1996). Sequence analysis of the genome of the unicellular cyanobacterium *Synechocystis* sp. strain PCC 6803. II. Sequence determination of the entire genome and assignment of potential protein-coding regions. DNA Res. *3*, 109–136.

Rippka, R., Deruelles, J., Waterbury, J.B., Herdman, M., and Stanier, R.Y. (1979). The Cyanobacteria: generic assignments, strain histories, and properties of pure cultures of cyanobacteria. J. Gen. Microbiol. *111*, 1–61.

Whitton, B.A. (2012). Ecology of Cyanobacteria II: Their Diversity in Space and Time (Springer-Verlag, Dordrecht, The Netherlands).

Whitton, B.A., and Potts, M. (2000). The Ecology of Cyanobacteria: Their Diversity in Time and Space (Dordrecht, The Netherlands: Kluwer Academic Publishers).

Cell Division in Cyanobacteria

2

Corinne Cassier-Chauvat and Franck Chauvat

Abstract

This review summarizes what is known regarding cell division in cyanobacteria, the fascinating microorganisms that are logically attracting a growing attention in various areas of basic and applied researches. Cyanobacteria, the only prokaryotes capable of oxygenic photosynthesis, colonize most water and soil environments of our planet, and provide a large part of the oxygenic atmosphere and biomass for the food chain. They display different morphologies ranging from unicellular (cylindrical, spherical and spirals) to complex multicellular (filamentous) forms that contain differentiated cells allowing the growth or survival of these organisms under adverse conditions. Furthermore, cyanobacteria divide in one or several successive planes, at right angles or in irregular planes so that the cells may appear singly or in aggregates of varying size. Nowadays, cyanobacteria are regarded as promising 'low-cost' microbial cell factories for carbon capture and storage, and for the sustainable production of secondary metabolites and biofuels, thanks to their simple nutritional requirements, their metabolic plasticity, and the powerful genetics of some model strains. In this chapter, we report that cyanobacteria (which are Gram-negative) share cytokinetic genes in common with both Gram-positive and Gram-negative bacteria, and/or the chloroplast and the nuclear genome of plants and algae. In agreement with cyanobacteria being regarded as the ancestor of the chloroplast, the stromal portion of the chloroplast division complex resemble the cyanobacterial cell division machinery, but many other components were lost after the endosymbiotic event.

Introduction

Cell division is a fundamental process tightly coordinated

Binary fission of a mother cell into two identical, or nearly identical, daughter cells is a widely conserved mechanism that is crucial for the proliferation of almost all cells. Dividing precisely at mid-cell presumably maximizes the collective fitness of the progeny by avoiding the generation of cells with unnecessarily low cytoplasmic mass or incomplete chromosomes. Hence, correct cell division requires the tight regulation and precise coordination between the key processes of cell growth, chromosome replication, chromosome segregation, septum formation and separation of the daughter cells. The temporal separation of these processes, which postpone septum formation until chromosome segregation has been completed, prevents the septum to come down onto and 'guillotine' the chromosomes (de Boer, 2010; Lutkenhaus, 2012; Wu and Errington, 2012).

Cell division in the rod-shaped model bacterium *Escherichia coli*

Because of their relative simplicity, bacteria are excellent models to characterize genes and proteins that operate in cell division (Margolin, 2009). Furthermore, as they come in a variety of shapes (straight or curved rods with or without branches, spheres, helix, ovoids) which are maintained during vegetative growth, bacteria are excellent models to study how cells grow and divide, and pass their morphology on to their progeny (Typas *et al.*, 2012).

In the 1960s, researchers set out to characterize many of the cell division processes of the unicellular enterobacterium *E. coli* by obtaining a large collection of temperature sensitive mutants defective in cytokinesis, which produced long filamentous cells at high temperature. The mutations allowed identification of a set of '*fts*' genes (filamentation thermosensitive), the products of which are essential for cell division. The first cell division protein thereby characterized, FtsZ, was shown to be abundant and to localize to mid-cell (Bi and Lutkenhaus, 1991). Then more than 10 cytokinetic proteins were identified and appeared to be recruited to the division site in the sequential order: FtsZ → FtsA/ZipA/ZapB → FtsK → FtsQ, FtsL/FtsB → FtsW → FtsI and FtsN (de Boer, 2010; Typas *et al.*, 2012). Cells lacking any of these core proteins fail to constrict. They elongate into very long filaments harbouring evenly spaced nucleoids and non-functional septal ring structures, before eventually dying. When complete, the division machine constricts in parallel with the synthesis of the septal cell wall (septal peptidoglycan), which matures into the new polar caps of the daughter cells. The overlap and interplay between the FtsZ systems and the peptidoglycan-synthesizing enzymes they recruit are the major driving forces of cylindrical shapes (Margolin, 2009; Typas *et al.*, 2012). During the cell cycle of *E. coli* (and the other rod-shaped model bacterium, *Bacillus subtilis*), the bacterial chromosome is replicated and segregated to the two cell halves before cell division can occur (Lenz and Sogaard-Andersen, 2011). Two molecular mechanisms – the Min system and the nucleoid occlusion (Noc) system – work synergistically to preclude Z ring formation near the nucleoid (i.e. the bacterial chromosome) thereby preventing the chromosome from being bisected and broken by the division septum (Kirkpatrick and Viollier, 2011).

By contrast to symmetrically placed Z-ring that partitions the cell into equally sized compartments, the Z-ring can also be naturally positioned asymmetrically to yield compartments of unequal size, as for the sporulation of *B. subtilis* (Kirkpatrick and Viollier, 2011; Wagner-Herman *et al.*, 2012).

The limitations of *the E. coli* and *B. subtilis* models

The 'historical' rod-shaped (cylindrical) models *E. coli* and *B. subtilis*, which possess a well-defined middle where constriction occurs, are not appropriate for the wealth of round-celled (spherical) bacteria (Margolin, 2009; Rippka *et al.*, 1979), which possess an infinite number of potential division planes at the point of greater cell diameter. Furthermore, cylindrical bacteria divide in one and the same plane perpendicular to the long axis of the cells, whereas spherical bacteria divide in alternating perpendicular planes (Mazouni *et al.*, 2004; Tzagoloff and Novick, 1977; Westling-Haggstrom *et al.*, 1977), as observed in *E. coli* mutants that became round after depletion of the cell wall synthesizing enzymes PBP2 (penicillin-binding protein 2) and RodA (Begg and Donachie, 1998; Margolin, 2009).

Why study cell division in cyanobacteria?

Cyanobacteria, the ancient Gram-negative prokaryotes, have a strong impact on the biosphere (Rippka *et al.*, 1979). They colonize most waters (fresh, brackish and marine) and terrestrial (including deserts) environments (Shi and Falkowski, 2008), and are regarded as the progenitors of the chloroplast (Archibald, 2009); the oxygen-rich atmosphere of our planet; and the biomass for the food chain (Partensky *et al.*, 1999; Scanlan *et al.*, 2009; Schopf, 2011; Zehr, 2011). In addition, cyanobacteria are receiving a growing attention as promising 'low-cost' microbial cell factories for the capture and storage of industrial CO_2 gas (Jansson and Northen, 2010), and the ecologically responsible production of important products (Abed *et al.*, 2009; Ducat *et al.*, 2011; Pearson *et al.*, 2010; Schirmer *et al.*, 2010; Wang *et al.*, 2011; Williams, 2009; Zhou and Li, 2010). Cyanobacteria display different cell sizes (1–10 μm) and forms (Bandyopadhyay *et al.*, 2011; Rippka *et al.*, 1979). They range from the unicellular spherical and cylindrical morphologies of *Synechocystis* PCC6803 (Mazouni *et al.*, 2004) and *Synechococcus* PCC7942 (Koksharova and Wolk, 2002), respectively, to complex multicellular (filamentous) forms of *Anabaena* or *Nostoc* species that differentiate specialized cells for growth or survival under adverse conditions (Chauvat *et al.*, 1982; Mariscal and Flores, 2010; Thiel and Wolk, 1983). Furthermore, cyanobacteria can divide in one, two or three successive perpendicular planes or in irregular planes, so that the cells may appear singly or in aggregates of varying size (Mazouni *et al.*, 2004; Rippka *et al.*, 1979), whereas *E. coli* and *B. subtilis* cells divide in one and the same plane. Moreover, cell cycle studies of the cyanobacteria *Synechocystis* PCC6803 and *Anabaena* PCC7120 show that the multiple copies of the polyploid chromosome (about 10 per cell; Hu *et al.*, 2007; Labarre *et al.*, 1989) segregate very late, i.e. just before complete closing of the division septum (Schneider *et al.*, 2007), thereby generating daughter cells with different DNA contents (Hu *et al.*, 2007). These findings indicate that cyanobacterial chromosomes may be segregated randomly and in a passive fashion, possibly through constriction of the division septum, unlike what is found in *B. subtilis* and *E. coli* (Lenz and Sogaard-Andersen, 2011; Lutkenhaus, 2012; Wu and Errington, 2012). Furthermore, the FtsZ rings of the cyanobacterium *Synechococcus* PCC7942 (which share the same rod shape as *B. subtilis* and *E. coli*) appeared to be formed at sites occupied by nucleoids (Miyagishima *et al.*, 2005), unlike what occurs in *B. subtilis* and *E. coli* (Kirkpatrick and Viollier, 2011). Collectively, these findings indicate that at least some aspects of the cell division process differ between cyanobacteria and both the *B. subtilis* and *E. coli* models.

In this chapter, we report on what is known concerning cell division in cyanobacteria, emphasizing on what has been learned with the fruitful models *Synechococcus* PCC7942 (unicellular, rod shaped) (Koksharova and Wolk, 2002; Miyagishima *et al.*, 2005); *Synechocystis* PCC6803 (unicellular, round shaped) (Marbouty *et al.*, 2009a–d; Mazouni *et al.*, 2004) and *Anabaena* PCC7120 (filamentous) (Hu *et al.*, 2007; Leganes *et al.*, 2005; Miyagishima *et al.*, 2005; Wilk *et al.*, 2011). We also report on what can be inferred by mining the information provided by the 70 sequenced genomes of morphologically and physiologically diverse cyanobacteria. We show that cyanobacteria (Gram-negative) share cytokinetic proteins in common with not only Gram-negative but also Gram-positive bacteria, as well as with the chloroplast- and nuclear-encoded plant proteins.

Peptidoglycan synthesis and hydrolysis

Much of what we know about the interplay between cell wall, cell shape and cytokinesis has come from studies with the rod-shaped unicellular bacteria *E. coli* (Gram-negative) and *B. subtilis* (Gram-positive) (de Boer, 2010; Kirkpatrick and Viollier, 2011; Lutkenhaus, 2012; Mannik *et al.*, 2012; Wu and Errington, 2012). Bacteria are designated Gram-positive or Gram-negative based on their ability or inability to retain the dye crystal violet (Gram stain). This difference is determined by the distinctive composition of the cell wall, the exoskeleton structure that largely dictates cell morphology and provides mechanical strength to resist intracellular pressure, and mechanical and osmotic challenges (Margolin, 2009; Typas *et al.*, 2012).

The major component of the cell wall is the peptidoglycan, a glycan polymer of *N*-acetylglucosamine and *N*-acetylmuramic acid strands cross-linked to each other via peptide chains, which is synthesized by various enzymes including the penicillin-binding proteins and forms a net-like structure covering the cell surface. In Gram-negative bacteria, an outer membrane surrounds a thin peptidoglycan (murein) layer with a periplasmic space between the inner and outer membranes (Typas *et al.*, 2012). In contrast, Gram-positive bacteria have only an inner membrane and a thicker layer of peptidoglycan (Reith and Mayer, 2011). The spatial and temporal synthesis, deposition and degradation of the peptidoglycan layer of the cell wall is responsible primarily for determination of cell shape and changes in cellular morphology. In newborn *E. coli* cells that have just divided, peptidoglycan is synthesized along the sidewall, through the insertion of newly synthesized glycan chains, or patches of them, and their attachment into the existing sacculus (a bag-like macromolecule made of peptidoglycan chains cross-linked by short peptides). This promotes elongation of the cell to ultimately twice the length of newborn cells. When cell division begins, the synthesis apparatus switches from sidewall peptidoglycan synthesis to division septum synthesis, which allows separation of the daughter cells. These two modes of peptidoglycan growth probably compete with one another (Margolin, 2009; Typas *et al.*, 2012).

Peptidoglycan growth involves glycosyltransferases (GTases), DD-transpeptidases (DD-TPases) and hydrolases. GTases that polymerize the glycan chains and DD-TPases that crosslink glycan chains via their peptide side-chains, simply attach new material to the sacculus, thereby thickening it without elongating it. Thus, hydrolases that cleave covalent bonds in the sacculus are required to allow the newly attached material to insert into the layer without increasing its thickness (Reith and Mayer, 2011; Typas *et al.*, 2012). The synthesis of peptidoglycan precursors occurs in three overall stages. First, *N*-acetylglucosamine is transformed by the cytoplasmic enzymes MurA–F into the N-acetylmuramic acid with a cross-linked peptide side chain (Typas *et al.*, 2012). Second, at the inner leaflet of the inner membrane, *N*-acetylglucosamine and *N*-acetylmuramic acid harbouring a cross-linked peptide side chain are assembled with undecaprenyl phosphate by MraY and MurG to form the lipid-anchored disaccharide-pentapeptide monomer subunit (lipid II), and are flipped across the membrane by RodA and FtsW. Third, lipid II is polymerized, releasing undecaprenyl pyrophosphate, and the resulting glycan chains are inserted into the sacculus (Typas *et al.*, 2012). TPases are also called penicillin-binding proteins (PBPs) because they can covalently bind penicillin. There are three types of peptidoglycan synthases: bifunctional GTase–TPases (class A PBPs), monofunctional TPases (class B PBPs) and monofunctional GTases. *E. coli* has three bifunctional synthases (PBP1A, PBP1B and PBP1C), one GTase (MgtA) and two TPases, which are essential for either cell elongation (PBP2) or cell division (PBP3; also named FtsI). PBP1A and PBP1B are partially redundant; one of them is enough for cell viability. PBP1A, thought to have a role in cell elongation, interacts with PBP2 (Banzhaf *et al.*, 2012), while PBP1B, enriched at the septum, interacts with two essential division proteins FtsN and PBP3 (Typas *et al.*, 2012).

During cell elongation the essentially single-layered murein is continuously degraded by peptidoglycan degrading enzymes (Typas *et al.*, 2012). In *E. coli*, about 40–50% of the total peptidoglycan material of the cell is removed every generation, and the soluble fragments that are removed from the sacculus are recycled (Reith and Mayer, 2011; Typas *et al.*, 2012). The endopeptidases and carboxypeptidases respectively hydrolyse the peptide crosslinks and release terminal D-Ala residues of the peptide portion of the peptidoglycan. *E. coli* has at least six DD-carboxypeptidases (class C PBPs), with the membrane anchored PBP5 being the most active. The lytic transglycosylases cleave the MurNAc-GlcNAc bonds in the glycan backbone of the peptidoglycan releasing unique 1,6-anhydro-*N*-acetylmuramic acid (anhMurNAc)-containing peptidoglycan fragments (anhydromuropeptides). Peptidoglycan hydrolases are essential for the separation of daughter cells during cell division, but likely because of high redundancy, multiple hydrolase genes have to be simultaneously inactivated to impair cell septation in *E. coli*. The periplasmic amidases AmiA, B and C, all of which cleave MurNAc-l-Ala bonds in the murein, are also involved in splitting the septum during daughter cell separation (Reith and Mayer, 2011; Typas *et al.*, 2012).

Cyanobacterial cell walls belong to the group of Gram-negative bacteria in lacking the teichoic acids that are characteristic of Gram-positive bacterial walls (Leganes *et al.*, 2005). However, the cyanobacterial peptidoglycan layer and the degree of cross-linking between the peptidoglycan chains within the murein resemble those of Gram-positive bacteria in being higher than their counterparts in most Gram-negative bacteria (Hoiczyk and Hansel, 2000; Typas *et al.*, 2012). Cyanobacteria possess several genes encoding proteins homologous to *E. coli* PBP proteins (Table 2.1). One of the at least 12 PBP-encoding genes of *Anabaena* PCC7120 (the *alr5101* gene encoding the class A PBP named PBP2), was found to be required for heterocyst differentiation and aerobic nitrogen fixation (Lazaro *et al.*, 2001). In *Synechocystis* PCC6803, which possesses at least nine PBP-encoding genes (Leganes *et al.*, 2005; Marbouty *et al.*, 2009a; Urbach *et al.*, 2008): sll0002 (PBP1; class A no. 1; PBP1A-1), slr1710

Table 2.1 Distribution and genomic organization of cyanobacterial genes operating in peptidoglycan synthesis and elongation

Species	mreBCD	rodA	PBPA1			PBP4	ampH	murA	murB	murC	murD	murE	murF	murG	mraY	bolA	penB	amiA	amiC	tolB	erzA	NlpD
			1	*2*	*3*																	
Acaryochloris marina MBIC11017	+++a	+b	+	+	3	2	–	+	+	+f	+	+h*	+	+	+j	+k*	+l	+	2	+	–	+o
Cyanothece sp. ATCC51142	+++a	+b	+	+	+	2	+	+	+	+f*	+	+	+	+	+j	+k	+	+	2m	4n	–	+o*
Cyanothece sp. PCC7424	+++a	+b	+	+	+	2	2	+	+	+f	+	+h*	+	+	+	+k	+l	+	2m	2n	+	+o*
Cyanothece sp. PCC7425	+++a	+	–	+	+	2	5	+	+	+f	+	+	+	+	+j	+k*	2l	+	2	–	–	+o*
Cyanothece sp. PCC7822	+++a	+b	+	+	+	2	2	+	+	+f	+	+h*	+	+	+	+	+l	+	+m	2n	+	+o
Cyanothece sp. PCC8801	+++a	+b	+	+	+	2	+	+	+	+f	+	+	+	+	+j	+k	+	+	2m	4n	+	+o*
Cyanothece sp. PCC8802	+++a	+b	+	+	+	2	2	+	+	+f	+	+	+	+	+j	+k	+	+	2m	4n	+	+o*
Microcystis aeruginosa NIES-843	++–a	+b	+	+	+	2	–	+	+	+f*	+	+	+	+	+j	+	+l	+	+m	4n	–	+o*
Synechococcus elongatus PCC6301	+++a	+b	+	+	+	+	–	+	+	+f*	+	+h	+	+	+j	+k	+l	+	+m*	2n	–	+o*
Synechococcus elongatus PCC7942	+++a	+b	+	+	+	+	–	+	+	+f*	+	+h	+	+	+j	+k	+l	+	2m*	2n	–	+o*
Synechococcus sp. CC9311	+++a*	+b*	+c	+d	–	+	+	+ e	+	+f	+	+h	+i	+	+j	+k**	+l*	–	+m*	2n	–	+p
Synechococcus sp. CC9605	+++a*	+b*	+c	–	–	+	+	+	+	+f	+	+h	+i	+	+j	+k**	+l*	–	+m*	2n	–	2p
Synechococcus sp. CC9902	+++a*	+b*	+c	+d	–	+	–	+ e	+	+f	+	+	+i	+	+j	+k**	+l*	–	+m*	2n	–	+p
Synechococcus JA-2–3B' a(2–13)	+++a	+	–	+	+	–	+	+	+	+f*	+	+	+	+	+	+	+	–	+m	+	–	–
Synechococcus JA-3–3B' Ab	+++a	+	–	+	+	–	+	+	+	+f*	+	+	+	+	+	+	+	–	+m	+	–	–

Table 2.1 (Continued)

Species	mreBCD	rodA	PBPA1 1	PBPA1 2	PBPA1 3	PBP4	ampH	murA	murB	murC	murD	murE	murF	murG	mraY	bolA	penB	amiA	amiC	tolB	erzA	NlpD
Synechococcus sp. PCC7002	+++a	+	+	+	+	+	+	+	+	+f	+	+	+	+	+	+k	+	–	+m	4n	–	+o
Synechococcus sp. RCC307	+++a	+b*	+c	+d	–	+	–	+e	+	+f	+	+h	+i	+	+j	+k**	+l*	–	+m*	2n	–	+p
Synechococcus sp. WH 7803	+++a*	+b*	+c	+d	–	+	–	+e	+	+f	+	+h	+i	+	+j	+k**	+l*	–	+m*	2n	–	+p
Synechococcus sp. WH 8102	– – –	+b*	+c	–	–	+	–	+e	+	+f	+	+	+i	+	+j	+k**	+l*	–	+m*	2n	–	+p
Synechocystis sp. PCC6803	+++a	–	+	+	+	2	2	+	+	+	+	+	+	+	+	+k	+l	+	2m	2n	–	+o
Thermosynechococcus elongatus BP-1	– – –	+	–	+	+	+	–	+	+	+f*	+	+	+	+	+j	+k	+l	+	+	–	–	+o
Cyanobacterium UCYN-A	2++a	+b	–	–	+	+	–	+	+	+f*	+	+	+	+	+j	+k*	+	–	–	–	–	+o*
Gloeobacter violaceus PCC7421	+++a	+	+	–	+	–	+	+	+	+f*	+	+	+	+	+j	+	+	+	+	–	+	–
Anabaena variabilis ATCC29413	+++a	+b	+	+	+	+	+	+	+	+f	+	+h*	+	+	+j	+k	2l	+	2m	2n	+	+o
Nostoc punctiforme PCC73102	+++a	+b	+	+d*	4	+	2	+	+	+f	+	+h*	+	+	+j	+k*	3l	+	2m	2n	+	+o*
Anabaena sp. PCC7120	+++a	+b	+	+d*	3	+	2	+	+	+f	+	+h*	+	+	+j	+k	2l	+	2m	2n	+	+o
Nostoc azollae 0708	+++a	+b	+	+d*	+	–	–	+	+	+f	+	+h*	+	+	+j	+k*	+l	+	2m	2n	–	+o*
Trichodesmium erythraeum ISM101	+++a	+	+c	+	+	+	+	+	+	+f	+	+h*	+	+	+j	+k*	+l*	+	+	–	5	+o*
Prochlorococcus marinus AS9601	+++a*	+b*	+c	–	–	+	–	+	+	+f	+g	+h	+i	+	+j	–	–	–	+m*	–	–	+p
Prochlorococcus marinus MIT 9211	+++a**	+b*	+c	–	–	+	–	+	+	+f	+g	+h	+i	+	+j	–	+l**	–	+m*	+	–	2p
Prochlorococcus marinus MIT 9215	2++a**	+b*	+c	–	–	+	–	+	+	+f	+g	+h	+i	+	+j	+k**	–	–	+m*	–	–	+p

Prochlorococcus marinus MIT 9301	+++a**	+b*	+c	–	–	+	–	+	+	+f	+g	+h	+i	+	+j	+k**	–	–	+m*	–	–	+p
Prochlorococcus marinus MIT 9303	+++a*	+b*	+c	–	–	+	–	+	+	+f	+	+h	+i	+	+j	+k**	+l	–	+m*	2n	–	+p
Prochlorococcus marinus MIT 9312	+++a**	+b*	+c	–	–	+	–	+	+	+f	+g	+h	+i	+	+j	+k**	–	–	+m*	–	–	+p
Prochlorococcus marinus MIT 9313	+++a*	+b*	+c	–	–	+	–	+	+	+f	+g	+h	+i	+	+j	+k**	+l*	–	+m*	2n	–	+p
Prochlorococcus marinus MIT 9515	+++a**	+b*	+c	–	–	+	–	+	+	+f	+g	+h	+i	+	+j	+k**	–	–	+m*	–	–	+p
Prochlorococcus marinus NATL1A	+++a**	+b*	–	–	–	+	–	+	+	+f	+g	+h	+i	+	+j	+k*	+l**	–	+m*	–	–	+p
Prochlorococcus marinus NATL2A	+++a**	+b*	–	–	–	+	–	+	+	+f	+g	+h	+i	+	+j	+k*	+l**	–	+m*	–	–	+p
Prochlorococcus marinus SS120	+++a**	+b*	+c	–	–	+	+	+	+	+f	+g	+h	+i	+	+j	+k**	+l**	–	+m*	–	–	2p
Prochlorococcus marinus MED4	+++a**	+b*	+c	–	–	+	–	+	+	+f	+g	+h	+i	+	+j	+k**	–	–	+m*	–	–	+p

All relevant genes, presented here or in the text, are indicated along with their presence (+, or indicated numbers) or absence (–) in each strain. Hypothetical genes are indicated as 'ho'.

Clusters: a = mreBCD; a* = mreBCD-ugpB tcr-lysS; a** = mreBCD-tcr-lysS; b = mrp-rodA; b* = mrp-rodA-tcshk; c = chlG-PBP1A-1; d = bioY-lspA-PBP1A-2; d* = bioY-lspA-one small unknown gene-PBP1A-2; e = murA-tRNAleu-folC; f = murC-murB-ho; f* = murC-murB-ho; g = murD-math; h = mpf-trx-murE; h* = mpf-trx-murE; i = menD-menB-glgA-murF; j = hoDU3134-mraY; k = bolA-grx; k* = ho-bolA-grx; k** = plsC-ho-bolA-grx; l = trmH-penB; l* = cytM-trmH-penB-ho-chlI-ruvc; l** = cytM-trmH-penB-ho-chlI-ruvc; m = amiC-murI; m* = comB-nrl-amiC-murI-thpt; n = psbX-ycf66-tolB-tolB; o = nlpD-pep; o* = nlpD-pep-tRNAmet; p = ald-nlpD.

Gene names: *ugpB*, sugar ABC transporter periplasmic protein; *tcr*, two component regulator; *lysS*, lysyl-tRNA synthetase; *mrp*, chromosome partitioning ATPase; *tcshk*, two-component sensor histidine kinase; PBP1A genes are classified according to the *Synechocystis* genes as follows: PBP1A-1, sll0002; PBP1A-2, slr1710; PBP1A-3, slr1434; PBP4, slr0646 and slr0804 (PBP5 and PBP8 in *Synechocystis*); *bioY*, biotin transporter; *lspA*, lipoprotein; *sgtpp*, small GTP-binding domain-containing protein; *ampH*, sll1167 and slr1924 (PBP6 and PBP7 in *Synechocystis*); *murA*, UDP-GlcNAc enolpyruvyl transferase; *murB*, UDP-*N*-acetylenolpyruvoylglucosamine reductase; *murC*, UDP-N-acetylmuramate-L-alanine ligase; *murD*, UDP-N-acetylmuramoyl-L-alanyl-D-glutamate synthetase; *ho*, Putative membrane protein insertion efficiency factor haemolytic domain; *math*, MATH domain-containing protein; *murE*, UDP-N-acetylmuramoylalanyl-D-glutamate-2, 6-diaminopimelate ligase; *trx*, thioredoxin/glutaredoxin superfamily; *murF*, UDP-*N*-acetylmuramoylalanyl-D-glutamyl-2, 6-diaminopimelate-D-alanyl-D-alanine ligase; *menD*, menaquinone synthesis protein 2-succinyl-6-hydroxy-2,4-cyclohexadiene-1-carboxylate synthase; *menB*, naphthoate synthase; *glgA*, glycogen synthase; *murG*, UDP-*N*-acetylglucosamine-*N*-acetylmuramyl-(pentapeptide)-pyrophosphoryl-undecaprenol N-acetylglucosamine transferase; *mraY*, phospho-*N*-acetylmuramoyl-pentapeptide-transferase; *murI*, aspartate and glutamate racemases; *grx*, CGFS-type monothiol glutaredoxin; *plsC*, phospholipid/glycerol acetyltransferase; *trmH*, RNA methyltransferase; *chlI*, protoporphyrin IX Mg-chelatase subunit ChlI; *ruvC*, Holliday junction resolvase; *comB*, 2-phosphosulfolactate phosphatase; *nrl*, nitrilase; murI, glutamate racemase; thpt, trans-hexaprenyltranstransferase; psbX, photosystem II protein X; *ycf66*, protein found in plant chloroplast and cyanobacteria; *pep*, peptidase; *ald*, aldehyde dehydrogenase.

(PBP1A-2), sll1434 (PBP1A-3), sll1833 (PBP4; class B PBP; FtsI), slr0646 (PBP5; class C PBP type 4), sll1167 (PBP6; class C PBP type ampH), slr1924 (PBP7; class C PBP type ampH), slr0804 (PBP8; class C PBP type 4) and slr0121 (PBP9; PBP homologous to class A beta-lactamase, also named PenB), we have found the following results. *Synechocystis* PCC6803 can survive in the absence of one but not two class A PBPs. PBP2 is more important than PBP1 and PBP3, and it likely operates in the synthesis of sidewall (not septal) peptidoglycan since the PBP2-less mutant displays minicells, whereas mutants lacking either PBP1 or PBP3 have no obvious phenotype (Marbouty *et al.*, 2009a). As a link between our basic work on cell division and applied researches oriented towards the sustainable production of biofuels it has been reported that the deletion of the PBP2-encoding gene facilitates the secretion of free fatty acids (Liu *et al.*, 2011). Returning to the *in vivo* analysis of the *Synechocystis* PCC6803 PBPs, we found that the unique class B PBP protein, PBP4 (FtsI), is indispensable to cell viability (Marbouty *et al.*, 2009a). The PBP4 depleted mutant displayed giant cells and cloverleaf-like clusters of four large un-separated cells that likely result from the delayed septation of daughter cells, which was not completed before the initiation of the second round of division. These findings suggest that FtsI might operate in the inward synthesis (or incorporation) of septal peptidoglycan that allows the separation of the daughter cells (Marbouty *et al.*, 2009a), as observed in *E. coli* (Margolin, 2009; Typas *et al.*, 2012). In contrast to FtsI, but similar to what observed for class A PBPs we found that *Synechocystis* can survive without one but not the two class C PBPs of type 4 (PBP5 and PBP8) and that the same is true for class C PBPs of type AmpH (PBP6 and PBP7). The mutants depleted of PBP5 and PBP8 on one hand, and PBP6 and PBP7 on the other hand grew slowly and displayed giant cells. Interestingly, the mutant depleted of both PBP5 and PBP8 (not the mutant depleted of both PBP6 and PBP7) exhibited a high proportion of cloverleaf-like four-cell clusters, like the FtsI-depleted strain. These findings suggest that PBP5 and PBP8, but not PBP6 and PBP7, are involved in completion of the septation enabling the separation of daughter cells, as FtsI does (Marbouty *et al.*, 2009a). Furthermore, we found that PBP1 and PBP3 physically interact with FtsI and PBP2 (Marbouty *et al.*, 2009a).

The analysis of cyanobacterial PBPs shows that the number of Class A PBPs tends to increase with genome size (Table 2.1). The unicellular cyanobacterium UCYN-A, which has the smallest genome within the analysed cyanobacteria, possesses only one PBP1A-encoding gene. By contrast, the unicellular cyanobacteria *Synechocystis* PCC6803 and *Synechococcus* PCC7942 have three Class A PBPs related to PBP1a or PB1b from *E. coli*. One possible way to explain this fact is that an ancestral gene was duplicated that encoded a PBP related to *E. coli* PBP1a in the unicellular cyanobacteria *Synechocystis* PCC6803 and *Synechococcus* PCC7942.

All cyanobacteria, including the small genome strain UCYN-A, possess the genes *murA-G, mraY* and *nlpD,* in conserved genome organization contexts (Table 2.1). By contrast, UCYN-A has none of the genes *amiA, amiC* and *tolB,* which occur in a few (*amiA*) or most (*amiC* and *tolB*) cyanobacteria (Table 2.1). The AmiC2 protein of *Nostoc punctiforme* ATCC 29133 is important for normal size and peptidoglycan structures, as well as cell–cell communication (Lehner *et al.*, 2011). The amiC2 mutant cells, which show altered septal peptidoglycan structures, have a larger size and aberrant morphology as compared to wild-type cells, and they cannot perform cellular differentiation (i.e. induction of hormogonia, heterocysts and akinetes).

Except *Synechocystis* PCC6803, all cyanobacteria have a presumptive *rodA* gene, located downstream of *mrp,* another presumptive gene possibly encoding a chromosome partitioning ATPase (Table 2.1). It will be important to investigate the role of these genes in the other model strains *Anabaena* PCC7120 and *Synechococcus* PCC7942.

The cytoskeleton and peptidoglycan synthesis

As mentioned above, peptidoglycan synthesis and insertion are guided by distinct elements of the bacterial cytoskeleton at different phases in the cell cycle (Typas *et al.*, 2012). One of the key factors in the elongation mode of peptidoglycan synthesis in rod-shaped bacteria is the actin-like protein MreB, which acts in conjunction with the proteins MreCD, PBP2, RodA, RodZ and others (Reith and Mayer, 2011; Typas *et al.*, 2012). In *E. coli*, MreB is encoded by the tricistronic *mreBCD* operon where *mreCD* are also expressed from an internal promoter (Wachi *et al.*, 2006). Two lines of evidence indicate that MreB is required for rod-shaped growth (Typas *et al.*, 2012). First, MreB is present in almost all rod-shaped bacteria, whereas it is absent from most spherical types (Carballido-Lopez, 2006; Margolin, 2009). Second, when MreB is depleted, cells rapidly stop elongating, increase their diameter and grow with spherical morphology (Typas *et al.*, 2012). MreB forms small filament patches, which move around the cell circumference perpendicularly to its long axis, and interact with the inner-membrane proteins MreC, MreD and RodZ, as well as with the lipid II synthesis enzymes MraY and MurG. As MreB lacks an obvious membrane-anchoring domain, the other proteins may at the very least help to tether MreB cables to the membrane (Margolin, 2009). Interestingly, inhibition or depletion of the enzymes PBP2 (in *E. coli*) or PBP2A and PBPH (in *B. subtilis*) eliminates MreB movement in both organisms, suggesting that peptidoglycan synthesis itself is the motor that drives MreB movement (Typas *et al.*, 2012). As MreB probably acts to shape the cell wall via MreC, it is not surprising that essentially all species that have MreB also have MreC and MreD, since both MreC and MreD control peripheral peptidoglycan synthesis by affecting the activity or localization of PBP1A (Typas *et al.*, 2012). Exceptions include Wolbachia and *Thermotoga maritima*, which have MreB but no MreC (Margolin, 2009); *Staphylococcus aureus*, which lacks MreB but has both MreC and MreD; and *Helicobacter pylori*, which has both MreB and MreC but lacks MreD (Margolin, 2009). Similarly, RodZ, which operates in MreB assembly, is usually only present in species that have MreB, but is somewhat less conserved than MreB (Margolin, 2009). Challenging the notion that MreB, MreC, MreD and RodZ are crucial for the rod-shape morphology of bacterial cell, the rod-shaped gamma-proteobacterium *Francisella tularensis* lacks all four genes (Margolin, 2009).

The mreBCD operon is also found in all cyanobacteria, to the noticeable exception of the unicellular strains UCYN-A, which has the smallest genome, and *Synechocystis* PCC6803, which also lacks *rodA* (Table 2.1). The absence of the *rodA* gene and *mreBCD* operon in the round-celled cyanobacterium *Synechocystis* PCC6803 (Table 2.1) is consistent with the round morphology of mutants of *E. coli* (normally rod-shaped) lacking either the *rodA* gene or the *mreBCD* operon (Begg and Donachie, 1998; Margolin, 2009; Typas *et al.*, 2012). In *Anabaena* PCC7120, the mreBCD operon is important to maintain the normal rod-shaped morphology of the cells and *mreCD* are also expressed from an internal promoter (Hu *et al.*, 2007), as occurs in *E. coli* (Wachi *et al.*, 2006). Both the cyanobacterial *mreB* gene and the mreBCD operon are dispensable to cell growth (Hu *et al.*, 2007), whereas they are essential in *E. coli* (Margolin, 2009).

In *E. coli*, the *mreB* gene is negatively regulated by the BolA protein (Freire *et al.*, 2009), which is widely conserved (Li *et al.*, 2012) and also regulates the PBP5- and PBP6-encoding genes (Freire *et al.*, 2009; Guinote *et al.*, 2011) that operate in peptidoglycan synthesis (Typas *et al.*, 2012). *E. coli* cells overexpressing *bolA* are spherical in shape, like *mreB* depleted mutants (Freire *et al.*, 2009; Guinote *et al.*, 2011). As *bolA* is present in almost all cyanobacterial genomes (Table 2.1), it will be possible to test BolA relations with murein-synthesizing components, including the possible BolA–MreB antagonism in MreB-possessing cyanobacteria.

Z-ring assembly

In nearly all bacteria, cytokinesis is initiated by the polymerization into a ring-like structure (the Z-ring) at mid-cell (the future division site) of the GTPase protein FtsZ, the likely ancestor of eukaryotic tubulin (Kirkpatrick and Viollier, 2011; Lowe and Amos, 1998). The construction of the Z ring on the cytoplasmic face of the inner cell membrane (IM) starts with the accumulation of FtsZ monomers, which arrange in a head-to-tail configuration to create a GTP-hydrolytic centre at the subunit interface, which facilitates FtsZ polymerization and Z-ring formation. FtsZ occurs in between 3200 and 15,000 molecules per cell in *E. coli* (Galli and Gerdes, 2012). FtsZ is also present in some archaea (Wang and Lutkenhaus, 1996), plastids and mitochondria (Margolin, 2005; Miyagishima, 2011), and all cyanobacteria (Doherty and Adams, 1995; Mazouni *et al.*, 2004; Miyagishima *et al.*, 2005; Mori and Johnson, 2001; Zhang *et al.*, 1995; see Table 2.2). As observed in *E. coli* (Bi and Lutkenhaus, 1991), FtsZ was found to be indispensable to cell viability and division in cyanobacteria. Indeed, the *ftsZ* gene could not be inactivated in *Anabaena* PCC7120 (Zhang *et al.*, 1995), *Synechocystis* PCC6803 (Mazouni *et al.*, 2004) or *Synechococcus* PCC7942 (Miyagishima *et al.*, 2005), irrespectively of the technique employed. The attempted deletion of *ftsZ* from the chromosome of *Synechocystis* PCC6803, which is polyploid (Griese *et al.*, 2011; Labarre *et al.*, 1989), yielded an heteroploid mutant harbouring two un-segregated types of chromosome copies, with and without *ftsZ* (Mazouni *et al.*, 2004). Furthermore, the GTPase activity of FtsZ was found to be crucial to the formation of the septal Z-ring at mid-cell (Mazouni *et al.*, 2004), as occurs in *E. coli* (de Boer, 2010). Moreover, the non-dividing differentiated heterocyst cells devoted to nitrogen fixation in some filamentous cyanobacteria such as *Anabaena* PCC7120 do not produce FtsZ proteins (Kuhn *et al.*, 2000), which appeared to be a target of specific proteolysis in this strain (Lopes Pinto *et al.*, 2011). By contrast, some families of bacteria, such as the Chlamydia and Planctomycetes lack FtsZ and, consequently, must divide using another as yet unknown process (Margolin, 2005).

The FtsZ ring serves as a scaffold for the sequential recruitment of a dozen of proteins that constitute the macromolecular complex called divisome, which executes the construction and constriction of the septum and the separation of the daughter cells (de Boer, 2010; Kirkpatrick and Viollier, 2011). As FtsZ has no affinity for phospholipid, the early divisome members operate in Z-ring stabilization and tethering to the inner cell (cytoplasmic) membrane. In *E. coli* this process minimally requires the presence of either FtsA or ZipA, the membrane-associated proteins that interact with a small C-terminus domain of FtsZ (de Boer, 2010). FtsA, the most conserved actin-like ATPase protein, which tethers FtsZ to the cell membrane via a C-terminal amphipathic helix, is essential in *E. coli* but not in *B. subtilis* (Kirkpatrick and Viollier, 2011). By contrast, the ZipA protein, which provides a trans-membrane anchor, is restricted to the γ-proteobacteria, in which it is indispensable (Kirkpatrick and Viollier, 2011). Recently, ZipA was found to be required for FtsZ-dependent preseptal peptidoglycan synthesis prior to invagination during cell division (Potluri *et al.*, 2012).

In addition to the FtsA and ZipA proteins, the soluble FtsZ-binding proteins ZapA and ZapC also associate with early Z-ring assemblies (Kirkpatrick and Viollier, 2011). ZapA promotes the assembly and stability of the Z ring in vivo, and recruits the ZapB protein to the division site (Galli and Gerdes, 2012). Unlike FtsA and ZipA, the ZapA and ZapB proteins are not essential and *E. coli* single mutants show only modest phenotypes (de Boer, 2010). ZapC stabilizes FtsZ protofilaments by suppressing GTPase activity and consolidates them into bundles. ZapA is well conserved, while ZapB appears restricted to the γ-proteobacteria. ZapA possesses an N-terminal globular domain, likely interacting with FtsZ, and a long C-terminal coiled-coil domain, likely mediating ZapA dimerization. *In* vivo, it is possible that the local concentration of ZapA at the division site, where it likely interacts stoichiometrically with FtsZ, may be sufficient to induce a tetrameric conformation (Galli and Gerdes, 2012). ZapB structure consists of a unique coiled-coil domain, which could be involved in ZapB dimerization and polymerization, and interaction with the coiled-coil domain of ZapA. A fine balance between the levels of cell division proteins is likely required for cytokinesis to proceed normally, as the FtsZ, ZapA and ZapB proteins are present in the cells in similar amounts (5,000 to 15,000 per cell) (Galli and Gerdes, 2012).

In Gram-positive bacteria (Hamoen *et al.*, 2006; Kirkpatrick and Viollier, 2011), FtsZ stabilization involves SepF, a shaper of FtsZ that influences the structure of FtsZ bundles (Gundogdu *et al.*, 2011). Mutations in *B. subtilis* SepF (alternative name YlmF) abolish formation of the Z-ring and disrupt cell division. The cytokinesis of *B. subtilis* is more strongly distorted by the simultaneous absence of SepF and either FtsA or EzrA (Kirkpatrick and Viollier, 2011).

Reminiscent of *E. coli* FtsA, *B. subtilis* EzrA interacts directly with FtsZ, as well as with many cytoplasmic and periplasmic divisome components, including the peptidoglycan synthesis apparatus (PBP1) and the divisome, thus providing a link between the Z-ring in the interior of the cell and septum formation in the periplasm (Kirkpatrick and Viollier, 2011).

Cyanobacteria lack the *ftsA, zipA, zapA, zapB* and *zapC* genes (Lutkenhaus, 2007; Marbouty *et al.*, 2009c; Miyagishima *et al.*, 2005; and this study, Table 2.2). By contrast, a few cyanobacteria harbour *erzA* (Table 2.1), which has not been studied yet. Despite the absence of the FtsA protein, an FtsA-like divisome assembly function has been reported for the FtsZ-binding protein ZipN (Ftn2) of *Synechocystis* PCC6803, which recruits several divisome proteins (see below). In addition, cyanobacteria possess a SepF orthologue (alternative name Cdv2), which appeared to be indispensable in both *Synechococcus* PCC7942 and *Synechocystis* PCC6803. In both strains, SepF depletion generates giant cells, which remained rod-shaped in *Synechococcus* PCC7942 (Miyagishima *et al.*, 2005) and spherical in *Synechocystis* PCC6803 (Marbouty *et al.*, 2009b). The *Synechocystis* PCC6803 SepF protein localizes at the division site at mid-cell, and physically interacts with FtsZ to stimulate formation and/or stability of FtsZ polymers (Marbouty *et al.*, 2009b). These findings show that cyanobacteria (Gram-negative) share cytokinetic genes in common with Gram-positive bacteria.

Positioning of the Z-ring

In rod-shaped bacteria, such as *E. coli* and *B. subtilis*, FtsZ assembly is directed to the midpoint of the long axis of the mother cell, thereby generating daughter cells of equal size with a fully functional genome (de Boer, 2010; Mannik *et al.*, 2012; Wu and Errington, 2012). Two molecular mechanisms – the Min system and the nucleoid occlusion (Noc) system – work synergistically to restrict bacterial division to mid-cell, while ParA operates in chromosome (and plasmid) segregation (Lenz and Sogaard-Andersen, 2011; Lutkenhaus, 2012; Wu and Errington, 2012). The Min system prevents division nearby the cell poles, which would otherwise lead to the disadvantageous formation of nucleoid-free minicells, by inhibiting the polymerization of FtsZ. The Min system consists of three proteins MinC, MinD, and either MinE (as in *E. coli*) or DivIVA (as in *B. subtilis*). MinC inhibits FtsZ polymerization, and promotes the disassembly of FtsZ structures and their interactions with the Z-ring stabilizing proteins FtsA and ZipA. MinC activity is stimulated upon binding the MinD ATPase. While the ADP form (monomer) of MinD freely diffuses in the cytoplasm, the ATP form (dimer) binds cooperatively to the membrane via its C-terminal amphipathic helix that recruits MinC to the membrane. The MinCD complex is then directed to the cell poles by the topological determinants MinE or DivIVA. In *E. coli*, MinE accumulates in a MinE-ring that oscillates periodically from pole to pole in the cell, and is thus most frequently present at mid-cell, where it excludes the MinCD complex (Lenz and Sogaard-Andersen, 2011). MinE binding to MinD at the membrane stimulates the ATPase activity of MinD, releasing it from the membrane (Park *et al.*, 2012). The resulting membrane-associated MinE either dissociates or jumps to another membrane-bound MinD. This interplay between MinD, MinE and the membrane culminates in the pole-to-pole oscillation of MinCD (with a period of 1–2 min) through diffusion-association cycles between the cytoplasm and the cytoplasmic membrane (Kirkpatrick and Viollier, 2011; Lenz and Sogaard-Andersen, 2011; Lutkenhaus, 2012).

In contrast to the pure oscillatory system in *E. coli* and many other Gram-negative organisms, in *B. subtilis* the MinCD complex is fastened at the poles by the MinD-binding protein MinJ, which itself interacts with the pole-scouting protein DivIVA (Kirkpatrick and Viollier, 2011; Lutkenhaus, 2012). Thus, both MinE and DivIVA proteins enrich the MinCD complex at the poles, thereby preventing Z ring formation at these inappropriate sites, which would otherwise lead to the disadvantageous formation of nucleoid-free minicells.

All cyanobacteria possess the *minCDE* genes, which are clustered, excepted in the closely related strains *Synechococcus* PCC7942 and *Synechococcus* PCC6301 where *minC* is located at a distance from the adjacent *minD* and *minE* genes (Table 2.2). In *Anabaena* PCC7120, MinC

Table 2.2 Distribution and genomic organization of cyanobacterial genes involved in Z-ring assembly and positioning

Species	*ftsZ*	*ftsQ*	*zipN*	*sepF*	*ftsW*	*ftsI*	*ftsN*	*ftn6*	*minCDE*	*div4A*	*sulA*	*ylmD*	*ylmE*	*ylmG1*	*ylmG2*	*ftsE*
Acaryochloris marina MBIC11017	+	+a*	+d	+e*	+f	+g	–	+	+++i	+j	+	20	+e*	+	+	+n*
Cyanothece sp. ATCC51142	+	+	+d	+e	+	+g	+	+	+++i	+	+	18	+	+	+	–
Cyanothece sp. PCC7424	+	+b	+d	+e*	+	+g	–	+	+++i	+j	+k	13	+e*	+	+	–
Cyanothece sp. PCC7425	+	+a	+d	+e	+	+g	–	+	+++i	+j	+	11	+e*	+	+	+n*
Cyanothece sp. PCC7822	+	+b*	+d	+e*	+	+g	–	+	+++i	+j	+k	14	+e*	+	+	–
Cyanothece sp. PCC8801	+	+a	+	+e*	+	+g	+	+	+++i	+j	+k	9	+e*	+	+	–
Cyanothece sp. PCC8802	+	+a	+	+e*	+	+g	+	+	+++i	+j	+k	9	+e*	+	+	–
Microcystis aeruginosa NIES-843	+	+c	+	+e*	+	+g	–	+	+++i	–	+k	+	+e*	+	–	–
Synechococcus elongatus PCC6301	+	+c*	+d	+e*	+f*	+h	–	+	+++	+j	+k	7	+e*	+	+	–
Synechococcus elongatus PCC7942	+	+c*	+d	+e*	+f*	+h	–	+	+++	+j	+k	7	+e*	+	+	–
Synechococcus sp. CC9311	+	+c	+d*	+e**	+f*	+h	–	–	+++i*	+j	+l	2	+e**	+m	+	–
Synechococcus sp. CC9605	+	+c	+d*	+e**	+f*	+h	–	–	+++i*	+j	+l*	–	+e**	+m	+	–
Synechococcus sp. CC9902	+	+c	+d*	+e**	+f*	+h	–	–	+++i*	+j	+l*	–	+e**	+m	+	–
Synechococcus JA-2–3B′ a(2–13)	+	+	+	+e	+	+	–	+	+++i	+	+	–	+e**	+	+	+n
Synechococcus JA-3–3B′ Ab	+	+	+	+e	+	+	–	+	+++i	+	+	–	+e**	+	+	+n
Synechococcus sp. PCC7002	+	+b	+d	+e*	+	+g	–	+	+++i	+j	+	6	+e*	+	+	–
Synechococcus sp. RCC307	+	+c	+d*	+e**	+f*	+h	–	–	+++i*	+j	+l	–	+e**	+m	+	–
Synechococcus sp. WH 7803	+	+c	+d*	+e**	+f*	+h	–	–	+++i*	+j	+l	–	+e**	+m	+	–
Synechococcus sp. WH 8102	+	+c	+d*	+e**	+f*	+h	–	–	+++i*	+j	+l	–	+e**	+m	+	–
Synechocystis sp. PCC6803	+	+a	+	+	+	+g	+	+	+++i	+j	+	10	+	+	+	–
Thermosynechococcus elongatus BP-1	+	+	+	+	+	+g	–	+	+++i	+j	+	5	+	+	+	+n*
Cyanobacterium UCYN-A	+	+a	+d	+	+	+	–	–	+++i	–	–	–	+	+	+	–
Gloeobacter violaceus PCC7421	+	+a	+	+e*	+	+g	–	+	+++i	+j	–	–	+e*	+	+	+n*
Anabaena variabilis ATCC29413	+	+a**	+d	+e**	+	+g	–	+	+++i	+j	+	5	+e**	+	+	+n
Nostoc punctiforme PCC73102	+	+a**	+d	+e**	+	+g	–	+	+++i	+j	+	8	+e**	+	+	+n

Anabaena sp. PCC7120	+	+a	+d	+e**	+	+g	–	+	+++i	+j	+	6	+e**	+	+	+n
Nostoc azollae 0708	+	+a*	+d	+e*	+	+g	–	+	+++i	+j	+	2	+	+	+	+n
Trichodesmium erythraeum ISM101	+	+b*	+d	+	+	+g	–	+	+++i	+j	+	3	+e***	+	+	–
Prochlorococcus marinus AS9601	+	+c	+d**	+e***	+f**	+h	–	–	+++i*	–	+l**	–	+e***	+m	+	–
Prochlorococcus marinus MIT 9211	+	+c	+d**	+e***	+f**	+h	–	–	+++i*	–	+l**	–	+e***	+m	+	–
Prochlorococcus marinus MIT 9215	+	+c	+d**	+e***	+f**	+h	–	–	+++i*	–	+l**	–	+e***	+m	+	–
Prochlorococcus marinus MIT 9301	+	+c	+d**	+e***	+f**	+h	–	–	+++i*	–	+l**	–	+e***	+m	+	–
Prochlorococcus marinus MIT 9303	+	+c	+d**	+e**	+f**	+h	–	–	+++i*	+j	+l*	–	+e**	+m	+	–
Prochlorococcus marinus MIT 9312	+	+c	+d**	+e***	+f**	+h	–	–	+++i*	–	+l**	–	+e***	+m	+	–
Prochlorococcus marinus MIT 9313	+	+c	+d**	+e**	+f**	+h	–	–	+++i*	+j	+l*	–	+e**	+m	+	–
Prochlorococcus marinus MIT 9515	+	+c	+d**	+e***	+f**	+h	–	–	+++i*	–	+l**	–	+e***	+m	+	–
Prochlorococcus marinus NATL1A	+	+c	+d**	+e**	+f**	+h	–	–	+++i*	–	+l**	–	+e**	+m	+	–
Prochlorococcus marinus NATL2A	+	+c	+d**	+e**	+f**	+h	–	–	+++i*	–	+l**	–	+e**	+m	+	–
Prochlorococcus marinus SS120	+	+c	+d**	+e***	+f**	+h	–	–	+++i*	–	+l**	–	+e***	+m	+	–
Prochlorococcus marinus MED4	+	+c	+d**	+e***	+f**	+h	–	–	+++i*	–	+l**	–	+e**	+m	+	–

All relevant genes, presented here or in the text, are indicated along with their presence (+, or indicated numbers) or absence (–) in each strain.
Clusters: **a** = ftsQ-ftsZ; **a*** = ftsQ//ho//ftsZ; **a**** = ftsQ-ho-ftsZ; **b** = ftsQ-ftsZ-thiD; **b*** = ftsQ-ftsZ-thiD-ho; **c** = miaB-ddl-ho-ftsQ-ftsZ; **c*** = miaB-ddl-ho-ftsQ-ftsZ-thiD; **d** = pdhA//zipN; **d*** = pdhA//zipN-ffh-rpsP-phoH; **d**** = mnmA//yjef-rpoD-pdhA//zipN-ffh-rpsP-phoH; **e** = sepF-proC; **e*** = pIIx-ylmE-sepF-proC; **e**** = engA-cbiQ-pIIx-ylmE-sepF-proC; **e***** = aroQ-miaE-cobl-engA-cbiQ-pIIx-ylmE-sepF-procC; **f** = ftsW-ccdA-ccs1-glnB; **f*** = apcE-apcA-apcB-apcC-ftsW-ccdA-ccs1-glnB; **f**** = atp//ftsW-ccdA-ccs1-glnB; **i** = minCDE; **i*** = ctpA-dgt-minCDE; **j** = coaD-cdv3; **k** = psb28-sulA; **l** = cysK-cbpA-ho//cpa1-sulA; **l**** = cysK-cbpA-arnT-ho//cpa1-sulA; **m** = ilvAA-scpB-ylmG1; **n** = wecB-ftsE; **n*** = wecB-ftsE-perm. The symbol // indicate the oppositely oriented genes while hypothetical genes are indicated by ho.
Gene names: *thiD*, phosphomethylpyrimidine kinase; *miaB*, (dimethylallyl) adenosine tRNA methylthiotransferase; *ddl*, D-alanyl-alanine synthetase A; *pdhA/acoA*, pyruvate dehydrogenase E1 alpha subunit; *ffh*, signal recognition particle protein; *rpsP*, 30S ribosomal protein S16; *phoH*, PhoH-like phosphate starvation-inducible protein; *yjeF*, fused sugar kinase domain-containing protein; *rpoD*, type II alternative RNA polymerase sigma-70 factor; *mnmA*, tRNA-specific 2-thiouridylase; *MnmA*; SepF (alternative names cdv2 and ycf50); *proC*, pyrroline-5-carboxylate reductase; *engA*, GTP-binding protein engA (ribosome stability); *ylmE*, TIM-barrel fold family protein; *pIIX*, pII interaction protein X; *cbiQ*, cobalt ABC transporter permease; ccdA, c-type cytochrome biogenesis protein; *ccs1*, c-type cytochrome biogenesis protein; *glnB*, PII-protein encoding gene; apcABCE, phycobiliproteins; dgt, HD domain-containing protein; ctpA, carboxyl-terminal protease; *DivIVA* (alternative name cdv3A); *coaD*, phosphopantetheine adenylyl-transferase; *psb28*, photosystem II reaction centre protein Psb28; *cysK*, cysteine synthase; cbpA, dnaJ chaperone; *CPA1*, Na^+/H^+ antiporter family protein; arnT, glycosyl transferase family protein; YlmG1 = ycf19; *ilvA*, threonine dehydratase; *scpB*, segregation and condensation protein B; *wecB*, glycosyl transferase WecB/TagA/CpsF family protein; perm, permease.

is important for normal cell morphology and cytokinesis. The over-expression of *minC* generates giant spherical cells and blocks heterocysts differentiation (Sakr *et al.*, 2006). Similarly, in *Synechocystis* PCC6803 the MinCDE system, which is dispensable for survival, is important for normal cell morphology and cytokinesis (Mazouni *et al.*, 2004). The mutants lacking one to all three *minCDE* genes displayed distorted morphologies (spiralled or heart-shaped cells and minicells) likely resulting from an asymmetric invagination or septation, as well as abnormal shape and position of FtsZ-structures. Furthermore, MinD was found to be targeted to cell membranes in a process involving its C-terminal amphipathic helix, but not its ATP binding region (Mazouni *et al.*, 2004). By contrast to *E. coli* where *minCDE* deletions generate DNA-less minicells (de Boer *et al.*, 1989), *Synechocystis* Min-less minicells remained capable to divide, indicating that they possessed at least one copy of their chromosome (Mazouni *et al.*, 2004). In agreement with what occurs in *Synechocystis* PCC6803, the *minE* gene was also found to be dispensable to *Synechococcus* PCC7942 (Miyagishima *et al.*, 2005).

The cyanobacterial Cdv3 protein has also been studied, because it resembles DivIVA in size, amino-acid sequence, and position and length of the coiled-coil domain. The *cdv3/divIVA* gene is indispensable to both cell growth and division in *Synechococcus* PCC7942 and *Synechocystis* PCC6803. The depletion of Cdv3/DivIVA generates giant cells retaining an otherwise normal morphology, cylindrical in *Synechococcus* PCC7942 (Miyagishima *et al.*, 2005) or spherical in *Synechocystis* PCC6803 (Marbouty *et al.*, 2009c). Also, interestingly, the *cdv3/divIVA* gene is not present in a few cyanobacteria including UCYN-A, *Microcystis aeruginosa* NIES-843 and most marine strains (Table 2.2).

Although MinC and MinD were shown to be involved in cyanobacterial division (Mazouni *et al.*, 2004), it is still unknown whether they oscillate as in *E. coli*, due to the presence of MinE, or localize at the cell poles as in *B. subtilis*, because of Cdv3/DivIVA, or perhaps neither. Study of the localization of the MinCD complex, MinE and Cdv3/DivIVA in cyanobacteria should provide significant insight into how two different systems cooperate in cyanobacteria.

The *ylmG* gene, which in Gram-positive bacteria is located in the *dcw* (division and cell wall) cluster of the genes *ylmD*, *ylmE*, *ylmF/sepF*, *ylmG*, *ylmH*, and *divIVA* in that order (Fadda et al., 2003), has also been studied in cyanobacteria (Kabeya *et al.*, 2010). Except *Microcystis aeruginosa*, they all possess two *ylmG* genes (Table 2.2). In marine cyanobacteria, the *ylmG1* gene is invariably located downstream of *scpB*, the gene encoding a presumptive chromosome segregation and condensation protein. Both *ylmG* genes of *Synechocystis* PCC6803 appeared to be dispensable, like the *Synechococcus* PCC7942 *ylmG1* gene (*ylmG2* has not been studied). All corresponding knockout mutants displayed a normal cell shape and DNA content (Ishikawa *et al.*, 2009; Kabeya *et al.*, 2010; Marbouty *et al.*, 2009c). In contrast, the overexpression of the studied *Synechococcus* PCC7942 *ylmG* gene increases cell length and slightly affects both Z-ring positioning (it is biased towards a cell side) and nucleoid segregation (Kabeya *et al.*, 2010). By contrast to *ylmG*, the number of *ylmD* genes is strongly variable (Table 2.2). It is not found in UCYN-A, *Gloeobacter violaceus* and marine species, whereas other strains such as *Synechocystis* PCC6803, *Cyanothece* sp. and the symbiotic strain *Acaryochloris* harbour at least ten copies of *ylmD*. These findings suggest that the *ylmD* genes might operate in the adaptation of the cell morphology and division to changing environment. Concerning ylmE, which occurs in all cyanobacteria, we found that it is located upstream of *ylmF/sepF* in many cyanobacterial genomes, to the noticeable exception of *Synechocystis* PCC6803, *Thermosynechococcus elongatus* and UCYN-A (Table 2.2). By contrast, we found no *ylmH* genes in cyanobacteria. Collectively, these findings show that cyanobacteria, which are Gram-negative bacteria, share cytokinetic genes in common with not only Gram-negative but also Gram-positive bacteria.

In rod-shaped bacteria such as *B. subtilis* and *E. coli*, the Min system is normally sufficient for blocking cell division at mid-cell until the sister chromosomes have segregated. Nevertheless, *B. subtilis* and *E. coli* possess a second system, the nucleoid occlusion system (Noc), to preclude Z ring formation near the nucleoid (i.e. the bacterial chromosome) thereby preventing the chromosome from being bisected and broken by the division septum (Kirkpatrick and Viollier, 2011). Unlike the Min proteins, the nucleoid occlusion factors SlmA and Noc are bound to specific regions of the chromosome of *E. coli* and *B. subtilis*, respectively, and the DNA-bound form of SlmA has enhanced activity in depolymerizing FtsZ filaments (Mannik *et al.*, 2012; Wu and Errington, 2012). This Noc system is crucial in round bacteria that do not possess a Min system, such as the Gram-positive pathogen *Staphylococcus aureus*, otherwise phylogenetically close to *B. subtilis*. Indeed, deletion of Noc in *S. aureus* resulted in the formation of multiple Z rings and DNA breaks in about 15% of the mutant cells (Wu and Errington, 2012). We found no *slmA* genes in cyanobacteria. Another important FtsZ regulator in *E. coli* is the DNA damage (SOS)-induced inhibitor SulA, which is not well conserved beyond the γ-proteobacteria (Kirkpatrick and Viollier, 2011). SulA, which is dispensable to cell growth in absence of stress, directly sequesters FtsZ to prevent cytokinesis to occur before the damage has been repaired.

A *sulA* homologue is present in all cyanobacteria, to the noticeable exception of *Gloeobacter violaceus* and UCYN-A (Table 2.2). In *Synechocystis* PCC6803, *sulA* is indispensable to cell life and division. The SulA-depleted mutant displayed giant cells and cloverleaf-like clusters of four large unseparated cells, which likely result from the delayed septation of daughter cells, which was not completed before the initiation of the second round of division (Raynaud *et al.*, 2004).

Maturation of the Z-ring to a septal ring and cell constriction

In *E. coli* and *B. subtilis*, there is a substantial delay between assembly of the Z-ring and recruitment of the later assembling components. During this interval, the Z-ring directs cylindrical murein synthesis at midcell. The Z-ring is subsequently used as a scaffold for the hierarchical recruitment of downstream cytokinetic factors (ZapB → FtsK → FtsQ, FtsL/FtsB → FtsW → FtsI and FtsN) through an as yet unclear process likely involving specific protein-protein contacts (de Boer, 2010; Kirkpatrick and Viollier, 2011).

When divisome assembly is complete, synthesis of peptidoglycan allows the overall suprastructure to progressively curve in order to encircle the cell cylinder and drive constriction of the division site leading to the new cell poles of the future daughter cells (Typas *et al.*, 2012; Wu and Errington, 2012). Once constriction ensues, the septal ring (SR) needs to accomplish four main tasks in a coordinated manner: (i) inner-membrane (IM) invagination and fission, (ii) septal murein synthesis, (iii) septal murein splitting, and (iv) outer membrane (OM) invagination and fission. First, IM invagination is likely performed by the bending force of the Z-ring portion. Second, synthesis of septal murein, which is probably added perpendicular to the cylindrical murein, requires the transpeptidase PBP3 (also known as FtsI) and at least one other synthase, likely PBP1A, PBP1B, and/or MtgA. PBP3, recruited to the septum by the lipid II flippase FtsW, spans the cytoplasmic membrane at the division site, and interacts with the FtsQLB complex, PBP1B and FtsN (Grenga *et al.*, 2010). The carboxy-terminal peptidoglycan-binding SPOR domain of the essential protein FtsN may contribute to its mid-cell localization via specific recognition of septal peptidoglycan. The essential function of FtsN, mediated by a short (35 amino acids) periplasmic region that does not include its SPOR domain, may be related to its interactions with the peptidoglycan synthases PBP1B and/or PBP3 (Typas *et al.*, 2012). Third, splitting of septal murein from the outer membrane (OM)-proximal side, which shapes the two nascent cell poles and allow OM invagination, is accomplished by the three murein amidases AmiA, AmiB, and AmiC, and the LytM-domain proteins EnvC and NlpD (de Boer, 2010). The EnvC-FtsEX system directly stimulates the activities of the autoinhibited AmiA and AmiB enzymes while NlpD stimulates that of AmiC (Yang *et al.*, 2012). Fourth, OM invagination, which uses proton motive force, requires the action of the Tol–Pal complex that comprises three IM proteins (TolQAR), one periplasmic protein (TolB) and the OM lipoprotein Pal. Finally, the septal pore in the IM must close to compartmentalize the cytoplasm, while the OM must close to compartmentalize the periplasm and finalize the separation of daughter cells. This process involves FtsB and the DNA translocase FtsK, as well as other mechanisms of membrane fission (Typas *et al.*, 2012).

All cyanobacterial genomes harbour *fstQ* and *ftsW* single-copy genes in well-conserved genomic organization (Table 2.2), whereas very few strains possess *ftsE* and *ftsN*, which appeared to be dispensable in *Synechocystis* PCC6803 (Marbouty *et al.*, 2009c). In most cases, *ftsQ* is located upstream of *ftsZ*, while *ftsW* is located upstream of cytochrome-encoding genes (Table 2.2). We have analysed the FtsQ and FtsW proteins of *Synechocystis* PCC6803, which appeared to be indispensable to cell viability, like FtsI (Marbouty *et al.*, 2009a). The mutants depleted of FtsQ, FtsW or FtsI exhibit similar defects; i.e. they grow slowly and display giant cells. We think that the giant morphology of spherical cells occurs when their septation is slowed-down more severely than their growth. By contrast, in a rod-shaped bacterium as *E. coli* or *Synechococcus* PCC7942, when septation is more affected than cell growth, the corresponding *fts* mutants become filamentous (Bi and Lutkenhaus, 1991; Koksharova and Wolk, 2002). Furthermore, we showed that both FtsQ and FtsI are self-interacting proteins, which physically interact with FtsW, as observed in *E. coli* (Di Lallo *et al.*, 2003; Karimova *et al.*, 2005). Also interestingly, we found that both FtsI and FtsQ interact physically with both PBP1 and PBP3 that interact with PBP2 (Marbouty *et al.*, 2009a).

Conservation of cytokinetic proteins in cyanobacteria and the models *E. coli* and/or *B. subtilis*

Several cell division genes occurring in *E. coli* and/or *B. subtilis* are conserved in cyanobacteria (Table 2.1 and Table 2.2). Likely in agreement with their importance in cyanobacterial morphogenesis and/or cytokinesis these are found either in all strains (*ftsZ, minCDE, ftsI, ftsQ, ftsW, PBP, ylmE, ylmF/sepF, ylmG, mraY, murA-G, amiC* (to the exception of strain UCYN-A), *mreBCD* (except in both UCYN-A and *Synechocystis* PCC6803) and *rodA* (except in *Synechocystis* PCC6803)); in most strains (*amiA, divIVA, nlpD, sulA, tolB* and *ylmD*) or in a few strains (*erzA, ftsE* and *ftsN*). These findings show that cyanobacteria (Gram-negative, photosynthetic) share molecular mechanisms for cytokinesis in common with non-photosynthetic bacteria of not only the Gram-negative (*E. coli*) but also the Gram-positive (*B. subtilis*) types. In addition, many of the cyanobacterial cell division proteins (FtsZ, MinDE, SulA, FtsI, FtsQ and FtsW) are conserved in chloroplasts (Miyagishima, 2011; Raynaud *et al.*, 2004; Yang *et al.*, 2008), in agreement with cyanobacteria being their likely progenitors (Archibald, 2009; Grey, 1993).

Identification and analysis of cyanobacterial cell division proteins that are not present in *E. coli* and *B. subtilis*

The genes *ftsA, ftsB, zipA, ftsL, ftsX, zapA, zapB, zapC* and *yabQ*, which operate in *E. coli* and/or *B. subtilis* cytokinesis, occur neither in cyanobacteria (Lutkenhaus, 2007; Marbouty *et al.*, 2009a,c; Miyagishima *et al.*, 2005) nor in the plastid or nuclear genomes of plants and algae (Miyagishima, 2011; Miyagishima *et al.*, 2005), in agreement again with the endosymbiotic theory (Archibald, 2009; Gray, 1993). Consequently, it is important to not only rely on sequence homology with cell division genes from well-known model bacteria to search for cyanobacterial genes operating in cytokinesis. Two such genes, *ftn2* (also named ZipN) and *ftn6* (alternative name ZipS), have been identified in the rod-shaped cyanobacterium *Synechococcus* PCC7942 after transposon mutagenesis and selection of mutants with an aberrant morphology of the colonies (extensive spreading) and a giant size (Dolganov and Grossman, 1993; Koksharova and Wolk, 2002; Miyagishima *et al.*, 2005). These genes are present in all (*ftn2/zipN*, Table 2.2) or many cyanobacteria (*ftn6/ZipS*, to the exception of UCYN-A and the marine strains). That they have no orthologue in *E. coli* and *B. subtilis* indicates that some of the molecular mechanisms for cytokinesis in cyanobacteria are different from those in *E. coli* and *B. subtilis*. The *ftn2/zipN* gene is dispensable to cell life in *Synechococcus* PCC7942, whereas it is indispensable in both *Anabaena* PCC7120 (the heteroploid, i.e. unsegregated mutant, harbours enlarged cells) (Koksharova and Wolk, 2002) and *Synechocystis* PCC6803 (the heteroploid mutant display spiralled cells and minicells) (Mazouni *et al.*, 2004). By contrast, the *ftn6/ZipS* gene is dispensable to both *Synechococcus* PCC7942 and *Anabaena* PCC7120 (Koksharova and Wolk, 2002), whereas it is indispensable in *Synechocystis* PCC6803 (the depletion of Ftn6 generates giant cells remaining spherical) (Marbouty *et al.*, 2009b). Z rings were not observed in the *ftn2/zipN* and *ftn6/ZipS* inactivation mutants of *Synechococcus* PCC7942, suggesting that their elongated size resulted from a default in Z ring assembly (Miyagishima *et al.*, 2005). By contrast, aberrant FtsZ structures were observed in the *ftn2/zipN*- and *ftn6/ZipS*-depleted mutants of *Synechocystis* PCC6803 (Marbouty *et al.*, 2009b,c). The reasons for these discrepancies between the two unicellular cyanobacteria *Synechococcus* PCC7942 (rod-shaped) and *Synechocystis* PCC6803 (round-shaped) are unclear, but these findings emphasize the importance of parallel studies of cell division in these two-model cyanobacteria.

We have thoroughly analysed the Ftn2/ZipN and Ftn6/ZipS indispensable proteins of *Synechocystis* PCC6803, as well as the interactions network between them and the other cell division proteins. Both Ftn2/ZipN and Ftn6/ZipS are septal proteins that physically interact with FtsZ (Marbouty *et al.*, 2009b; Mazouni *et al.*, 2004). We showed that ZipN is a self-interacting protein that interacts with the other indispensable cytokinetic factors, FtsZ (Mazouni *et al.*, 2004), FtsI and FtsQ (Marbouty *et al.*, 2009a), Cdv3/DivIVA (also a self-interacting protein) (Marbouty *et al.*, 2009c), SepF/YlmF and ZipS/Ftn6 (Marbouty *et al.*, 2009b,c), as well as with the dispensable cell division protein YlmD (Marbouty *et al.*, 2009c). By contrast, we found no direct interaction between Cdv3, SepF and ZipS, and between ZipN and any of the dispensable factors AmiC, AmpH, FtsE, FtsN, SpoIID, YlmE and YlmG (Marbouty *et al.*, 2009c). Collectively, these findings show that ZipN plays a central role in the assembly of the *Synechocystis* PCC6803 divisome (Fig. 2.1). Hence, we propose that ZipN, which occurs in all cyanobacteria and the chloroplast (alternative name Arc6), is a functional analogue of the FtsA orchestrator of *E. coli* cell division, which is absent in cyanobacteria and chloroplasts (Marbouty *et al.*, 2009c; Miyagishima *et al.*, 2005). However, the presence in ZipN of a membrane anchor makes it also comparable to ZipA (Kirkpatrick and Viollier, 2011). Concerning the FtsZ interacting protein Ftn6/ZipS, we showed that it operates in the tolerance to DNA damages generated by UV

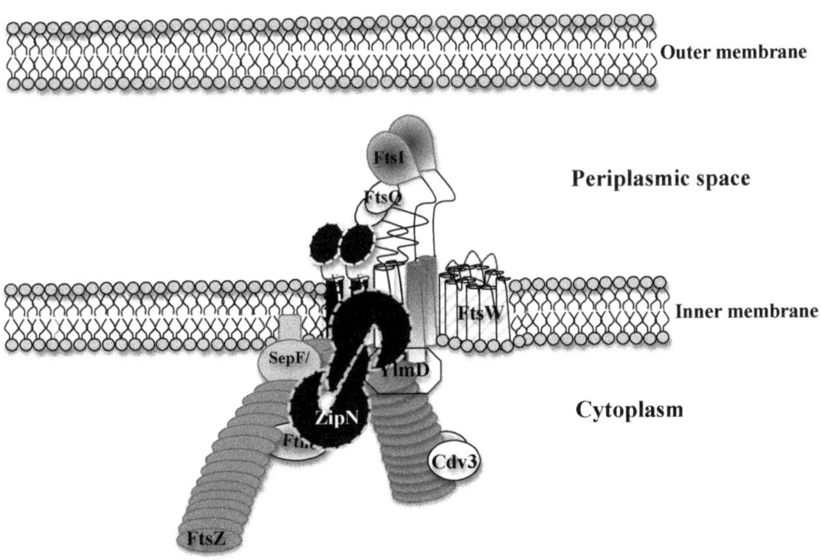

Figure 2.1 Topological model of the cyanobacterial divisome based on available information described in this review.

light, in agreement with the fact that it possesses a DnaD-like domain known in other proteins to physically interact with DNA. Hence, Ftn6/ZipS might be a molecular link between cell division and DNA metabolism (Marbouty *et al.*, 2009d).

Cyanobacteria-derived components of the chloroplast division machinery

Although plastids have retained some proteins of the cyanobacterial cell division machine, many other were lost after the endosymbiotic event (Miyagishima, 2011; Yang *et al.*, 2008). Whereas cyanobacterial genomes encode a single FtsZ protein, green algae and land plants have two types of nucleus-encoded FtsZ proteins, FtsZ1/FtsZ2, which are required for assembling the chloroplast Z-ring that enables chloroplast division. FtsZ2 contains a short C-terminal motif that mediates the interaction between FtsZ2 and ARC6, the nucleus-encoded protein descended from the cyanobacterial protein Ftn2/ZipN (Koksharova and Wolk, 2002; Maple *et al.*, 2005; Mazouni *et al.*, 2004; Miyagishima *et al.*, 2005; Vitha *et al.*, 2003), the orchestrator of the divisome assembly (Marbouty *et al.*, 2009c). FtsZ1, which likely emerged by duplication of FtsZ2 after the endosymbiotic event, does not possess the Arc6-interacting C-terminal motif, unlike Ftsz2 (Miyagishima, 2011). SepF is encoded in the plastid (cyanelle) genome of the glaucophyte *Cyanophora paradoxa*, but the function in plastid division has not been determined. Similarly, FtsW is encoded in the plastid genome of glaucophytes and a certain green algae. FtsI is encoded in the plastid genome of a certain green algae. MinC, absent in plants, is encoded in the nuclear genome of (some) certain algal lineages and the moss *P. patens*. The nucleus-encoded MinD and MinE protein regulate positioning of the chloroplast FtsZ ring.

Concluding remarks

Cyanobacteria that display a variety of shapes (rods, spheres, ovoids, helixes), which are maintained during vegetative growth, are excellent models to study how cells grow, divide, and pass their morphology on to their progeny. In this chapter, we have shown that these processes require the tight regulation and precise coordination of several sophisticated mechanisms, as occurs in the 'historical' bacterial models *Escherichia coli* (Gram-negative bacterium) and *Bacillus subtilis* (Gram-negative bacterium). Consistently, cyanobacteria (which are Gram-negative bacteria) share cell division genes in common with both Gram-positive and Gram-negative bacteria. Cyanobacteria also share cytokinetic genes with the chloroplast and the nuclear genome of plants and algae, in agreement with the endosymbiotic theory. The stromal portion of the chloroplast division complex resemble the cyanobacterial cell division machinery, but many other components were lost after the endosymbiotic event that transformed a cyanobacterial ancestor into a chloroplast.

Much remains to be learned about cell division in cyanobacteria. How all the activities of septal ring proteins are controlled to achieve an efficient and coordinated invagination of the cell envelope layers? How the whole constriction process is coordinated with other cell cycle events, such as chromosome replication and cell growth? Further construction and/or investigations of cell division mutants to better characterize cytokinetic factors, and possibly identify missing players, combined to efforts to reconstitute parts of the cyanobacterial division machinery and its regulators with purified components *in vitro* should shed light on some of these questions.

References

Abed, R.M., Dobretsov, S., and Sudesh, K. (2009). Applications of cyanobacteria in biotechnology. J Appl Microbiol *106*, 1–12.

Archibald, J.M. (2009). The puzzle of plastid evolution. Curr. Biol. *19*, R81–88.

Bandyopadhyay, A., Elvitigala, T., Welsh, E., Stockel, J., Liberton, M., Min, H., Sherman, L.A., and Pakrasi, H.B. (2011). Novel metabolic attributes of the genus *Cyanothece*, comprising a group of unicellular nitrogen-fixing *Cyanothece*. mBio 2, e00214–11.

Banzhaf, M., van den Berg van Saparoea, B., Terrak, M., Fraipont, C., Egan, A., Philippe, J., Zapun, A., Breukink, E., Nguyen-Disteche, M., den Blaauwen, T., *et al.* (2012). Cooperativity of peptidoglycan synthases active in bacterial cell elongation. Mol. Microbiol. *85*, 179–194.

Begg, K.J., and Donachie, W.D. (1998). Division planes alternate in spherical cells of *Escherichia coli*. J. Bacteriol. *180*, 2564–2567.

Bi, E.F., and Lutkenhaus, J. (1991). FtsZ ring structure associated with division in *Escherichia coli*. Nature *354*, 161–164.

de Boer, P.A. (2010). Advances in understanding *E. coli* cell fission. Curr. Opin. Microbiol. *13*, 730–737.

de Boer, P.A., Crossley, R.E., and Rothfield, L.I. (1989). A division inhibitor and a topological specificity factor coded for by the minicell locus determine proper placement of the division septum in *E. coli*. Cell *56*, 641–649.

Carballido-Lopez, R. (2006). Orchestrating bacterial cell morphogenesis. Mol. Microbiol. *60*, 815–819.

Chauvat, F., Corre, B., Herdman, M., and Joset-Espardellier, J. (1982). Energetic and metabolic requirements for the germination of akinetes of the cyanobacterium *Nostoc* PCC7524. Arch. Microbiol. *133*, 44–49.

Di Lallo, G., Fagioli, M., Barionovi, D., Ghelardini, P., and Paolozzi, L. (2003). Use of a two-hybrid assay to study the assembly of a complex multicomponent protein machinery: bacterial septosome differentiation. Microbiology *149*, 3353–3359.

Doherty, H.M., and Adams, D.G. (1995). Cloning and sequence of *ftsZ* and flanking regions from the cyanobacterium *Anabaena* PCC 7120. Gene *163*, 93–96.

Dolganov, N., and Grossman, A.R. (1993). Insertional inactivation of genes to isolate mutants of *Synechococcus* sp. strain PCC 7942: isolation of filamentous strains. J. Bacteriol. *175*, 7644–7651.

Ducat, D.C., Way, J.C., and Silver, P.A. (2011). Engineering cyanobacteria to generate high-value products. Trends Biotechnol. *29*, 95–103.

Fadda, D., Pischedda, C., Caldara, F., Whalen, M.B., Anderluzzi, D., Domenici, E., and Massidda, O. (2003). Characterization of divIVA and other genes located in the chromosomal region downstream of the dcw cluster in *Streptococcus pneumoniae*. J. Bacteriol. *185*, 6209–6214.

Freire, P., Moreira, R.N., and Arraiano, C.M. (2009). BolA inhibits cell elongation and regulates MreB expression levels. J. Mol. Biol. *385*, 1345–1351.

Galli, E., and Gerdes, K. (2012). FtsZ-ZapA-ZapB interactome of *Escherichia coli*. J. Bacteriol. *194*, 292–302.

Gray, M.W. (1993). Origin and evolution of organelle genomes. Curr. Opin. Genet. Dev. *3*, 884–890.

Grenga, L., Guglielmi, G., Melino, S., Ghelardini, P., and Paolozzi, L. (2010). FtsQ interaction mutants: a way to identify new antibacterial targets. New Biotechnol. *27*, 870–881.

Griese, M., Lange, C., and Soppa, J. (2011). Ploidy in cyanobacteria. FEMS Microbiol. Lett. *323*, 124–131.

Guinote, I.B., Matos, R.G., Freire, P., and Arraiano, C.M. (2011). BolA affects cell growth, and binds to the promoters of penicillin-binding proteins 5 and 6 and regulates their expression. J. Microbiol. Biotechnol. *21*, 243–251.

Gundogdu, M.E., Kawai, Y., Pavlendova, N., Ogasawara, N., Errington, J., Scheffers, D.J., and Hamoen, L.W. (2011). Large ring polymers align FtsZ polymers for normal septum formation. EMBO J. *30*, 617–626.

Hamoen, L.W., Meile, J.C., de Jong, W., Noirot, P., and Errington, J. (2006). SepF, a novel FtsZ-interacting protein required for a late step in cell division. Mol. Microbiol. *59*, 989–999.

Hoiczyk, E., and Hansel, A. (2000). Cyanobacterial cell walls: news from an unusual prokaryotic envelope. J. Bacteriol. *182*, 1191–1199.

Hu, B., Yang, G., Zhao, W., Zhang, Y., and Zhao, J. (2007). MreB is important for cell shape but not for chromosome segregation of the filamentous cyanobacterium *Anabaena* sp. PCC 7120. Mol. Microbiol. *63*, 1640–1652.

Ishikawa, M., Fujiwara, M., Sonoike, K., and Sato, N. (2009). Orthogenomics of photosynthetic organisms: bioinformatic and experimental analysis of chloroplast proteins of endosymbiont origin in *Arabidopsis* and their counterparts in *Synechocystis*. Plant Cell. Physiol. *50*, 773–788.

Jansson, C., and Northen, T. (2010). Calcifying cyanobacteria – the potential of biomineralization for carbon capture and storage. Curr. Opin. Biotechnol. *21*, 365–371.

Kabeya, Y., Nakanishi, H., Suzuki, K., Ichikawa, T., Kondou, Y., Matsui, M., and Miyagishima, S.Y. (2010). The YlmG protein has a conserved function related to the distribution of nucleoids in chloroplasts and cyanobacteria. BMC Plant Biol. *10*, 57.

Karimova, G., Dautin, N., and Ladant, D. (2005). Interaction network among *Escherichia coli* membrane proteins involved in cell division as revealed by bacterial two-hybrid analysis. J. Bacteriol. *187*, 2233–2243.

Kirkpatrick, C.L., and Viollier, P.H. (2011). New(s) to the (Z-)ring. Curr. Opin. Microbiol. *14*, 691–697.

Koksharova, O.A., and Wolk, C.P. (2002). A novel gene that bears a DnaJ motif influences cyanobacterial cell division. J. Bacteriol. *184*, 5524–5528.

Kuhn, I., Peng, L., Bedu, S., and Zhang, C.C. (2000). Developmental regulation of the cell division protein FtsZ in *Anabaena* sp. strain PCC 7120, a cyanobacterium capable of terminal differentiation. J. Bacteriol. *182*, 4640–4643.

Labarre, J., Chauvat, F., and Thuriaux, P. (1989). Insertional mutagenesis by random cloning of antibiotic resistance genes into the genome of the cyanobacterium *Synechocystis* strain PCC 6803. J. Bacteriol. *171*, 3449–3457.

Lazaro, S., Fernandez-Pinas, F., Fernandez-Valiente, E., Blanco-Rivero, A., and Leganes, F. (2001). pbpB, a gene coding for a putative penicillin-binding protein, is required for aerobic nitrogen fixation in the cyanobacterium *Anabaena* sp. strain PCC7120. J. Bacteriol. *183*, 628–636.

Leganes, F., Blanco-Rivero, A., Fernandez-Pinas, F., Redondo, M., Fernandez-Valiente, E., Fan, Q., Lechno-Yossef, S., and Wolk, C.P. (2005). Wide variation in the cyanobacterial complement of presumptive penicillin-binding proteins. Arch. Microbiol. *184*, 234–248.

Lehner, J., Zhang, Y., Berendt, S., Rasse, T.M., Forchhammer, K., and Maldener, I. (2011). The morphogene *AmiC2* is pivotal for multicellular development in the cyanobacterium *Nostoc punctiforme*. Mol. Microbiol. 79, 1655–1669.

Lenz, P., and Sogaard-Andersen, L. (2011). Temporal and spatial oscillations in bacteria. Nat. Rev. Microbiol. *9*, 565–577.

Li, H., Mapolelo, D.T., Randeniya, S., Johnson, M.K., and Outten, C.E. (2012). Human glutaredoxin 3 forms [2Fe–2S]-bridged complexes with human BolA2. Biochemistry *51*, 1687–1696.

Liu, X., Sheng, J., and Curtiss, R., 3rd (2011). Fatty acid production in genetically modified cyanobacteria. Proc. Natl. Acad. Sci. U.S.A. *108*, 6899–6904.

Lopes Pinto, F., Erasmie, S., Blikstad, C., Lindblad, P., and Oliveira, P. (2011). FtsZ degradation in the cyanobacterium *Anabaena* sp. strain PCC 7120. J. Plant Physiol. *168*, 1934–1942.

Lowe, J., and Amos, L.A. (1998). Crystal structure of the bacterial cell-division protein FtsZ. Nature *391*, 203–206.

Lutkenhaus, J. (2007). Assembly dynamics of the bacterial MinCDE system and spatial regulation of the Z ring. Annu. Rev. Biochem. *76*, 539–562.

Lutkenhaus, J. (2012). The ParA/MinD family puts things in their place. Trends Microbiol. *20*, 411–418.

Mannik, J., Wu, F., Hol, F.J., Bisicchia, P., Sherratt, D.J., Keymer, J.E., and Dekker, C. (2012). Robustness and accuracy of cell division in *Escherichia coli* in diverse cell shapes. Proc. Natl. Acad. Sci. U.S.A. *109*, 6957–6962.

Maple, J., Aldridge, C., and Moller, S.G. (2005). Plastid division is mediated by combinatorial assembly of plastid division proteins. Plant J. *43*, 811–823.

Marbouty, M., Mazouni, K., Saguez, C., Cassier-Chauvat, C., and Chauvat, F. (2009a). Characterization of the *Synechocystis* strain PCC 6803 penicillin-binding proteins and cytokinetic proteins FtsQ and FtsW and their network of interactions with ZipN. J. Bacteriol. *191*, 5123–5133.

Marbouty, M., Saguez, C., Cassier-Chauvat, C., and Chauvat, F. (2009b). Characterization of the FtsZ-interacting septal proteins SepF and Ftn6 in the spherical-celled cyanobacterium *Synechocystis* strain PCC 6803. J. Bacteriol. *191*, 6178–6185.

Marbouty, M., Saguez, C., Cassier-Chauvat, C., and Chauvat, F. (2009c). ZipN, an FtsA-like orchestrator of divisome assembly in the model cyanobacterium *Synechocystis* PCC6803. Mol. Microbiol. *74*, 409–420.

Marbouty, M., Saguez, C., and Chauvat, F. (2009d). The cyanobacterial cell division factor Ftn6 contains an N-terminal DnaD-like domain. BMC Struct. Biol. *9*, 54.

Margolin, W. (2005). FtsZ and the division of prokaryotic cells and organelles. Nature Rev. Mol. Cell Biol. *6*, 862–871.

Margolin, W. (2009). Sculpting the bacterial cell. Curr. Biol. *19*, R812–822.

Mariscal, V., and Flores, E. (2010). Multicellularity in a heterocyst-forming cyanobacterium: pathways for intercellular communication. Adv. Exp. Med. Biol. *675*, 123–135.

Mazouni, K., Domain, F., Cassier-Chauvat, C., and Chauvat, F. (2004). Molecular analysis of the key cytokinetic components of cyanobacteria: FtsZ, ZipN and MinCDE. Mol. Microbiol. *52*, 1145–1158.

Miyagishima, S.Y. (2011). Mechanism of plastid division: from a bacterium to an organelle. Plant Physiol. *155*, 1533–1544.

Miyagishima, S.Y., Wolk, C.P., and Osteryoung, K.W. (2005). Identification of cyanobacterial cell division genes by comparative and mutational analyses. Mol. Microbiol. *56*, 126–143.

Mori, T., and Johnson, C.H. (2001). Independence of circadian timing from cell division in cyanobacteria. J. Bacteriol. *183*, 2439–2444.

Park, K.T., Wu, W., Lovell, S., and Lutkenhaus, J. (2012). Mechanism of the asymmetric activation of the MinD ATPase by MinE. Mol. Microbiol. *85*, 271–281.

Partensky, F., Hess, W.R., and Vaulot, D. (1999). *Prochlorococcus*, a marine photosynthetic prokaryote of global significance. Microbiol. Mol. Biol. Rev. *63*, 106–127.

Pearson, L., Mihali, T., Moffitt, M., Kellmann, R., and Neilan, B. (2010). On the chemistry, toxicology and genetics of the cyanobacterial toxins, microcystin, nodularin, saxitoxin and cylindrospermopsin. Marine Drugs *8*, 1650–1680.

Potluri, L.P., Kannan, S., and Young, K.D. (2012). ZipA is required for FtsZ-dependent pre-septal peptidoglycan synthesis prior to invagination during cell division. J. Bacteriol. *194*, 5334–5342.

Raynaud, C., Cassier-Chauvat, C., Perennes, C., and Bergounioux, C. (2004). An *Arabidopsis* homolog of the bacterial cell division inhibitor SulA is involved in plastid division. Plant Cell *16*, 1801–1811.

Reith, J., and Mayer, C. (2011). Peptidoglycan turnover and recycling in Gram-positive bacteria. Appl. Microbiol. Biotechnol. *92*, 1–11.

Rippka, R., Deruelles, J., Waterbury, J.B., Herdman, M., and Stanier, R.Y. (1979). Generic assignments, strains histories and properties of pure cultures of cyanobacteria. J. Gen. Microbiol. *111*, 1–61.

Sakr, S., Thyssen, M., Denis, M., and Zhang, C.C. (2006). Relationship among several key cell cycle events in the developmental cyanobacterium *Anabaena* sp. strain PCC 7120. J. Bacteriol. *188*, 5958–5965.

Scanlan, D.J., Ostrowski, M., Mazard, S., Dufresne, A., Garczarek, L., Hess, W.R., Post, A.F., Hagemann, M., Paulsen, I., and Partensky, F. (2009). Ecological genomics of marine picocyanobacteria. Microbiol. Mol. Biol. Rev. *73*, 249–299.

Schirmer, A., Rude, M.A., Li, X., Popova, E., and del Cardayre, S.B. (2010). Microbial biosynthesis of alkanes. Science *329*, 559–562.

Schneider, D., Fuhrmann, E., Scholz, I., Hess, W.R., and Graumann, P.L. (2007). Fluorescence staining of live cyanobacterial cells suggest non-stringent chromosome segregation and absence of a connection between cytoplasmic and thylakoid membranes. BMC Cell Biol. *8*, 39.

Schopf, W.J. (2011). The paleobiological record of photosynthesis. Photosynthesis Res. *107*, 87–101.

Shi, T., and Falkowski, P.G. (2008). Genome evolution in cyanobacteria: the stable core and the variable shell. Proc. Natl. Acad. Sci. U.S.A. *105*, 2510–2515.

Thiel, T., and Wolk, C.P. (1983). Metabolic activities of isolated akinetes of the cyanobacterium *Nostoc spongiaeforme*. J. Bacteriol. *156*, 369–374.

Typas, A., Banzhaf, M., Gross, C.A., and Vollmer, W. (2012). From the regulation of peptidoglycan synthesis to bacterial growth and morphology. Nat. Rev. Microbiol. *10*, 123–136.

Tzagoloff, H., and Novick, R. (1977). Geometry of cell division in *Staphylococcus aureus*. J. Bacteriol. *129*, 343–350.

Urbach, C., Fastrez, J., and Soumillion, P. (2008). A new family of cyanobacterial penicillin-binding proteins. A missing link in the evolution of class A beta-lactamases. J. Biol. Chem. *283*, 32516–32526.

Vitha, S., Froehlich, J.E., Koksharova, O., Pyke, K.A., van Erp, H., and Osteryoung, K.W. (2003). ARC6 is a J-domain plastid division protein and an evolutionary descendant of the cyanobacterial cell division protein Ftn2. Plant Cell *15*, 1918–1933.

Wachi, M., Osaka, K., Kohama, T., Sasaki, K., Ohtsu, I., Iwai, N., Takada, A., and Nagai, K. (2006). Transcriptional analysis of the *Escherichia coli mreBCD* genes responsible for morphogenesis and chromosome segregation. Biosci. Biotechnol. Biochem. *70*, 2712–2719.

Wagner-Herman, J.K., Bernard, R., Dunne, R., Bisson-Filho, A.W., Kumar, K., Nguyen, T., Mulcahy, L., Koullias, J., Gueiros-Filho, F.J., and Rudner, D.Z. (2012). RefZ Facilitates the Switch from Medial to Polar Division during Spore Formation in *Bacillus subtilis*. J. Bacteriol. *194*, 4608–4618.

Wang, H., Fewer, D.P., and Sivonen, K. (2011). Genome mining demonstrates the widespread occurrence of gene clusters encoding bacteriocins in cyanobacteria. PloS One *6*, e22384.

Wang, X., and Lutkenhaus, J. (1996). FtsZ ring: the eubacterial division apparatus conserved in archaebacteria. Mol. Microbiol. *21*, 313–319.

Westling-Haggstrom, B., Elmros, T., Normark, S., and Winblad, B. (1977). Growth pattern and cell division in *Neisseria gonorrhoeae*. J. Bacteriol. *129*, 333–342.

Wilk, L., Strauss, M., Rudolf, M., Nicolaisen, K., Flores, E., Kuhlbrandt, W., and Schleiff, E. (2011). Outer membrane continuity and septosome formation between vegetative cells in the filaments of *Anabaena* sp. PCC 7120. Cell. Microbiol. *13*, 1744–1754.

Williams, P.G. (2009). Panning for chemical gold: marine bacteria as a source of new therapeutics. Trends Biotechnol. *27*, 45–52.

Wu, L.J., and Errington, J. (2012). Nucleoid occlusion and bacterial cell division. Nat. Rev. Microbiol. *10*, 8–12.

Yang, D.C., Tan, K., Joachimiak, A., and Bernhardt, T.G. (2012). A conformational switch controls cell wall-remodelling enzymes required for bacterial cell division. Mol. Microbiol. *85*, 768–781.

Yang, Y., Glynn, J.M., Olson, B.J., Schmitz, A.J., and Osteryoung, K.W. (2008). Plastid division: across time and space. Curr. Opin. Plant Biol. *11*, 577–584.

Zehr, J.P. (2011). Nitrogen fixation by marine cyanobacteria. Trends Microbiol. *19*, 162–173.

Zhang, C.C., Huguenin, S., and Friry, A. (1995). Analysis of genes encoding the cell division protein FtsZ and a glutathione synthetase homologue in the cyanobacterium *Anabaena* sp. PCC 7120. Res. Microbiol. *146*, 445–455.

Zhou, J., and Li, Y. (2010). Engineering cyanobacteria for fuels and chemicals production. Protein Cell *1*, 207–210.

The Cell Envelope

3

Alexander Hahn and Enrico Schleiff

Abstract

Cyanobacteria are prokaryotes with cell envelopes typical for Gram-negative bacteria. The cell envelope consists of four distinct layers, the plasma membrane, the peptidoglycan layer, the outer membrane and, in some cases, the surface or S-layer. Often, the latter three are referred to as the cell wall. The functionality of the cell envelope is defined by the cooperative action of lipids and membrane-embedded proteins. The membranes of cyanobacterial species contain two types of lipids, phosphoglycerolipids and galactolipids. The proteins of the plasma membrane show the typical α-helix based membrane domain architecture, whereas the proteins in the outer membrane have a β-barrel shaped membrane domain. The proteins perform many distinct functions ranging from solute transport to signal transduction. Thus, several features are indeed comparable between the cyanobacterial and the proteobacterial systems investigated so far. However, some properties are unique for the cyanobacterial branch. In this chapter, we summarize the current knowledge on composition, structure and function of the cell envelope including information obtained from different cyanobacterial strains. We also compare the properties of the cyanobacterial envelope to those of non-photosynthetic Gram-negative bacteria.

The structural properties of the cell envelope

Cyanobacteria are generally classified as Gram-negative bacteria. The cell envelope has a well-defined architecture containing a plasma membrane, which is sometimes also termed cytoplasmic membrane, a peptidoglycan layer, an outer membrane and, in some cases, also an S-layer (Fig. 3.1A). They all together are annotated as the cell envelope, while the term cell wall often refers to the peptidoglycan layer and the outer membrane (and in some cases the S-layer). In addition, cell-type specific surface structures exist in special cases. For example, the heterocysts of filamentous cyanobacteria are surrounded by a specific glycolipid layer (HGL) and by a specific polysaccharide layer (Hep). In this section we describe the overall structure of the cell envelope.

The continuity of the outer membrane in filamentous cyanobacteria

Some cyanobacteria grow as filaments of more than 100 cells including vegetative, normally photosynthetically active cells and more specialized cells such as heterocysts. Early electron microscopic analyses suggested that a structural element corresponding to the outer membrane extends into the periphery of such filaments, but does not enter the septum between two consecutive cells (Ris and Singh, 1961). More recent studies, e.g. on *Phormidium uncinatum* (Hoiczyk and Baumeister, 1995) or *Anabaena* sp. PCC 7120 (Bauer *et al.*, 1995; Flores *et al.*, 2006), are consistent with this observation. By electron tomography on entire cells of *Anabaena* sp. PCC 7120 it was demonstrated that the outer membrane does not penetrate the septum between two consecutive cells in a filament (Fig. 3.1; Wilk *et al.*, 2011). In contrast, the peptidoglycan layer surrounds

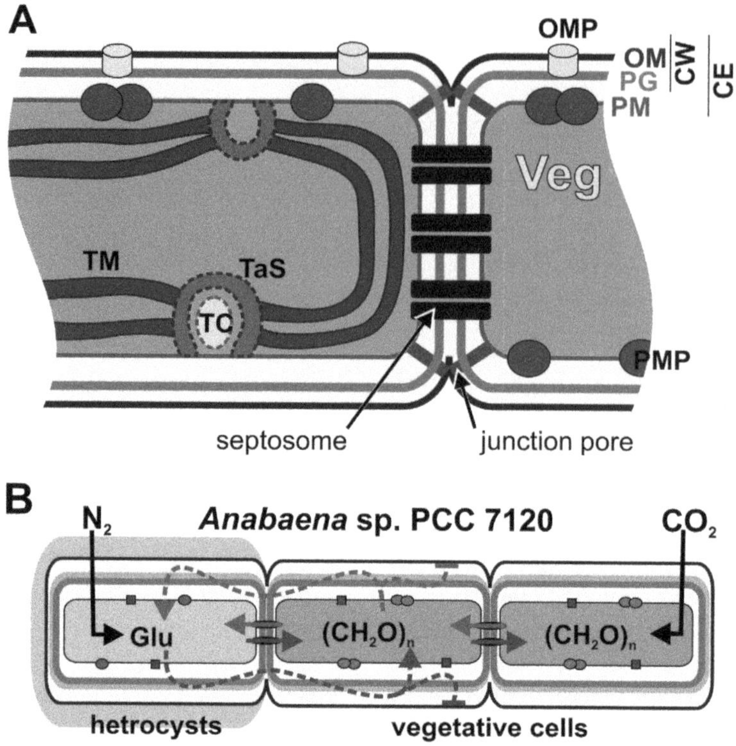

Figure 3.1 The membrane organization of cyanobacteria. (A) The structural composition of the cell envelope of cyanobacteria is shown as scheme. The outermost barrier is attached to the outer membrane (dark green) and the peptidoglycan layer (brown) and the plasma membrane (light green). At the septum border the periplasm is crossed by junction pores (dark brown), and in the septum by septosomes (blue). Outer membrane proteins (OMP, yellow) and plasma membrane proteins (red) are shown to indicate their existence. The inner-cellular structures formed by the thylakoid membranes (TM) are likely associated with the plasma membrane by the thylakoid associated semi-circular structures (TaS), which in some cases are associated with the thylakoid centres (TC). The figure is adapted from Wilk *et al.* (2011). (B) The N_2-fixing heterocysts and the CO_2-fixing vegetative cells co-exist in a single filament. Amino acids such as glutamine are nitrogen carriers that are transported to vegetative cells. In turn, sugars are carbon carriers produced by vegetative cells, which need to be transferred to heterocysts. Two possible transport routes are currently discussed. One considers the periplasm as a possible transport path for metabolites (dashed arrows), while another considers the septosomes as major structures for the transport of metabolites (solid arrows). The figure is adapted from Flores *et al.* (2006). A colour version of this figure is available in the plate section at the back of the book.

each cell in the filament (Wilk *et al.*, 2011). However, in the septum the peptidoglycan layers of the adjacent cells might be in physical contact (Lehner *et al.*, 2011). Thus, Wilk and co-workers suggested that cells of *Anabaena* sp. PCC 7120 are like unicellular Gram-positive bacteria with a plasma membrane and a thick peptidoglycan layer arranged in an outer membrane sack. This proposal has many implications for the function of the plasma membrane and for the required complexity of the peptidoglycan layer, which are discussed below.

The proteomes of the outer membrane of vegetative cells and heterocysts of *Anabaena* sp. PCC 7120 are rather comparable (Moslavac *et al.*, 2005, 2007), which is consistent with the idea that a single outer membrane surrounds a filament. However, a final conclusion based on only these results is not justified, as these studies did not detect the entire possible outer membrane proteome that can be anticipated based on bioinformatic identification of β-barrel proteins (discussed in 'The protein composition of the cyanobacterial cell envelope'). Remarkably, in *Anabaena* sp. PCC 7120 one Omp85 family protein (Alr0075) was observed specifically localized in the outer membrane of vegetative cells (Nicolaisen *et al.*, 2009b). Thus, a mechanism for proper localization and spatial fixation of outer membrane proteins must exist. This could either be the anchoring of the corresponding proteins to the peptidoglycan layer or the presence of a diffusion barrier for outer membrane proteins between adjacent but functionally specialized cells. The latter might be formed by proteinaceous components, e.g. those that have been suggested to form the junction pore in *Phormidium uncinatum* and *Oscillatoria princeps* (Fig. 3.1A; Hoiczyk and Baumeister, 1995).

A continuous outer membrane results in a continuous periplasmic space along the filament. This space is discussed to allow the exchange of a large collection of molecules (Fig. 3.1B). By use of the green fluorescent protein (GFP) as a marker it was demonstrated that solutes can be exchanged between cells via the periplasmic space of *Anabaena* sp. PCC 7120 (Mariscal *et al.*, 2007), although the periplasm is thought to be crowded with molecules and forms a gel-like texture (Hobot *et al.*, 1984). In the filamentous cyanobacteria the periplasmic space might thus 'connect' heterocysts and vegetative cells, and this 'connection' might be required for the transfer of reduced carbon from vegetative cells to heterocysts and of fixed nitrogen in the reverse direction (Wolk, 1968; Wolk *et al.*, 1974). Because heterocysts are separated by 10–20 vegetative cells (see, for example, Adams, 2000; Golden and Yoon, 2003; Herrero *et al.*, 2004), fixed nitrogen transport has to bridge up to 10 cells to reach the vegetative cell most distant from the heterocyst (Picossi *et al.*, 2005). However, not only metabolite transport but also signal transduction might require a continuous periplasmic space. Regulatory molecules such as the penta-peptide Arg-Gly-Ser-Gly-Arg (the C-terminal fragment of PatS) might be transported through the periplasmic space in order to repress heterocyst differentiation of adjacent cells and ensure correct spacing (Yoon and Golden, 1998).

Several prerequisites exist for this hypothetical mode of transport. First, cell type specific transporters are needed for the recognition, uptake or secretion of the corresponding molecules. Indeed, most of the cyanobacterial genomes encode many putative plasma membrane proteins, which might account for such function. Second, scavenger molecules for the specific transport of the metabolites have to be present in the periplasm. Currently not much is known about the proteome of the periplasm, although it has been analysed for the unicellular cyanobacterium *Synechocystis* sp. PCC 6803 (Fulda *et al.*, 2000). However, only very few of the identified proteins have been experimentally approached. Furthermore, one would expect that the periplasmic proteome of filamentous cyanobacteria is significantly different from the one of unicellular bacteria if the periplasmic space is used for communication between cells. Thus, periplasmic signal molecules might exist in heterocyst-forming cyanobacteria. Third, the outer membrane has to constitute a barrier for the passive diffusion of the molecules into the extracellular space. The original hypothesis of the molecular transfer between vegetative cells and heterocysts through the periplasmic space was based on the assumption that cyanobacteria do not contain porin-like outer membrane transporters (Flores *et al.*, 2006). However, experimental (Moslavac *et al.*, 2005, 2007) and bioinformatic (e.g. Nicolaisen *et al.*, 2009a) analysis of the outer membrane proteome revealed the existence of several outer membrane proteins, which might function in transport or diffusion of solutes across the outer membrane (see 'The protein composition of the cyanobacterial cell envelope'). Nonetheless, electrophysiological measurements on cyanobacterial outer membranes yielded a lower conductance across this membrane than expected for classical proteobacterial porins (Benz *et al.*, 1985), suggesting that cyanobacterial porins are either rather small or their opening is actively regulated. Thus, even in the presence of outer membrane β-barrel proteins, the outer membrane might still represent a selectivity barrier as previously suggested (Nicolaisen *et al.*, 2009a).

The concept of a continuous periplasmic space (Fig. 3.1B) is, however, still discussed. In contrast to the work described above, Zhang and co-workers (2008) concluded that GFP diffuses within the periplasm of the producing cell, while a diffusion barrier exists at the cell border impeding the exchange of GFP between two neighbouring cells. In line with such a diffusion barrier between cell borders, each cell is surrounded by an individual peptidoglycan layer (Wilk *et al.*, 2011), which might separate the periplasm of two cells. In this respect, the peptidoglycan layer could represent a diffusion barrier for molecules as large as GFP; especially as the cyanobacterial peptidoglycan layer shows properties significantly different from the proteobacterial one (see 'The peptidoglycan layer of cyanobacteria' below). Thus, although very attractive, the hypothesis considering diffusion through the periplasmic space of the filament should be worked out further.

The peptidoglycan layer of cyanobacteria

The first reports on the existence of a peptidoglycan layer in cyanobacteria date back to the fifties and sixties of the twentieth century (Frank, 1962; Drews and Meyer, 1964, and references therein). Based on electron-microscopy and compound analysis of the cell wall of, for example, *Phormidium uncinatum*, *Anacystis nidulans* (*Synechococcus*), *Microcoleus vaginatus* or *Chlorogloea fritschii*, it was concluded that cyanobacteria contain a 'muco-polymer' between the outer and inner membranes, reminiscent of what was observed for other Gram-negative bacteria such as *E. coli* (Frank, 1962; Drews and Meyer, 1964; Lamont, 1969). The subsequent isolation and visualization of the sacculus (peptidoglycan layer, older name: 'L II layer') confirmed this concept (e.g. Golecki, 1977).

Composition

Based on the determined molar ratio between glucosamine (GlcN), muramic acid (MurN), alanine (Ala), glutamate (Glu), and diaminopimelic acid (A_2pm) of about 1:1:2:1:1 (Table 3.1), the cyanobacterial peptidoglycan was classified as A1-type peptidoglycan (Schleifer and Kandler, 1972), although the ratio between GlcN and MurN varies between 1:1 and 2:1 (Jones and Yopp, 1979; Schrader *et al.*, 1981). The analysis of the peptidoglycan structure of the unicellular cyanobacterium *Synechocystis* sp. PCC 6714 confirmed this attribution (Jürgens *et al.*, 1983; Fig. 3.2A). The width of the peptidoglycan layer of this bacterium was estimated based on electron microscopic images to be in the range of 12 nm (Jürgens *et al.*, 1985; Jürgens and Weckesser, 1986). This value is comparable to the one found for the peptidoglycan layer by electron tomography of *Anabaena* sp. PCC 7120 cells (Wilk *et al.*, 2011). Considering the width of 1.5 nm for a single peptidoglycan layer (Braun *et al.*, 1973), it was suggested that the peptidoglycan of *Synechocystis* sp. PCC 6714 (and therefore also of *Anabaena* sp. PCC 7120) is composed of 8–10 layers (Jürgens *et al.*, 1985). However, more recent investigations suggest a dimension of a single peptidoglycan layer in *E. coli* of 6 nm (Labischinski *et al.*, 1991; Yao *et al.*, 1999; Matias *et al.*, 2003; Gan *et al.*, 2008), which would account for two layers in *Synechocystis* sp. PCC 6714 or *Anabaena* sp. PCC 7120. Anyhow, cyanobacteria appear to have more layers than the model system *E. coli*. In addition, the dimensions of the peptidoglycan layers of different

Table 3.1 The molar ratio of important peptidoglycan compounds identified in various cyanobacteria

		Anacystis nidulans	*Phormidium uncinatum*[c]	*Synechococcus* PCC 6716[d]	*Prochlorothrix hollandica*[e]	*E. coli*[f]
Glucosamine	GlcN	1.34[a]/1.50[b]	1.00	2.1	1.76	1.0
Muramic acid	MurN	1.07/0.90	0.63	0.9	0.95	1.1
Diaminopimelic acid	**(meso-)A_2pm**	**1.00**	**1.00**	**1.0**	**1.0**	**1.0**
Glutamate	D-Glu and L-Glu	1.52/1.10	0.94	1.4	0.98	1.6
Alanine	D-Ala and L-Ala	2.40/1.90	2.00	2.3	1.78	2.9
Glycine	Gly	0.41/0.30	0.11	0.10		0.2

Molar ratio with respect to A_2pm; [a]Drews and Meyer (1964); [b]Golecki (1977); [c]Frank *et al.* (1962); [d]Jürgens *et al.* (1983); [e]Jürgens and Burger-Wiersma (1989); [f]Martin and Frank (1962). *E. coli* included for comparison.

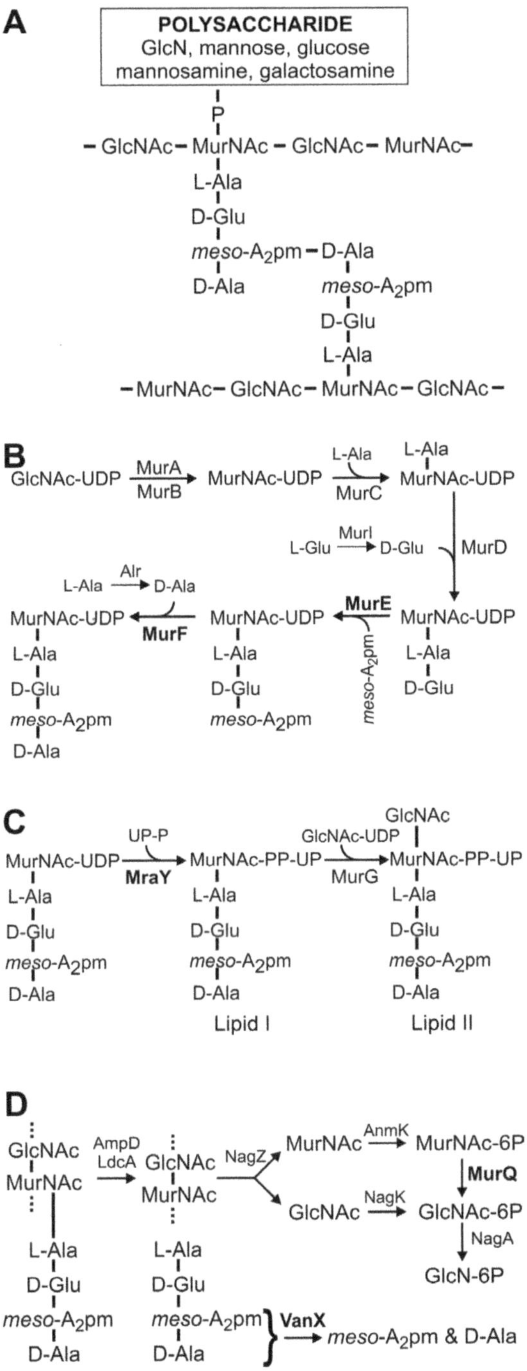

Figure 3.2 The peptidoglycan layer of cyanobacteria. (A) The basic structure of the peptidoglycan is of Gram-negative type with the exception of the cross-linked sugar moieties (Jürgens *et al.*, 1983; Jürgens and Weckesser, 1986). (B) The synthesis path of the peptidoglycan precursor as established in *E. coli* is depicted (Typas *et al.*, 2012). (C) The synthesis of the final peptidoglycan precursors at the cytosolic side of the plasma membrane is shown. (D) The general path of peptidoglycan layer degradation is depicted. Enzymes written in bold have been experimentally approached. The abbreviations are: A_2pm, diaminopimelic acid; Alr, Ala racemase; AmpD, anhMurNAc-l-Ala amidase; AnmK, anhMurNAc kinase; D-Ala, D-alanine; D-Glu, D-glutamate; GlcN, glucosamine; GlcNAc, N-acetyl glucosamine; L-Ala, L-alanine; *meso*-A_2pm, *meso*-diaminopimelic acid; LdcA, L,D-carboxypeptidase; MraY, UDP-MurNAc-pentapeptide phosphotransferase; MurA, UDP-GlcNAc enolpyruvyl transferase; MurB, UDP-MurNAc dehydrogenase; MurC, UDP-MurNAc–L-Ala ligase; MurD, UDP-MurNAc-L-Ala–D-Glu ligase; MurE, UDPMurNAc-L-Ala-D-Glu–meso-A_2pm ligase; MurF, UDP-MurNAc-tripeptide–D-Ala ligase; MurG, UDP-GlcNAc-undecaprenoyl-pyrophosphoryl-MurNAc-pentapeptide transferase; MurI, Glu racemase; MurN, muramic acid; MurNAc, N-acetyl muramic acid; MurQ, N-acetylmuramic acid 6-phosphate etherase; NagA, GlcNAc-6-P deacetylase; NagK, GlcNAc kinase; NagZ, β-N-acetylglucosaminidase; UDP, uridine diphosphate, UP-P, undecaprenol phosphate.

cyanobacterial species vary quite significantly. For example, for the large cyanobacterium *Oscillatoria princeps* a peptidoglycan layer of up to 700 nm has been reported (Hoiczyk and Baumeister, 1995).

While analysing the molecular composition of the peptidoglycan layer of *Synechocystis* sp. PCC 6714 it was found that a polysaccharide composed of glucosamine, mannosamine, galactosamine, mannose, and glucose is directly linked to the muramic acid-6-phosphate of the peptidoglycan of this bacterium forming a PG–PS complex (Jürgens and Weckesser, 1986), which is rather typical for Gram-positive bacteria. In addition, the degree of covalent crosslinks between peptidoglycan chains is described to be in the range of 60% (Jürgens *et al.*, 1983). Again, this exceeds the degree of crosslinks typically found in Gram-negative bacteria (~30%; e.g. Glauner *et al.*, 1988), and is rather comparable to that in the layer of Gram-positive bacteria. However, other components typical for Gram-positive peptidoglycan layers such as teichoic acid or L-lysine are not found in the cyanobacterial peptidoglycan (Hoiczyk and Hansel, 2000). Thus, it can be suggested that the cyanobacterial peptidoglycan layer is an intermediate between the typical Gram-negative and the typical Gram-positive layers.

The peptidoglycan layer of cyanobacteria is perforated by two structures (Fig. 3.1A). First, septosomes between cells of filamentous cyanobacteria have been discovered by conventional microscopy (Wildon and Mercer, 1963; Lang and Fay, 1971), freeze fracture electron microscopy (Giddings and Staehelin, 1978, 1981) or electron tomography (Wilk *et al.*, 2011). Approximately 300–500 of such structures penetrate the peptidoglycan layer to connect two adjacent cells, most likely to enable the exchange of small solutes (Mullineaux *et al.*, 2008; Zhang *et al.*, 2008). The length of the septosome is in the range of 25 nm, its outer diameter determined by ruthenium red staining is about 15 nm and its inner diameter determined by potassium permanganate staining has a dimension of about 6 nm (Wilk *et al.*, 2011).

Second, structures annotated as junction pores were observed to perforate the peptidoglycan layer (Lamont, 1969; Hoiczyk and Baumeister, 1995). For *Microcoleus vaginatus* it was found that the diameter and the spacing of the junctional pore increase from 5 nm and 30 nm during early cell division, respectively, to 15 nm and 60 nm between mature cells (Lamont, 1969). They bridge the outer and the inner membranes and have been suggested to form pores for slime secretion (Fig. 3.1A; Hoiczyk and Baumeister, 1995, 1998). In addition, it was speculated that these structures are involved in membrane bending in the circumference of the junction (Wilk *et al.*, 2011). The latter was suggested as these structures are regularly distributed at the circumferential junction and not within the septum.

Synthesis and degradation

Although the existence and molecular composition of the peptidoglycan is well known, its path of synthesis has not yet been described in cyanobacteria. In *E. coli*, many proteins are involved in the synthesis of peptidoglycan precursors, in the final formation of the peptidoglycan layer, and in its crosslinking and anchoring to the outer membrane (Typas *et al.*, 2012). Cytoplasmic precursor synthesis requires the action of eight different enzymes (Fig. 3.2B) for which homologues can be found in the cyanobacterial genomes. However, only MurE and MurF have been experimentally approached thus far. Plastid targeted MurE from *Anabaena* sp. PCC 7120 complements the phenotype caused by a mutation of the plastidal MurE in the moss *Physcomitrella patens* (Garcia *et al.*, 2008), but a relation of the cyanobacterial protein to peptidoglycan precursor synthesis has not yet been confirmed. For the cyanobacterial MurF a direct relation to this process could be established, because it was observed that MurF from *Synechocystis* sp. PCC 6803 restores the wild-type behaviour of a temperature sensitive *murF* mutant in *E. coli* (Malakhov *et al.*, 1995). The functional relation of the other enzymes detected by bioinformatic means has still to be approached.

The next two steps of synthesis producing the precursors annotated as Lipid I and Lipid II occur on the inner membrane (Fig. 3.2C). These two precursors are subsequently transferred to the periplasm by the action of the flipases FtsW and RodA. The first enzyme is an UDP-MurNAc-pentapeptide phosphotransferase (MraY) that links the precursor to the membrane-localized undecaprenol phosphate. Although the enzyme activity has not been demonstrated in cyanobacteria, a protein with clear homology to MraY from *Anabaena* sp. PCC 7120 complements the phenotype of the plastidal MraY in the moss *Physcomitrella patens* (Homi *et al.*, 2009). The second enzyme (MurG) links N-acetyl glucosamine to the precursor. Again, homologues can be found in all cyanobacterial genomes, but the functions of these homologues are not yet confirmed.

A glycosyltransferase catalyses the polymerization of the peptidoglycan chain in the periplasm. It is thought that this occurs while the chain is still anchored to the inner membrane (Typas *et al.*, 2012). The crosslinking of the emerging chain to the existing sacculus is catalysed by so called DD-transpeptidases, which are also annotated as penicillin-binding proteins (PBPs) and have been identified e.g. in *Synechococcus* sp. PCC 7942 by labelling of proteins with radioactive benzylpenicillin (Stone *et al.*, 1988). They are generally divided into class A, B and C (Goffin and Ghuysen, 1998). Class A enzymes have transglycosylase and transpeptidase activities and are involved in peptidoglycan polymerization. Class B enzymes have transpeptidase activity only and are involved in the modification of the peptidoglycan layer during cell elongation and division. Class C PBP's are involved in the regulation of the peptidoglycan synthesis by hydrolysing acyl-groups and break down the peptidoglycan. The ability to hydrolyse covalent acyl-enzyme intermediates makes them general β-lactamases (Goffin and Ghuysen, 1998; Typas *et al.*, 2012).

Genes coding for PBPs have been identified and classified subsequent to the sequencing of several cyanobacterial genomes (Lazaro *et al.*, 2001; Leganés *et al.*, 2005). By experimental investigations of PBPs it was documented that Class A enzymes of the filamentous *Anabaena* sp. PCC 7120 encoded by *alr5101*, *all2981* and *alr4579* are specifically important for heterocyst differentiation (Lazaro *et al.*, 2001, Leganés *et al.*, 2005), suggesting that these particular enzymes are involved in the modification of the peptidoglycan layer during differentiation. The analysis of all eight PBPs in *Synechocystis* sp. PCC 6803 showed that the different Classes (A–C) are essential for cell survival (Marbouty *et al.*, 2009). Although three Class A PBPs enzymes exist in *Synechocystis*, deletion of two Class A enzymes already results in a lethal phenotype (Marbouty *et al.*, 2009). In addition, the enzyme annotated as PBP2 may be involved in the synthesis of the peptidoglycan precursor as its deletion results in a reduced cell size (Marbouty *et al.*, 2009). Only one Class B enzyme was found in the genome of *Synechocystis* sp. PCC 6803 that was annotated as FtsI. It is not surprising that this enzyme was found to be essential and it has been suggested to be involved in the incorporation of peptidoglycan at the septum during separation of daughter cells (Marbouty *et al.*, 2009). The four Class C enzymes were further subclassified in two type 4 (PBP5 and PBP8) and two type AmpH-like PBPs (PBP6 and PBP7; Marbouty *et al.*, 2009). Again, one, but not both enzymes of each subclass was found to be dispensable. The analysis of heteroploid double mutants (PBP5 and PBP8; PBP6 and PBP7) suggested that PBP5 and PBP8 are involved in the completion of the septation during daughter cell separation. These data are a first step into the understanding of cyanobacterial peptidoglycan synthesis and maintenance during division. However, it is evident that further experimental approaches are required to fully understand this process and its relation to the different cellular functions (see Chapter 2).

Similarly to its synthesis pathway, not much is known about peptidoglycan degradation (Fig. 3.1D). The initial steps are catalysed by an anhMurNAc-l-Ala amidase (AmpD), a β-N-acetylglucosaminidase (NagZ), and an L,D-carboxypeptidase (LdcA). Although cyanobacterial genomes contain genes with similarity to these enzymes, their role has not yet been established. After digestion, GlcNAc and N-acetylmuramic acid (MurNAc) are phosphorylated by NagK and AnmK, respectively. For these two kinases, homologous genes can be found in cyanobacterial genomes, but again, no experimental evidence for their involvement in peptidoglycan degradation exists. Interestingly, *N*-acetylmuramic acid 6-phosphate is converted into *N*-acetylglucosamine 6-phosphate by the *N*-acetylmuramic acid 6-phosphate etherase (MurQ; Uehara *et al.*, 2006; Fig. 3.1D), and this enzyme was identified in *Synechocystis* sp. PCC 6803 (Sll0861), *Anabaena* sp. PCC 7120 (Alr2432) and *Microcystis aeruginosa* PCC 7806 (Mlla-1; protein ID: CAO90902.1). The genes encoding these proteins were found to complement each other. Moreover, *sll0861* and *mlla-1* complement *murQ* function in *E. coli*, suggesting that their products function in a similar pathway and are able to convert N-acetylmuramic acid 6-phosphate into *N*-acetylglucosamine 6-phosphate (Jiang *et al.*, 2010). Interestingly, the mutants had a growth delay under low light conditions. For the mutant of *Anabaena* sp. PCC 7120 a reduced heterocyst development rate was observed under low light as well (Jiang *et al.*, 2010). Based on this observation the authors speculated that recycling of the peptidoglycan might be an important pathway to increase the pool of carbon metabolites under low light conditions, a notion that needs to be confirmed, especially as the function of the putative cyanobacterial GlcNAc-6-P deacetylase (NagA) is not yet established. This enzyme converts N-acetylglucosamine 6-phosphate into glucosamine 6-phosphate, which can either be used as precursor for peptidoglycan layer synthesis, or which can enter the carbohydrate metabolism (e.g. Plumbridge, 1991). Another enzyme described with putative function in the decomposition of the peptidoglycan precursor is the cytoplasmic VanX (Lessard *et al.*, 1998). However, experimentally only the cleavage of the dipeptides D-Ala-D-Ala, L-Ala-D-Ala and aminoisobutyryl-D-Ala by the VanX enzyme isolated from *Synechocystis* sp. PCC 6803 could be confirmed (Lessard *et al.*, 1998). Therefore, the final proof that the enzyme is directly related to peptidoglycan degradation remains to be established.

Thus, after the primary structure of the cyanobacterial peptidoglycan was discovered to be comparable to the one of other Gram-negative bacteria, the synthesis and degradation pathways were not analysed in much detail. It is rather assumed that the pathways in cyanobacteria are similar to those described for *E. coli* (Typas *et al.*, 2012). However, given the importance of the peptidoglycan for the functionality of the proteins associated with this structure and the periplasm in general, the investigation of this cellular structure should experience a renaissance. For instance, it is still not known how solutes and proteins traverse this layer. Considering the higher complexity when compared, for example, to *E. coli*, the general assumption that the peptidoglycan function in cyanobacteria can be approximated from the knowledge on this model bacterium might not be justified.

The plasma membrane of cyanobacteria

The carotenoids in the plasma membrane

In general, the plasma membrane is composed of three different components, namely lipids (see 'The outermost-layer of the cyanobacterial cell'), proteins (see 'The protein composition of the cyanobacterial cell envelope'), and carotenoids. In *Anacystis nidulans* the ratio (in weight) of the three components is 57:40:3 (Omata and Murata, 1983). In the same bacterium the carotenoid composition (based on weight) is made of 1% cryptoxanthin, 3% β-carotene, 11% nostoxanthin, 25% caloxanthin and 60% zeaxantin (Omata and Murata, 1983). In *Synechocystis* sp. PCC 6714 a very similar ratio was found (1% cryptoxanthin, 1% echinenone, 4% β-carotene, 31% zeaxantin and 63% xanthophylls; Omata and Murata, 1984). The comparable values observed indicate that the carotenoids are indeed compounds of the plasma membrane and not contaminants from the thylakoid membranes.

The importance of carotenoids for the non-photosynthetic membranes can be discussed in two ways. On the one hand, carotenoids increase the osmotic tolerance of membranes by decreasing water permeability (e.g. Lazrak *et al.*, 1987, Berglund *et al.*, 1999). This might be an important feature for especially the seawater cyanobacteria. On the other hand, carotenoids can protect membrane lipids against reactive oxygen species (Cogdell and Frank, 1987; Gruszecki and Strzalka, 2005). The unsaturated fatty acid chains are targets of oxidation by reactive oxygen species (e.g. Halliwell, 1991), which can cleave acyl-ester bonds in membrane lipids (McKersie *et al.*, 1990). Therefore, carotenoids are central components of the cytoplasmic membrane to ensure its stability and functionality as a barrier.

Architecture and dynamics of the plasma membrane

The plasma membrane surrounds the individual cells of cyanobacteria irrespective of whether they are unicellular or filamentous. However, most cyanobacteria contain an additional internal membrane system, the thylakoids (see Chapter 4). It has been widely discussed whether both membranes are directly connected or not. Initially, most reports pointed to an independence of the two membrane systems (e.g. Golecki, 1977, 1979; Giddings and Staehelin, 1979) and connections observed were discussed as produced by 'inadequate preservation' (Kunkel, 1982). Subsequently it was suggested that a 'thylakoid centre' with functional relevance for thylakoid biogenesis exists (Fig. 3.1A), which, however, was discussed not to be connected with the plasma membrane (Kunkel, 1982). In contrast, recent studies by electron tomography showed a possible connection between this thylakoid centres and the plasma membrane (van de Meene *et al.*, 2006) – an observation that would explain why the initial assembly of the photosystems occurs not in the thylakoid membranes, but in the plasma membrane (e.g. Zak *et al.*, 2001). Interestingly, the periplasmic biogenesis factor PratA of

Synechocystis sp. PCC 6803 was co-purified with a membrane fraction that was distinct from typical thylakoids and plasma membranes (Schottkowski *et al.*, 2009). The authors termed it 'PratA defined membrane (PDM) subfraction'. By immuno-gold labelling of PratA and D1 and electron-microscopic analysis of the ultrastructure of wild-type and a *pratA* mutant of *Synechocystis* sp. PCC 6803, it was further documented that a 'semi-circular structure' surrounding the thylakoid centre exists, to which the thylakoids are associated and which is a continuation of the plasma membrane (Fig. 3.1A). Although these structures are also connected to the thylakoids, the authors suggested that the lumen of these structures and that of the thylakoids are not continuous (Stengel *et al.*, 2012). The emerging picture suggests that plasma membrane inserted proteins can migrate into the thylakoid membrane, which poses the question, how the sorting of the thylakoid proteins is achieved (see Chapter 5).

Gloeobacter violaceus is the only known cyanobacterium that does not contain internal thylakoid membranes. Instead, this cyanobacterium possesses specialized domains in its plasma membrane (Fig. 3.3A), in which the photosynthetic and the respiratory complexes co-localize (Rexroth *et al.*, 2011). These domains can be separated by mechanical forces from the remaining membrane system. This strongly suggests that their physical properties are significantly different from that of the remaining plasma membrane. In line with this

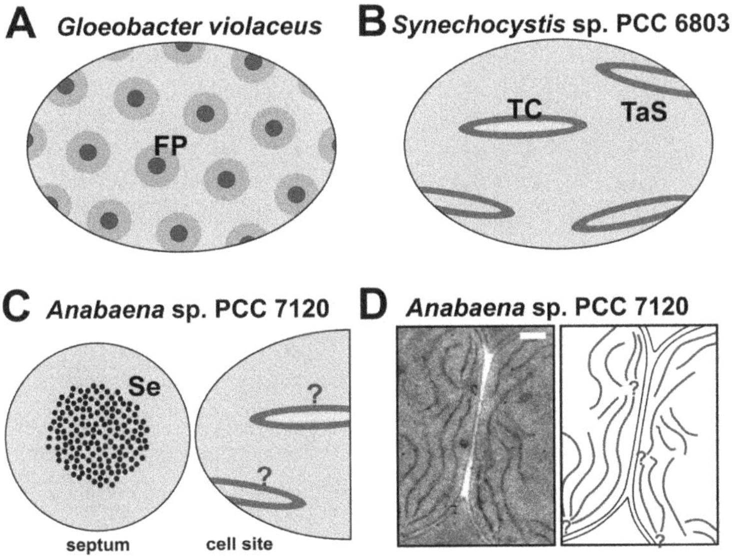

Figure 3.3 The plasma membrane of cyanobacteria contains specialized regions. (A) Shown is the distribution of the fluorescent patches (FP) observed in the plasma membrane of the unicellular cyanobacterium *Gloeobacter violaceus*, which contains the photosynthetic and respiratory machines. Per cell, 46 fluorescent patches have been found. The illustration is adapted from Rexroth *et al.* (2011). (B) The thylakoid membrane sheets of *Synechocystis* sp. PCC 6803 converge at sites adjacent to the plasma membrane ($N=3$ to 10 per cell; van de Meene *et al.*, 2006), which are likely coincident with the thylakoid associated semicircular structures (TaS, Stengel *et al.*, 2012). In addition, these sites are in $N=0$ to 4 cases associated with the described thylakoid centres (van de Meene *et al.*, 2006). (C) The septum of the filamentous cyanobacterium *Anabaena* sp. PCC 7120 contains up to 500 septosomes (Se). In this cyanobacterium, the thylakoid membranes converge at the plasma membrane in some cases. Although thylakoid-associated semicircular structures have not been described for this cyanobacterium, they might exist as well. (D) The image of the septum between two cells of *Anabaena* sp. PCC 7120 is shown to illustrate the convergence of the thylakoid membranes. On the right side the thylakoid membranes are drawn to scale, and question marks indicate possible positions of convergence close to the plasma membrane. The image was taken by Laura Wilk, MPI of Biophysics, Frankfurt, Germany. The cells were prepared as described in Wilk *et al.* (2011) by the method employing potassium permanganate. Scale bar, 200 nm. A colour version of this figure is available in the plate section at the back of the book.

notion, the photosystem hosting fraction contains (as expected) chlorophyll, while the remaining membrane shows high contents of carotenoids (Rexroth *et al.*, 2011). In contrast, a distinction of the lipid composition of either of the membranes was not observed, which led to the conclusion that these specialized domains are formed by protein complexes. It has been discussed that the specialized domains found in *Gloeobacter violaceus* might have been the evolutionary precursors for the development of the thylakoid membranes (Rexroth *et al.*, 2011).

Specialized membrane domains are not unique for *Gloeobacter violaceus* (Fig. 3.3A). It was reported that the thylakoid membranes of *Synechocystis* sp. PCC 6803 converge at the plasma membrane (van de Meene *et al.*, 2006), which might be linked to thylakoid associated semicircular structures (Fig. 3.3B; Stengel *et al.*, 2012). These thylakoid-associated semicircular structures might represent or contain a thylakoid centre (van de Meene *et al.*, 2006) and might again represent highly specialized regions of the plasma membrane. Consistently, a convergence of the thylakoid and plasma membrane can be observed in some regions of the cells of the filamentous cyanobacterium *Anabaena* sp. PCC 7120 (e. g. Wilk *et al.*, 2011; Fig. 3.3D). However, in filamentous cyanobacteria such as *Anabaena* sp. PCC 7120 an additional highly specialized region of the plasma membrane exists, namely at the septum where the septosomes are localized (Fig. 3.3C). In summary, it can be proposed that the plasma membrane of cyanobacteria contains specific domains of defined functionality. It might be speculated that additional specialized domains might exist, e.g. for the transport of metabolites, which, however, has not yet been experimentally approached.

The outermost layer of the cyanobacterial cell wall

S-layer

The surface layer (S-layer) is a non-membranous lattice of protein or glycoprotein on top of the cell envelope of some archaea and eubacteria. In archaea and Gram-positive bacteria, the S-layer is linked to the pseudomurein or the peptidoglycan layer respectively, whereas in Gram-negative-bacteria it is attached to the lipopolysaccharide (LPS) of the outer membrane (Fig. 3.4A). The function of the S-layer is very diverse and rather environment dependent (Engelhardt, 2007). It consists of a single species-specific protein, which in the corresponding bacterium accounts for up to 15% of total protein content (Messner *et al.*, 1994). The protein varies in size, ranging from 40 to 200 kDa, and forms a paracrystalline, two-dimensional monolayer of identical lattice or unit 'cells' (Sleytr *et al.*, 1994). These 'cells' form a small repetitive unit of oblique (p2), tetragonal (p4) or hexagonal (p3, p6) symmetry (Sara 2000). The distance between two 'cell' centres varies from 3 to 35 nm. However, the S-layer is rather fragile and binds to the cell surface by weak hydrophobic or anionic interactions (Koval and Murray, 1984) and can be released by treatment with a chelator such as EDTA.

Although the S-layer has been identified in more than 60 cyanobacterial strains, the knowledge about cyanobacterial S-layer proteins is still very scarce (Smarda, 2002). By combination of freeze-etching techniques and electron microscopy in the early 1980s, several groups demonstrated that the S-layer throughout cyanobacteria predominantly form the p6 symmetry with a distance of 13–16 nm between the centre of two units. More specifically, in unicellular cyanobacteria, p1, p3 and p6 symmetry has been observed, whereas in filamentous species only p1 and p4 could be found. Interestingly, no S-layer could be detected in the *Nostocales* and *Stigonematales*. The presence or absence as well as architecture between different species of the same genus are conserved (e.g. Lounatmaa *et al.*, 1980; Vaara, 1982).

Recently, the S-layer of the marine cyanobacterium *Synechococcus* sp. WH8102 was analysed in some detail. The 130 kDa glycoprotein SwmA was found to be important for movement because a mutation in *swmA* renders the bacteria non-motile (Brahamsha *et al.*, 1996; McCarren, 2009). *Synechococcus* generally has no flagella, and no obvious cell shape alterations by the mutation could be observed by light microscopy. Thus, the function of SwmA is most likely the formation of surface distortions. These distortions are suggested to be

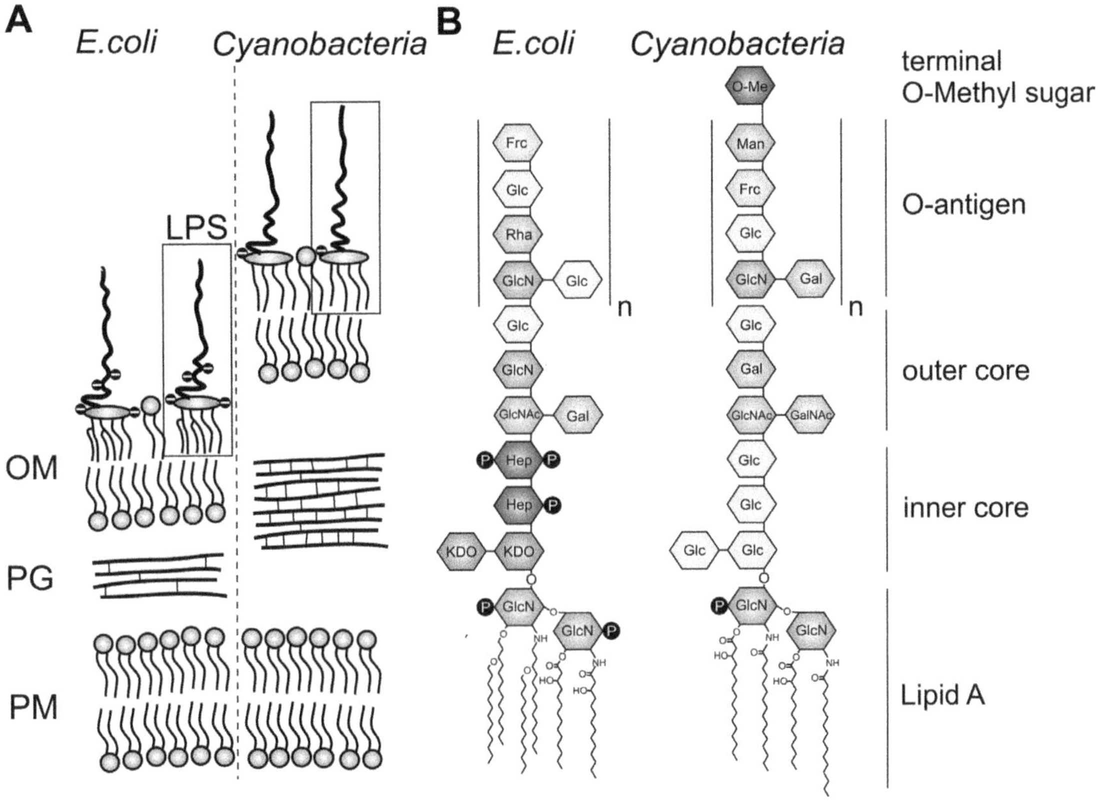

Figure 3.4 The lipopolysaccharide of cyanobacteria. (A) Cyanobacteria have a typical Gram-negative cell envelope composed of plasma membrane (PM) peptidoglycan layer (PG) and outer membrane (OM). The outer leaflet of the OM of Gram-negative bacteria, including cyanobacteria, is formed by lipopolysaccharides that serve as a barrier for hydrophobic substances. (B) Lipopolysaccharide consists of 3 parts, the O-antigen, the core region (including inner- and outer core) and the Lipid A membrane anchor. The highly diverse O-antigen is formed by repetitive units of polysaccharide. The sugar composition of the units is highly diverse and varies among cyanobacteria. Information about structural composition of the unit cells is scarce; nonetheless, typically found sugars are depicted. The O-antigen of cyanobacteria is terminated by rare O-methyl sugars. In some strains O-methyl sugars are abundant and incorporated in the unit cells. The LPS core region of enterobacteria consists of the conserved heptose and KDO and is highly phosphorylated. In contrast cyanobacteria lack heptose while KDO is only rarely found and is replaced by glucose. The level of phosphorylation is much lower than in enterobacteria. The most conserved part of LPS is the Lipid A. While the D-glycosamine disaccharide backbone of enterobacteria is mono- or polyphosphorylated in position 1 and 4', cyanobacterial Lipid A has a low phosphate level. The D-glycosamine disaccharide is either ester- or amide-bound to fatty acids of variable chain length.

required for the movement of *Synechococcus* by a mechanism annotated as 'helical rotor mechanism' (Nan *et al.*, 2011), where a cargo protein moves along a continuous looped helical track. This movement creates a travelling helical wave along the S-layer, and this surface wave is thought to propel the unicellular cyanobacteria (Ehlers and Oster, 2012). (For further details on motility, see Chapter 11.)

One characteristic feature of S-layer proteins is the surface layer homology (SLH) domain that mediates the binding to the sugar molecules of LPS. Interestingly, a similar domain is found at the N-terminus of cyanobacterial porins (Hansel, 1998). This domain could serve as an anchor to connect the S-layer to the LPS and porins of the outer membrane to the peptidoglycan (see 'Porins in cyanobacteria').

Lipopolysaccharides

Endotoxins were first described by the German physician Richard Pfeiffer in the late 19th century. He discriminated between proteinogenic exotoxins, which can be inactivated by heat, and the cell wall-associated endotoxins, which keep their pathological activity after heating (Schletter *et al.*, 1995). Further analysis indicated that endotoxins are composed of lipids and polysaccharides and were termed lipopolysaccharide (LPS). Lipopolysaccharides are abundant amphipathic heteropolymers, which form the outer leaflet monolayer of the outer membrane of Gram-negative bacteria. LPS consist of three parts: (i) the O-specific antigen, (ii) the R core and (iii) the lipid A membrane anchor, which will be described in the following sections (Fig. 3.4B).

The LPS has been best described for *E. coli* and *Salmonella*. Isolated endotoxin lysate is well known for induction of a strong immune response or even septic shock in mammals (Schletter *et al.*, 1995). Other symptoms in mammalian hosts are gastrointestinal and respiratory diseases. LPS is considered to contribute to outer membrane integrity, surface adhesion, and protection against phagocytosis, but also to mediate sensitivity to bacteriophages (Xu *et al.*, 1997).

LPS is present throughout Gram-negative bacteria including cyanobacteria and other photosynthetic bacteria (Table 3.2; Schmidt *et al.*, 1980a, b). The diversity of cyanobacteria and their inhabited ecosystems (sea, fresh water, tropical forests, hot springs or polar regions) are reflected by an adapted composition of the LPS when compared to enterobacteria. Nevertheless, our knowledge of the latter might serve as a basis for the understanding of the cyanobacterial LPS.

The O-specific antigen

The O-specific antigen is the most diverse part of the LPS. It is a repetitive polymer of homo- or hetero-oligosaccharide units of variable length (Schletter *et al.*, 1995). They define the antigenic specificity by interaction with the immune response system but also serve as receptor for bacteriophages. In *Anabaena* sp. PCC 7120 it was shown that deletion of O-antigen by mutation of genes required for O-antigen synthesis leads to spontaneous resistance to bacteriophages (Xu *et al.*, 1997). The structural sugar composition of the linked polysaccharide units and their chain length can be extremely variable. Typical components of *E. coli* O-antigens are D-glucose, L-fructose, D-galactose as well as D-glycosamine. In contrast, the O-antigen of *P. aeruginosa* consists mostly of D-rhamnose, which forms a homopolymer (King *et al.*, 2009).

The LPS from *Synechococcus* sp. PCC 6301 contains a polysaccharide comprising predominantly 1–3 linked D-mannose trisaccharides, which are 1–4 interlinked in the O-antigen (Weckesser *et al.*, 1979). Additionally small amounts of 3-O-methyl-D-mannose and 4-O-methyl-D-mannose are present (Table 3.3). In contrast, the dominant sugar components of strains of *Synechococcus* and *Synechocystis* are fucose, mannose, galactose, glucose and glucosamine, which are supplemented with other strain specific sugars (Schmidt *et al.*, 1980a,b). The composition is very flexible and differs significantly not only between cyanobacterial species but also between strains of the same genus or species. The *Synechococcus* sp. PCC 6908 and PCC 6311 have a mannose content of more than 30% of dry weight of the isolated lipopolysaccharide fraction. In contrast, *Synechococcus* sp. PCC 6907 and PCC 6307 possess only 2% mannose but more than 12% of rhamnose, which seems to be entirely absent in *Synechococcus* sp. PCC 6908 and PCC 6311 (Table 3.2). The O-antigen of *Anabaena variabilis* PCC 7118 contains larger amounts of L-rhamnose, O-mannose, O-glucose, and O-galactose. In addition, the O-methyl sugar L-acofriose was identified (Weckesser *et al.*, 1974). O-methyl sugars are rare in LPS of Gram-negative bacteria but are found frequently in that of photosynthetic bacteria. The low abundant O-methyl sugars such as 3-O-methyl-D-mannose and 4-O-methyl-D-mannose in *Synechococcus* sp. PCC 6301 are preferentially incorporated at the terminal positions of the O-antigen chain (Katz *et al.*, 1977). However, the high content of 3-O-methyl-L-rhamnose (L-acrofriose) (18% of dry weight) in *Anabaena variabilis* (Weckesser *et al.*, 1974) is a good example that O-methyl sugars can also be present in the repeating unit of the O-antigen. The exact composition of the linked hetero-oligosaccharide units cannot completely be deduced from the summarized studies.

Table 3.2 The sugar composition of the lipopolysaccharides in cyanobacteria (% dry weight)

	3,6-Dideoxy-D-arabino-hexose	3,6-Dideoxy-D-arabnose	Fructose	Mannose	Glucose	Galactose	Arabinose	Xylose	Heptose	Rhamnose	Tyvelose	Glycosamine	Mannosamine	2-Keto-3-deoxyoctonate
Anabaena variabilis[1]				3.9	34.3	2.3				19.3		2.1		
Anacystis nidulans[2]			+	33	3.4	5.3				+		6.4		1.5
Schizothrix calcicola[3]				+	+	+		+		+		+		
Synechococcus PCC 6907[4]	3.8		2.6	2.1	3.1	6.6	+			12.2	+	1.8	0.4	
Synechococcus PCC 6307[4]	4.6		2.4	2.2	4	6.9	+			12.9	+	2	0.4	
Synechococcus PCC 6911[4]	2.6		2.7	2.1	1	6.2	+			9.5	+	1.3	0.3	
Synechococcus PCC 6603[4]		0.2	8.3	4.4	1.9	3.1	+					1.8		
Synechococcus PCC 6908[4]			4.7	35.3	1.2	5						2.9	0.5	1.8
Synechococcus PCC 6301[5]				+	+	+								
Synechococcus PCC 6311[6]			5.2	30.1	1.3	5.4						2.4	0.5	1.3
Synechococcus PCC 6312[6]			8.5	4.6	2.2	2.8	+					1.5		
Synechococcus PCC 6910[6]			4.4	35.7	1.8	5						4	1	1.2
Synechocystis PCC 6714[6]			1.1	7.5	33.9	4.5						+		
Synechocystis PCC 6803[6]			3.7	3.4	21.8	3.4						+		
Synechocystis PCC 6807[6]			4.1	7.9	4.5	3.5			3	0.5		+		
Synechocystis PCC 6308[6]			6.4	3.8	6.1	3.8			1.8	1		+		
Anabaena cylindrica[7]				+	+	+		+	+			+		

[1]Weckesser (1974); [2]Weiss (1970); [3]Keleti (1979); [4]Schmidt (1980a); [5]Katz (1977); [6]Schmidt (1980b); [7]Keleti (1982).
+, detected in traces.

Nevertheless, the cyanobacterial O-specific antigens are at least partially equivalent and therefore reminiscent of those from *E. coli*.

The R-core

The R-core of the LPS connects the polysaccharide of the O-antigen with the Lipid A. It can be differentiated in outer and inner core oligosaccharides. The inner core of enterobacteria consists of an acidic hetero-oligosaccharide that contains a phosphorylated 2-keto-3-deoxyoctonate (KDO) and phosphorylated L-glycero-D-mannoheptose (heptose; Tables 3.2 and 3.3). The KDO and heptose are highly conserved. The

Table 3.3 The composition of the O-methyl sugar in cyanobacteria (% dry weight)

Anabaena variabilis[1]	18.4									+
Synechococcus PCC 6603[2]		0.2								
Synechococcus PCC 6908[2]								+	+	
Synechococcus PCC 6301[3]								+	+	
Synechococcus PCC 6311[4]								+	+	
Synechococcus PCC 6312[4]		+								
Synechococcus PCC 6910[4]								+	+	
Synechocystis PCC 6803[4]			0.4	3	+	2.8				
Synechocystis PCC 6807[4]							0.5			0.6
Synechocystis PCC 6308[4]							+		0.4	
Anabaena cylindrica[5]								0.4		

[1]Weckesser (1974); [2]Schmidt (1980a); [3]Katz (1977); [4]Schmidt (1980b); [5]Keleti (1982). +, detected in traces.

outer core consists of D-glucose, D-galactose, and *N*-acetylglucosamine. The KDO of the R-core forms an acid-labile glycosidic linkage to 5′ of Lipid A that can be split by mild acidic treatment. Although in Gram-negative bacteria the R-core is more conserved than the O-antigen, the cyanobacterial core region shows several important differences when compared with that of enterobacteria. The most important difference is the lack of the conserved heptose and, to a large extent, of the KDO (Weckesser *et al.*, 1979). As a consequence, the Lipid A cannot be removed from the oligosaccharide by mild acidic hydrolysis. The missing L-glycero-D-mannoheptose is frequently replaced by glucose which makes cyanobacterial LPS energetically 'cheaper' to produce. Based on the comprehensive knowledge of enterobacterial LPS, they are often considered as the 'raw model'. However, considering the abundance and ancient character of cyanobacteria (Cavalier-Smith, 2006), it is conceivable that cyanobacterial LPS is the more ancient and 'original LPS' form, from which the more complex LPS of enterobacteria derived (Snyder *et al.*, 2009).

Lipid A

E. coli Lipid A is the most conserved part of the LPS and capable of inducing a strong immune response. Lipid A of most Gram-negative bacteria is closely related to Lipid A of *E. coli*. In mammalian cells Lipid A is recognized by the Toll-like receptor 4 (TLR4), which induces an inflammatory response (Alexander and Rietschel, 2001; Raetz 2007). Lipid A of *E. coli* consists of a β(1–6)-linked D-glycosamine disaccharide with phosphoryl groups at position 1 and 4′. Together with KDO of the inner core, the negative charges of the phospho groups present a diffusion barrier for hydrophobic molecules. These two phosphates are also needed to trigger full TLR4 response in human cells (Rietschel, 1994). The positions 2, 2′ and 3, 3′ of the disaccharide are ester or amid bound to mostly saturated hydroxy fatty acids, which anchor the Lipid A in the outer membrane. Although Lipid A is the most conserved part of the LPS, the diversity of the fatty acids vary among different Gram-negative bacteria and are also prone to modification in order to adapt to different environmental conditions as described for membrane lipids (see below).

It is noteworthy that Lipid A of cyanobacteria again shows some structural differences when compared with its enterobacterial counterpart. In most cyanobacteria the glucosamine–disaccharide backbone of Lipid A is not phosphorylated. Also the hydroxyl fatty acid chains are different. The dominant fatty acid of strains of *Synechococcus* and *Synechocystis* is beta-hydroxy palmimitic acid (Table 3.4). In *Anabaena variabilis* β-hydroxymyristic acid, β-hydroxypalmitic acid, and β-hydroxystearic acid are bound to the amino group whereas palmitic acid is ester bound (Weckesser *et al.*, 1974).

Table 3.4 The composition of the fatty acids within Lipid A of cyanobacteria (% dry weight)

	Beta-hydroxylauric	Beta-hydroxymyristic	Beta-Hyroxypentadecanoic	Beta-hydroxypalmitic	beta-Hydroxymargaric	beta-Hydroxymargaric	Beta-hydroxystearic	Lauric	Myristic	Pentadecanoic	Palmitic	Palmitoleic	Heptadecanoic	Heptadec-9-enoic	Stearic	Oleic	Linoleic	Nonadecanoic	Arachidic	Behenic	Lignoceric
		β-C_{14}OH	β-C_{15}OH	β-C_{16}OH	Iso-β-$C_1$7OH	Anteiso b-C_{17}OH	β-C_{18}OH	12:0	14:0	15:0	16:0	16:1	17:0	17:1	18:0	18:1	18:2	19:0	20:0	22:0	24:0
Anabaena variabilis[1]		+		+			+				+										
Anacystis nidulans[2]		+						+			+				+	+				+	
Phormidium spp.											+				+	+					
Schizothrix calcicola[3]	+			+					+	+	+				+	+	+				
Synechococcus PCC 6907[4]		1.3	0.8	2.5		0.4			5.5		11.7	0.8			0.3						
Synechococcus PCC 6307[4]		1.4	0.4	2.8		0.7			4.4		9.6	1.1			0.4						
Synechococcus PCC 6911[4]		1.9	1.1	3.5		0.7			1.1		6.2	1.5			0.2						
Synechococcus PCC 6603[4]				5.5	1.3				tr.	0.4	tr.		0.3								
Synechococcus PCC 6908[4]		0.3	0.3	5.4							0.3	0.2			0.2						
Synechococcus PCC 6301[5]																					
Synechococcus PCC 6311[6]		0.3	0.2	5.4							0.4	tr.		0.3							0.3
Synechococcus PCC 6312[6]		0.2		5.4	0.7						0.7	0.2			0.2						
Synechococcus PCC 6910[6]		0.3	0.2	6							0.4	0.2			0.3						
Synechocystis PCC 6714[6]		1.8	0.1	2.8							0.6	0.2				0.3					
Synechocystis PCC 6803[6]		1.8	0.2	3.6							0.8	0.2				0.3					
Synechocystis PCC 6807[6]		1.8	0.4	4.1					0.5		0.4	tr.			0.5						
Synechocystis PCC 6308[6]		1.6		4.7					tr.	0.5	tr.			0.5							1.6
Anabaena cylindrica[7]	+	+	+	+		+	+	+	+	+	+	+	+	+	+	+	+	+	+	+	+

[1]Weckesser (1974); [2]Weiss (1970); [3]Keleti (1979); [4]Schmidt (1980a); [5]Katz (1977); [6]Schmidt (1980b); [7]Keleti (1982). +, detected in traces.

No lipopolysaccharides were found in *Anabaena flos-aquae* A-37 (Wang and Hill, 1977). Analysis of phenol–water extracts (Wastphal and Jann, 1965) showed that isolated material consists exclusively of polysaccharides without lipids attached. The absence of KDO, heptose and lipid fatty acids suggests that no classical LPS are present. Glucose and mannose are the major sugars of the envelope polysaccharide. However, no further work has been conducted to understand the nature of this unusual LPS although a similar case has been reported for *Borrelia burgdorferi* (Takayama *et al.*, 1987).

Lipopolysaccharides – true endotoxins?

LPS by itself is a relatively inactive and nontoxic molecule (Henderson, 1996). Its inflammatory effect depends on a cascade of host-mediated signals. LPS binds to an extracellular lipopolysaccharide binding protein (LBP), which mediates the interaction with membrane bound CD14 (cluster of differentiation protein 14) on the surface of phagocytes. This interaction activates the Toll-like receptor 4/myeloid differentiation protein 2 (TLR4-MD-2) complex and stimulates the inflammatory response through activation of cytokines, including interleukin 1 (IL-1), IL-6 and tumour necrosis factor-α (TNF-α) enforcing a broad range of cellular reactions (e.g. Heumann and Roger, 2002; Re and Strominger, 2002; Steward *et al.*, 2006; Noreen *et al.*, 2012). Under normal physiological conditions patients only suffer from mild fever with antibacterial and antiviral effects. However, high concentrations of LPS can lead to a strong and unregulated immune response that can result in thrombosis, a septic shock and multiple organ failure (e.g. Alexander and Rietschel, 2001).

Because of the high potential of enterobacterial LPS to trigger an acute inflammatory response, cyanobacterial LPS are often assumed to be endotoxic. However, there is little evidence to support this assumption (Stewart *et al.*, 2006). In contrast, some reports suggest that humans and animals with a normal gut flora and intact intestines would be little if at all affected by drinking cyanobacterial contaminated water (Heaney, 1971; Stewart *et al.*, 2006). Interestingly, the immunodominant sugar of the O-antigen of *Salmonella* is tyvelose, which is also present in the O-antigen of some strains of *Synechococcus* (Table 3.2; Jörbeck, 1979; Keleti *et al.*, 1982). Still, haemagglutination assays shows significantly reduced aggregation when heat-inactivated LPS of *Synechococcus* sp. is used instead of *Salmonella* LPS (Schmidt, 1980). Additionally, after injection of LPS into rabbits, the LPS of *Synechococcus* sp. induces only a mild fever while the pyrogenicity of *Salmonella* LPS was 1000-fold higher (Galanos *et al.*, 1972). The low immune reactivity against cyanobacterial LPS can be explained by the missing phosphorylation at positions 1–4′ of the glycosamine disaccharide in the cyanobacterial Lipid A backbone. Both phosphate groups are important to trigger a full immune response as they are major determinants of the Lipid A conformation (Rietschel, 1994), which in turn influences the binding affinity of Lipid A to LBD and its ability to activate the inflammatory response via TRL4-receptor (Schromm, 1998). Thus, although structurally related, cyanobacterial LPS does not posses true endotoxic effects harming human health.

The lipid composition of cyanobacterial membranes

Membranes of the proteobacterium *E. coli* have a very simple phospholipid composition. They consist of 75% phosphatidylethanolamine (PE), 20% phosphatidylglycerol (PG), and 5% of the PG-derivative cardiolipin (CL; Cronan, 2003). All lipids are synthesized at (or in) the inner leaflet of the plasma membrane. The transport of the lipids to the outer membrane is not well understood. The first step of lipid transfer into the periplasmic leaflet of the membrane could occur by four different modes: (i) spontaneous, or catalysed by (ii) P-type flipases, (iii) ABC-type flipases, or (iv) the so-called scramblases (Sharom *et al.*, 2011). However, P-type flipases and the scramblases are only found in eukaryotes. The subsequent lipid transport to the outer membrane may be catalysed by a mechanism similar to that established for lipopolysaccharide transport (Ruiz *et al.*, 2009; see 'The protein composition of the cyanobacterial cell envelope').

The membrane lipid composition of cyanobacteria is quite different to that of the proteobacteria, whereas it resembles the lipid composition of chloroplasts of higher-plants and algae. The most abundant lipids are the two neutral galactolipids monogalactosyl diglyceride (MGDG) and digalactosyl diglyceride (DGDG). Two charged lipids, the anionic glycolipid sulfoquinovosyldiacylglycerol (SQDG) and the acidic phospholipid phosphatidylglycerol (PG; Nichols *et al.*, 1965), are minor components, while traces of other lipids such as monoglucosyl diacylglycerol (GlcDG) have been found as well (in *Anabaena variabilis*; Sato and Murata, 1982). Analysis of the lipid distribution between the cell envelope and thylakoid membranes of *Anacystis nidulans* showed that the envelope is dominated by MGDG (55%), while it contains 24% of DGDG, 18% PG and only 5% SQDG (Murata *et al.*, 1981). Whereas the MGDG content is similar in both membrane systems, thylakoids contain significantly more PG (21%) and SQDG (14%), but less DGDG (14%), than the cell envelope. The higher level of DGDG in the cell envelope fraction is most likely present in the outer membrane, because only 12% DGDG was observed in the plasma membrane (Omata and Murata, 1983). The concentration of PG (18%) and SQDG (7%) in the plasma membrane is rather comparable to the concentration observed in the entire cell envelope. This clearly speaks for an enrichment of MGDG in the plasma membranes that makes up to 63% of the lipids of this membrane of *Anacystis nidulans* (Omata and Murata, 1983). A very similar lipid distribution was found in several other cyanobacteria. For example, the plasma membrane of the terrestrial cyanobacterium *Nostoc flagelliforme* contains 62% MGDG as well (Huang *et al.*, 2004). In contrast, the concentration of PG is significantly reduced to 2%, which results in an increased concentration of SQDG (17%) and DGDG (19%). This, however, is logically consistent, as a further increase of the non-bilayer lipid MGDG would disturb the membrane integrity. Whether the reduced PG content reflects a lower phosphate usage by *Nostoc flagelliforme* remains speculative.

Cyanobacteria are very diverse and, thus, it is not unexpected that other lipids and different lipid concentrations can be found in them. For example, the genus *Chroococcidiopsis* contains, in addition to the major lipids, the non-phosphorous betaine lipid *N,N,N*-trimethylhomoserin-4-*O*-yl-diacylglycerol (DGTS; Rezanka, 2003). Furthermore, PG, phosphatidylcholin (PC) and sulfolipid (SL), but not MGDG or DGDG, have been identified in the cell envelope of *Synechocystis* sp. PCC 6714. Remarkably, MGDG, DGDG, SL and PG could be detected in the plasma membrane and thylakoid membrane of the same cyanobacterium (Jürgens *et al.*, 1985).

The difference in lipid composition as compared to enterobacteria, on the one hand, and the lipid distribution between the three membrane systems of cyanobacteria, on the other hand, could be explained by the need and availability of environmental phosphate in their respective habitats. Hence, phosphate might be limiting for some bacteria, and thus, alternative compositions might have evolved.

Phosphate availability as determinant of the lipid composition?

Phosphate starvation is a serious growth-limiting factor. Enterobacteria, for example, are supplied with phosphate by their host or environment and can deposit large quantities of phosphate in phospholipid membranes. In contrast, the habitat of most cyanobacteria is not particularly rich in phosphate. In addition, photosynthetic prokaryotes possess large ramified internal membrane systems, the thylakoids. If these would largely consist of phospholipids, tremendous amounts of phosphate would be required for their synthesis. Hence, phosphate shortage or temporary environmental fluctuations in phosphate availability would pose the high risk of starvation and malfunction of photosystems. In turn, carbon fixation via photosynthesis provides large quantities of sugar precursors, which offers the unique opportunity to build up high amounts of low-cost lipids with sugar head groups.

The acidic lipid PG and its derivative CL are indispensable for the maturation of lipoprotein in *E. coli* (Suzuki, 2002; Kikuchi, 2000) or the stability of the respiratory chain supercomplexes in mitochondrial membranes (Pfeiffer *et al.*, 2003). Remarkably, PG is found in cyanobacteria such as *Synechocystis* sp. PCC 6803, where it is essential for biogenesis as demonstrated by the fact that a strain with a mutation in the CDP-diacylglycerol synthase gene required PG supplementation for growth (Sato *et al.*, 2002), and in the terrestrial cyanobacterium *Nostoc flagelliforme* (Hagio *et al.*, 2000; Huang *et al.*, 2004). Interestingly, SQDG can replace PG although the efficiency varies among different cyanobacterial species. Thus, loss of SQDG in *Synechococcus* sp. PCC 7492 by mutation of the UDP-sulfoquinovose synthase (*sqdB*) or the sulfoquinovosyldiacylglycerol synthase (*sqdX*) gene can be complemented by PG (Frentzen, 2004) and the mutant does not show a significant decrease in PSII efficiency. In contrast, the *sqd* mutant of *Synechocystis* sp. PCC 6803 requires SQDG for growth and shows defects in photosynthesis suggesting that PG cannot complement for SQDG requirement (Aoki *et al.*, 2004). The different phenotypes might be related to differences of the UDP-sulfoquinovosyltransferase with respect to molecular composition and activity, as it was found that the one from *Synechocystis* is more similar to the plant orthologue than to the one from *Synechococcus* (Aoki *et al.*, 2004). However, a recent study demonstrated that SQDG is required for the replication of chromosomal DNA in *Synechocystis* sp. PCC 6803 but not in *Synechococcus* sp. PCC 7492, which might explain the differential dependence on SQDG as well (Aoki *et al.*, 2012).

Nevertheless, based on these observations, it can be suggested that the acidic galactolipid SQDG has replaced the acidic PG in the curse of evolution, but SQDG is not able to serve all functions performed by the acidic phospholipid. The remaining need for PG is most likely related to the composition of the photosystems. PG binds to the D1 protein in the PSII reaction centre (Guskov *et al.*, 2009), facilitating formation of the PSII heterodimer complex (Kruse *et al.*, 2000, Sozer *et al.*, 2011), and is a constituent of PSI (Jordan *et al.*, 2001). Thus, it can be concluded that the low environmental availability of phosphate provided the evolutionary pressure for adaptation of the cell envelope membrane composition towards a high content of the non-phosphorous galactolipids MGDG, DGDG and SQDG, although PG is still indispensable for proper photosystem biogenesis and function.

Fatty acid alterations as protective mechanism against stress

The majority of the fatty acids that are incorporated in the membrane lipids is made up by palmitic acid (16:0), palmitoleic acid (16:1), hexadecadienoic acid (16:2), stearic acid (18:0), oleic acid (18:1), linoleic acid (18:2) and α- and γ-linolenic acid (18:3) (see, for example, Nichols *et al.*, 1965; Vigh *et al.*, 1993; Murata *et al.*, 1981; Omata and Murata, 1983; Huang *et al.*, 2004). However, traces of lauric acid (12:0), myristic acid (14:0), myristolic acid (14:1) and ω3,6,9,12-octadecatetraenoic acid (18:4) have been found as well. The distribution is exemplified in Table 3.5. In general, the fatty acid composition of the different lipids is largely comparable between the different cyanobacteria included in Table 3.5. Notably, the number of saturated C16 fatty acids exceeds the number of unsaturated ones, whereas the opposite is true for C18 fatty acids. For MGDG and DGDG, the portion of linoleic acid and linolenic acid exceeds the one of oleic acid, while for SQDG and PG the content of oleic acid is larger or at least similar to the other two.

A comparison of the overall fatty acid content of the membrane lipids of outer and plasma membranes in different cyanobacteria shows that the portion of C18 fatty acids is rather low (e.g. in *Anacystis nidulans* (*Synechococcus*); see Table 3.5). Interestingly, comparison of the fatty acid distribution found in the plasma membrane of the freshwater cyanobacterium *Anacystis nidulans* with that of the terrestrial cyanobacterium *Nostoc flagelliforme* shows a clear distinction for MGDG and DGDG. In *Anacystis nidulans* the palmitic acid is the predominant C16 fatty acid and almost no C18 fatty acids are found. In contrast, unsaturated palmitoleic acid is the dominant C16 fatty acid in *Nostoc flagelliforme*, and high concentrations of linolenic acid were found for these two lipids. SQDG and PG are predominantly composed of saturated palmitic acid in both cyanobacteria. This comparison of the two strains might suggest that the fatty acid composition of cyanobacterial membranes, particularly of the lipids dominating the cell envelope, depends on the environmental habitat where these bacteria are found or have been grown; a notion which has to be challenged by the analysis of other cyanobacteria.

Table 3.5 The molar ratio of lipid-incorporated fatty acids

	MGDG						DGDG						SQDG						PG					
	Total			CE	CM		Total			CE	CM		Total			CE	CM		Total			CE	CM	
	1	2	3	4	5	6	1	2	3	4	5	6	1	2	3	4	5	6	1	2	3	4	5	6
12:0	0	ND	0	ND	ND	ND	0	ND	0	ND	ND	ND	0	ND	4	ND	ND	ND	6	ND	4	ND	ND	ND
14:0	0	ND	1	2	1	ND	0	ND	2	4	2	ND	0	ND	3	2	2	ND	8	ND	8	3	Trace	ND
14:1	ND	ND	Trace	5	3	ND	ND	ND	0	10	2	ND	ND	ND	0	7	1	ND	ND	ND	2	3	Trace	ND
16:0	27	52	43	45	48	10	26	50	51	42	51	9	53	66	43	52	58	39	31	51	35	49	54	61
16:1	28	3	34	46	44	36	26	3	28	42	41	42	5	4	28	36	35	12	9	1	17	40	41	Trace
18:0	Trace	Trace	4	1	1	2	2	Trace	4	1	1	Trace	2	2	4	2	1	3	6	5	8	1	Trace	2
18:1	12	10	20	1	3	5	9	7	16	1	3	3	19	18	16	1	3	14	21	20	22	3	5	34
18:2	18	22	ND	ND	ND	11	20	21	ND	ND	ND	10	13	9	ND	ND	ND	5	12	22	ND	ND	ND	1
18:3	15	11	ND	ND	ND	38	17	17	ND	ND	ND	35	7	Trace	ND	ND	ND	28	6	1	ND	ND	ND	2

Values given in mol% for (1) *Anabaena variabilis* (Nichols *et al.*, 1965), (2) *Synechocystis* PCC6803 (Vigh *et al.*, 1993); (3) *Anacystis nidulans* (Nichols *et al.*, 1965); (4) *Anacystis nidulans* (Murata *et al.*, 1981); (5) *Anacystis nidulans* (Omata and Murata, 1983); (6) *Nostoc flagelliforme* (Huang *et al.*, 2004); ND, not determined; Trace, less than 1%; CE, cell envelope; CM, cytoplasmic membrane.

It has long been known that the fatty acid composition of membrane lipids changes with the growth temperature (Sato and Murata, 1980). The changes are related to the adaptation of the phase behaviour of the entire membrane (Murata, 1989; Nishida and Murata, 1996). In general, desaturation of the palmitoleic acid is observed under low temperatures, although the alteration of the degree of saturation of the C18 fatty acids was found to be different for the four strains investigated (*Anacystis nidulans; Anabaena variabilis; Synechococcus lividus; Synechocystis* sp. PCC 6803; Murata, 1989). In *Anabaena variabilis* and *Synechocystis* sp. PCC 6803 linolenic acid accumulates, whereas the amounts of oleic and linoleic acid are reduced when the temperature is decreased. This clearly documents that desaturation contributes to the adaptation of the membrane to changing environmental conditions. A similar desaturation of C18 fatty acids has been found for the two *Synechococcus* strains. However, in contrast to the other two strains, the amount of stearic acid was reduced and that of oleic acid was increased (at least for *Synechococcus lividus*) after reduction of the temperature. These differences can be explained by a different capacity of the cyanobacteria to desaturate the fatty acids. For a detailed description of the effects and of the fatty acid synthesis pathway, please see the excellent review by Nishida and Murata (1996).

Chilling-induced desaturation is regulated by a central plasma membrane inserted kinase, Hik33, which in *Synechocystis* sp. PCC 6803 regulates the expression of the ω3-desaturase DesB (Suzuki *et al.*, 2000). It was demonstrated that the regulation of Hik33-dependent gene expression is related to the membrane fluidity and further regulated by light (Mironov *et al.*, 2012). The mode of regulation of Hik33 is thought to be similar to that of the temperature sensor DesK in *Bacillus subtilis* (Sakamoto and Murata, 2002). For DesK, it was found that the regulatory function is dependent on its transmembrane domains (Albanesi *et al.*, 2004), which holds true for Hik33 as well (Shimura *et al.*, 2012). The transmembrane domain senses the dimension of the bilayer (Cybulski *et al.*, 2010), and it is thought that a cluster of hydrophilic amino acids, which is usually positioned above the bilayer, changes the positioning of the transmembrane domain upon bilayer size increase in response to the enhanced lipid order during chilling. The information of this change in the transmembrane domain position is likely transferred, through an unknown mechanism, to the kinase domain resulting in a structural change that induces an active state.

The function of Hik33 is not restricted to the chilling stress response, but regulates gene expression in many environmental situations affecting membrane fluidity, such as high-salt stress (Marin *et al.*, 2003), hyperosmotic stress (Paithoonrangsarid *et al.*, 2004) and oxidative stress (Kanesaki *et al.*, 2007). This documents that the lipid and fatty acid composition of the membranes is a prime sensor and a prime target for the perception of, and reaction to, environmental changes.

First hints on the lipid transport system

To date, not much is known about the fatty acid and lipid transport pathways in cyanobacteria. Recent studies provided a first hint at a possible lipid transport pathway in *Synechocystis* sp. PCC 6803. In a screen of mutants of putative ABC transporters for sensitivity against acidic stress, a mutant of *slr1045* was identified (Tahara *et al.*, 2012). The Slr1045 protein has a high similarity to MlaE from *E. coli* (Malinverni and Silhavy, 2009) and Tgd1 from *A. thaliana* (Xu *et al.*, 2005), which are both involved in lipid transport across a membrane. Based on this similarity and a reduced PG content in the *slr1045* mutant under acidic growth conditions, it was suggested that Slr1045 might be involved in phospholipid transport (Tahara *et al.*, 2012). Accordingly, the protein was named '*Synechocystis* sp. PCC 6803 lipid transport protein A' (SltA1). Moreover, it was proposed that the gene product of *slr1344* (SltA2) might perform a similar function (Tahara *et al.*, 2012). Both proteins might function together with the gene products of *sll1002* and *sll0751*, which might act as substrate binding factors and are annotated as SltB1 and SltB2, respectively (Tahara *et al.*, 2012). Finally, the proteins encoded by *sll1001* (SltC1) and *sll1041* (SltC2) have been proposed as ATP binding proteins of the transport system (Tahara *et al.*, 2012). However, experimental evidence is still lacking for these proposed roles, especially a lipid transport function has not been yet demonstrated for SltA1.

The protein composition of the cyanobacterial cell envelope

With respect to the protein content of the cell envelope, cyanobacteria are globally comparable to other Gram-negative bacteria. In the last decade, several proteomic studies have been undertaken to decipher the protein composition of the individual compartments. By proteomics and bioinformatic approaches more than 100 proteins have been identified in the outer membrane/cell wall (Table 3.6). The majority of these proteins are outer membrane-associated proteins, whereas many β-barrel membrane proteins have only been detected via bioinformatics and at the transcript level, but not at the protein level (e.g. Mirus *et al.*, 2009). Among the proteins identified by mass spectrometry, proteins important for the biogenesis of the outer membrane (see 'The biogenesis of the cyanobacterial membranes'), TonB-dependent outer membrane transporters (see 'Iron uptake systems in cyanobacteria', below), a TolC-like protein (see 'The TolC- a central factor for cell survival'), porins (see 'Porins – more than just holes in the wall') and proteins associated with these factors have been identified.

In the plasma membrane of *Synechocystis* sp. PCC 6803, 196 proteins have been identified in proteomic studies, 80 of which have been found in at least two independent studies (Huang *et al.*, 2002, 2006; Pisareva *et al.*, 2007, 2011; Zhang *et al.*, 2008). The large number of proteins observed in one study can be explained by the plasticity of the membrane proteome in response to environmental changes such as salt or pH stress (Huang *et al.*, 2006; Zhang *et al.*, 2009; Li *et al.*, 2012). In general, subunits of the NADH dehydrogenase, the cytochrome b_6f complex, the ATP synthase, the photosystems I and II, and the respiratory terminal oxidases have been found (Table 3.7). In addition, several enzymes for carotenoid, cobalamin, haem and porphyrin synthesis, and proteins involved in the formation of structural elements such as pili have been identified. However, the largest family that has been found consists of proteins involved in the transport of solutes across the plasma membrane (see 'The plasma membrane proteome'). Remarkably, most of these have been exclusively identified in the plasma membrane, while many of the other proteins have also been found in other cellular compartments (Table 3.6). In addition, 32 plasma membrane proteins of unknown function have been detected, which strongly suggests that we do not yet fully understand the properties of the cyanobacterial proteome.

Interestingly, a recent study has dissected the protein content generally found in the plasma membrane, in the regions close to the thylakoid centres and in thylakoids (Pisareva *et al.*, 2011). The authors showed that most of the components of the photosystems, the NADH dehydrogenase, the cytochrome b_6f complex and the ATP synthase were found in both membranes, whereas the subunits of the respiratory terminal oxidases, the proteins involved in chemotaxis and most of the proteins involved in transport of solutes are restricted to the plasma membrane. Even more, some of the transporters were not identified in the regions of the plasma membrane that are in close proximity with the thylakoid centres (Table 3.6; Pisareva *et al.*, 2011). This observation clearly speaks in favour of the presence of functionally distinct regions in the plasma membrane.

Remarkably, two porins and the outer membrane protein TolC have been found among the proteins that are considered likely contaminants often co-purified with the cytoplasmic membrane (uncertain localization; Table 3.6). For these three proteins experimental evidence exists that they are true outer membrane proteins.

The biogenesis of the cyanobacterial membranes

Protein translocation across and insertion into the cytoplasmic membrane is catalysed by the SEC translocon (Bohnsack and Schleiff, 2010; for a further discussion of this topic, see Chapter 5). The outer membrane of cyanobacteria contains two central components, outer membrane lipids (including the LPS) and outer membrane proteins (OMPs). Thus, factors for lipopolysaccharide layer formation (Sperandeo *et al.*, 2009) and for the integration of OMPs are central for the

Table 3.6 The protein content of the outer membrane identified by proteomics and bioinformatic predictions in *Synechocystis* sp. PCC 6803 and *Anabaena* sp. PCC 7120

Synechocystis sp. PCC 6803	*Anabaena* sp. PCC 7120	Code	Putative function	Reference					
				1	2	3	4	5	6
OMPA/MOTB									
n.f.	All3289								+
n.f	Alr5049								+
OM biogenesis and secretion									
Sll0350	Alr1278	LptD	Organic solvent tolerance protein OstA	+					+
n.f	All0406		Auto transporter						+
Slr1661	All5116	FhaC1	TpsB transporter						+
	Alr1659	FhaC2	TpsB transporter						+
Slr1227	Alr2269	Omp85	Outer membrane biogenesis factor	+	+	+	+		+
	Alr0075	Omp85–2	Outer membrane biogenesis factor						+
	Alr4893	Omp85–3	Outer membrane biogenesis factor						+
Slr1270	Alr2887	TolC	Outer membrane efflux protein	+	+	+			+
Sll1581	All0495	Oma	Outer Membrane Auxiliary (OMA) Protein Family						
	Alr2294	Oma	Outer Membrane Auxiliary (OMA) Protein Family						
Slr0019	Alr3592	PapC	Pilus biogenesis						+
Porins									
Slr1841	Alr4550	OprB1	Carbohydrate-selective porin like	+	+	+			+
Slr1908	All4499	OprB2	Carbohydrate-selective porin like	+	+	+			+
Sll0772	All5191	OprB3	Carbohydrate-selective porin like						+
Sll1550	Alr2231	OprB4	Carbohydrate-selective porin like						+
Slr0042	Alr4741	OprB5	Carbohydrate-selective porin like						+
Sll1271	All7614	OprB6	Carbohydrate-selective porin like						
Slr1163		OprB7	Carbohydrate-selective porin like						
	Alr0834	Porin1	Porin-like	+	+				+
Slr1272	Alr3608	Porin2	Porin-like	+		+	+		
	Alr3917	Porin3	Porin-like						+
TonB-dependent transporter									
Sll1206	Alr0397	IutA/SchT	TonB-dependent transporter	+	+			+	+
	Alr2209	IutA2	TonB-dependent transporter					+	+
	Alr2581	IutA3	TonB-dependent transporter					+	+
Sll1406	All1101	FhuA1	TonB-dependent transporter	+		+	+	+	+
Sll1409	All2148	FhuA2	TonB-dependent transporter					+	+
Slr1490	All2158	FhuA3	TonB-dependent transporter					+	+
	All2236	FhuA4	TonB-dependent transporter					+	+
	All2610	FhuA5	TonB-dependent transporter					+	+
	All2674	FhuA6	TonB-dependent transporter					+	+
	All4924	FhuA7	TonB-dependent transporter					+	+
	Alr2175	FhuA8	TonB-dependent transporter					+	+
	Alr2185	FhuA9	TonB-dependent transporter					+	+
	Alr2211	FhuA10	TonB-dependent transporter					+	+
	Alr2588	FhuA11	TonB-dependent transporter	+				+	+
	Alr2592	FhuA12	TonB-dependent transporter					+	+

Synechocystis sp. PCC 6803	Anabaena sp. PCC 7120	Code	Putative function	Reference 1	2	3	4	5	6
	Alr2596	FhuA13	TonB-dependent transporter					+	+
	Alr2626	FhuA14	TonB-dependent transporter					+	+
n.f	All3310	BtuB	TonB-dependent transporter	+				+	+
n.f	All4026	lacT	TonB-dependent transporter	+				+	+
n.f	Alr2153	HutA1	TonB-dependent transporter					+	+
	Alr3242	HutA2	TonB-dependent transporter					+	+
TBDT-associated factors									
Slr1484	Alr0248	TonB1	TBDT-associated energy transducer						*
	All3585	TonB2	TBDT-associated energy transducer						*
	All5036	TonB3	TBDT-associated energy transducer	+					*
	Alr5329	TonB4	TBDT-associated energy transducer						*
Slr1295	Alr1382	FutA1	Periplasmic iron binding protein			+			*
Slr0513		FutA2	Periplasmic iron binding protein						*
n.f	All0388	FhuD	Periplasmic iron binding protein						*
Sll1202	Alr2583	FecB1	Periplasmic iron binding protein						*
Slr1319	Alr3243	FecB2	Periplasmic iron binding protein						*
Slr1491	Alr4031	FecB3	Periplasmic iron binding protein						*
Slr1492		FecB4	Periplasmic iron binding protein						*
Peptidoglycan-binding proteins									
Sll0068	All1861		Peptidoglycan-binding protein	+	+				
Sll1360	All3826		Peptidoglycan-binding protein	+	+				
Slr0646	Alr1666	LdcA	D-Alanyl-D-alanine carboxypeptidase	+					
Slr0891	All4999	AmpD	*N*-acetylmuramoyl-L-alanine amidase	+					
Slr1734	All4294	AmpD	*N*-acetylmuramoyl-L-alanine amidase	+					
Slr1734	Alr1690	AmpD	*N*-acetylmuramoyl-L-alanine amidase	+					
Slr1744	Alr0092	AmpD	*N*-acetylmuramoyl-L-alanine amidase	+					
Slr1910	All1140	AmpD	*N*-acetylmuramoyl-L-alanine amidase	+					
Periplasmic proteins									
Sll0180	All3144		Multidrug efflux transporter fusion protein	+		+			
Sll0680	All4575		Phosphate-binding periplasmic protein	+					
Sll0842	Alr0663		Neopullulanase	+					
Sll0915	All1940	pqqE	Periplasmic protease			+			
Sll1358	All7319		Periplasmic oxalate decarboxylate			+			
Sll1481	All5347		ABC-transporter membrane fusion protein	+					
Sll1491	All2352		Periplasmic WD-repeat protein	+					
Sll1581	All4388	GumB	Periplasmic polysaccharide export	+	+	+	+		
Slr0250	Alr4067	LptA	Organic solvent tolerance protein OstA	+					
Slr0447	All1951		Periplasmic ABC-type transport system subunit	+					
Slr0559	Alr1834		Periplasmic ABC-type branched-chain amino acid transp. comp.	+					
Secreted proteins									
Sll0736	All1776		fenl S-layer protein	+					

Table 3.6 (Continued)

Synechocystis sp. PCC 6803	*Anabaena* sp. PCC 7120	Code	Putative function	Reference					
				1	2	3	4	5	6
Sll1483	Alr1819		Secreted and surface protein with fasciclin-like repeats	+					
Slr1704	All3983		Secreted and surface protein	+					
	All3984			+	+				
Uncertain localization									
Sll0173	–	Vgb	Virginiamycin b hydrolase			+			
Sll0359	Alr0946		Regulators of stationary/sporulation	+					
Sll0789	All7606		OmpR homologue	+					
Sll1908	Alr1890		D-3-phosphoglycerate dehydrogenase	+					
Slr0319			β-Lactamase				+		
Slr1204	Alr0702		Protease HtrA	+		+	+		
Slr1277	Alr1031		General secretion protein pathway D			+	+		
Slr1751	All2500		Carboxyl terminal protease CtpC	+		+			
Slr2005	All0268		Predicted Zn-dependent hydrolases	+	+				
Slr8030	All8079		Predicted phosphatase	+					
Hypothetical									
Sll0319			Hypothetical			+			
Sll0470	All7598		Hypothetical	+					
Sll0470	Alr3411		Hypothetical	+					
Sll0471	All0405		Hypothetical	+					
Sll0552			Hypothetical				+		
Sll0749	Alr5055		Hypothetical		+				
Sll1025	Alr0668		Hypothetical		+				
Sll1307			Hypothetical			+			
Sll1338			Hypothetical			+	+		
Sll1586	All2430		Hypothetical protein	+					+
Sll1586	All7597		Hypothetical protein	+					
Sll1638			Hypothetical			+			
Sll1835			Hypothetical			+			
Sll1940			Hypothetical			+	+		
Slr0172	All2571		hypothetical protein	+					
Slr0306			Hypothetical				+		
Slr0356			Hypothetical			+	+		
Slr0400	All3585		Hypothetical	+	+				
Slr0431	All0089		Hypothetical	+	+	+			
Slr0695			Hypothetical			+			
Slr1150			Hypothetical				+		
Slr1457	Alr4740		unknown protein	+					
Slr1457	Alr7326		hypothetical protein	+					
Slr1506			Hypothetical			+			
Slr1753			Hypothetical			+	+		
Slr6044	All7611		hypothetical protein	+					

1, Moslavac *et al.* (2005); 2, Moslavac *et al.* (2007); 3, Huang *et al.*, (2004b); 4, Rajalahti *et al.* (2007); 5, Mirus *et al.* (2009); 6 +, Nicolaisen *et al.* (2009a); 6*, Stevanovic *et al.* (2012).

Table 3.7 The protein content of the cytoplasmic membrane identified by proteomic analysis of *Synechocystis* sp. PCC 6803

	Accession no.	Name	Putative function	Reference				
				1	2	3	4	5
Photosystems								
1	Sll0226	Ycf4	PSI assembly					PM1/2
2	Sll0258	PsbV	PSII subunit					PM1
3	Sll0427	PsbO	PSII subunit,	+	+			PM1/2
4	Sll0819	PsaF	PSI subunit	+	+		+	PM1/2
5	Sll0851	PsbC	PSII subunit					PM1
6	Sll1194	PsbU	PSII subunit					PM1
7	Sll1298	PsbW	PSII subunit					PM1
8	Sll1867	PsbA3	PSII subunit					PM1/2
9	Sll1638	PsbQ	PSII subunit	+	+			PM1/2
10	Slr0399	Ycf39	PSII subunit					PM1
11	Slr0737	PsaD	PSI subunit	+				PM1/2
12	Slr0823	Ycf3	PSI assembly protein		+			
13	Slr0906	PsbB	PSII subunit					PM1
14	Slr1311	PsbA2	PSII subunit					PM1/2
15	Slr1645	PsbZ	PSII subunit					PM1/2
16	Slr1655	PsaL	PSI subunit					PM1/2
17	Slr1834	PsaA	PSI subunit					PM1/2
18	Slr1835	PsaB	PSI subunit					PM1/2
19	Slr2034	Hcf136	PSII assembly protein					PM1
20	Ssl0563	PsaC	PSI subunit	+				PM1/2
21	Ssr2831	PsaE	PSI subunit					PM1/2
22	Smr0006	PsbF	PSII subunit					PM1
23	Smr0007	PsbL	PSII subunit					PM1
Cytochrome b_6f complex								
24	Sll1812	PetC3	Cytochrome b6f complex iron–sulfur subunit					PM1/2
25	Sll1316	PetC1	Cytochrome b6f complex iron–sulfur subunit					PM1
26	Sll1317	PetA	Cytochrome b6f complex iron–sulfur subunit					PM1
NADH dehydrogenase								
27	Sll0520	NdhI	NADH dehydrogenase subunit					PM1
28	Sll1484	NdbC	NADH-dehydrogenase type II,			+		**PM2**
29	Sll1262	NfhN	NADH dehydrogenase subunit					PM2
30	Slr1280	NdhK	NADH dehydrogenase subunit					PM1
31	Slr1281	NdhJ	NADH dehydrogenase subunit					PM1
32	Slr1623	NfhM	NADH dehydrogenase subunit					PM2
33	Slr1743	NdbB	NADH dehydrogenase				+	PM1/2
34	Ssl1690	NfhO	NADH dehydrogenase subunit					PM2
Phycobilisome								
35	Sll1471	CpcG	Phycobilisome rod-core linker polypeptide			+		
36	Slr0335	ApcE	Phycobilisome -core membrane linker polypeptide			+		

Table 3.7 (Continued)

	Accession no.	Name	Putative function	Reference				
				1	2	3	4	5
Phycocyanin								
37	Sll1577	CpcB	Phycocyanin b chain		+	+		
38	Sll1578	CpcA	Phycocyanin chain	+	+		+	
39	Sll1579	CpcC	Phycocyanin			+		
40	Sll1580	CpcC2	Phycocyanin-associated linker protein			+		
41	Slr2067		Allophycocyanin alpha subunit		+			
ATP synthase								
42	Sll1323	AtpG	ATP synthase subunit	+	+			PM1/2
43	Sll1324	AtpF	ATP synthase subunit b	+	+			PM2
44	Sll1326	AtpA	ATP synthase chain	+	+			PM1/2
45	Sll1327	AtpC	ATP synthase chain					PM1/2
46	Slr1329	AtpB	ATP synthase beta subunit,		+			PM1/2
47	Slr1330	AtpE	ATP synthase epsilon subunit,		+			PM1/2
Respiratory terminal oxidases								
48	Sll0813	CtaC	Cytochrome *c* oxidase subunit II	+	+	+	+	**PM2**
49	Sll1184		Haem oxygenase I				+	
50	Slr2082	CtaDII	Haem-copper quinol oxidase subunit I					**PM2**
Chemotaxis								
51	Sll0041	PixJ1	Methyl-accepting chemotaxis protein			+		**PM1/2**
52	Sll1294	Mcp	Methyl-accepting chemotaxis protein			+		**PM2**
53	Sll1694	PilA1	Pilin			+		PM1/2
54	Slr0162	PilC	Twitching motility protein					**PM2**
55	Slr1044	PilJ	Methyl-accepting chemotaxis protein			+		PM1/2
56	Slr1274	PilM	Pilus assembly protein	+	+			**PM1/2**
57	Slr1275	PilN	Pilus assembly protein	+	+			**PM2**
58	Slr1276	PilO	Pilus assembly protein	+				PM1/2
59	Slr1928	PilA5	Pilin-like protein					**PM1/2**
60	Slr2016	PilA10	Pilin-like protein					**PM2**
61	Slr2017	PilA11	Pilin-like protein					**PM2**
Enzymes								
62	Sll0034	VanY	Putative carboxypeptidase	+	+	+		
63	Sll0646	Cya2	Guanylyl cyclases				+	
64	Sll1091	ChlP	Geranylgeranyl hydrogenase					PM1/2
65	Sll1178		Probable carbamoyl transferase			+		
66	Sll1363	IlvC	Ketol-acid reductoisomerase	+				
67	Sll1713		Histidinol-phosphate aminotransferase				+	
68	Slr0194		Ribose-5-phophate isomerase		+			
69	Slr0394	PgK	Phosphoglycerate kinase,			+		
70	Slr0506	Por	NADPH-protochlorophyllide oxidoreductase					PM1/2
71	Slr0839	ScpA	Ferrochelatase HemH					PM1
72	Slr0940	CrtQ	Zeta-carotene desaturase					PM1
73	Slr0963		Ferredoxin sulfite reductase			+		

	Accession no.	Name	Putative function	Reference				
				1	2	3	4	5
74	Slr1377	LepB	Leader peptidase type I	+	+			**PM1/2**
75	Slr1609	FadD	Long-chain-fatty-acid CoA ligase			+		
76	Slr1751	CtpC	Carboxyl-terminal protease	+				
77	Slr1943		Putative glycosyltransferase			+		
78	Ssl0707	GlnB	Nitrogen regulatory protein P-II	+	+			
Transport and binding proteins								
79	Sll0180		Membrane fusion protein	+	+	+	+	PM1/2
80	Sll0141		Membrane fusion protein	+				PM2
81	Sll0289	MinD	Septum site-determining protein				+	
82	Sll0374	UrtE	ABC urea transporter ATP-binding subunit					**PM1/2**
83	Sll0415		ABC transporter ATP-binding protein					**PM1/2**
84	Sll0540	PstS	ABC Phosphate transporter					**PM1**
85	Sll0679		ABC transporter; phosphate-binding protein					**PM2**
86	Sll0680	PstS1	ABC transporter; phosphate-binding protein		+			**PM1/2**
87	Sll0683	PstB	Phosphate transport ATP-binding protein				+	**PM2**
88	Sll0684	PstB	Phosphate transport ATP-binding protein				+	**PM2**
89	Sll0923	EpsB	Exopolysaccharide export protein			+		
90	Sll1041		ATP-binding protein of ABC transporter	+				**PM2**
91	Sll1053	MtrC	Membrane fusion protein	+				**PM1/2**
92	Sll1080		ABC transporter; substrate binding protein		+			PM1/2
93	Sll1082		ABC-transport system ATP-binding protein				+	**PM2**
94	Sll1180	HlyB	Toxin secretion ABC transporter ATP-binding protein			+		**PM1/2**
95	Sll1181	HlyD			+			PM1/2
96	Sll1270	BgtB	ABC glutamine transporter substrate binding					**PM2**
97	Sll1404	TolQ	Biopolymer transport protein, ExbB			+		**PM2**
98	Sll1405	TolR	ExbD protein	+				**PM2**
99	Sll1450	NrtA	ABC transporter; nitrate/nitrite binding protein		+	+	+	PM1/2
100	Sll1481		HlyD family protein					**PM2**
101	Sll1482		ABC-transporter permease protein				+	**PM2**
102	Sll1699		Periplasmic oligopeptide-binding protein	+	+		+	**PM2**
103	Sll1835		periplasmic curli production component CsgG family	+	+			
104	Sll1864		Probable chloride channel					PM1/2
105	Sll1878	FutC	ABC iron(III) transp., ATP-binding protein					**PM1/2**
106	Slr0040	CmpA	Bicarbonate transporter substrate-binding protein	+		+	+	PM1/2
107	Slr0043	CmpC	Bicarbonate transporter ATP-binding protein					**PM1/2**
108	Slr0044	CmpD	Bicarbonate transporter ATP-binding protein					**PM1**
109	Slr0074	ycf24	ABC-transporter subunit				+	**PM2**
110	Slr0369		Cation / multidrug efflux system protein			+		**PM1/2**
111	Slr0513	FutA2	Periplasmic iron-binding protein	+	+		+	PM1/2
112	Slr0529	GgtB	ABC transporter; glucosylglycerol binding protein		+			PM2
113	Slr0559	NatB	ABC neutral amino acids transport, substrate binding					PM1/2
114	Slr0677	TolQ	ExbB protein	+	+	+		**PM2**
115	Slr0678	TolR	ExbD protein					**PM2**

Table 3.7 (Continued)

	Accession no.	Name	Putative function	Reference				
				1	2	3	4	5
116	Sll0679		Periplasmic phosphate-binding protein				+	
117	Slr0765	MscS	Mechanosensitive ion channel,			+		**PM2**
118	Slr0798	ZiaA	Zinc-transporting P-type ATPase			+		**PM2**
119	Slr0875	MscL	Large conductance mechanosensitive channel	+				PM1/2
120	Slr1106		Prohibitin	+	+			
121	Slr1128		Stomatin-like protein	+		+		
122	Slr1207		HlyD family secretion protein					PM1/2
123	Slr1216		Magnesium transport protein					PM1
124	Slr1224		ABC sugar transporter ATP-binding protein					**PM2**
125	Slr1295	FutA1	Periplasmic iron-binding protein	+	+	+	+	PM1/2
126	Slr1319	FecB	Iron(III)dicitrate periplasmic binding protein	+		+		**PM2**
127	Slr1416		MorR protein		+			
128	Slr1452	SbpA	Sulfate transport substrate-binding protein					**PM1**
129	Slr1512	SbtA	Sodium-dependent bicarbonate transporter			+		PM1/2
130	Slr1513		Periplasmic protein	+		+	+	PM2
131	Slr1730	KdpC	Potassium-transporting ATPase C chain	+				
132	Slr1735	BgtA	ABC glutamine transporter ATP-binding subunit					**PM1/2**
133	Slr1740		ABC transporter; oligopeptide binding protein		+			PM1/2
134	Slr1768		Prohibitin	+	+			
135	Slr1794		Probable anion transporting ATPase				+	**PM2**
136	Slr1881	BraG	Branched-chain amino acid transporter LivF	+				
137	Slr1897	SrrA	ABC transporter; sugar-binding protein,	+	+			
138	Slr1962		Probable extracellular solute-binding protein					PM1
139	Slr2000		S-layer homology domain proteins		+		+	
140	Slr2024		CheY superfamily		+			
141	Slr2057	AqpZ	Water channel protein					**PM2**
142	Slr2131		Cation/multidrug efflux system protein			+		**PM1/2**
Proteins of unknown relation to the PM								
143	Sll0020	Clp	ATP-dependent protease regulatory subunit		+			
144	Sll0227		Peptidyl-prolyl cis–trans isomerase B		+			
145	Sll0314		Periplasmic TPR repeat-containing protein		+			
146	Sll0617	Im30	Vipp1	+	+			PM1/2
147	Sll0947	LrtA	Light-repressed protein	+	+			
148	Sll1679		HhoA protease		+			
149	Slr0151		TPR domain-containing protein	+	+	+	+	
150	Slr1390	FtsH2	Protease			+		
151	Slr1604	FtsH4	Protease			+		
Unknown proteins								
152	Sll0325		Hypothetical		+			
153	Sll0443		Hypothetical		+			
154	Sll0565		Hypothetical	+	+			
155	Sll0606		Hypothetical	+	+			
156	Sll0749		Hypothetical	+	+		+	

	Accession no.	Name	Putative function	Reference 1	2	3	4	5
157	Sll0752		Hypothetical	+		+	+	
158	Sll1021		Hypothetical	+	+	+		
159	Sll1106		Hypothetical				+	
160	Sll1118		Hypothetical	+	+			
161	Sll1239		Hypothetical				+	
162	Sll1628		Hypothetical		+			
163	Sll1665		Hypothetical			+		
164	Sll1696		Hypothetical		+			
165	Sll1757		Hypothetical			+		
166	Sll5132		Hypothetical				+	
167	Slr0013		Hypothetical	+	+	+		
168	Slr0362		Hypothetical		+			
169	Slr0431		Hypothetical	+				
170	Slr0455		Hypothetical		+			
171	Slr0483		Hypothetical		+			
172	Slr0516		Hypothetical			+		
173	Slr0695		Hypothetical lipoprotein	+				
174	Slr0848		Hypothetical			+		
175	Slr0872		Hypothetical	+	+			
176	Slr1220		Hypothetical	+	+			
177	Slr1258		Hypothetical	+	+		+	
178	Slr1506		Hypothetical	+				
179	Slr1590		Hypothetical		+			
180	Slr1612		Hypothetical				+	
181	Slr1624		Hypothetical	+				
182	Slr1721		Hypothetical			+		
183	Slr2105		Hypothetical			+		
184	Slr6071		Hypothetical			+		
Uncertain localization								
185	Sll0416	GroEL-2	60 kDa chaperonin	+	+			
186	Sll1028	CcmK	Carbon dioxide-concentrating mechanism protein	+				
187	Sll1029	CcmK	Carbon dioxide-concentrating mechanism protein	+				
188	Sll1031	CcmD	Carbon dioxide concentrating mechanism protein			+		
189	Sll1099	TufA	Elongation factor EF-Tu	+	+		+	
190	Sll1626	LexA					+	
191	Slr1270	TolC		+			+	PM2
192	Slr1272		Probable porin				+	
193	Slr1463	EF-G	Elongation factor				+	
194	Slr1841		Putative porin	+	+	+		
195	Slr1908		Putative porin	+	+	+	+	

1, Huang *et al.*, 2002; 2, Huang *et al.*, 2006; 3, Pisareva *et al.* (2007); 4, Zhang *et al.* (2009); 5, Pisareva *et al.* (2011). PM1, plasma membrane section close to the thylakoid centre; PM2, general plasma membrane. Bold indicates proteins which have not be found in the thylakoid fraction of this analysis.

outer membrane function (Mirus *et al.*, 2010). Lipid A is synthesized on the cytoplasmic side of the plasma membrane (see 'The lipid composition of cyanobacterial membranes') and has to be transported to the outer leaflet of the outer membrane by the β-barrel-type protein LptD (formerly known as Imp/OstA, see 'The LptD-dependant Lipid A transport' below). Protein insertion into the outer membrane is mediated by Omp85 proteins, which are a subclass of the polypeptide transporting β-barrel (PTB) protein superfamily. Both systems are related, as the open reading frame encoding Omp85 in *Anabaena* sp. PCC 7120, *alr2269*, is the first gene of an operon that also includes genes for Lipid A biosynthesis (Nicolaisen *et al.*, 2009b).

The LptD-dependent Lipid A transport

The central outer membrane protein involved in Lipid A transport is the Lipid A transport protein D (LptD). This protein is essential in *E. coli* (Wu *et al.*, 2006), but not in *Neisseria meningitidis* (Bos *et al.*, 2004). Mutation of the gene encoding this protein in *N. meningitidis* leads to a 90% reduction of LPS as compared to the wild-type (Bos *et al.*, 2004), and the remaining LPS is not exported to the cell surface resulting in an altered lipid/protein ratio of the outer membrane (Braun and Silhavy, 2002). In *E. coli*, LptD forms a complex with the outer membrane-anchored lipoprotein LptE (formerly known as RlpB). Depletion of LptE resulted in a similar phenotype as depletion of LptD (Wu *et al.*, 2006). However, LptE is not generally found in bacteria and does not exist in cyanobacteria (Haarmann *et al.*, 2010). In contrast, LptD and the periplasmic factor involved in the delivery of Lipid A, LptA, are present in cyanobacteria (Haarmann *et al.*, 2010).

In *Anabaena* sp. PCC 7120 the LptD orthologue is encoded by *alr1278* (Haarmann *et al.*, 2010). Although nothing is known about the *in vivo* function of LptD, the *in vitro* properties are comparable to those of LptD from *E. coli*. By electrophysiological measurements an interaction with Lipid A could be demonstrated, which suggests that anaLptD functions similarly in Lipid A export as established for the proteobacterial system.

The other components involved in the transport of Lipid A from its place of synthesis in the cytosolic leaflet of the plasma membrane to the outer leaflet of the outer membrane have only been described by bioinformatic means. From the 7 components described in *E. coli* (LptA-C, LptE-G, MsbA; Ruiz *et al.*, 2009), four have been detected in the genome of *Anabaena* sp. PCC 7120, namely LptA, LptB, LptF and MsbA (Haarmann *et al.*, 2010).

MsbA is required for the transfer of Lipid A from the inner leaflet to the outer leaflet of the plasma membrane. The protein encoded in the genome of *Anabaena* sp. PCC 7120 with highest similarity to MsbA is HepA (formerly termed HetA, encoded by *alr2835*; Holland and Wolk, 1990); the closest homologue in *Synechocystis* sp. PCC 6803 is Sll1276. The protein was found to be essential for the formation of the heterocysts-specific polysaccharide layer (Huang *et al.*, 2005), but not for cell survival per se. This might be explained by the similarity of MsbA to the protein encoded by *alr5199*, for which no experimental information is yet available and which might in part complement the function of MsbA.

Alr4069 is assigned as LptF and shares similarity to LptG as well, which suggests that in contrast to the situation in *E. coli* a homodimer functions as the plasma membrane-inserted unit facilitating the release of Lipid A from the plasma membrane in an LptB (Alr4068)-dependent manner. LptA (Alr4067) recognizes Lipid A in the periplasm and transfers it to the outer membrane component LptD. Remarkably, LptA was detected in a proteomic study of the cell wall (Table 3.6), which suggests a tight interaction with LptD that has not yet been experimentally confirmed.

Thus, cyanobacterial genes encoding proteins homologous to proteins of the Lipid A transport pathway to the outer leaflet of the outer membrane have been identified. However, the function of these genes in cyanobacteria and the transfer mode of Lipid A into the outer leaflet remain to be established.

Omp85/FhaC system of cyanobacteria

Omp85-like proteins were initially discovered in *H. influenzae* as D15 protective surface antigens

(Thomas *et al.*, 1990), and subsequently in all branches of life representing one of the most ancient outer membrane proteins (see, for example, Bredemeier *et al.*, 2007). In bacteria, these proteins are essential for insertion of porins into the outer membrane (Genevrois *et al.*, 2003; Voulhoux *et al.*, 2003; Wu *et al.*, 2005). The protein is composed of a 16-stranded β-barrel and an N-terminal domain with polypeptide transport-associated (POTRA) repeats (Schleiff *et al.*, 2011). The function of the β-barrel domain remains partially unknown, while the POTRAs appear to serve as a receptor for incoming outer membrane proteins and as a docking site for Omp85 complex components (Schleiff *et al.*, 2011).

In cyanobacteria, Omp85 was first identified in *Synechocystis* sp. PCC 6803 (Slr1227; Bölter *et al.*, 1998; Reumann *et al.*, 1999), in which only one Omp85 is present. In contrast, the genome of the filamentous cyanobacterium *Anabaena* sp. PCC 7120 encodes three different Omp85 homologues (Alr0075, Alr2269, Alr4893; Bredemeier *et al.*, 2007), but only one of these proteins, namely Alr2269, which is annotated as anaOmp85, was identified in proteomic studies of the outer membranes isolated from heterocysts or vegetative cells (Moslavac *et al.*, 2005, 2007). The other two Omp85-like proteins have been detected by immunodecoration (Nicolaisen *et al.*, 2009b), which suggests a lower abundance of All0075 and Alr4893 than of Alr2269. Moreover, Alr0075 was only found in vegetative cells, which argues for a rather specialized function of this protein.

Similar to proteobacterial Omp85 proteins, the N-terminal domain of Alr2269 is composed of three POTRA elements (Koenig *et al.*, 2010) and is involved in substrate recognition and Omp85 oligomerization in vitro (Ertel *et al.*, 2005). However, by electrophysiological analysis of reconstituted proteins it was demonstrated that cyanobacterial Omp85 proteins have a five times larger pore diameter than proteobacterial Omp85 proteins (Bölter *et al.*, 1998, Bredemeier *et al.*, 2007), which explains why anaOmp85 could not complement the *omp85* mutant in *E. coli* (Wunder *et al.*, 2009). Remarkably, the two other identified homologues in *Anabaena* sp. PCC 7120 possess smaller pore diameters than anaOmp85, which however are still about 2–3 times larger than that of the *E. coli* Omp85 (Nicolaisen *et al.*, 2009b). The three *Anabaena* proteins may perform different functions, which are supported by the observation that all three homologues are essential proteins, since chromosomes mutated in their corresponding genes could not be segregated.

The importance of Omp85 proteins for the outer membrane integrity has been concluded from several lines of evidence. On one hand, the (not fully segregated) mutant of anaOmp85 showed an increased sensitivity to erythromycin, SDS, lysozyme and proteinase K suggesting that the outer membrane has an increased permeability (Nicolaisen *et al.*, 2009b). On the other hand, the anaOmp85 mutant has an increased amino acid uptake rate, which also documents the importance of Omp85 function for the formation of the outer membrane barrier. In addition, the structure of the outer membrane is largely altered in the mutant (Nicolaisen *et al.*, 2009b; Tripp *et al.*, 2012). However, a direct evidence for a catalytic function of cyanobacterial Omp85 in outer membrane protein insertion is still lacking.

In proteobacterial systems, OMPs are delivered to Omp85 by the periplasmic chaperone SurA, after they have been translocated across the plasma membrane (Sklar *et al.*, 2007). However, a SurA homologue is not encoded in any cyanobacterial genome analysed thus far (Bohnsack and Schleiff, 2010). Interestingly, anaOmp85 was found to interact with All0114 in *Anabaena* sp. PCC 7120 (Tripp *et al.*, 2012). All0114 encodes a homologue of Tic22, which in plants is involved in protein translocation into chloroplasts (e.g. Schleiff and Becker, 2011). The mutant of anaTic22 shows a defect in the ultrastructure of the outer membrane comparable to that of the anaOmp85 mutant (Tripp *et al.*, 2012). In addition, the anaTic22 mutant is also sensitive to SDS or lysozyme. Thus, one might speculate that the cyanobacterial Tic22 has a function similar to that of the proteobacterial SurA (Tripp *et al.*, 2012). Again, many lines of indirect evidence point to a role of anaTic22 in the insertion of outer membrane proteins, but a final proof is still missing.

Iron uptake systems in cyanobacteria

The TonB-dependent transport systems

Cyanobacteria require large amounts of iron, generally 10 times more than non-photosynthetic prokaryotes (Nicolaisen and Schleiff, 2010). Iron is taken up bound to siderophores, which are Fe(III)-chelators secreted by microorganisms. Siderophore production and secretion occurs under iron starvation to foster iron uptake (see, for example, Estep *et al.*, 1975; Murphy *et al.*, 1976; Simpson and Neilands, 1976; Armstrong and van Baalen, 1979; McKnight and Morel, 1980; Lammers and Sanders-Loehr, 1982; Goldman *et al.*, 1983; Brown and Trick, 1992; Wilhelm and Trick, 1994; Beiderbeck *et al.*, 2000), but it also occurs when the cells are exposed to high concentrations of other metal ions, such as copper, to protect the organism from such toxic compounds (Clarke *et al.*, 1987).

The molecular mode of siderophore secretion into the cellular surrounding has been studied in *Anabaena* sp. PCC 7120, which produces the siderophore schizokenin (Lammers and Sanders-Loehr, 1982). A major facilitator superfamily (MFS) plasma membrane protein, SchE (All4025), was proposed to be involved in the export of schizokinen, since a mutant of *schE* showed a reduced secretion of siderophore (Nicolaisen *et al.*, 2010). The TolC-like outer membrane protein HgdD (see 'The cyanobacterial TolC') appears to be required for siderophore secretion as well (Nicolaisen *et al.*, 2010).

In general, Gram-negative bacteria contain TonB-dependent-transporters (TBDTs) in their outer membranes, which bind siderophores with high affinity (Faraldo-Gómez and Sansom, 2003). TBDTs are generally composed of a 22-stranded membrane-embedded β-barrel and an N-terminal plug domain that is located within the barrel (Faraldo-Gómez and Sansom, 2003). Substrates bind to the extracellular pocket of the transporter, which is formed by the plug and the barrel domain and does not exhibit a characteristic conserved sequence, consistent with the different substrate specificities of the different transporters (Chimento *et al.*, 2005).

In cyanobacteria various genes coding for putative TBDTs have been annotated by genome analyses (Mirus *et al.*, 2009). However, most of them are yet to be characterized experimentally. In *Synechocystis* sp. PCC 6803, four genes (*sll1206*, *sll1406*, *sll1409*, *slr1490*) code for putative TBDTs. Mutants of any of these genes showed no altered growth phenotype in the absence of iron and no reduction in Fe(II) or Fe(III) uptake activity (Katoh *et al.*, 2001). Whether, as these observations suggest, iron is not the substrate of these TBDTs, which in general are also involved in uptake of other metals (Schauer *et al.*, 2008), or whether the putative TBDTs are only important under not yet identified conditions remains to be explored.

To date, the siderophore recognized could be defined only for one cyanobacterial TBDT, namely for the schizokinen transporter SchT (Alr0379) from *Anabaena* sp. PCC 7120 (Fig. 3.5A; Nicolaisen *et al.*, 2008). The mutation in *schT* drastically reduces the uptake of schizokinen (Nicolaisen *et al.*, 2008), although two additional TBDTs encoded in the genome of *Anabaena* sp. PCC 7120 (Alr2209 and Alr2581) might have similar substrate specificities (Mirus *et al.*, 2009). Nevertheless, 19 additional TBDTs have been identified in the genome of this cyanobacterium for which the substrates still have to be defined (Mirus *et al.*, 2009). Indeed, they might be involved in the transport of other siderophores, as multiple gene clusters encoding proteins involved in siderophore synthesis have been observed as well (Jeanjean *et al.*, 2008; Nicolaisen *et al.*, 2008), and they might also serve to acquire siderophores produced by other bacteria. Indeed, *Anabaena* sp. PCC 7120 (as well as *Synechocystis* sp. PCC 6803) is able to utilize ferric-aerobactin from *E. coli* (Goldman *et al.*, 1983, Kranzler *et al.*, 2011). Alternatively or additionally, the TBDTs might be required for uptake of other metals (Schauer *et al.*, 2008).

Iron uptake by *Anabaena* sp. PCC 7120 was found to be related to copper uptake. On one hand, schizokinen can bind to copper, but this complex is not taken up by the cyanobacterium (Clarke *et al.*, 1987). On the other hand, initial information on the TBDT IacT (iron and copper transporter; All4026), which is homologous to

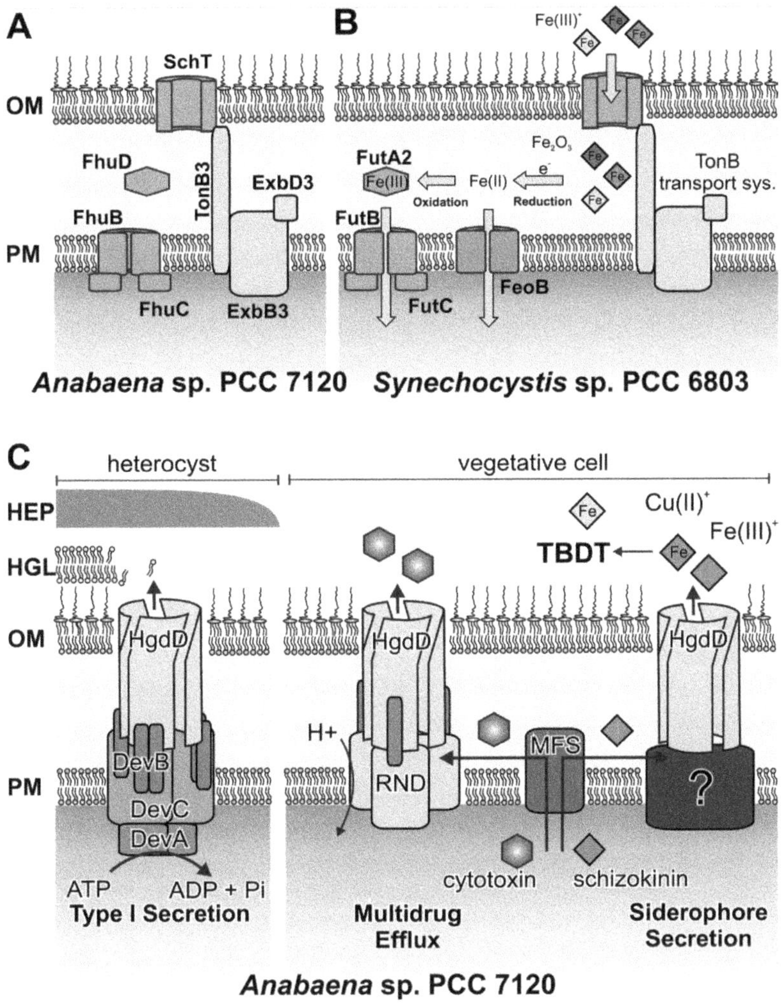

Figure 3.5 Iron uptake and metabolite export strategies. (A) Siderophore uptake: recognition and transport through the outer membrane of the siderophore schizokinen is mediated by the TonB-dependent-transporter SchT energized by TonB3 in complex with ExbB3 and ExbD3. The final uptake is regulated by a Fhu system in the periplasm and the plasma membrane. (B) Reductive uptake: In dilute heterogeneous environments iron can be free, bound to a variety of organic ligands (squares) or form ferric oxide colloids. In *Synechocystis* sp. PCC 6803 the mode of transport through the OM is not known. Once inside the periplasm, all iron species are reduced and unchelated Fe ions are transported through the PM. Free iron in the Fe(II) form can be taken up by transporters such as FeoB or can be re-oxidized to Fe(III) and taken up by transporters like FutABC. (C) The HgdD-dependent secretion systems. The homotrimeric TolC-like HgdD (yellow) is the outer membrane factor for many secretion processes in *Anabaena* sp. PCC 7120. HgdD is expressed in both vegative cells and heterocysts but shows a transient enhanced expression with a peak at nine hours after nitrogen deprivation. Together with the DevA, DevB and DevC (left), HgdD is involved in the secretion of heterocyst specific glycolipids to form the heterocyst specific glycolipid (HGL)-layer that is needed to maintain the micro-oxic environment for nitrogen fixation. DevB is the membrane fusion protein (MFP), DevC the permease and DevA the ATP-hydrolysing subunit of an ABC-transporter. Together they form a classical Type I secretion system. HgdD is additionally involved in the secretion of small molecular weight substances such as antibiotics, dyes (middle) and siderophores (right). The secretion of antibiotics and dyes depends on the membrane potential of the plasma membrane and likely functions in cooperation with an RND-family efflux pump protein such as *E. coli* AcrAB. For siderophore secretion both HgdD and SchE, a protein of the major facilitator superfamily (MFS), are required. Unlike the ABC transporter and RND-like proteins, MFS proteins are not encoded next to a membrane MFP and are unlikely to interact directly with HgdD. Therefore, a third component might be involved to charge HgdD in siderophore secretion. A colour version of this figure is available in the plate section at the back of the book.

the catecholate-type transporter ViuA from *Vibrio cholerae* (Mirus *et al.*, 2009), suggests a cross-talk between both pathways. The mutant of *iacT* shows an enhanced secretion of siderophores and reduced iron and copper transport and cellular levels when compared to the wild-type (Nicolaisen *et al.*, 2010). Based on the observations done for SchT and IacT, a model for regulation was proposed, which, however, might be specific for *Anabaena* sp. PCC 7120 (Nicolaisen *et al.*, 2010). It is assumed that for this cyanobacterium schizokinen is the major siderophore for iron uptake. However, under high copper concentrations, it can bind copper as well. This interaction leads to square planar conformation of the siderophore–copper complex instead of the octahedral structure of the ferric siderophore complex explaining why the copper-loaded siderophore cannot be transported (Clarke *et al.*, 1987). As a consequence, an alternative transport system for iron is required under high copper concentrations. A system co-transporting iron and copper would be of evolutionary advantage, because it would ensure iron and copper uptake in situations where copper is largely bound to schizokinen. Consistently with this model, two genes (*all7619* and *all7632*) coding for homologues to CusB, which is involved in copper homeostasis in *E. coli* (Rensing and Grass, 2003), are repressed in an IacT overexpression mutant (Nicolaisen *et al.*, 2010).

The energy for the transport of the ferric–iron–siderophore complexes by TBDTs is supplied by the energy-transducing unit TonB (Faraldo-Gómez and Sansom, 2003). TonB is anchored to the cytoplasmic membrane and associated with the integral membrane proteins ExbB and ExbD. The interaction between TBDT and TonB is manifest by the recognition of the TonB-box positioned in front of the N-terminal plug domain of TBDTs. Knowledge on the Exb/TonB system in cyanobacteria is rather limited. The identification of *tonB* genes in the cyanobacterial genomes based on comparisons with the sequences of *E. coli* TonBs proved to be rather difficult, perhaps because of differences in the peptidoglycan layers with respect to size and composition (see 'The peptidoglycan layer of cyanobacteria'). However, four putative *tonB* genes and three *exbB/D* gene clusters have been found in the genome of *Anabaena* sp. PCC 7120 (Stevanovic *et al.*, 2012). By expression profiling and mutant analysis, a possible role of *tonB3* and the *exbB3/D3* gene cluster for the siderophore-dependent iron uptake has been assigned. The expression of these genes increases under iron limiting conditions, while genes coding for siderophore synthesis and secretion are up-regulated in the *tonB3* and *exbB3/D3* gene cluster mutants (Stevanovic *et al.*, 2012). The function and functional relevance of the other identified proteins remains unknown.

Siderophore-independent transport

Siderophore-mediated iron uptake may not account for iron uptake by all cyanobacteria, since not all cyanobacterial genomes contain genes coding for siderophore synthesis machineries, e.g. non-ribosomal peptide synthetases (NRPS), modular polyketide synthases or other known synthases (Ehrenreich *et al.*, 2005; Hopkinson and Morel, 2009). Thus, these cyanobacteria either (i) use the siderophores produced by other bacteria in the community or (ii) use an alternative Fe uptake mode as found in plants, yeast and algae (Allen *et al.*, 2007; Morrissey and Guerinot, 2009). In line with the first proposal, typical components of the siderophore-based iron uptake system have been identified in *Synechocystis* sp. PCC 6803 (Katoh *et al.*, 2001). An alternative path, on the other hand, would depend on extracellular reduction of Fe(III) or Fe(III)-chelates and uptake of inorganic non-chelated free hydrated iron species (Fe') present as e.g. $Fe(OH)_2^+$, $Fe(OH)_3$ and $Fe(OH)_4^-$ (Sunda *et al.*, 2005).

Consistent with the alternative mode of iron uptake, it could be demonstrated that extracellular iron reduction facilitated by a superoxide dismutase enhances Fe uptake in the marine cyanobacterium *Lyngbya majuscula* (Rose and Waite, 2005). Similarly, based on iron uptake experiments it was proposed that a natural *Synechococcus* population from the Gulf of Aqaba, *Synechococcus* sp. WH8102, *Synechocystis* sp. PCC 6803 and *Anabaena flos-aquae* acquire organic iron after extracellular reduction of Fe' (Lis and Shaked, 2009; Wirtz *et al.*, 2010; Kranzler *et al.*, 2011). Thus, a general path of Fe' uptake involves iron reduction either on the cell surface or in the periplasmic space, in which redox reactions occur as it has been shown for *Synechocystis* sp. PCC 6803 (Pils and Schmetterer, 2001). Because it is not yet known whether the Fe(II) trapping agents can cross the outer membrane of cyanobacteria, reduction in the periplasm is an interesting possibility (Fig. 3.5B). Thus, iron uptake would not require ferrisiderophore transporters in the outer membrane and might be facilitated by porin-like proteins (Table 3.6). After reduction, iron might be transported across the plasma membrane either as Fe(II) or in the re-oxidized form Fe(III). Consistently, a plasma membrane-localized cation diffusion facilitator protein was found to be involved in Fe(III) transport in *Synechocystis* sp. PCC 6803, which is unique for photosynthetic bacteria (Jiang *et al.*, 2012). Additionally, genes encoding FeoB-like proteins typically involved in Fe(II) transport have also been identified in this cyanobacterium (Katoh *et al.*, 2001).

Summarizing, the current state of the description of cyanobacterial iron (metal) uptake is only at its beginning. For a detailed understanding, the experimental dissection of the function of the different TBDTs and their energizing systems is required. The same holds true for our understanding of the diversity of the uptake systems in terms of composition and complexity found in different cyanobacteria, which is not yet solved. Two different modes of iron uptake might exist, which, however, are not fully understood at the molecular and structural levels. The co-existence of chelator-dependent and -independent systems would be of evolutionary advantage as cyanobacteria have a very high demand on iron.

TolC – a central factor for cell survival

TolC (tolerance to colicin) is an outer membrane secretion channel involved in multidrug resistance (MDR; Nagel de Zwaig and Luria, 1967). The crystal structure of *E. coli* TolC and of the *Pseudomonas aeruginosa* homologue OprM (Koronakis *et al.*, 2000; Akama *et al.*, 2004) revealed remarkably similar three-dimensional structures although both proteins show very little sequence similarity. TolC consists of three protomers that form a homotrimeric protein channel-tunnel of 140 Å with 'six-fold symmetry', as each TolC protomer itself were a product of gene duplication. Each protomer contributes four anti-parallel β-sheets to form the outer membrane-inserted β-barrel and four α-helices (H3, H4, H7, H8) to shape the α-helical channel-tunnel. The α-helices are twisted and can perform an iris-like movement to facilitate an open/close state of the channel. The helices H1 and H5 are part of the equatorial domain. The β-barrel of TolC has a diameter of 30 Å and is inserted in the outer membrane with the help of BamA, the *E. coli* Omp85 homologue (Masi *et al.*, 2009). The α-helical barrel part faces the periplasm. The combination of β-barrel and α-helical elements to form a channel-tunnel is a unique design for outer membrane proteins.

TolC functions in the secretion of a broad range of structurally different components. To achieve this versatility, it is part of at least two different secretion systems: (i) the Type I secretion systems and (ii), together with RND (resistance–nodulation–division) efflux pumps, the tripartite multidrug efflux system.

TolC and its function in Type I secretion and multidrug efflux

TolC interacts with ATP-dependent permeases (ABC-type transporters) to form a 'Type I secretion system' that facilitates a Sec-independent secretion of unfolded proteins (see, for example, Piddock, 2006). TolC and the ABC-transporter are connected by a multimeric membrane fusion protein (MFP), which is inserted in the plasma membrane via an N-terminal transmembrane domain. The ATP-hydrolysing part of the ABC-transporter can either be the cytosol-facing domain of the permease itself or a subunit that is encoded as a separate protein (Holland, 2005; Fiedler *et al.*, 1998). Typically, the genes encoding the permease, the membrane fusion protein and the substrate form an operon. Protein substrates vary in size (10–800 kDa) and have a C-terminal secretion signal of approximately 50 amino acids. The signal is poorly conserved on primary sequence level and is not cleaved off after translocation. Within the post-translational transport process, the ABC-transporter, but not TolC confers substrate specificity.

The best-known example of Type I protein secretion system is the haemolysin A (HlyA) export system of *E. coli* (see, for example, Gentschev *et al.*, 2002). HlyB forms a dimer that is stabilized upon ATP loading. The pre-existing complex of the permease HlyB and the membrane fusion protein HlyD interact with the substrate haemolysin A (HlyA) (Zaitseva *et al.*, 2005). The C-terminal signal sequence of HlyA binds to the ATPase domain of HlyB, which leads to the recruitment of TolC (Holland, 2005). The binding of ATP releases the signal sequence and ATP hydrolysis causes the translocation of HlyA across the cell wall (Zaitseva, 2005). Molecular dynamic simulations suggest that TolC facilitates the substrate translocation by a peristaltic motion of the periplasmic domain (Vaccaro *et al.*, 2008).

A tripartite complex comprising TolC, a membrane fusion protein and a plasma membrane-localized RND forms multidrug efflux pumps. Together with the low permeability of the outer membrane, this complex confers resistance to lipophilic antibiotics such as β-lactams and macrolides, organic solvents, cytotoxic dyes and divalent metal ions (Goldberg, 1999). In addition, TolC is also involved in the transport of cytoplasmic produced siderophores (see 'Iron uptake systems in cyanobacteria'). Strains with constitutive expression of RND proteins acquire a high potential for multidrug export (Poole *et al.*, 1993) and pose a serious medical threat because of their multiple resistance against broad-spectra antibiotics. Although multidrug efflux pumps have a similar organization as Type I secretion systems, their mode of function is quite different.

The best-characterized RND-TolC multidrug efflux system is the AcrAB-TolC system of *E. coli* (Pagès *et al.*, 2005). The AcrB consists of a trimer that is driven by the proton motive force (pmf) across the plasma membrane and acts as a proton-drug antiporter. The deduced working principle of the RND is remarkably similar to that of an F_1F_O ATPase (Murakami *et al.*, 2002, 2006; Seeger *et al.*, 2006; Nakashima *et al.*, 2011). The asymmetric trimer combines three different conformational states: loose (L), tight (T) and open (O), which perform a functional cycling (Murakami *et al.*, 2006; Seeger *et al.*, 2006). The substrate binds at the periplasmic side to a low affinity-binding pocket of the L-conformation, which then undertakes a conformational change to the T-state. The T-state converts to the O-state and releases the substrate to TolC (Eicher *et al.*, 2009). The MFP AcrA stabilizes the interaction between AcrB and TolC. In contrast to MFP from the ABC-system, AcrA has no transmembrane domain, but it bears a lipidated N-terminus that anchors the protein to the plasma membrane.

The cyanobacterial TolC

To date, the best-characterized TolC homologue in cyanobacteria is HgdD (Alr2887) from *Anabaena* sp. PCC 7120 (Fig. 3.5C). This protein was first identified in proteomic analyses of isolated outer membranes from vegetative cells (Moslavac *et al.*, 2005) and heterocysts (Moslavac *et al.*, 2007a). HgdD has only 20% sequence identity to *E. coli* TolC but is highly conserved at the structural level as suggested by homology modelling (Moslavac *et al.*, 2007b). A similar protein, Slr1270, was previously identified in the outer membrane of *Synechocystis* sp. PCC 6803 (Huang *et al.*, 2004; Table 3.5). In *Anabaena* sp. PCC 7120, the *hgdD* gene is up-regulated under nitrogen deprivation and is essential for the formation of mature heterocysts (Moslavac *et al.*, 2007b). A mutant of *hgdD* lacks the heterocyst-specific glycolipid layer, which forms a diffusion barrier against oxygen for protection of the highly oxygen-sensitive nitrogenase from inactivation. Nonetheless, the synthesis of heterocyst-specific glycolipids is not impaired in the mutant, which suggests that HgdD is involved in the secretion of glycolipids, leading to the name Heterocyst glycolipid deposition protein D (HgdD). As a consequence, mutants of *hgdD* are incapable of nitrogen fixation in the presence of oxygen (Fox$^-$ mutant phenotype) and cannot survive without a source of fixed nitrogen. The same phenotype could be observed for the ABC-transporter DevBCA (Fiedler *et al.*, 1998), for which a physical interaction with HgdD has been observed (Staron *et al.*, 2011). Thus, DevBCA and HgdD likely form a Type I secretion system to export glycolipids, especially as glycolipid extracts induce a slight increase in ATPase activity of DevBCA.

HgdD is the only TolC homologue found to be encoded in the genome of *Anabaena* sp. PCC 7120. Similarly, most other cyanobacteria encode a single TolC. Thus, in order to handle other substrates, HgdD has to be as versatile as its homologue in *E. coli*. As described above, cyanobacteria have a high demand for iron as an indispensable cofactor for photosynthesis and nitrogen fixation. As described in 'Iron uptake systems in cyanobacteria', schizokinen is the central siderophore secreted by *Anabaena* sp. PCC 7120 for iron uptake. HgdD, which acts together with the plasma membrane protein SchE (Alr4025), is the central outer membrane component for the secretion of this siderophore (Nicolaisen *et al.*, 2010). Because SchE is a MFS protein, a direct physical interaction between SchE and HgdD is unlikely. As a consequence, a third component not yet identified must be involved, which would serve as an interaction partner for HgdD-dependent siderophore secretion.

HgdD of *Anabaena* sp. PCC 7120 has a classical function in secretion of cytotoxic dyes and antibiotics as well (Hahn *et al.*, 2012). The deletion mutant of HgdD is highly sensitive to low concentrations of ethidium bromide and macrolide antibiotics such as erythromycin. The efflux of ethidium across the outer membrane by the wild-type cyanobacterium has a Michaelis Menten constant of 170 μM, while the transport in the *hgdD* mutant is diffusion-driven. The rate obtained for the transport is at least two orders of magnitude lower in the mutant than in wild-type, indicating that HgdD specifically export cytotoxic dyes. In contrast to the glycolipid transport, the export of the toxins is dependent on the proton gradient of the plasma membrane (Hahn *et al.*, 2012).

Summarizing, the TolC of *Anabaena* sp. PCC 7120, and likely in other cyanobacteria as well, plays a major role in iron homoeostasis, heterocyst differentiation and detoxification. To facilitate this function, TolC interacts with different plasma membrane-anchored machineries. Whether TolC switches dynamically between these machineries or whether stable complexes are formed remains to be investigated, as is also the case for the interaction of TolC with the peptidoglycan layer.

Porins – more than just holes in the wall

The outer membrane is an effective diffusion barrier for hydrophobic toxins and other types of molecules. Therefore, mechanisms and components for the exchange of solutes required for cellular function have to exist. In Gram-negative bacteria, integral outer membrane β-barrel-shaped channels generally annotated as porins turn the outer membrane into a selective filter (Mirus *et al.*, 2010). The proteins of the porin family usually consist of 8–24 β-strands that can be tilted 30–60°. The β-strands of the membrane-inserted barrel have an alternating hydrophilic-hydrophobic amino acid pattern to form the aliphatic exterior and the hydrophilic interior of the barrel (Fairman *et al.*, 2011). They are connected by short periplasmic turns (D-loops) and rather long loops on the extracellular side (L-loops; Mirus *et al.*, 2010; Fairman *et al.*, 2011). The L-loops are in direct contact with the environment and make Gram-negative bacteria susceptible to bacteriophages or (in case of pathogens) the immune response system of the host organism. However, the L-loops show a very high sequence diversity, which contributes to a general low sequence identity of porins. The latter makes the comparison of porins by their amino acid sequence rather difficult. In order to discuss recent advances in understanding porins of cyanobacteria, the 'classical porins' (OmpF, OmpC, PhoE), the 'slow porins' (OprF, OmpA) and carbohydrate specific porins (LamB, ScrY and OprB) of *E. coli* and *P. aeruginosa* will be introduced first.

Classical and slow porins

Classical porins such as OmpF, OmpC and PhoE from *E. coli* are considered to be unspecific diffusion channels with a size exclusion limit of approximately 600 Da (see, for example, Nikaido and Vaara, 1985). Porins form a stable homotrimer in which each monomer forms a β-barrel of 16 β-strands as determined by crystallization, first attained for OmpF (Weiss *et al.*, 1991; Cowan *et al.*, 1992). Consistent with the dimensions found by structural analysis, OmpF, OmpC and PhoE have a conductance of ~1–2 nS (1 M KCl) reflecting an internal diameter of 1–1.2 nm, determined by reconstitution and single channel measurements (Table 3.8; Benz *et al.*, 1985; Buehler *et al.*, 1991). In the structure, two loops have been found to be important. The L-loop 2 bends over to the edge of the neighbouring

Table 3.8 The properties of porins present in Gram-negative bacteria

Porin	Organism	No. of β-strands	Oligomerization	Channel conductance at 1 M KCl (nS)*	Pore diameter (nm)	Reference
OmpF	*Escherichia coli*	16	3	0.70 and 2.10	1.2	1, 2
OmpC	*Escherichia coli*	16	3	1.50–2.80	1.0	1, 3
PhoE	*Escherichia coli*	16	3	1.80	1.1	1
LamB	*Escherichia coli*	18	3	0.16	0.6	2, 4–6
OprB	*Pseudomonas aeruginosa*	18	3	0.04	0.6	1
OprF	*Pseudomonas aeruginosa*	8/16	1/3	0.36	1.1/1.6	7–9
OmpA	*Escherichia coli*	8/16	1/3	0.05	1–2.6	10–12
Isolated porin fraction	*Anabaena variabilis*	?	?	3.50	1.6	13
Isolated porin fraction	*Synechocystis* sp. PCC 6714	?	?	0.30 and 2.20	?	14, 15
SomA	*Synechococcus* sp. PCC 6301	?	3	0.40	?	16
SomB	*Synechococcus* sp. PCC 6301	?	3	0.90	?	16

1, Benz *et al.* (1985); 2, Benz *et al.* (1987); 3, Biró *et al.* (2010); 4, Benz *et al.* (1986); 5, Saravolac *et al.* (1991); 6, Schirmer *et al.* (1995); 7, Brinkman *et al.* (2000); 8, Nestorovich *et al.* (2006); 9, Woodruff *et al.* (1986); 10, Bond *et al.* (2002); 11, Pautsch and Schluz (1998); 12, Sugawara and Nikaido (1992); 13, Benz and Böhme (1985); 14, Hansel and Schmid (1994); 15, Jürgens and Benz (1989); 16, Umeda *et al.* (1996).

barrel and stabilizes the homotrimer, whereas the long L-loop 3, also called 'eyelet', reaches into the barrel forming a constriction to narrow the pore size and to regulate gating and ion selectivity (Weiss *et al.*, 1991; Cowan *et al.*, 1992; Bainbridge *et al.*, 1998; Schirmer, 1998; Fairman *et al.*, 2011). Thus, OmpF, OmpC and PhoE share a high structural similarity, but the L-loop 3 defines unique properties. Thereby, OmpF and OmpD prefer cations while PhoE shows a slight preference for anions (Benz *et al.*, 1985; Bauer *et al.*, 1989).

The expression and abundance of the three porins in the outer membrane depends on different environmental conditions. Hence, OmpF and OmpC are counter-regulated: low osmolarity or temperature induces *ompF* expression, while an increase of the temperature or osmolarity enhances the expression of *ompC* (Lugtenberg *et al.*, 1976; Alphen and Lugtenberg, 1977; Bassford *et al.*, 1977). This counter-regulation can be explained by the smaller transport rate found for OmpC than for OmpF (Schulz, 2002). Higher osmolarity or temperature would increase the transport rate and thus, the reduction of the transport rate by decreasing the OmpF/C ratio balances the influx rate. Regarding PhoE, it is expressed only under phosphate starvation (Tommassen and Lugtenberg, 1982).

The major porin of *Pseudomonas aeruginosa*, OprF, and its *E. coli* homologue, OmpA, are annotated as 'slow porins', as they show a 50-fold slower diffusion rate for small molecules than OmpF. However, the pore diameter is wide enough to allow passive diffusion of oligosaccharides larger than 600 Da (Table 3.8; Bellido *et al.*, 1992; Sugawara *et al.*, 1992). A two-domain model was proposed to explain these contradicting results. In the closed (majority) state, the N-terminal domain forms an eight-stand β-barrel with a pore diameter too small for the transport of hydrophilic molecules, while the α-helical C-terminus is interacting with the peptidoglycan layer (Rawling, 1998). This view has been supported by the crystallization of the N-terminal domain, which yielded an eight-stranded β-barrel (Pautsch and Schulz, 1998, 2000). In the so-called open (minority) state, the whole protein forms a 16-stranded β-barrel, similar to OmpF, which allows the diffusion of large oligosaccharides. By introducing cysteine residues that could be captured by biotin the open conformation of OprF could be enriched and it was found to tend to form trimers (Sugawara *et al.*, 2012). In addition, in a mutant of *Pseudomonas aeruginosa* lacking the periplasmic chaperone Skp, an increase of the open form was observed (Sugawara *et al.*, 2012). This, however, could also be interpreted in a different way: if the C-terminal portion is not properly folded it could insert into the membrane enforcing the formation of an artificial, non-native channel. Thus, the issue of rate of diffusion versus pore size has not yet been solved.

Carbohydrate-specific porins

Maltose is an important carbon source for enterobacteria. LamB, a maltodextrin-inducible porin, facilitates the uptake of maltose and malto-oligosaccharides across the outer membrane, but at the same time prevents the diffusion of glucose (Luckey, 1980). Under conditions of carbohydrate-starvation, LamB becomes induced and facilitates the uptake of lactose, arabinose, and glycerol (Death, 1993). LamB forms a β-barrel with 18 β-strands and oligomerizes to a trimer state (Table 3.8; Schirmer *et al.*, 1995). Similar to OmpF, the L-loop 3 inserts itself into the barrel and forms a charged constriction zone leaving a pore diameter of less than 6 Å with a low conductance of 0.16 nS (1 M KCl; Saravolac *et al.*, 1991). The channel is characterized by a line of aromatic amino acids called 'greasy slide', which form a helical path that leads the maltose through the barrel. Polar residues on both sides of the barrel form 'ionic tracks', which can establish hydrogen bonds with the sugar to stabilize its orientation (Dutzler *et al.*, 2002; Dumas *et al.*, 2000b).

ScrY and OprB are homologous to LamB and similar in its function and structure (Dumas, 2000a). ScrY enables bacteria to use a variety of sugars (glucose, fructose, arabinose, maltose, lactose, raffinose, sucrose and maltodextrins) as a carbon source (Ulmke *et al.*, 1999, Van Gelder *et al.*, 2001). OprB was described as a carbohydrate-selective porin that is induced by glucose and other sugars and repressed by organic acids such as succinate and citrate (Death *et al.*, 1993; Wylie *et al.*, 1995). OprB is slightly cation-selective and has a single-channel conductance of 35 pS (1 M KCl), which is five times lower than that of LamB (155 pS, 1 M KCl; Saravolac *et al.*, 1991). However the channel activity of OprB can be blocked by glucose in a concentration dependent manner, which parallels the behaviour of the maltodextrin-specific transporter LamB of *E. coli* (Saravolac *et al.*, 1991). Although LamB and OprB facilitate transport of multiple carbohydrates, it seems that OprB is the porin for small sugars and LamB for big sugars (Nikaido, 2003).

Porins in cyanobacteria

The first cyanobacterial outer membrane fractions with porin-forming proteins were isolated from *Anabaena variabilis* by Benz and Böhme (1985). The analysis of the electrophysiological properties of a purified fraction after insertion into an artificial lipid bilayer revealed a single pore conductance of 3.5 nS in 1 M KCI, which led to the calculation of a diameter of 1.6 nm (Table. 3.8). Based on this observation, the authors proposed the existence of a cyanobacterial porin (Benz and Böhme, 1985). The existence of such outer membrane protein is consistent with the dependence of the photoheterotrophic growth of *Anabaena variabilis* on large disaccharide molecules in the media (Rippka *et al.*, 1978). Similarly, analysis of outer membrane fractions from *Synechocystis* sp. PCC 6714 revealed the existence of two activities with 2.2 nS and 0.3 nS, respectively (Jürgens and Benz, 1989), whereas a similar approach using *Synechococcus* sp. PCC 6301 revealed a single-channel conductance of 5.5 nS (in 1 M KCl; Hansel *et al.*, 1994). The latter activity was initially assigned to a single protein of 45 kDa based on its amino acid composition, which appeared to be strongly associated with the peptidoglycan layer (Hansel *et al.*, 1994). However, subsequently it was shown that the fraction contained two proteins migrating with almost the same molecular weight, which form a 140 kDa trimer under native conditions (Hansel and Tadros, 1998). Both proteins show partial sequence similarity to the *Synechococcus* sp. PCC 7942 SomA (*Synechococcus* outer membrane protein A; Umeda *et al.*, 1996) and were therefore annotated as SomA and SomB (Hansel *et al.*, 1998). The single-channel conductance of SomA (0.4 nS) and SomB (0.9 nS; Table 3.8; Hansel and Tadros, 1998) are significantly lower than the conductance determined for the partially purified proteins (Hansel *et al.*, 1994). They are also lower than the conductance of OmpF from *E. coli* (Benz *et al.*, 1985), but higher than that of OprF from *Pseudomonas aeruginosa* (Yoshimura *et al.*, 1983) or LamB (Benz and Hancock, 1987). SomA from *Synechococcus* sp. PCC 7942 was shown to contain a typical N-terminal signal sequence that could target OmpF to the outer membrane in *E. coli* (Umeda *et al.*, 1996). This signal sequence is followed by a domain similar to one found in S-layer proteins (Gupta *et al.*, 1999), where it binds to the LPS layer (Olabarria *et al.*, 1996).

Since then, significant progress has been made by genome and proteome studies. In the genome of *Synechocystis* sp. PCC 6803, based on similarity to SomA and SomB, seven ORFs (*sll0772, sll1271, sll1550, slr0042, slr1272, slr1841, and slr1908*) have been proposed to encode proteins of the cyanobacterial porin (CBP) family (Kaneko *et al.*, 1996; Ren *et al.*, 2007). The proteins encoded by six of these genes (*sll0772, sll1271, sll1550, slr0042, slr1841, slr1908*) as well as by *slr1163* have additionally been assigned to the OprB family (Ren *et al.*, 2007). Three of these proteins (Slr1908, Slr1841, Slr1272) have been identified in proteomic studies of isolated outer membrane (Table 3.6; Huang *et al.*, 2004). Similarly, nine proteins encoded by *Anabaena* sp. PCC 7120 have been assigned as porins (Table 3.6; Ren *et al.*, 2007, Nicolaisen *et al.*, 2009a), four of which (Alr4550, All4499, Alr0834, Alr3608) could be identified in outer membrane proteome analysis (Moslavac *et al.*, 2005).

Based on sequence similarity, most of the identified cyanobacterial porins have been annotated as carbohydrate-selective porin OprB (Ren *et al.*, 2007). This is rather surprising. The conductance of 0.4–0.9 nS measured for purified cyanobacterial porins (Table 3.8; Hansel and Tadros, 1998) argues for classical- or slow-type porins. Very recently it has been shown that ethidium can readily pass the outer membrane of *Anabaena* sp. PCC 7120 in a concentration-dependent and passive manner. Pre-incubation of cells with polyamines, a typical inhibitor of porins such as OmpF in *E. coli* (Dela Vega *et al.*, 1996, Iyer *et al.*, 2000) could significantly reduce the uptake rate of ethidium (Hahn *et al.*, 2012). This observation suggests a conventional function of porins in cyanobacteria with an exclusion limit not smaller than 300 Da, which would be large enough for the passage of monosaccharides. Furthermore, it was recently demonstrated that the inactivation of ORF *Npun_R5320* encoding an OprB-like porin in *Nostoc punctiforme* indeed impaired glucose and fructose uptake (Ekman *et al.*, 2013).

Cyanobacterial porins characterized to date have an N-terminal S-layer homology (SLH)-domain that probably binds to the peptidoglycan layer and stabilizes the outer membrane. A similar architecture has been recently observed for OmpM1, a putative porin of *Mitsuokella multacida* (Kalmokoff, 2009). Remarkably, the bacterium *Thermus thermophilus*, which is considered as very ancient, contains a primitive outer membrane-like structure (Castan *et al.*, 2002) composed of the porin-like protein SlpA. The protein is discussed to form a hexagonal S-layer (Caston *et al.*, 1988, 1993) and to interact with the peptidoglycan layer through its SLH domain (Olabarria *et al.*, 1996). Considering the Gram-positive and Gram-negative features of the cyanobacterial envelope (see 'The peptidoglycan layer of cyanobacteria') and the early branching of the cyanobacterial clade from the bacterial tree (Cavalier-Smithet *et al.*, 2006), it is tempting to speculate that cyanobacteria still possess this mode of anchoring to the peptidoglycan layer by the SLH-domains of outer membrane porins.

Thus, it appears that cyanobacterial porins fulfil a classical porin function, like, for example, OmpF. The electrophysiological data, however, suggests that these porins are smaller than those from proteobacteria. Larger cyanobacterial porins might not be required, because their autotrophic lifestyle only requires the uptake of small inorganic compounds. In addition, such small porins would not allow uptake of harmful substances. However, the smaller pore size contradicts the observed ability of (at least some) cyanobacteria to grow heterotrophically on sugars. Thus, one might speculate that cyanobacterial porins follow the two-state system described for the slow porins OprF and OmpA, if one admits that the SLH-domain might not only interact with the peptidoglycan layer. Unfortunately, however, this speculation lacks any experimental support.

The plasma membrane proteome

The proteome of the plasma membrane can be dissected in three large categories: (i) proteins controlling uptake and export of solutes and proteins; (ii) proteins anchored to the membrane for structural purposes; and (iii) enzymes involved in metabolism or signal transduction. All of these functions are essential but might be of distinct evolutionary age. It is obvious that the uptake and export of all classes of substances must have evolved as early as the membrane system to enable the supply of essential cellular ingredients, although some particular transporters may have been acquired only recently. In turn, the existence of the membrane provided a platform for the spacial two-dimensional concentration of enzymatic reactions, which likely represents an advantage over reactions performed in the three-dimensional cell interior. The last components required were probably the structural components anchoring the peptidoglycan layer or joining neighbouring cells. The latter would have been important after evolution of filamentous cyanobacteria, but their classification is somewhat arbitrary, as they might also be seen as transporters or enzymes.

Proteins involved in solute exchange across the plasma membrane

A large family of plasma membrane proteins is involved in the uptake or export of all kinds of substances across the plasma membrane. These substances range from proteins translocated by the Sec system (see Chapter 5) to lipids (see 'The LptD-dependent Lipid A transport') or metabolites utilizing the secretion system (see 'The TolC- a central factor for cell survival') to nutrients transported by permeases or transporters.

A large plasma membrane protein family with diverse functions in substrate transport is that of the ABC-type transporters. These transporters are characterized by the presence of two ABC-binding domains (which reside in the cytoplasmic face of the plasma membrane) and two integral membrane domains, 'transmembrane permeases' (which may reside in a single polypeptide or in two independent polypeptides). Additionally, bacterial uptake ABC-transporters usually bear a periplasmic substrate binding protein. Some cyanobacterial ABC-transporters are presented schematically in Fig. 3.6. An example of an exporter is the *Microcystis aeruginosa* McyH involved in microcystin production (Pearson *et al.*, 2004). Examples of importers are systems involved in the uptake of (i) nitrate in *Anabaena* sp. PCC 7120, *Phormidium laminosum* and *Synechococcus* sp. PCC 7942 (NrtABCD; Merchán *et al.*, 1995; Frias *et al.*, 1997); (ii) nitrite and cyanate (NCO^-) in *Synechococcus elongatus* PCC 7942 and *Synechococcus* sp. UTEX625 (cynABD; Espie *et al.*, 2007; Maeda and Omata, 2009); (iii) bicarbonate in *Synechococcus* sp. PCC 7942, *Synechocystis* sp. PCC 6803 and *Anabaena* sp. PCC 7120 (Cmp or BCT1; Omata *et al.*, 1999, 2001; López-Igual *et al.*, 2012); (iv) ethyl viologen in *Synechocystis* sp. PCC 6803 (ErvA-C; Prosecka *et al.*, 2009); (v) vanadate in *Anabaena variabilis* ATCC 29413 (vupABC; Pratte and Thiel, 2006); (vi) urea in *Synechocystis* sp. PCC 6803 and *Anabaena* sp. PCC 7120 (urtABCDE, Valladares *et al.*, 2002); (vii) glucosylglycerol in *Synechocystis* sp. PCC 6803 (ggtBCD; Mikkat and Hagemann, 2000), (viii) phosphate in *Synechocystis* sp. PCC 6803 and *Anabaena* sp. PCC 7120 (pstABB'CS1 and pstABCS2; Pitt *et al.*, 2010; Muñoz-Martín *et al.*, 2011); (ix) neutral amino acids in *Anabaena* sp. PCC 7120 (NatA-E and NatFGH-BgtA; Picossi *et al.*, 2005; Pernil *et al.*, 2008); and (x) basic amino acids in *Synechocystis* sp. PCC 6803 and *Anabaena* sp. PCC 7120 (Bgt; Quintero *et al.*, 2001; Pernil *et al.*, 2008).

As mentioned, several ABC transporters are involved in amino acid uptake. In the 1990s it was observed that cyanobacteria are capable of taking up amino acids (Montesinos *et al.*, 1995, 1997). *Synechocystis* sp. PCC 6803, for example, contains four different amino acid transport systems (Labarre *et al.*, 1987, Flores and Muro-Pastor, 1990; Quintero *et al.*, 2001). The first is an ABC transporter composed of the ATP binding subunit BgtA (Slr1735) and the composite protein BgtB (Sll1270), which bears a periplasmic substrate binding and an integral membrane protein domain, and is involved in the transport of basic amino acids (arginine/lysine and glutamine) (Quintero *et al.*, 2001). A second ABC transporter is involved in the uptake of neutral amino acids

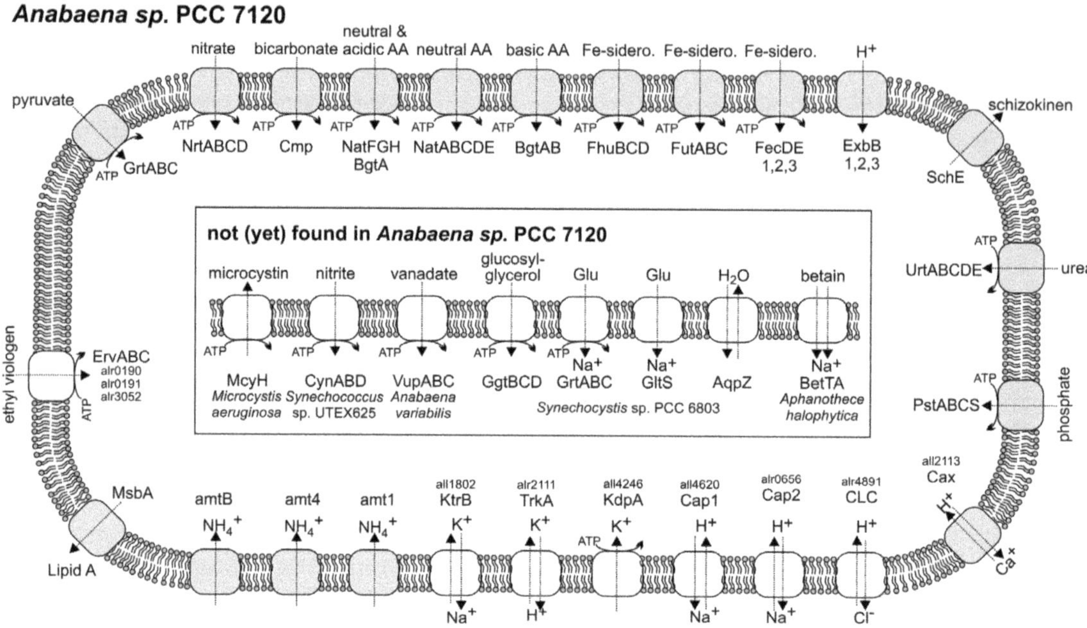

Figure 3.6 The transporters of the cyanobacterial plasma membrane. The different transporters found in the cyanobacterial plasma membrane are shown in the example of *Anabaena* sp. PCC 7120. For all transporters the abbreviation and the confirmed substrate are given. In case that the transporter from other cyanobacterial species is characterized and the transporter in *Anabaena* sp. PCC 7120 is only identified by homology, the gene identification number is given as well. Eight transporters (given in the box) have been described in different cyanobacterial species but have not yet been identified in *Anabaena* sp. PCC 7120. For these transporters, the gene name and the species where the transporter was identified are indicated.

and is composed of the ATP-binding subunits NatA (slr0467) and NatE (slr1881), the periplasmic substrate-binding protein NatB (slr0559), and the transmembrane permeases NatC (sll0146) and NatD (slr0949). The third system, which is specific for glutamate transport, however, does not belong to the ABC-transporter family. The protein encoded by *slr1145* shares significant homology to the Na^+/glutamate permease GltS of *E. coli* (Deguchi *et al.*, 1990). Indeed, the rate of glutamate uptake by *Synechocystis* sp. PCC 6803 is dependent on the sodium concentration and the mutant of *slr1145* shows significantly reduced glutamate uptake rates (Quintero *et al.*, 2001). In parallel, a second Na^+-dependent glutamate transport system composed of Sll1102 (GrtA), Sll1103 (GrtB), and Sll1104 (GrtC) has been described (Quintero *et al.*, 2001). This system is similar to TRAP permeases, which are binding protein-dependent secondary transport systems (Rabus *et al.*, 1999). The homologous system in *Anabaena* sp. PCC 7120 (composed of Alr3026, Alr3027, and All3028) appears to function as a pyruvate/monocarboxylate 2-oxoacid transporter rather than as a glutamate transporter (Pernil *et al.*, 2010). Thus, it seems that these systems perform species-specific functions, which are mainly determined by the periplasmic binding protein of the system in each cyanobacterium (Pernil *et al.*, 2010).

Cyanobacteria contain several proton-dependent transporters, which in many cases have been described for salt stress tolerance. For example, the chloride channel (CLC) is a Cl^-/H^+ antiporter identified in *Synechocystis* sp. PCC 6803. The protein is able to transport 20 chloride ions per second (Jayaram *et al.*, 2011), which is about 100 fold slower than what is found for the homologous protein from *E. coli* (Picollo *et al.*, 2009). However, it is likely that the cyanobacterial CLC has a similar function as the proteobacterial protein, which is involved in the regulation of the salt homeostasis. In addition, cyanobacteria contain a plasma membrane localized Ca^{2+}/H^+

antiporter, which appears to be important for salt tolerance. Genes coding for such transporter are found in the genome of *Synechocystis* sp. PCC 6803 (*synCAX; slr1336*), *Anabaena* sp. PCC 7120 (*anaCAX; all0626*), *Aphanothece halophytica* (*apCAX*) and *Thermosynechococcus elongatus* BP-1 (*telCAX; tll1783*). The proteins from *Synechocystis* sp. PCC 6803 and *Aphanothece halophytica* were found to complement a Ca^{2+}-sensitive *E. coli* mutant (Waditee *et al.*, 2004). A mutant of *Synechocystis* sp. PCC 6803 with disrupted *synCAX* has a lower Ca^{2+} efflux activity and a salt-sensitive phenotype (Waditee *et al.*, 2004). A third type of proton driven transporter was discovered in the halotolerant cyanobacterium *Aphanothece halophytica*, where a Na^+/H^+ antiporter was identified (Waditee *et al.*, 2001). The expression of this protein in *Synechococcus* sp. PCC 7942 resulted in a phenotype of tolerance to salt stress (Waditee *et al.*, 2002) demonstrating the importance of this membrane protein for the salt homeostasis. Remarkably, seven genes with signatures of a Na^+/H^+ antiporter have been identified by bioinformatic means in the genome of *Synechococcus* sp. PCC 7942 (Billini *et al.*, 2008). Furthermore, one of them, annotated as Nha3, was able to complement a salt sensitive *E. coli* strain, strongly supporting a function as a Na^+/H^+ antiporter (Billini *et al.*, 2008).

Besides chloride, calcium and sodium, potassium is also transported in a proton-dependent manner. However, potassium is transported by several different mechanisms (Matsuda and Uozumi, 2006). The KdpA encoded by *slr1728* in the genome of *Synechocystis* sp. PCC 6803 transports potassium in an ATP-dependent manner (Berry *et al.*, 2003); TrkA (Slr0773) transports potassium in a proton-dependent manner; the KtrAB (KtrA: Sll0493; KtrB = NtpJ: Slr1509) transports potassium in a sodium-dependent manner (Berry *et al.*, 2003), and potassium-efflux, Kef-like potassium channels can be found as well (Slr1575, Sll0536, Sll0993).

The smallest molecule transported by plasma membrane proteins is water. In *Synechocystis* sp. PCC 6803, a so-called aquaporin (AqpZ), encoded by *slr2057*, has been identified. In a mutant of this gene, water uptake was significantly reduced, and the mutant cells showed reduced growth under hyperosmotic conditions (Akai *et al.*, 2011; Azad *et al.*, 2011).

Ammonium is another important nutrient for cyanobacteria, as it is the preferred nitrogen source. Ammonium transporters have been identified in different cyanobacterial species (Montesinos *et al.*, 1998; Vázquez-Bermúdez *et al.*, 2002; Paz-Yepes *et al.*, 2007, 2008). In *Synechocystis* sp. PCC 6803, three ammonium transporters encoded by *amt* genes have been characterized. Amt1 (Sll0108) is the major high-affinity transporter, while Amt2 (Sll1017) and Amt3 (Sll0537) make minor contributions to uptake at low substrate concentrations (Montesinos *et al.*, 1998). Similarly, two Amt transporters have been found in *Synechococcus elongatus* PCC 7942. However, only one (Amt1; Vázquez-Bermúdez *et al.*, 2002) was found to be similar to an ammonium transporter from *Synechocystis* sp. PCC 6803, while the other (AmtB) is more similar to a proteobacterial ammonium transporter (Paz-Yepes *et al.*, 2007). The uptake of methylammonium was extremely low in the *amt1* mutant and only partially reduced in the *amtB* mutant leading to the suggestion that Amt1 is required for transport whereas AmtB somehow supports full methylammonium uptake (Paz-Yepes *et al.*, 2007). *Anabaena* sp. PCC 7120 bears three *amt* genes encoding Amt4 (Alr0990), Amt1 (Alr0991) and AmtB (Alr0992), but these genes have not been inactivated individually (Paz-Yepes *et al.*, 2008).

The transporter content and the properties of the transporters differ between cyanobacterial species. As a final example, in the halotolerant alkaliphilic cyanobacterium *Aphanothece halophytica* a Na^+-betaine symporter has been discovered, which has been annotated as BetTA and which belongs to the betaine–choline–carnitine transporter (BCCT) family (Laloknam *et al.*, 2006). As expected for this cyanobacterium, the transport activity is particularly high at alkaline pHs.

The enzymes of the plasma membrane

These enzymes can be classified in components of the photosynthetic and respiratory processes, metabolic enzymes, and proteins involved in signal transduction. (The latter will be separately described below.) It has been a longstanding discussion whether the assembly of the photosystems occurs in the plasma membrane (see 'Architecture and dynamics of the plasma membrane'; Table 3.7), but with the exception of *Gloeobacter violaceus*, it is clear that the photosystems function in the specialized membrane system of the thylakoids.

Although respiration occurs in the thylakoids, enzymes of the respiratory chain seem to be localized in the plasma membrane as well (Table 3.7), although to a different extent in different cyanobacterial strains. For example, *Synechococcus* sp. PCC 6301 harbours high levels of cytochrome *c* oxidase in the plasma membrane, while in *Synechocystis* sp. PCC 6803 the activity was found in both, plasma membrane and thylakoids (Peschek *et al.*, 1989, 1994). Nevertheless, recent studies document that a rudimentary electron transfer complex exists in the plasma membrane of *Synechocystis* sp. PCC 6803, which might be composed of a type 2 NADH dehydrogenase (NDH-2), a quinol oxidase (Cyd; Kufryk and Vermaas, 2006) and most likely an alternative terminal oxidase (CtaII) as another putative quinol oxidase (Pils and Schmetterer, 2001; Schultze *et al.*, 2009). However, this incomplete respiratory chain would not produce a proton gradient (Schultze *et al.*, 2009). The proton gradient could be either formed by plastoquinone reduction at the cytoplasmic side and oxidation in the periplasm, a mode that has been described for the cytochrome *bd* oxidase of *E. coli* (Puustinen *et al.*, 1991), or by plasma membrane ATPase, which in the presence of high levels of ATP could pump protons into the periplasm (Teuber *et al.*, 2001). Based on this, it has been discussed that the plasma membrane-localized 'electron transfer complexes could facilitate the oxidation of a surplus of NAD(P)H' (Schultze *et al.*, 2009). However, the exact function of these complexes remains to be further explored.

As stated above, there are also metabolic enzymes that are inserted or associated with the plasma membrane (Table 3.6). For example, lipid synthesis occurs at this membrane (see 'The outermost-layer of the cyanobacterial cell wall'). Similarly, the recently identified plasma membrane-associated or integrated acyl-acyl carrier protein (ACP) synthetase (SynAas) is involved in the transfer of fatty acids across the plasma membrane (von Berlepsch *et al.*, 2012). Another example of enzymes that are found associated with the plasma membrane is the protease FtsH2. In *Synechocystis* sp. PCC 6803, the mutant of this protein shows a defect in osmoregulation in response to salt-stress, but not to non-ionic osmotic stress (Stirnberg *et al.*, 2007). Proteomic analysis of the cellular response of the wild-type and the *ftsH2* mutant suggests that the sensing mechanism involves the release of FtsH2 from the membrane, or, alternatively, the targeted transfer of cytosolic proteins to the membrane surface for degradation. Glycosyltransferases, essential for the synthesis of the cell wall, have been discussed in 'The peptidoglycan layer of cyanobacteria'.

Signal transduction proteins

Similarly to the membrane-inserted or associated enzymes, many components of signal transduction cascades are present in the plasma membrane. The complexity of these systems cannot be fully explored in this article, but two major processes, which are related to membrane and cellular organization and are guided by plasma membrane proteins involved in signalling, will be highlighted. On one hand, cellular differentiation in filamentous cyanobacteria requires plasma membrane protein based-signalling. In bacteria, signalling is often based on phosphorylation cascades involving two-component regulatory systems (Stock *et al.*, 2000), kinases and phosphatases (Kennelly, 2001). In *Anabaena* sp. PCC 7120, 13 genes encode proteins containing a C-terminal histidine kinase and a N-terminal Ser/Thr kinase domain, and thus belong to the histidine and Ser/Thr kinase hybrid class. This protein class is defined by a fusion between the upstream acting component and the Ser/Thr kinase (Wang *et al.*, 2002). Four histidine kinases have been described to be involved in the regulation of heterocyst formation. The deletion of the two kinases encoded by *pkn30* (*all3691*) and *pkn44* (*all1625*) causes a defect in the synthesis of the 'minor species' of the HGL (Shi *et al.*, 2007). Although the downstream cascade has not yet been explored, this phenotype supports a function of these two kinases in the regulation of cell differentiation.

Another two-component system regulates the synthesis of HepA, an ABC transporter essential for Hep layer formation (Zhu *et al.*, 1998). This system involves the kinase HepK (All4496), which contains a sensor and transmitter domain, and the response regulator DevR (Alr0442; Wang *et al.*, 2002). Deletion of either gene impedes the formation of the HEP-layer (e.g. Zhou and Wolk, 2003). The Ser/Thr kinase HepS might be the cytoplasmic signal carrier, as *hepK* and *hepS* (*all2760*) mutants affect the expression of a common set of genes (Lechno-Yossef *et al.*, 2007). Direct evidence for this cascade is however still missing. HepN (Alr0117) is the fourth histidine kinase involved in cell differentiation (Fan *et al.*, 2005; Lechno-Yossef *et al.*, 2007). This protein does not contain a sensor domain and is thus distinct from HepK. However, regulation by HepN is at least partially overlapping with the one exerted by HepK as determined by global expression analysis (Lechno-Yossef *et al.*, 2007). To finally assign the function of these two pathways, biochemical approaches and the description of the phospho-signalling and phosphor–protein network regulated by all of these cascades are required.

To date, only two plasma membrane-localized phosphatases have been described to be involved in the process of heterocyst differentiation, namely PrpJ1 (All1731; Jang *et al.*, 2007) and PrpJ2 (All2470; Jang *et al.*, 2009). Mutation of *prpJ1* causes the loss of HGL1 during heterocyst formation, which likely results from the down-regulation of *hglE* and *nifH* (Jang *et al.*, 2007). The cascade regulated by these two factors, however, has to be further explored.

Membrane reshaping in the course of osmoprotein involves a signalling system as well, which involves some histidine kinases. Although Hik33 is best investigated as regulator of membrane fluidity, (see 'Fatty acid alterations as a mechanism against stress'), six different Hik proteins (Hik2, Slr1147; Hik10, Slr0533; Hik16, Slr1805; Hik33, Sll0698; Hik34, Slr1285; Hik41, Sll1229) and the cognate response regulators (Rre1, Slr1783; Rre3, Sll0649; Rre17, Sll1708; Rre31, Slr0115) have been described to be involved in the response to hyperosmotic stress in *Synechocystis* sp. PCC 6803 (Marin *et al.*, 2003; Paithoonrangsarid *et al.*, 2004; Shoumskaya *et al.*, 2005; Kanesaki *et al.*, 2007). However, only Hik33, Hik16 and Hik10 have been assigned as plasma membrane-inserted/associated (Paithoonrangsarid *et al.*, 2004). It has been discussed that Hik33 transduces the signal to the response regulator Rre31, Hik10 to response regulator Rre3 and Hik16 via Hik41 to Rre17. However, it is unknown by which mechanism Hik34 and Hik2, the two cytosolic kinases, are activated, although it has been suggested that both transmit the signal to Rre1 (Paithoonrangsarid *et al.*, 2004; Shoumskaya *et al.*, 2005; Kanesaki *et al.*, 2007).

Another pair of histidine kinases and response regulators is the Hik31-Rre34 system, also known as CopS-CopR, which is involved in copper-resistance in *Synechocystis* sp. PCC 6803 (Giner-Lamia *et al.*, 2012). It is suggested that the system can sense both the periplasmic and the cytoplasmic copper level and regulates the expression of the heavy-metal RND (resistance, nodulation, and division)-type copper efflux system CopBAC. It has also been proposed that CopS localizes to the thylakoid membranes as well, and this might suggest that it can sense the copper levels also in the thylakoid lumen (Giner-Lamia *et al.*, 2012). Thus, this system globally senses the copper status in the different cellular compartments.

These examples, the regulation of cell differentiation, osmolarity and copper resistance, document that plasma membrane-inserted kinases are involved in the regulation of cell function and fate. However, the molecular cascades are still under investigations and, thus, the full spectrum of reaction paths initiated at the plasma membrane has still to be established.

Proteins involved in shaping the plasma membrane

The major structural elements involving plasma membrane proteins are described in 'The plasma membrane of cyanobacteria'. Nothing is currently known about the composition of the junction pore and, thus, the proteinaceous components involved in shaping this structure remain to be established. Regarding the pilus complex, some components have been identified by mass spectrometric analysis (Table 3.6). In *Synechocystis* sp. PCC6803, two pilus morphotypes have been described, which are annotated as thin and thick pili (Bhaya *et al.*, 2000). Inactivation of the plasma membrane component PilD required for pili assembly leads to the absence of both pili types and the loss of mobility (Bhaya *et al.*, 2000). Subsequent mutagenesis dissected the components of the two pili: Sll1694 (PilA1), Slr0063 (PilB1), Slr1274 (PilM), Slr1275 (PilN), Slr1276 (PilO), Slr0161 (PilT1) and Slr1277 (PilQ) are required for the thick pili formation, while Sll1694 is required for thin pili formation as well (Yoshihara *et al.*, 2001). A similar system has been identified in *Microcystis aeruginosa* PCC 7806 (Nakasugi *et al.*, 2007) or *Nostoc punctiforme* (Duggan *et al.*, 2007). In the latter, pilus formation is important for plant infection. For further information on this topic, see Chapter 11.

Proteins involved in septosome formation have been discussed in the case of *Anabaena* sp. PCC 7120. These proteins have been classified as 'Cell–Cell joining' proteins and mutants of the corresponding genes have been isolated in the early 1990s (Buikema and Haselkorn, 1991; Ernst *et al.*, 1992). These mutants typically show a so-called filament fragmentation phenotype, as the cell–cell connections are disrupted (Bauer *et al.*, 1995). The detailed analysis of these mutants revealed a protein annotated as SepJ/FraG encoded by *alr2338* (Flores *et al.*, 2007; Nayar *et al.*, 2007). The protein was localized to the septum between adjacent cells (Flores *et al.*, 2007) and the positioning of the protein was found to be influenced by two factors annotated as FraC (encoded by *alr2392*) and FraD (*alr2393*), which in turn are also localized in the intercellular septa (Merino-Puerto *et al.*, 2010). The SepJ protein is encoded in all analysed filamentous cyanobacteria (Mariscal *et al.*, 2011). By mutagenesis it was demonstrated that its C-terminal portion (which is inserted in the plasma membrane) is required for the anchoring of the protein (Mariscal *et al.*, 2011), and the absence of the protein reduces the distance between two cells in the septum (Wilk *et al.*, 2011). FraC and FraD are membrane-inserted proteins as well as documented by prediction and fractionation studies (Merino-Puerto *et al.*, 2011). Remarkably, the rate of transfer of the fluorescent probe 5-carboxylfluorescein between cells is more drastically impaired in the mutants of *fraC* and *fraD* than in the mutant of *sepJ* (Merino-Puerto *et al.*, 2011). This observation suggests that these two factors are not only required for the localization of SepJ but must be involved in the formation of the septosome as well. However, further work is required to define the role of the three plasma membrane proteins in septosome formation.

Taking these examples, it becomes clear that some structural elements define the shape of the cell, the cell–cell connections and the cell envelope. However, in some cases we even do not know the components of these structures, which documents that the analysis is still at an early stage.

Concluding remarks

Many if not most components of the cell envelope of cyanobacteria have been detected in the past, and this holds true for the (poly)saccharides, lipids and proteins. The same is valid for many structural features of the cell envelope, which have been described at the structural level in the past. In addition, many reactions to environmental situations have been documented at global structural or transcriptomic levels. However, the understanding of the molecular processes taking place at the cell envelope is limited, and this is even more notorious for the function of the peptidoglycan layer. Additionally, the cyanobacterial species are rather diverse, and, thus, dissecting the general principles from species-specific variation is required. In this chapter, this difference has been highlighted, for example, when discussing the functional domains in the plasma membrane, which ranges from the presence of photosystems in this membrane in *Gloeobacter violaceus* to different functional regions in filamentous cyanobacteria such as *Anabaena* sp. PCC 7120. Thus, although unified by the ability to perform oxygenic photosynthesis, it is clear that cyanobacterial behaviour and function – even of the cell envelope – can be different in different cyanobacteria. This also involves nutrition uptake strategies as exemplified by the discussion of iron uptake strategies by siderophore-producing and non-producing strains.

In order to understand the cyanobacterial behaviour and the function of the cell envelope in more detail two strategies of investigation should be considered. On one hand, the comparative analysis of mechanisms in different cyanobacteria at molecular and structural levels will be needed to dissect the diversity of strategies in one organism and in the cyanobacterial world. On the other hand, it will be important to study the behaviour of different cyanobacteria in communities and natural environments, as the laboratory conditions currently used are optimized for rapid cell proliferation and are often significantly different from those in the native habitats. However, modern technologies and techniques do no longer require massive amounts of biological material, and the question arises: are the pathways and cell envelope functions found in the laboratory relevant for the cyanobacteria in their natural environments? The answer to this question cannot be foreseen, however, because it is frequent that some components appear to be not essential under laboratory conditions. Thus, much is known but still more remains to be explored, as our understanding of cell envelope function at the molecular level is just at its beginning and its importance in the exchange of solutes, exchange of information, protection, symbiosis and defence makes the envelope a central component of the cell.

Acknowledgements

We are grateful to Eva-Maria Brouwer, Mara Stevanovic and Mareike Rudolf (Frankfurt) for critical reading of the manuscript. In particular, we thank Enrique Flores for helpful discussions and comments during manuscript preparation. The work was supported by grants from Deutsche Forschungsgemeinschaft to ES (DFG SCHL 585 2-1 and DFG SCHL 585 7-1).

References

Adams, D.G. (2000). Heterocyst formation in cyanobacteria. Curr. Opin. Microbiol. *3*, 618–624.

Akai, M., Onai, K., Kusano, M., Sato, M., Redestig, H., Toyooka, K., Morishita, M., Miyake, H., Hazama, A., Checchetto, V., *et al.* (2011). Plasma membrane aquaporin AqpZ protein is essential for glucose metabolism during photomixotrophic growth of *Synechocystis* sp. PCC 6803. J. Biol. Chem. *286*, 25224–25235.

Akama, H., Kanemaki, M., Yoshimura, M., Tsukihara, T., Kashiwagi, T., Yoneyama, H., Narita, S., Nakagawa, A., and Nakae, T. (2004). Crystal structure of the drug discharge outer membrane protein, OprM, of *Pseudomonas aeruginosa*: dual modes of membrane anchoring and occluded cavity end. J. Biol. Chem. *279*, 52816–52819.

Albanesi, D., Mansilla, M.C., and de Mendoza, D. (2004). The membrane fluidity sensor DesK of *Bacillus subtilis* controls the signal decay of its cognate response regulator. J. Bacteriol. *186*, 2655–2663.

Alexander, C., and Rietschel, E.T. (2001). Bacterial lipopolysaccharides and innate immunity. J. Endotoxin Res. *7*, 167–202.

Allen, M.D., del Campo, J.A., Kropat, J., and Merchant, S.S. (2007). FEA1, FEA2, and FRE1, encoding two homologous secreted proteins and a candidate ferrireductase, are expressed coordinately with FOX1 and FTR1 in iron-deficient *Chlamydomonas reinhardtii*. Eukaryotic Cell *6*, 1841–1852.

Alphen, W.V., and Lugtenberg, B. (1977). Influence of osmolarity of the growth medium on the outer membrane protein pattern of *Escherichia coli*. J. Bacteriol. *131*, 623–630.

Aoki, M., Sato, N., Meguro, A., and Tsuzuki, M. (2004). Differing involvement of sulfoquinovosyl diacylglycerol in photosystem II in two species of unicellular cyanobacteria. Eur. J. Biochem. *271*, 685–693.

Aoki, M., Tsuzuki, M., and Sato, N. (2012). Involvement of sulfoquinovosyl diacylglycerol in DNA synthesis in *Synechocystis* sp. PCC 6803. BMC Res. Notes *5*, 98.

Armstrong, J.E., and van Baalen, C. (1979). Iron transport in microalgae: The isolation and biological activity of a hydroxamate siderophore from the blue-green algae *Agmenellum quadruplicatum*. J. General Microbiol. *111*, 253–262.

Azad, A.K., Sato, R., Ohtani, K., Sawa, Y., Ishikawa, T., and Shibata, H. (2011). Functional characterization and hyperosmotic regulation of aquaporin in *Synechocystis* sp. PCC 6803. Plant Sci. *180*, 375–382.

Bainbridge, G., Mobasheri, H., Armstrong, G.A., Lea, E.J., and Lakey, J.H. (1998). Voltage-gating of *Escherichia coli* porin: a cystine-scanning mutagenesis study of loop 3. J. Mol. Biol. *275*, 171–176.

Bassford, P.J. Jr., Diedrich, D.L., Schnaitman, C.L., and Reeves, P. (1977). Outer membrane proteins of *Escherichia coli*. VI. Protein alteration in bacteriophage-resistant mutants. J. Bacteriol. *131*, 608–622.

Bauer, C.C., Buikema, W.J., Black, K., and Haselkorn, R. (1995). A short-filament mutant of *Anabaena* sp. strain PCC 7120 that fragments in nitrogen-deficient medium. J. Bacteriol. *177*, 1520–1526.

Bauer, K., Struyvé, M., Bosch, D., Benz, R., and Tommassen, J. (1989). One single lysine residue is responsible for the special interaction between polyphosphate and the outer membrane porin PhoE of *Escherichia coli*. J. Biol. Chem. *264*, 16393–16398.

Beiderbeck, H., Taraz, K., Budzikiewicz, H., and Walsby, A.E. (2000). Anachelin, the siderophore of the cyanobacterium *Anabaena cylindrica* CCAP 1403/2A. Z. Naturforsch *55*, 681–687.

Bellido, F., Martin, N.L., Siehnel, R.J., and Hancock, R.E. (1992). Reevaluation, using intact cells, of the exclusion limit and role of porin OprF in *Pseudomonas aeruginosa* outer membrane permeability. J. Bacteriol. *174*, 5196–5203.

Benz, R., and Böhme, H. (1985). Pore formation by an outer membrane protein of the cyanobacterium

Anabaena variabilis. Biochim. Biophys. Acta *812*, 286–292.

Benz, R., and Hancock, R.E. (1987). Mechanism of ion transport through the anion-selective channel of the *Pseudomonas aeruginosa* outer membrane. J. Gen. Physiol. *89*, 275–295.

Benz, R., Schmid, A., and Hancock, R.E. (1985). Ion selectivity of Gram-negative bacterial porins. J. Bacteriol. *162*, 722–727.

Benz, R., Schmid, A., Nakae, T., and Vos-Scheperkeuter, G.H. (1986). Pore formation by LamB of *Escherichia coli* in lipid bilayer membranes. J. Bacteriol. *165*, 978–986.

Berglund, A.H., Nilsson, R., and Liljenberg, C. (1999). Permeability of large unilamellar digalactosyldiacylglycerol vesicles for protons and glucose – influence of α-tocopherol, α -carotene, zeaxanthin and cholesterol. Plant Physiol. Biochem. *37*, 179–186.

von Berlepsch, S., Kunz, H.H., Brodesser, S., Fink, P., Marin, K., Flügge, U.I., and Gierth, M. (2012). The acyl-acyl carrier protein synthetase from *Synechocystis* sp. PCC 6803 mediates fatty acid import. Plant Physiol. *159*, 606–617.

Berry, S., Esper, B., Karandashova, I., Teuber, M., Elanskaya, I., Rögner, M., and Hagemann, M. (2003). Potassium uptake in the unicellular cyanobacterium *Synechocystis* sp. strain PCC 6803 mainly depends on a Ktr-like system encoded by slr1509 (ntpJ). FEBS Lett. *548*, 53–58.

Bhaya, D., Bianco, N.R., Bryant, D., and Grossman, A. (2000). Type IV pilus biogenesis and motility in the cyanobacterium *Synechocystis* sp. PCC6803. Mol. Microbiol. *37*, 941–951.

Billini, M., Stamatakis, K., and Sophianopoulou, V. (2008). Two members of a network of putative Na+/H+ antiporters are involved in salt and pH tolerance of the freshwater cyanobacterium *Synechococcus elongatus*. J. Bacteriol. *190*, 6318–6329.

Biró, I., Pezeshki, S., Weingart, H., Winterhalter, M., and Kleinekathöfer, U. (2010). Comparing the temperature-dependent conductance of the two structurally similar *E. coli* porins OmpC and OmpF. Biophys. J. *98*, 1830–1839.

Bohnsack, M.T., and Schleiff, E. (2010). The evolution of protein targeting and translocation systems. Biochim. Biophys. Acta *1803*, 1115–1130.

Bölter, B., Soll, J., Schulz, A., Hinnah, S., and Wagner, R. (1998). Origin of a chloroplast protein importer. Proc. Natl. Acad. Sci. U.S.A. *95*, 15831–15836.

Bond, P.J., Faraldo-Gómez, J.D., and Sansom, M.S. (2002). OmpA: a pore or not a pore? Simulation and modeling studies. Biophys. J. *83*, 763–775.

Bos, M.P., Tefsen, B., Geurtsen, J., and Tommassen, J. (2004). Identification of an outer membrane protein required for the transport of lipopolysaccharide to the bacterial cell surface. Proc. Natl. Acad. Sci. U.S.A. *101*, 9417–9422.

Brahamsha, B. (1996). An abundant cell-surface polypeptide is required for swimming by the nonflagellated marine cyanobacterium *Synechococcus*. Proc. Natl. Acad. Sci. U.S.A. *93*, 6504–6509.

Braun, M., and Silhavy, T.J. (2002). Imp/OstA is required for cell envelope biogenesis in *Escherichia coli*. Mol. Microbiol. *45*, 1289–1302.

Braun, V., Gnirke, H., Henning, U., and Rehn, K. (1973). Model for the structure of the shape-maintaining layer of the *Escherichia coli* cell envelope. J. Bacteriol. *114*, 1264–1270.

Bredemeier, R., Schlegel, T., Ertel, F., Vojta, A., Borissenko, L., Bohnsack, M.T., Groll, M., von Haeseler, A., and Schleiff, E. (2007). Functional and phylogenetic properties of the pore-forming beta-barrel transporters of the Omp85 family. J. Biol. Chem. *282*, 1882–1890.

Brinkman, F.S., Bains, M., and Hancock, R.E. (2000). The amino terminus of *Pseudomonas aeruginosa* outer membrane protein OprF forms channels in lipid bilayer membranes: correlation with a three-dimensional model. J. Bacteriol. *182*, 5251–5255.

Brown, C.M., and Trick, C.G. (1992). Response of the cyanobacterium, *Oscillatoria tenuis*, to low iron environments: the effect on growth rate and evidence for siderophore production. Arch. Microbiol. *157*, 349–354.

Buehler, L.K., Kusumoto, S., Zhang, H., and Rosenbusch, J.P. (1991). Plasticity of *Escherichia coli* porin channels. Dependence of their conductance on strain and lipid environment. J. Biol. Chem. *266*, 24446–24450.

Buikema, W.J., and Haselkorn, R. (1991). Isolation and complementation of nitrogen fixation mutants of the cyanobacterium *Anabaena* sp. strain PCC 7120. J. Bacteriol. *173*, 1879–1885.

Castan, P., Zafra, O., Moreno, R., de Pedro, M.A., Valle´s, C., Cava, F., Caro, E., Schwarz, H., and Berenguer, J. (2002). The periplasmic space in *Thermus thermophilus*: evidence from a regulation-defective S-layer mutant overexpressing an alkaline phosphatase. Extremophiles *6*, 225–232.

Caston, J., Carrascosa, J., de Pedro, M., and Berenguer, J. (1988). Identification of a crystalline layer on the cell envelope of the thermophilic eubacterium *Thermus thermophilus*. FEMS Lett. *51*, 225–230.

Caston, J.R., Berenguer, J., de Pedro, M.A., and Carrascosa, J.L. (1993). S-layer protein from *Thermus thermophilus* HB8 assembles into porin like structures. Mol. Microbiol *9*, 65–75.

Cavalier-Smith, T. (2006). Rooting the tree of life by transition analyses. Biol. Direct. *1*, 19.

Chimento, D.P., Kadner, R.J., and Wiener, M.C. (2005). Comparative structural analysis of TonB-dependent outer membrane transporters: Implications for the transport cycle. PROTEINS: Struc. Func. Bioinform. *59*, 240–251.

Clarke, S.E., Stuart, J., and Sanders-Loehr, J. (1987). Induction of siderophore activity in Anabaena spp. and moderation of copper toxicity. Appl. Environ. Microbiol. *53*, 917–922.

Cogdell, R.J., and Frank, H.A. (1987). How carotenoids function in photosynthetic bacteria. Biochim. Biophys. Acta. *895*, 63–79.

Cowan, S.W., Schirmer, T., Rummel, G., Steiert, M., Ghosh, R., Pauptit, R.A., Jansonius, J.N., and Rosenbusch, J.P. (1992). Crystal structures explain functional properties of two *E. coli* porins. Nature *358*, 727–733.

Cronan, J.E. (2003). Bacterial membrane lipids: where do we stand? Annu. Rev. Microbiol. *57*, 203–224.

Cybulski, L.E., Martín, M., Mansilla, M.C., Fernández, A., and de Mendoza, D. (2010). Membrane thickness cue for cold sensing in a bacterium. Curr. Biol. *20*, 1539–1544.

Death, A., Notley, L., and Ferenci, T. (1993). Derepression of LamB protein facilitates outer membrane permeation of carbohydrates into *Escherichia coli* under conditions of nutrient stress. J. Bacteriol. *175*, 1475–1483.

Deguchi, Y., Yamato, I., and Anraku, Y. (1990). Nucleotide sequence of gltS, the Na+/glutamate symport carrier gene of *Escherichia coli* B. J. Biol. Chem. *265*, 21704–21708.

Dela Vega, A.L., and Delcour, A.H. (1996). Polyamines decrease *Escherichia coli* outer membrane permeability. J. Bacteriol. *178*, 3715–3721.

Drews, G., and Meyer, H. (1964). Untersuchungen zum chemischen Aufbau der Zellwände von *Anacystis nidulans* und *Chlorogloea fritschii*. Arch. Mikrobiol. *48*, 259–267.

Duggan, P.S., Gottardello, P., and Adams, D.G. (2007). Molecular analysis of genes in *Nostoc punctiforme* involved in pilus biogenesis and plant infection. J. Bacteriol. *189*, 4547–4551.

Dumas, F., Frank, S., Koebnik, R., Maillet, E., Lustig, A., and Van Gelder, P. (2000a). Extended sugar slide function for the periplasmic coiled coil domain of ScrY. J. Mol. Biol. *300*, 687–695.

Dumas, F., Koebnik, R., Winterhalter, M., and Van Gelder, P. (2000b). Sugar transport through maltoporin of *Escherichia coli*. Role of polar tracks. J. Biol. Chem. *275*, 19747–19751.

Dutzler, R., Schirmer, T., Karplus, M., and Fischer, S. (2002). Translocation mechanism of long sugar chains across the maltoporin membrane channel. Structure *10*, 1273–1284.

Ehlers, K., and Oster, G. (2012). On the mysterious propulsion of *Synechococcus*. PLoS One 7, e36081.

Ehrenreich, I.M., Waterbury, J.B., and Webb, E.A. (2005). Distribution and diversity of natural product genes in marine and freshwater cyanobacterial cultures and genomes. Appl. Environ. Microbiol. *71*, 7401–7413.

Eicher, T., Brandstätter, L., and Pos, K.M. (2009). Structural and functional aspects of the multidrug efflux pump AcrB. Biol. Chem. *390*, 693–699.

Ekman, M., Picossi, S., Campbell, E., Meeks, J.C., and Flores, E. (2013). A *Nostoc punctiforme* sugar transporter necessary to establish a cyanobacterium–plant symbiosis. Plant Physiol. *161*, 1984–1992.

Engelhardt, H. (2007). Are S-layers exoskeletons? The basic function of surface protein layers revisited. J. Struct. Biol. *160*, 115–124.

Ernst, A., Black, T., Cai, Y., Panoff, J.M., Tiwari, D.N., and Wolk, C.P. (1992). Synthesis of nitrogenase in mutants of the cyanobacterium *Anabaena* sp. strain PCC 7120 affected in heterocyst development or metabolism. J. Bacteriol. *174*, 6025–6032.

Ertel, F., Mirus, O., Bredemeier, R., Moslavac, S., Becker, T., and Schleiff, E. (2005).The evolutionarily related beta-barrel polypeptide transporters from *Pisum sativum* and *Nostoc* PCC7120 contain two distinct functional domains. J. Biol. Chem. *280*, 28281–28289.

Espie, G.S., Jalali, F., Tong, T., Zacal, N.J., and So, A.K. (2007). Involvement of the cynABDS operon and the CO_2-concentrating mechanism in the light-dependent transport and metabolism of cyanate by cyanobacteria. J. Bacteriol. *189*, 1013–1024.

Estep, M., Armstrong, J.E., and van Baalen, C. (1975). Evidence for the occurrence of specific iron (III)-binding compounds in near-shore marine ecosystems. Applied Microbiol. *30*, 186–188.

Fairman, J.W., Noinaj, N., and Buchanan, S.K. (2011). The structural biology of β-barrel membrane proteins: a summary of recent reports. Curr. Opin. Struct. Biol. *21*, 523–531.

Fan, Q., Huang, G., Lechno-Yossef, S., Wolk, C.P., Kaneko, T., and Tabata, S. (2005). Clustered genes required for synthesis and deposition of envelope glycolipids in *Anabaena* sp. strain PCC 7120. Mol. Microbiol. *58*, 227–243.

Faraldo-Gómez, J.D., and Sansom, M.S.P. (2003). Acquisition of siderophores in Gram-negative bacteria. Nat. Rev. Mol. Cell Biol. *4*, 105–116.

Fiedler, G., Arnold, M., Hannus, S., and Maldener, I. (1998). The DevBCA exporter is essential for envelope formation in heterocysts of the cyanobacterium *Anabaena* sp. strain PCC 7120. Mol. Microbiol. *27*, 1193–1202.

Flores, E., and Muro-Pastor, A.M. (1990). Mutational and kinetic analysis of basic amino acid transport in the cyanobacterium *Synechocystis* sp. PCC 6803. Arch. Microbiol. *154*, 521–527.

Flores, E., Herrero, A., Wolk, C.P., and Maldener, I. (2006). Is the periplasm continuous in filamentous multicellular cyanobacteria? Trends Microbiol. *14*, 439–443.

Flores, E., Pernil, R., Muro-Pastor, A.M., Mariscal, V., Maldener, I., Lechno-Yossef, S., Fan, Q., Wolk, C.P., and Herrero, A. (2007). Septum-localized protein required for filament integrity and diazotrophy in the heterocyst-forming cyanobacterium *Anabaena* sp. strain PCC 7120. J. Bacteriol. *189*, 3884–3890.

Frank, H., Lefort, M., and Martin, H.H. (1962). Elektronenoptische und chemische Untersuchungen an Zellwänden der Blaualge *Phormidium uncinatum*. Z. Naturforsch *17b*, 262–268.

Frentzen, M. (2004). Phosphatidylglycerol and sulfoquinovosyldiacylglycerol: anionic membrane lipids and phosphate regulateon. Curr. Opin. Plant Biol. *7*, 270–276.

Frías, J.E., Flores, E., and Herrero, A. (1997). Nitrate assimilation gene cluster from the heterocyst-forming cyanobacterium *Anabaena* sp. strain PCC 7120. J. Bacteriol. *179*, 477–486.

Fulda, S., Huang, F., Nilsson, F., Hagemann, M., and Norling, B. (2000). Proteomics of *Synechocystis* sp. strain PCC 6803. Identification of periplasmic proteins in cells grown at low and high salt concentrations. Eur. J. Biochem. *267*, 5900–5907.

Galanos, C., Rietschel, E.T., Lüderitz, O., Westphal, O., Kim, Y.B., and Watson, D.W. (1972). Biological activities of lipid A complexed with bovine-serum albumin. Eur. J. Biochem. *31*, 230–233.

Gan, L., Chen, S., and Jensen, G.J. (2008). Molecular organization of Gram-negative peptidoglycan. Proc. Natl. Acad. Sci. U.S.A. *105*, 18953–18957.

Garcia, M., Myouga, F., Takechi, K., Sato, H., Nabeshima, K., Nagata, N., Takio, S., Shinozaki, K., and Takano, H. (2008). An *Arabidopsis* homolog of the bacterial peptidoglycan synthesis enzyme MurE has an essential role in chloroplast development. Plant J. *53*, 924–934.

Genevrois, S., Steeghs, L., Roholl, P., Letesson, J.J., and van der Ley, P. (2003). The Omp85 protein of Neisseria meningitidis is required for lipid export to the outer membrane. EMBO J. *22*, 1780–1789.

Gentschev, I., Dietrich, G., and Goebel, W. (2002). The *E. coli* alpha-hemolysin secretion system and its use in vaccine development. Trends Microbiol. *10*, 39–45.

Giddings, T.H. Jr., and Staehelin, L.A. (1978). Plasma membrane architecture of *Anabaena cylindrica*: occurrence of microplasmodesmata and changes associated with heterocyst development and the cell cycle. Cytobiology *16*, 235- 249.

Giddings, T.H., and Staehelin, L.A. (1979). Changes in thylakoid structure associated with the differentiation of heterocysts in the cyanobaeterium *Anabaena cylindrica*. Biochim. Biophys. Acta *546*, 373- 382.

Giddings, T.H. Jr., and Staehelin, L.A. (1981). Observation of microplasmodesmata in both heterocyst-forming and non-heterocyst forming filamentous cyanobacteria by freeze-fracture electron microscopy. Arch. Microbiol. *129*, 295–298.

Giner-Lamia, J., López-Maury, L., Reyes, J.C., and Florencio, F.J. (2012). The CopRS two-component system is responsible for resistance to copper in the cyanobacterium *Synechocystis* sp. PCC 6803. Plant Physiol. *159*, 1806–1818.

Glauner, B., Höltje, J.V., and Schwarz, U. (1988). The composition of the murein of *Escherichia coli*. J. Biol. Chem. *263*, 10088–10095.

Goldberg, M., Pribyl, T., Juhnke, S., and Nies, D.H. (1999). Energetics and topology of CzcA, a cation/proton antiporter of the resistance-nodulation-cell division protein family. J. Biol. Chem. *274*, 26065–26070.

Goffin, C., and Ghuysen, J.M. (1998). Multimodular penicillin-binding proteins: an enigmatic family of orthologs and paralogs. Microbiol. Mol. Biol. Rev. *62*, 1079–1093.

Golden, J.W., and Yoon, H.S. (2003). Heterocyst development in *Anabaena*. Curr. Opin. Microbiol. *6*, 557–563.

Goldman, S.J., Lammers, P.J., Berman, M.S., and Sanders-Loehr, J. (1983). Siderophore-mediated uptake in different strains of *Anabaena* sp. J. Bacteriol. *156*, 1144–1150.

Golecki, J.R. (1977). Studies on ultrastructure and composition of cell walls of the cyanobacterium *Anacystis nidulans*. Arch. Microbiol. *114*, 35–41.

Golecki, J.R. (1979). Ultrastructure of cell wall and thylakoid membranes of the thermophilic cyanobacterium *Synechococcus lividus* under the influence of temperature shifts. Arch. Microbiol. *120*, 125–133.

Gruszecki, W.I., and Strzalka, K. (2005). Carotenoids as modulators of lipid membrane physical properties. Biochim. Biophys. Acta *1740*, 108–115.

Gupta, R.S., Mukhtar, T., and Singh, B. (1999). Evolutionary relationships among photosynthetic prokaryotes (*Heliobacterium chlorum, Chloroflexus aurantiacus, cyanobacteria, Chlorobium tepidum* and *proteobacteria*): implications regarding the origin of photosynthesis. Mol. Microbiol. *32*, 893–906.

Guskov, A., Kern, J., Gabdulkhakov, A., Broser, M., Zouni, A., and Saenger, W. (2009). Cyanobacterial photosystem II at 2.9-Å resolution and the role of quinones, lipids, channels and chloride. Nat. Struct. Mol. Biol. *16*, 334–342.

Haarmann, R., Ibrahim, M., Stevanovic, M., Bredemeier, R., and Schleiff, E. (2010). The properties of the outer membrane localized Lipid A transporter LptD. J. Physics: Condensed Matter. *22*, 454124.

Hagio, M., Gombos, Z., Várkonyi, Z., Masamoto, K., Sato, N., Tsuzuki, M., and Wada, H. (2000). Direct evidence for requirement of phosphatidylglycerol in photosystem II of photosynthesis. Plant Physiol. *124*, 795–804.

Hahn, A., Stevanovic, M., Mirus, O., and Schleiff, E. (2012). The TolC-like protein HgdD of the cyanobacterium *Anabaena* sp. PCC 7120 is involved in secondary metabolite export and antibiotic resistance. J. Biol. Chem. *287*, 41126–41138.

Halliwell, B. (1991). Reactive oxygen species in living systems: source, biochemistry and role in human disease. Am. J. Med. *91*, 14–22.

Hansel, A., and Tadros, M.H. (1998). Characterization of two pore-forming proteins isolated from the outer membrane of *Synechococcus* PCC 6301. Curr. Microbiol. *36*, 321–326.

Hansel, A., Schmid, A., Tadros, M.H., and Jürgens, U.J. (1994). Isolation and characterisation of porin from the outer membrane of *Synechococcus* PCC 6301. Arch. Microbiol. *161*, 163–167.

Hansel, A., Pattus, F., Jürgens, U.J., and Tadros, M.H. (1998). Cloning and characterization of the genes coding for two porins in the unicellular cyanobacterium *Synechococcus* PCC 6301. Biochim. Biophys. Acta *1399*, 31–39.

Heaney, S.I. (1971). The toxicity of *Microcystis aeruginosa* Kutz from some English reservoirs. Water Treat. Exam. *20*, 235–244.

Henderson, B., Poole, S., and Wilson, M. (1996). Microbial/host interactions in health and disease: who controls the cytokine network? Immunopharmacology *35*, 1–21.

Herrero, A., Muro-Pastor, A.M., Valladares, A., and Flores, E. (2004). Cellular differentiation and the NtcA transcription factor in filamentous cyanobacteria. FEMS Microbiol. Rev. *28*, 469–487.

Heumann, D., and Roger, T. (2002). Initial responses to endotoxins and Gram-negative bacteria. Clin. Chim. Acta. *323*, 59–72.

Hoiczyk, E., and Baumeister, W. (1995). Envelope structure of four gliding filamentous cyanobacteria. J. Bacteriol. *177*, 2387–2395.

Hoiczyk, E., and Baumeister, W. (1998). The junctional pore complex, a prokaryotic secretion organelle, is the molecular motor underlying gliding motility in cyanobacteria. Curr. Biol. *8*, 1161–1168.

Hoiczyk, E., and Hansel, A. (2000). Cyanobacterial cell walls: news from an unusual prokaryotic envelope. J. Bacteriol. *182*, 1191–1199.

Hobot, J.A., Carlemalm, E., Villiger, W., and Kellenberger, E. (1984). Periplasmic gel: new concept resulting from the reinvestigation of bacterial cell envelope ultrastructure by new methods. J. Bacteriol. *160*, 143–152.

Holland, D., and Wolk, C.P. (1990). Identification and characterization of hetA, a gene that acts early in the process of morphological differentiation of heterocysts. J. Bacteriol. *172*, 3131–3137.

Holland, K.A., and Holland, I.B. (2005). Adventures with ABC-proteins: highly conserved ATP-dependent transporters. Acta Microbiol. Immunol. Hung. *52*, 309–322.

Homi, S., Takechi, K., Tanidokoro, K., Sato, H., Takio, S., and Takano, H. (2009). The peptidoglycan biosynthesis genes MurA and MraY are related to chloroplast division in the moss *Physcomitrella patens*. Plant Cell Physiol. *50*, 2047–2056.

Hopkinson, B.M., and Morel, F.M. (2009). The role of siderophores in iron acquisition by photosynthetic marine microorganisms. Biometals *22*, 659–669.

Huang, F., Parmryd, I., Nilsson, F., Persson, A.L., Pakrasi, H.B., Andersson, B., and Norling, B. (2002). Proteomics of *Synechocystis* sp. strain PCC 6803: identification of plasma membrane proteins. Mol. Cell Proteom. *1*, 956–966.

Huang, F., Hedman, E., Funk, C., Kieselbach, T., Schröder, W.P., and Norling, B. (2004). Isolation of outer membrane of *Synechocystis* sp. PCC 6803 and its proteomic characterization. Mol. Cell Proteom. *3*, 586–595.

Huang, F., Fulda, S., Hagemann, M., and Norling, B. (2006). Proteomic screening of salt-stress-induced changes in plasma membranes of *Synechocystis* sp. strain PCC 6803. Proteomics *6*, 910–920.

Huang, G., Fan, Q., Lechno-Yossef, S., Wojciuch, E., Wolk, C.P., Kaneko, T., and Tabata, S. (2005). Clustered genes required for the synthesis of heterocyst envelope polysaccharide in *Anabaena* sp. strain PCC 7120. J. Bacteriol. *187*, 1114–1123.

Huang, H., Zhong, Z.P., Wang, K.B., Bai, K.Z., Li, L.B., and Kuang, T.Y. (2004). Isolation and Characterization of the Cytoplasmic Membrane from the Terrestrial Cyanobacterium – *Nostoc flagelliforme*. Acta Botanica Sinica *46*, 1186–1191.

Iyer, R., Wu, Z., Woster, P.M., and Delcour, A.H. (2000). Molecular basis for the polyamine-ompF porin interactions: inhibitor and mutant studies. J. Mol. Biol. *297*, 933–945.

Jang, J., Wang, L., Jeanjean, R., and Zhang, C.C. (2007). PrpJ, a PP2C-type protein phosphatase located on the plasma membrane, is involved in heterocyst maturation in the cyanobacterium *Anabaena* sp. PCC 7120. Mol. Microbiol. *64*, 347–358.

Jang, J., Shi, L., Tan, H., Janicki, A., and Zhang, C.C. (2009). Mutual regulation of ntcA and hetR during heterocyst differentiation requires two similar PP2C-type protein phosphatases, PrpJ1 and PrpJ2, in *Anabaena* sp. strain PCC 7120. J. Bacteriol. *191*, 6059–6066.

Jayaram, H., Robertson, J.L., Wu, F., Williams, C., and Miller, C. (2011). Structure of a slow CLC $Cl^-/H+$ antiporter from a cyanobacterium. Biochem. *50*, 788–794.

Jeanjean, R., Talla, E., Latifi, A., Havaux, M., Janicki, A., and Zhang, C.C. (2008). A large gene cluster encoding peptide synthetases and polyketide synthases is involved in production of siderophores and oxidative stress response in the cyanobacterium *Anabaena* sp. PCC 7120. Environ. Microbiol. *10*, 2574–2585.

Jiang, H., Kong, R., and Xu, X. (2010). The N-acetylmuramic acid 6-phosphate etherase gene promotes growth and cell differentiation of cyanobacteria under light-limiting conditions. J. Bacteriol. *192*, 2239–2245.

Jiang, H.B., Lou, W.J., Du, H.Y., Price, N.M., and Qiu, B.S. (2012). Sll1263, a unique cation diffusion facilitator protein that promotes iron uptake in the cyanobacterium *Synechocystis* sp. Strain PCC 6803. Plant Cell Physiol. *53*, 1404–1417.

Jones, J.H., and Yopp, J.H. (1979). Cell wall constituents of *Aphanothece halopytica* (Cyanophyta). J. Phycol. *15*, 62–66.

Jörbeck, H., Carlsson, H.E., Svenson, S.B., Lindberg, A.A., Alfredsson, G., Garegg, P.J., Svensson, S., and Wallin, N.H. (1979). Immunochemistry of Salmonella O-antigens. Specificity and cross-reactivity of factor O9 serum and of antibodies against tyvelose (Formula: see text) mannose coupled to bovine serum albumin. Int. Arch. Allergy Appl. Immunol. *58*, 11–19.

Jordan, P., Fromme, P., Witt, H.T., Klukas, O., Saenger, W., and Krauss, N. (2001). Three-dimensional structure of cyanobacterial photosystem I at 2.5 Å resolution. Nature *411*, 909–917.

Jürgens, U.J., and Benz, R. (1989). Pore-forming activity of outer membrane extracts from the unicellular cyanobacteria *Synechocystis* sp. PCC 6714. Z. Naturforsch *44c*, 147–151.

Jürgens, U.J., and Burger-Wiersma, T. (1989). Peptidoglycan–polysaccharide complex in the cell wall of the filamentous prochlorophyte *Prochlorothrix hollandica*. J. Bacteriol. *171*, 498–502.

Jürgens, U.J., and Weckesser, J. (1986). Polysaccharide covalently linked to the peptidoglycan of the cyanobacterium *Synechocystis* sp. strain PCC6714. J. Bacteriol. *168*, 568–573.

Jürgens, U.J., Drews, G., and Weckesser, J. (1983). Primary structure of the peptidoglycan from the unicellular cyanobacterium *Synechocystis* sp. strain PCC 6714. J. Bacteriol. *154*, 471–478.

Jürgens, U.J., Golecki, J.R., and Weckesser, J. (1985). Characterization of the cell wall of the unicellular cyanobacterium *Synechocystis* PCC 6714. Arch. Microbiol. *142*, 168–174.

Kalmokoff, M.L., Austin, J.W., Cyr, T.D., Hefford, M.A., Teather, R.M., and Selinger, L.B. (2009). Physical and genetic characterization of an outer-membrane protein (OmpM1) containing an N-terminal S-layer-like homology domain from the phylogenetically Gram-positive gut anaerobe *Mitsuokella multacida*. Anaerobe *15*, 74–81.

Kaneko, T., Sato, S., Kotani, H., Tanaka, A., Asamizu, E., Nakamura, Y., Miyajima, N., Hirosawa, M., Sugiura, M., Sasamoto, S., *et al.* (1996). Sequence analysis of the genome of the unicellular cyanobacterium *Synechocystis* sp. strain PCC6803. II. Sequence determination of the entire genome and assignment of potential protein-coding regions. DNA Res. *3*, 109–136.

Kanesaki, Y., Yamamoto, H., Paithoonrangsarid, K., Shoumskaya, M., Suzuki, I., Hayashi, H., and Murata, N. (2007). Histidine kinases play important roles in the perception and signal transduction of hydrogen peroxide in the cyanobacterium, *Synechocystis* sp PCC 6803. Plant J. *49*, 313–324.

Katoh, H., Hagino, N., Grossman, A.R., and Ogawa, T. (2001). Genes essential to iron transport in the cyanobacterium *Synechocystis* sp. strain PCC 6803. J. Bacteriol. *183*, 2779–2784.

Katz, A., Weckesser, J., and Drews, G. (1977). Chemical and Biological Studies on the Lipopolysaccharide (O-Antigen) of *Anacystis nidulans*. Arch. Microbiol. *113*, 247–256.

Keleti, G., and Sykora, J.L. (1982). Production and properties of cyanobacterial endotoxins. Appl. Environ. Microbiol. *43*, 104–109.

Keleti, G., Sykora, J.L., Lippy, E.C., and Shapiro, M.A. (1979). Composition and biological properties of lipopolysaccharides isolated from *Schizothrix calcicola* (Ag.) Gomont (Cyanobacteria). Appl. Environ. Microbiol. *38*, 471–477.

Kennelly, P.J. (2001). Protein phosphatases – a phylogenetic perspective. Chem. Rev. *101*, 2291–2312.

Kikuchi, S., Shibuya, I., and Matsumoto, K. (2000). Viability of an *Escherichia coli* pgsA null mutant lacking detectable phosphatidylglycerol and cardiolipin. J. Bacteriol. *182*, 371–376.

King, J.D., Kocíncová, D., Westman, E.L., and Lam, J.S. (2009). Lipopolysaccharide biosynthesis in *Pseudomonas aeruginosa*. Innate Immun.*15*, 261–312.

Koenig, P., Mirus, O., Haarmann, R., Sommer, M.S., Sinning, I., Schleiff, E., and Tews, I. (2010). Conserved properties of polypeptide transport-associated (POTRA) domains derived from cyanobacterial Omp85. J. Biol. Chem. *285*, 18016–18024.

Koronakis, V., Sharff, A., Koronakis, E., Luisi, B., and Hughes, C. (2000). Crystal structure of the bacterial membrane protein TolC central to multidrug efflux and protein export. Nature *405*, 914–919.

Koval, S.F., and Murray, R.G. (1984). The isolation of surface array proteins from bacteria. Can. J. Biochem. Cell Biol. *62*, 1181–1189.

Kranzler, C., Lis, H., Shaked, Y., and Keren, N. (2011). The role of reduction in iron uptake processes in a unicellular, planktonic cyanobacterium. Environ. Microbiol. *13*, 2990–2999.

Kruse, O., Hankamer, B., Konczak, C., Gerle, C., Morris, E., Radunz, A., Schmid, G.H., and Barber, J. (2000). Phosphatidylglycerol is involved in the dimerization of photosystem II. J. Biol. Chem. *275*, 6509–6514.

Kufryk, G.I., and Vermaas, W.F.J. (2006). Sll1717 affects the redox state of the plastoquinone pool by modulating quinol oxidase activity in thylakoids. J. Bacteriol. *188*, 1286–1294.

Kunkel, D.D. (1982). Thylakoid Centers: Structures Associated with the Cyanobacterial Photosynthetic Membrane System. Arch. Microbiol. *133*, 97–99.

Labarre, J., Thuriaux, P., and Chauvat, F. (1987). Genetic analysis of amino acid transport in the facultatively heterotrophic cyanobacterium *Synechocystis* sp. strain 6803. J. Bacteriol. *169*, 4668–4673.

Labischinski, H., Goodell, E.W., Goodell, A., and Hochberg, M.L. (1991). Direct proof of a 'more-than-single-layered' peptidoglycan architecture of *Escherichia coli* W7: a neutron small-angle scattering study. J. Bacteriol. *173*, 751–756.

Laloknam, S., Tanaka, K., Buaboocha, T., Waditee, R., Incharoensakdi, A., Hibino, T., Tanaka, Y., and Takabe, T. (2006). Halotolerant cyanobacterium *Aphanothece halophytica* contains a betaine transporter active at alkaline pH and high salinity. Appl. Environ. Microbiol. *72*, 6018–6026.

Lammers, P.J., and Sanders-Loehr, J. (1982). Active transport fo ferric schizokinen in *Anabaena* sp. J. Bacteriol. *151*, 288–294.

Lamont, H.C. (1969). Sacrificial cell death and trichome breakage in an oscillatoriacean Blue-Green alga: the role of murein. Arch. Mikrobiol. *69*, 237–259.

Lang, N.J., and Fay, P. (1971). The heterocysts of blue-green algae II. Details of ultrastructure. Proc. Roy. Soc. Lond. B. *178*, 193–203.

Lázaro, S., Fernández-Piñas, F., Fernández-Valiente, E., Blanco-Rivero, A., and Leganés, F. (2001). pbpB, a gene coding for a putative penicillin-binding protein, is required for aerobic nitrogen fixation in the cyanobacterium *Anabaena* sp. strain PCC7120. J. Bacteriol. *183*, 628–636.

Lazrak, T., Milon, A., Wolff, G., Albrecht, A.M., Miehé, M., Ourisson, G., and Nakatani, Y. (1987). Comparison of the effects of inserted C40- and C50-terminally dihydroxylated carotenoids on the mechanical properties of various phospholipid vesicles. Biochim. Biophys. Acta. *903*, 132–141.

Lechno-Yossef, S., Fan, Q., Ehira, S., Sato, N., and Wolk, C.P. (2006). Mutations in four regulatory genes have interrelated effects on heterocyst maturation in *Anabaena* sp. strain PCC 7120. J. Bacteriol. *188*, 7387–7395.

Leganés, F., Blanco-Rivero, A., Fernández-Piñas, F., Redondo, M., Fernández-Valiente, E., Fan, Q., Lechno-Yossef, S., and Wolk, C.P. (2005). Wide variation in the cyanobacterial complement of presumptive penicillin-binding proteins. Arch. Microbiol. *184*, 234–248.

Lehner, J., Zhang, Y., Berendt, S., Rasse, T.M., Forchhammer, K., and Maldener, I. (2011). The morphogene AmiC2 is pivotal for multicellular development in the cyanobacterium Nostoc punctiforme. Mol. Microbiol. 79, 1655–1669.

Lessard, I.A., Pratt, S.D., McCafferty, D.G., Bussiere, D.E., Hutchins, C., Wanner, B.L., Katz, L., and Walsh, C.T. (1998). Homologs of the vancomycin resistance D-Ala-D-Ala dipeptidase VanX in *Streptomyces toyocaensis*, *Escherichia coli* and *Synechocystis*: attributes of catalytic efficiency, stereoselectivity and regulation with implications for function. Chem. Biol. *5*, 489–504.

Li, T., Yang, H.M., Cui, S.X., Suzuki, I., Zhang, L.F., Li, L., Bo, T.T., Wang, J., Murata, N., and Huang, F. (2012). Proteomic study of the impact of Hik33 mutation in *Synechocystis* sp. PCC 6803 under normal and salt stress conditions. J. Proteome Res. *11*, 502–514.

Lis, H., and Shaked, Y., (2009). Probing the bioavailability of organically bound iron: a case study in the *Synechococcus*-rich waters of the Gulf of Aqaba. Aquatic Microbial. Ecology. *56*, 241–253.

López-Igual, R., Picossi, S., López-Garrido, J., Flores, E., and Herrero, A. (2012). N and C control of ABC-type bicarbonate transporter Cmp and its LysR-type transcriptional regulator CmpR in a heterocyst-forming cyanobacterium, *Anabaena* sp. Environ. Microbiol. *14*, 1035–1048.

Lounatmaa, K., Vaara, T., Osterlund, K., and Vaara, M. (1980). Ultrastructure of the cell wall of a *Synechocystis* strain. Can. J. Microbiol. *26*, 204–208.

Luckey, M., and Nikaido, H. (1980). Specificity of diffusion channels produced by lambda phage receptor protein of *Escherichia coli*. Proc Natl. Acad. Sci. U.S.A. *77*, 167–171.

Lugtenberg, B., Peters, R., Bernheimer, H., and Berendsen, W. (1976). Influence of cultural conditions and mutations on the composition of the outer membrane proteins of *Escherichia coli*. Mol. Gen. Genet. *147*, 251–262.

Maeda, S., and Omata, T. (2009). Nitrite transport activity of the ABC-type cyanate transporter of the cyanobacterium *Synechococcus elongatus*. J. Bacteriol. *191*, 3265–3272.

Malakhov, M.P., Los, D.A., Wada, H., Semenenko, V.E., and Murata, N. (1995). Characterization of the murF gene of the cyanobacterium *Synechocystis* sp. PCC 6803. Microbiology *141*, 163–169.

Malinverni, J.C., and Silhavy, T.J. (2009). An ABC transport system that maintains lipid asymmetry in the Gram-negative outer membrane. Proc. Natl. Acad. Sci. U.S.A. *106*, 8009–8014.

Marbouty, M., Mazouni, K., Saguez, C., Cassier-Chauvat, C., and Chauvat, F. (2009). Characterization of the *Synechocystis* strain PCC 6803 penicillin-binding proteins and cytokinetic proteins FtsQ and FtsW and their network of interactions with ZipN. J. Bacteriol. *191*, 5123–5133.

Marin, K., Suzuki, I., Yamaguchi, K., Ribbeck, K., Yamamoto, H., Kanesaki, Y., Hagemann, M., and Murata, N. (2003). Identification of histidine kinases that act as sensors in the perception of salt stress in *Synechocystis* sp. PCC 6803. Proc. Natl. Acad. Sci. U.S.A. *100*, 9061–9066.

Mariscal, V., Herrero, A., and Flores, E. (2007). Continuous periplasm in a filamentous, heterocyst-forming cyanobacterium. Mol. Microbiol. *65*, 1139–1145.

Mariscal, V., Herrero, A., Nenninger, A., Mullineaux. C.W., and Flores, E. (2011). Functional dissection of the three-domain SepJ protein joining the cells in cyanobacterial trichomes. Mol. Microbiol. *79*, 1077–1088.

Martin, H.H., and Frank, H. (1962). Quantitative Bausteinanalyse der Stützmembran in der Zellwand von *E. coli* B. Z. Naturforsch. *17b*, 190–196.

Masi, M., Duret, G., Delcour, A.H., and Misra, R. (2009). Folding and trimerization of signal sequence-less mature TolC in the cytoplasm of *Escherichia coli*. Microbiology *155*, 1847–1857.

Matias, V.R., Al-Amoudi, A., Dubochet, J., and Beveridge, T.J. (2003). Cryo-transmission electron microscopy of frozen-hydrated sections of Escherichia coli and *Pseudomonas aeruginosa*. J. Bacteriol. *185*, 6112–6118.

Matsuda, N., and Uozumi, N. (2006). Ktr-mediated potassium transport, a major pathway for potassium uptake, is coupled to a proton gradient across the membrane in *Synechocystis* sp. PCC 6803. Biosci. Biotechnol. Biochem. *70*, 273–275.

McCarren, J., and Brahamsha, B. (2009). Swimming motility mutants of marine *Synechococcus* affected in production and localization of the S-layer protein SwmA. J. Bacteriol. *191*, 1111–1114.

McKersie, B.D., Hoekstra, F.A., and Krieg, L.C. (1990). Differences in the susceptibility of plant membrane lipids to peroxidation. Biochim. Biophys. Acta. *1030*, 119–126.

McKnight, M.D., and Morel, F.M.M. (1980). Copper complexation by siderophores from filamentous blue-green algae. Limnol. Oceanog. *25*, 62–71.

van de Meene, A.M., Hohmann-Marriott, M.F., Vermaas, W.F., and Roberson, R.W. (2006). The three-dimensional structure of the cyanobacterium *Synechocystis* sp. PCC 6803. Arch. Microbiol. *184*, 259–270.

Merchán F., Kindle, K.L., Llama, M.J., Serra, J.L., and Fernández, E. (1995). Cloning and sequencing of the nitrate transport system from the thermophilic, filamentous cyanobacterium *Phormidium laminosum*: comparative analysis with the homologous system from Synechococcus sp, PCC 7942. Plant Mol. Biol. *28*, 759–766.

Merino-Puerto, V., Mariscal, V., Mullineaux, C.W., Herrero, A., and Flores, E. (2010). Fra proteins influencing filament integrity, diazotrophy and localization of septal protein SepJ in the heterocyst-forming cyanobacterium *Anabaena* sp. Mol. Microbiol. *75*, 1159–1170.

Merino-Puerto, V., Schwarz, H., Maldener, I., Mariscal, V., Mullineaux, C.W., Herrero, A., and Flores, E. (2011). FraC/FraD-dependent intercellular molecular exchange in the filaments of a heterocyst-forming cyanobacterium, Anabaena sp. Mol. Microbiol. *82*, 87–98.

Messner, P., and Sleytr, U.B. (1992). Crystalline bacterial cell-surface layers. Adv. Microb. Physiol. *33*, 213–275.

Mikkat, S., and Hagemann, M. (2000). Molecular analysis of the ggtBCD gene cluster of *Synechocystis* sp. strain PCC6803 encoding subunits of an ABC transporter for osmoprotective compounds. Arch. Microbiol. *174*, 273–282.

Mironov, K.S., Sidorov, R.A., Trofimova, M.S., Bedbenov, V.S., Tsydendambaev, V.D., Allakhverdiev, S.I., and Los, D.A. (2012). Light-dependent cold-induced fatty acid unsaturation, changes in membrane fluidity, and alterations in gene expression in *Synechocystis*. Biochim. Biophys. Acta *1817*, 1352–1359.

Mirus, O., Strauss, S., Nicolaisen, K., von Haeseler, A., and Schleiff, E. (2009). TonB-dependent transporters and their occurrence in cyanobacteria. BMC Biol. *7*, 68.

Mirus, O., Hahn, A., and Schleiff, E. (2010). Outer membrane proteins. In Prokaryotic Cell Wall Compounds: Structure and Biochemistry, König, H., Claus, H., and Varma, A., eds. (Springer Verlag), pp. 175–230.

Montesinos, M.L., Herrero, A., and Flores, E. (1995). Amino acid transport systems required for diazotrophic growth in the cyanobacterium Anabaena sp. strain PCC 7120. J. Bacteriol. *177*, 3150–3157.

Montesinos, M.L., Herrero, A., and Flores, E. (1997). Amino acid transport in taxonomically diverse cyanobacteria and identification of two genes encoding elements of a neutral amino acid permease putatively involved in recapture of leaked hydrophobic amino acids. J. Bacteriol. *179*, 853–862.

Montesinos, M.L., Muro-Pastor, A.M., Herrero, A., and Flores, E. (1998). Ammonium/methylammonium permeases of a Cyanobacterium. Identification and analysis of three nitrogen-regulated amt genes in *Synechocystis* sp. PCC 6803. J. Biol. Chem. *273*, 31463–31470.

Morrissey, J., and Guerinot, M.L. (2009). Iron uptake and transport in plants: the good, the bad, and the ionome. Chem. Rev. *109*, 4553–4567.

Moslavac, S., Bredemeier, R., Mirus, O., Granvogl, B., Eichacker, L.A., and Schleiff, E. (2005). Proteomic analysis of the outer membrane of Anabaena sp. strain PCC 7120. J. Proteome Res. *4*, 1330–1338.

Moslavac, S., Reisinger, V., Berg, M., Mirus, O., Vosyka, O., Plöscher, M., Flores, E., Eichacker, L.A., and Schleiff, E. (2007a). The proteome of the heterocyst cell wall in Anabaena sp. PCC 7120. Biol. Chem. *388*, 823–829.

Moslavac, S., Nicolaisen, K., Mirus, O., Al Dehni, F., Pernil, R., Flores, E., Maldener, I., and Schleiff, E. (2007b). A TolC-like protein is required for heterocyst development in Anabaena sp. strain PCC 7120. J. Bacteriol. *189*, 7887–7895.

Mullineaux, C.W., Mariscal, V., Nenninger, A., Khanum, H., Herrero, A., Flores, E., and Adams, D.G. (2008). Mechanism of intercellular molecular exchange in heterocystforming cyanobacteria. EMBO J. *27*, 1299–1308.

Muñoz-Martín, M.A., Mateo, P., Leganés, F., and Fernández-Piñas, F. (2011). Novel cyanobacterial bioreporters of phosphorus bioavailability based on alkaline phosphatase and phosphate transporter genes of Anabaena sp. PCC 7120. Anal. Bioanal. Chem. *400*, 3573–3584.

Murakami, S., Nakashima, R., Yamashita, E., and Yamaguchi, A. (2002). Crystal structure of bacterial multidrug efflux transporter AcrB. Nature *419*, 587–593.

Murakami, S., Nakashima, R., Yamashita, E., Matsumoto, T., and Yamaguchi, A. (2006). Crystal structures of a multidrug transporter reveal a functionally rotating mechanism. Nature *443*, 173–179.

Murata, N. (1989). Low-temperature effects on cyanobacterial membranes. J. Bioenergetics Biomem. *21*, 61–75.

Murata, N., Sato, N., Omata, T., and Kuwabara, T. (1981). Separation and characterization of thylakoid and cell envelope of the blue-green alga (cyanobacterium) *Anacystis nidulans*. Plant Cell. Physiol. *22*, 855–866.

Murphy, T.P., Lean, D.R., and Nalewajko, C. (1976). Blue-green algae: their excretion of iron-selective chelators enables them to dominate other algae. Science *192*, 900–902.

Nagel de Zwaig, R., and Luria, S.E. (1967). Genetics and physiology of colicin-tolerant mutants of *Escherichia coli*. J Bacteriol. *94*, 1112–1123.

Nakashima, R., Sakurai, K., Yamasaki, S., Nishino, K., and Yamaguchi, A. (2011). Structures of the multidrug exporter AcrB reveal a proximal multisite drug-binding pocket. Nature *480*, 565–569.

Nakasugi, K., Alexova, R., Svenson, C.J., and Neilan, B.A. (2007). Functional analysis of PilT from the toxic cyanobacterium *Microcystis aeruginosa* PCC 7806. J. Bacteriol. *189*, 1689–1697.

Nan, B., Chen, J., Neu, J.C., Berry, R.M., Oster, G., and Zusman, D.R. (2011). Myxobacteria gliding motility requires cytoskeleton rotation powered by proton motive force. Proc. Natl. Acad. Sci. U.S.A. *108*, 2498–2503.

Nayar, A.S., Yamaura, H., Rajagopalan, R., Risser, D.D., and Callahan, S.M. (2007). FraG is necessary for filament integrity and heterocyst maturation in the cyanobacterium Anabaena sp. strain PCC 7120. Microbiology *153*, 601–603.

Nestorovich, E., Sugawara, E., Nikaido, H., and Bezrukov, S.M. (2006). *Pseudomonas aeruginosa* Porin OprF: properties of the channel. J. Biol. Chem. *281*, 16230–16237.

Nichols, B.W., Harris, R.V., and James, A.T. (1965). The lipid metabolism of blue-green algae. Biochem. Biophys. Res. Commun. *20*, 256–262.

Nicolaisen, K., and Schleiff, E. (2010). Iron dependency of and transport by cyanobacteria. In Iron Uptake in Microorganisms. Andrews, S., and Cornelis, P., eds. (Caister Academic Press, Norfolk, UK), pp. 203–229.

Nicolaisen, K., Moslavac, S., Samborski, A., Valdebenito, M., Hantke, K., Maldener, I., Muro-Pastor, A.M., Flores. E., and Schleiff, E. (2008). Alr0397 is an outer membrane transporter for the siderophore schizokinen in Anabaena sp. strain PCC 7120. J. Bacteriol. *190*, 7500–7507.

Nicolaisen, K., Hahn, A., and Schleiff, E. (2009a). The cell wall in heterocyst formation by Anabaena sp. PCC 7120. J. Basic. Microbiol. *49*, 5–24.

Nicolaisen, K., Mariscal, V., Bredemeier, R., Pernil, R., Moslavac, S., López-Igual, R., Maldener, I., Herrero, A., Schleiff, E., and Flores, E. (2009b). The outer membrane of a heterocyst-forming cyanobacterium is a permeability barrier for uptake of metabolites that are exchanged between cells. Mol. Microbiol. *74*, 58–70.

Nicolaisen, K., Hahn, A., Valdebenito, M., Moslavac, S., Samborski, A., Maldener, I., Wilken, C., Valladares, A., Flores, E., Hantke, K., and Schleiff, E. (2010).The interplay between siderophore secretion and coupled iron and copper transport in the heterocyst-forming cyanobacterium Anabaena sp. PCC 7120. Biochim. Biophys. Acta *1798*, 2131–2140.

Nikaido, H. (2003). Molecular basis of bacterial outer membrane permeability revisited. Microbiol. Mol. Biol. Rev. *67*, 593–656.

Nikaido, H., and Vaara, M. (1985). Molecular basis of bacterial outer membrane permeability. Microbiol. Rev. *49*, 1–32.

Nishida, I., and Murata, N. (1996). Chilling sensitivity in plants and cyanobacteria: The crucial contribution of membrane lipids. Annu. Rev. Plant Physiol. Plant Mol. Biol. *47*, 541–568.

Noreen, M., Shah, M.A., Mall, S.M., Choudhary, S., Hussain, T., Ahmed, I., Jalil, S.F., and Raza, M.I. (2012). TLR4 polymorphisms and disease susceptibility. Inflamm. Res. *61*, 177–188.

Olabarria, G., Carrascosa, J.L., de Pedro, M.A., and Berenguer, J. (1996). A conserved motif in S-layer proteins is involved in peptidoglycan binding in *Thermus thermophilus*. J. Bacteriol. *178*, 4765–4772.

Omata, T., and Murata, N. (1983). Isolation and characterization of the cytoplasmic membranes from the blue-green alga (cyanobacterium) *Anacystis nidulans*. Plant Cell Physiol. *24*, 1101–1112.

Omata, T., and Murata, N. (1984). Isolation and characterization of three types of membranes from the cyanobacterium (blue-green alga) *Synechocystis* PCC 6714. Arch. Microbiol. *139*, 113–116.

Omata., T., Price, G.D., Badger, M.R., Okamura, M., Gohta, S., and Ogawa, T. (1999). Identification of an ATP-binding cassette transporter involved in bicarbonate uptake in the cyanobacterium *Synechococcus* sp. strain PCC 7942. Proc. Natl. Acad. Sci. U.S.A. *96*, 13571–13576.

Omata, T., Gohta, S., Takahashi, Y., Harano, Y., and Maeda, S. (2001). Involvement of a CbbR homolog in low CO_2- induced activation of the bicarbonate transporter operon in cyanobacteria. J. Bacteriol. *183*, 1891–1898.

Pagès, J.M., Masi, M., and Barbe, J. (2005). Inhibitors of efflux pumps in Gram-negative bacteria. Trends Mol. Med. *11*, 382–389.

Paithoonrangsarid, K., Shoumskaya, M.A., Kanesaki, Y., Satoh, S., Tabata, S., Los, D.A., Zinchenko, V.V., Hayashi, H., Tanticharoen, M., Suzuki, I., and Murata, N. (2004). Five histidine kinases perceive osmotic stress and regulate distinct sets of genes in *Synechocystis*. J. Biol. Chem. *279*, 53078–53086.

Pautsch, A., and Schulz, G.E. (1998). Structure of the outer membrane protein A transmembrane domain. Nat. Struct. Biol. *5*, 1013–1007.

Pautsch, A., and Schulz, G.E. (2000). High-resolution structure of the OmpA membrane domain. J. Mol. Biol. *298*, 273–282.

Paz-Yepes, J., Herrero, A., and Flores, E. (2007). The NtcA-regulated amtB gene is necessary for full methylammonium uptake activity in the cyanobacterium *Synechococcus elongatus*. J. Bacteriol. *189*, 7791–7798.

Paz-Yepes, J., Merino-Puerto, V., Herrero, A., and Flores, E. (2008). The amt gene cluster of the heterocyst-forming cyanobacterium Anabaena sp. strain PCC 7120. J. Bacteriol. *190*, 6534–6539.

Pearson, L.A., Hisbergues, M., Börner, T., Dittmann, E., and Neilan, B.A. (2004). Inactivation of an ABC transporter gene, mcyH, results in loss of microcystin production in the cyanobacterium *Microcystis aeruginosa* PCC 7806. Appl. Environ. Microbiol. *70*, 6370–6378.

Pernil, R., Picossi, S., Mariscal, V., Herrero, A., and Flores, E. (2008). ABC-type amino acid uptake transporters Bgt and N-II of Anabaena sp. strain PCC 7120 share an ATPase subunit and are expressed in vegetative cells and heterocysts. Mol. Microbiol. *67*, 1067–1080.

Pernil, R., Herrero, A., and Flores, E. (2010). A TRAP transporter for pyruvate and other monocarboxylate 2-oxoacids in the cyanobacterium Anabaena sp. strain PCC 7120. J. Bacteriol. *192*, 6089–6092.

Peschek, G.A., Wastyn, M., Trnka, M., Molitor, V., Fry, I.V., and Packer, L. (1989). Characterization of the cytochrome c oxidase in isolated and purified plasma membranes from the cyanobacterium *Anacystis nidulans*. Biochemistry *28*, 3057–3063.

Peschek, G.A., Obinger, C., Fromwald, S., and Bergman., B. (1994). Correlation between immuno-gold labels and activities of the cytochrome-c oxidase (aa3-type) in membranes of salt stressed cyanobactria. FEMS Microbiol. Lett. *124*, 431–437.

Pfeiffer, K., Gohil, V., Stuart, R.A., Hunte, C., Brandt, U., Greenberg, M.L., and Schägger, H. (2003). Cardiolipin stabilizes respiratory chain supercomplexes. J. Biol. Chem. *278*, 52873–52880.

Picollo, A., Malvezzi, M., Houtman, J.C., and Accardi, A. (2009). Basis of substrate binding and conservation of selectivity in the CLC family of channels and transporters. Nature Struct. Mol. Biol. *16*, 1294–1301.

Picossi, S., Montesinos, M.L., Pernil, R., Lichtlé, C., Herrero, A., and Flores, E. (2005). ABC-type neutral amino acid permease N-I is required for optimal diazotrophic growth and is repressed in the heterocysts of *Anabaena* sp. strain PCC 7120. Mol. Microbiol. *57*, 1582–1592.

Piddock, L.J. (2006). Multidrug-resistance efflux pumps – not just for resistance. Nat. Rev. Microbiol. *4*, 629–636.

Pils, D., and Schmetterer, G. (2001). Characterization of three bioenergetically active respiratory terminal oxidases in the cyanobacterium *Synechocystis* sp. strain PCC 6803. FEMS Microbiol. Lett. *203*, 217–222.

Pisareva, T., Shumskaya, M., Maddalo, G., Ilag, L., and Norling, B. (2007). Proteomics of *Synechocystis* sp. PCC 6803. Identification of novel integral plasma membrane proteins. FEBS J. *274*, 791–804.

Pisareva, T., Kwon, J., Oh, J., Kim, S., Ge, C., Wieslander, A., Choi, J.S., and Norling, B. (2011). Model for membrane organization and protein sorting in the cyanobacterium *Synechocystis* sp. PCC 6803 inferred from proteomics and multivariate sequence analyses. J. Proteome Res. *10*, 3617–3631.

Pitt, F.D., Mazard, S., Humphreys, L., and Scanlan, D.J. (2010). Functional characterization of *Synechocystis* sp. strain PCC 6803 pst1 and pst2 gene clusters reveals a novel strategy for phosphate uptake in a freshwater cyanobacterium. J. Bacteriol. *192*, 3512–3523.

Plumbridge, J.A. (1991). Repression and induction of the nag regulon of *Escherichia coli* K-12: the roles of nagC and nagA in maintenance of the uninduced state. Mol. Microbiol. *5*, 2053–2062.

Poole, K., Heinrichs, D.E., and Neshat, S. (1993). Cloning and sequence analysis of an EnvCD homologue in *Pseudomonas aeruginosa*: regulation by iron and possible involvement in the secretion of the siderophore pyoverdine. Mol. Microbiol. *10*, 529–544.

Pratte, B.S., and Thiel, T. (2006). High-affinity vanadate transport system in the cyanobacterium *Anabaena variabilis* ATCC 29413. J. Bacteriol. *188*, 464–468.

Prosecka. J., Orlov, A.V., Fantin, Y.S., Zinchenko, V.V., Babykin, M.M., and Tichy, M. (2009). A novel ATP-binding cassette transporter is responsible for resistance to viologen herbicides in the cyanobacterium *Synechocystis* sp. PCC 6803. FEBS J. *276*, 4001–4011.

Puustinen, A., Finel, M., Haltia, T., Gennis, R.B., and Wikström, M. (1991). Properties of the two terminal oxidases of *Escherichia coli*. Biochem. *30*, 3936–3942.

Quintero, M.J., Montesinos, M.L., Herrero, A., and Flores, E. (2001). Identification of genes encoding amino acid permeases by inactivation of selected ORFs from the *Synechocystis* genomic sequence. Genome. Res. *11*, 2034–2040.

Rabus, R., Jack, D.L., Kelly, D.J., and Saier, M.H., Jr. (1999). TRAP transporters: An ancient family of extracytoplasmatic solute-receptor-dependent secondary active transporters. Microbiology *145*, 3431–3445.

Raetz, C.R.H., Reynolds, C.M., Trent, M.S., and Bishop, R.E. (2007). Lipid A modification systems in Gram-negative bacteria. Annu. Rev. Biochem. *76*, 295–329.

Rajalahti, T., Huang, F., Klement, M.R., Pisareva, T., Edman, M., Sjöström, M., Wieslander, A., and Norling, B. (2007). Proteins in different *Synechocystis* compartments have distinguishing N-terminal features: a combined proteomics and multivariate sequence analysis. J. Proteome Res. *6*, 2420–2434.

Rawling, E.G., Brinkman, F.S., and Hancock, R.E. (1998). Roles of the carboxy-terminal half of *Pseudomonas aeruginosa* major outer membrane protein OprF in cell shape, growth in low-osmolarity medium, and peptidoglycan association. J. Bacteriol. *180*, 3556–3562.

Re, F., and Strominger, J.L. (2002). Monomeric recombinant MD-2 binds toll-like receptor 4 tightly and confers lipopolysaccharide responsiveness. J. Biol. Chem. *277*, 23427–23432.

Ren, Q., Chen, K., and Paulsen, I.T. (2007). TransportDB: a comprehensive database resource for cytoplasmic membrane transport systems and outer membrane channels. Nucleic. Acids. Res. *35*, D274-D279.

Rensing, C., and Grass, G. (2003). *Escherichia coli* mechanisms of copper homeostasis in a changing environment. FEMS Microbiol. Rev. *27*, 197–213.

Reumann, S., Davila-Aponte, J., and Keegstra, K. (1999). The evolutionary origin of the protein-translocating channel of chloroplastic envelope membranes: identification of a cyanobacterial homolog. Proc. Natl. Acad. Sci. U.S.A. *96*, 784–789.

Rexroth, S., Mullineaux, C.W., Ellinger, D., Sendtko, E., Rögner, M., and Koenig, F. (2011). The plasma membrane of the cyanobacterium *Gloeobacter violaceus* contains segregated bioenergetic domains. Plant Cell *23*, 2379–2390.

Rezanka, T., Víden, I., Go, J.V., Dor, I., and Dembitsky, V.M. (2003). Polar lipids and fatty acids of three wild cyanobacterial strains of the genus Chroococcidiopsis. Folia. Microbiol. *48*, 781–786.

Rietschel, E.T., Kirikae, T., Schade, F.U., Mamat, U., Schmidt, G., Loppnow, H., Ulmer, A.J., Zähringer, U., Seydel, U., Di Padova, F., *et al.* (1994). Bacterial endotoxin: molecular relationships of structure to activity and function. FASEB J. *8*, 217–225.

Rippka. R., Deruelles, J., Waterbury, J.B., Herdman, M., and Stanier, R.Y. (1979). Generic assignments, strain histories and properties of pure cultures of cyanobacteria. J. Gen. Microbiol. *111*, 1–61.

Ris, H., and Singh, R.N. (1961). Electron microscope studies on blue-green algae. J. Biophys. Biochem. Cytol. *9*, 63–80.

Rose, A.L., and Waite, T.D. (2005). Reduction of organically complexed ferric iron by superoxide in simulated natural water. Environ. Sci. Technol. *39*, 2645–2650.

Ruiz, N., Kahne, D., and Silhavy, T.J. (2009). Transport of lipopolysaccharide across the cell envelope: the long road of discovery. Nat. Rev. Microbiol. *7*, 677–683.

Sakamoto, T., and Murata, N. (2002). Regulation of the desaturation of fatty acids and its role in tolerance to cold and salt stress. Curr. Opin. Microbiol. *5*, 208–210.

Saravolac, E.G., Taylor, N.F., Benz, R., and Hancock, R.E. (1991). Purification of glucose-inducible outer membrane protein OprB of *Pseudomonas putida* and reconstitution of glucose-specific pores. J. Bacteriol. *173*, 4970–4976.

Sato, N., and Murata, N. (1980). Temperature shift-unduces responses in lipids in the blue-green alga, *Anabaena variabilis*, the central role of diacylmonogalactosylglycerol in thermo-adaptation. Biochim. Biophys. Acta *619*, 353–366.

Sato, N., and Murata, N. (1982). Lipid biosynthesis in the blue-green alga, *Anabaena variabilis*. I. Lipid class. Biochim. Biophys. Acta *710*, 271–278.

Sato, N., Hagio, M., Wada, H., and Tsuzuki, M. (2002). Requirement of phosphatidylglycerol for photosynthetic function in thylakoid membranes. Proc. Natl. Acad. Sci. U.S.A. *97*, 10655–10660.

Schauer, K., Rodionov, D.A., and de Reuse, H. (2008). New substrates for TonB-dependent transport: do we only see the 'tip of the iceberg'? Trends. Biochem. Sci. *33*, 330–338.

Schirmer, T. (1998). General and specific porins from bacterial outer membranes. J. Struct. Biol. *121*, 101–109.

Schirmer, T., Keller, T.A., Wang, Y.F., and Rosenbusch, J.P. (1995). Structural basis for sugar translocation through maltoporin channels at 3.1 Å resolution. Science *267*, 512–514.

Schleifer, K.H., and Kandler, O. (1972). Peptidoglycan types of bacterial cell walls and their taxonomic implications. Bacteriol. Rev. *36*, 407–477.

Schleiff, E., and Becker, T. (2011). Common ground for protein translocation: access control for mitochondria and chloroplasts. Nat. Rev. Mol. Cell. Biol. *12*, 48–59.

Schleiff, E., Maier, U.G., and Becker, T. (2011). Omp85 in eukaryotic systems: one protein family with distinct functions. Biol. Chem. *392*, 21–27.

Schletter, J., Heine, H., Ulmer, A.J., and Rietschel, E.T. (1995). Molecular mechanisms of endotoxin activity. Arch. Microbiol. *164*, 383–389.

Schmidt, W., Drews, G., Weckesser, J., Fromme, I., and Borowiak, D. (1980a). Characterization of the lipopolysaccharides from eight strains of the cyanobacterium *Synechococcus*. Arch. Microbiol. *127*, 209–215.

Schmidt, W., Drews, G., Weckesser, J., and Mayer, H. (1980b). Lipopolysaccharides in four strains of the unicellular cyanobacterium *Synechocystis*. Arch. Microbiol. *127*, 217–222.

Schromm, A.B., Brandenburg, K., Loppnow, H., Zähringer, U., Rietschel, E.T., Carroll, S.F., Koch, M.H., Kusumoto, S., and Seydel, U. (1998). The charge of endotoxin molecules influences their conformation and IL-6-inducing capacity. J. Immunol. *161*, 5464–5471.

Schottkowski, M., Gkalympoudis, S., Tzekova, N., Stelljes, C., Schuenemann, D., Ankele, E., and Nickelsen, J. (2009). Interaction of the periplasmic PratA factor and the PsbA (D1) protein during biogenesis of photosystem II in *Synechocystis* sp. PCC 6803. J. Biol. Chem. *284*, 1813–1819.

Schrader, M., Drews, G., and Weckesser, J. (1981). Chemical analyses on cell wall constituents of the thermophilic cyanobacterium Synechococcus PCC6716. FEMS Microbiol. Lett. *11*, 37–40.

Schultze, M., Forberich, B., Rexroth, S., Dyczmons, N.G., Roegner, M., and Appel, J. (2009). Localization of cytochrome b6f complexes implies an incomplete respiratory chain in cytoplasmic membranes of the cyanobacterium *Synechocystis* sp. PCC 6803. Biochim. Biophys. Acta. *1787*, 1479–1485.

Schulz, G.E. (2002). The structure of bacterial outer membrane proteins. Biochim. Biophys. Acta *1565*, 308–317.

Seeger, M.A., Schiefner, A., Eicher, T., Verrey, F., Diederichs, K., and Pos, K.M. (2006). Structural asymmetry of AcrB trimer suggests a peristaltic pump mechanism. Science *313*, 1295–1298.

Sharom, F.J. (2011). Flipping and flopping – lipids on the move. IUBMB Life *63*, 736–746.

Shi, L., Li, J.H., Cheng, Y., Wang, L., Chen, W.L., and Zhang, C.C., (2007). Two genes encoding protein kinases of the HstK family are involved in synthesis of the minor heterocyst-specific glycolipid in the cyanobacterium *Anabaena* sp. strain PCC 7120. J. Bacteriol. *189*, 5075–5081.

Shimura, Y., Shiraiwa, Y., and Suzuki, I. (2012). Characterization of the subdomains in the N-terminal region of histidine kinase Hik33 in the cyanobacterium *Synechocystis* sp. PCC 6803. Plant Cell Physiol. *53*, 1255–1266.

Shoumskaya, M.A., Paithoonrangsarid, K., Kanesaki, Y., Los, D.A., Zinchenko, V.V., Tanticharoen, M., Suzuki, I., and Murata, N. (2005). Identical Hik-Rre systems are involved in perception and transduction of salt signals and hyperosmotic signals but regulate the expression of individual genes to different extents in *Synechocystis*. J. Biol. Chem. *280*, 21531–21538.

Simpson, F.B., and Neiland, J.B. (1976). *Siderochromesin cyanophyceae*: Isolation and characterization of schizokinen from *Anabaena* sp. J. Phycol. *12*, 44–48.

Sklar, J.G., Wu, T., Kahne, D., and Silhavy, T.J. (2007). Defining the roles of the periplasmic chaperones SurA, Skp, and DegP in *Escherichia coli*. Genes Dev. *21*, 2473–2484.

Sleytr, U.B., Sára, M., Messner, P., and Pum, D. (1994). Two-dimensional protein crystals (S-layers): fundamentals and applications. J. Cell Biochem. *56*, 171–176.

Smarda, J., Smajs, D., Komrska, J., and Krzyzánek, V. (2002). S-layers on cell walls of cyanobacteria. Micron *33*, 257–277.

Snyder, D.S., Brahamsha, B., Azadi, P., and Palenik, B. (2009). Structure of compositionally simple lipopolysaccharide from marine Synechococcus. J. Bacteriol. *191*, 5499–5509.

Sozer, O., Kis, M., Gombos, Z., and Ughy, B. (2011). Proteins, glycerolipids and carotenoids in the functional photosystem II architecture. Front. Biosci. *16*, 619–643.

Sperandeo, P., Dehò, G., and Polissi, A. (2009). The lipopolysaccharide transport system of Gram-negative bacteria. Biochim. Biophys. Acta *1791*, 594–602.

Staron, P., Forchhammer, K., and Maldener, I. (2011). Novel ATP-driven pathway of glycolipid export involving TolC protein. J. Biol. Chem. *286*, 38202–38210.

Stengel, A., Gügel, I.L., Hilger, D., Rengstl, B., Jung, H., and Nickelsen, J. (2012), Initial steps of photosystem II de novo assembly and preloading with manganese take place in biogenesis centers in *Synechocystis*. Plant Cell *24*, 660–675.

Stevanovic, M., Hahn, A., Nicolaisen, K., Mirus, O., and Schleiff, E. (2012). The components of the putative iron transport system in the cyanobacterium Anabaena sp. PCC 7120. Environ. Microbiol. *14*, 1655–1670.

Stewart, I., Schluter, P.J., and Shaw, G.R. (2006). Cyanobacterial lipopolysaccharides and human health – a review. Environ. Health. *5*, 7.

Stirnberg, M., Fulda, S., Huckauf, J., Hagemann, M., Krämer, R., and Marin, K. (2007). A membrane-bound FtsH protease is involved in osmoregulation in *Synechocystis* sp. PCC 6803: the compatible solute synthesizing enzyme GgpS is one of the targets for proteolysis. Mol. Microbiol. *63*, 86–102.

Stock, A.M., Robinson, V.L., and Goudreau, P.N. (2000). Two-component signal transduction. Annu. Rev. Biochem. *69*, 183–215.

Stone. B.J., Boussiba, S., and Gibson, J. (1988). Penicillin-binding proteins in the unicellular cyanobacterium

Synechococcus sp. strain R2 (PCC 7942): implications for membrane identification. J. General. Microbiol. *134*, 2951–2958.

Sugawara, E., and Nikaido, H. (1992). Pore-forming activity of OmpA protein of *Escherichia coli*. J. Biol. Chem. *267*, 2507–2511.

Sugawara, E., Nagano, K., and Nikaido, H. (2012). Alternative folding pathways of the major porin OprF of *Pseudomonas aeruginosa*. FEBS J. *279*, 910–918.

Sunda, W., Price, N.M., and Morel, F.M.M. (2005). Trace metal ion buffers and their use in culture studies. In Algal Culturing Techniques. Andersen, R., ed. (Burlington, MA: Academic Press), pp. 35–63.

Suzuki, I., Los, D.A., Kanesaki, Y., Mikami, K., and Murata, N. (2000). The pathway for perception and transduction of low-temperature signals in *Synechocystis*. EMBO J. *19*, 1327–1334.

Suzuki, M., Hara, H., and Matsumoto, K. (2002). Envelope disorder of *Escherichia coli* cells lacking phosphatidylglycerol. J. Bacteriol. *184*, 5418–5425.

Tahara, H., Uchiyama, J., Yoshihara, T., Matsumoto, K., and Ohta, H. (2012). Role of Slr1045 in environmental stress tolerance and lipid transport in the cyanobacterium *Synechocystis* sp. PCC6803. Biochim. Biophys. Acta *1817*, 1360–1366.

Takayama, K., Rothenberg, R.J., and Barbour, A.G. (1987). Absence of lipopolysaccharide in the Lyme disease spirochete, *Borrelia burgdorferi*. Infect. Immun. *55*, 2311–2313.

Teuber, M., Rögner, M., and Berry, S. (2001). Fluorescent probes for non-invasive bioenergetic studies of whole cyanobacterial cells. Biochim *1506*, 31–46.

Thomas, W.R., Callow, M.G., Dilworth, R.J., and Audesho, A.A. (1990). Expression in *Escherichia coli* of a high-molecular-weight protective surface antigen found in nontypeable and type b *Haemophilus influenzae*. Infect. Immun. *58*, 1909–1913.

Tommassen, J., and Lugtenberg, B. (1982). PHO-regulon of *Escherichia coli* K12: a minireview. Ann. Microbiol. *133*, 243–249.

Tripp, J., Hahn, A., Koenig, P., Flinner, N., Bublak, D., Brouwer, E.M., Ertel, F., Mirus, O., Sinning, I., Tews, I., and Schleiff, E. (2012). Structure and conservation of the periplasmic targeting factor Tic22 protein from plants and cyanobacteria. J. Biol. Chem. *287*, 24164–24173.

Typas, A., Banzhaf, M., Gross, C.A., and Vollmer, W. (2012). From the regulation of peptidoglycan synthesis to bacterial growth and morphology. Nature Rev. Microbiol. *10*, 123–136.

Uehara, T., Suefuji, K., Jaeger, T., Mayer, C., and Park, J.T. (2006). MurQ etherase is required by *Escherichia coli* in order to metabolize anhydro-n-acetylmuramic acid obtained either from the environment or from its own cell wall. J. Bacteriol. *188*, 1660–1662.

Ulmke, C., Kreth, J., Lengeler, J.W., Welte, W., and Schmid, K. (1999). Site-directed mutagenesis of loop L3 of sucrose porin ScrY leads to changes in substrate selectivity. J. Bacteriol. *181*, 1920–1923.

Umeda, H., Aiba, H., and Mizuno, T. (1996). SomA, a novel gene that encodes a major outer-membrane protein of *Synechococcus* sp. PCC 7942. Microbiology *142*, 2121–2128.

Vaara, T. (1982). The outermost surface structures in chroococcacean cyanobacteria. Can. J. Microbiol. *28*, 929–941.

Vaccaro, L., Scott, K.A., and Sansom, M.S. (2008). Gating at both ends and breathing in the middle: conformational dynamics of TolC. Biophys. J. *95*, 5681–5691.

Valladares, A., Montesinos, M.L., Herrero, A., and Flores, E. (2002). An ABC-type, high-affinity urea permease identified in cyanobacteria. Mol. Microbiol. *43*, 703–715.

Van Gelder, P., Dutzler, R., Dumas, F., Koebnik, R., and Schirmer, T. (2001). Sucrose transport through maltoporin mutants of *Escherichia coli*. Protein Eng. *14*, 943–948.

Vázquez-Bermúdez, M.F., Paz-Yepes, J., Herrero, A., and Flores, E. (2002). The NtcA-activated amt1 gene encodes a permease required for uptake of low concentrations of ammonium in the cyanobacterium Synechococcus sp. PCC 7942. Microbiology *148*, 861–869.

Vigh, L., Los, D.A., Horváth, I., and Murata, N. (1993). The primary signal in the biological perception of temperature: Pd-catalyzed hydrogenation of membrane lipids stimulated the expression of the desA gene in *Synechocystis* PCC6803. Proc. Natl. Acad. Sci. U.S.A. *90*, 9090–9094.

Voulhoux, R., Bos, M.P., Geurtsen, J., Mols, M., and Tommassen, J. (2003). Role of a highly conserved bacterial protein in outer membrane protein assembly. Science *299*, 262–265.

Waditee, R., Hibino, T., Tanaka, Y., Nakamura, T., Incharoensakdi, A., and Takabe, T. (2001). Halotolerant cyanobacterium *Aphanothece halophytica* contains an Na(+)/H(+) antiporter, homologous to eukaryotic ones, with novel ion specificity affected by C-terminal tail. J. Biol. Chem. *276*, 36931–36938.

Waditee, R., Hibino, T., Nakamura, T., Incharoensakdi, A., and Takabe, T. (2002). Overexpression of a Na+/H+ antiporter confers salt tolerance on a freshwater cyanobacterium, making it capable of growth in sea water. Proc. Natl. Acad. Sci. U.S.A. *99*, 4109–4114.

Waditee, R., Hossain, G.S., Tanaka, Y., Nakamura, T., Shikata, M., Takano, J., and Takabe, T. (2004). Isolation and functional characterization of Ca2+/H+ antiporters from cyanobacteria. J. Biol. Chem. *279*, 4330–4338.

Wang, A.W., and Hill, A. (1977). Chemical analysis of the phenol-water-extractable materials from Anabaena flos-aquae. J. Bacteriol. *130*, 558–560.

Wang, L., Sun, Y.P., Chen, W.L., Li, J.H., and Zhang, C.C. (2002). Genomic analysis of protein kinases, protein phosphatases and two-component regulatory systems of the cyanobacterium Anabaena sp. strain PCC 7120. FEMS Microbiol. Lett. *217*, 155–165.

Weckesser, J., Katz, A., Drews, G., Mayer, H., and Fromme, I. (1974). Lipopolysaccharide containing L-acofriose in the filamentous blue-green alga *Anabaena variabilis*. J. Bacteriol. *120*, 672–678.

Weckesser, J., Drews, G., and Mayer, H. (1979). Lipopolysaccharides of photosynthetic prokaryotes. Annu. Rev. Microbiol. *33*, 215–239.

Weiss, M.S., Kreusch, A., Schiltz, E., Nestel, U., Welte, W., Weckesser, J., and Schulz, G.E. (1991). The structure of porin from Rhodobacter capsulatus at 1.8 Å resolution. FEBS Lett. *280*, 379–382.

Wildon, D.C., and Mercer, F.V. (1963). The Ultrastructure of the heterocyst and akinete of the blue-green algae. Arch. Microbiol. *47*, 19–31.

Wilhelm, S.W., and Trick, C.G. (1994). Iron-limited growth of cyanobacteria: Multiple siderophore production is a common response. Limnol. Oceanog. *39*, 1979–1984.

Wilk, L., Strauss, M., Rudolf, M., Nicolaisen, K., Flores, E., Kühlbrandt, W., and Schleiff, E. (2011). Outer membrane continuity and septosome formation between vegetative cells in the filaments of Anabaena sp. PCC 7120. Cell Microbiol. *13*, 1744–1754.

Wolk, C.P. (1968). Movement of carbon from vegetative cells to heterocysts in Anabaena cylindrical. J. Bacteriol., *96*, 2138–2143.

Wolk, C.P., Austin, S.M., Bortins, J., and Galonsky, A. (1974). Autoradiographic localization of 13N after fixation of 13N-labeled nitrogen gas by a heterocyst-forming blue-green alga. J. Cell Biol. *61*, 440–453.

Woodruff, W.A., Parr, T.R., Hancock, R.E., Hanne, L.F., Nicas, T.I., and Iglewski, B.H. (1986). Expression in *Escherichia coli* and function of *Pseudomonas aeruginosa* outer membrane porin protein F. J. Bacteriol. *167*, 473–479.

Wirtz, N.L., Treble, R.G., and Weger, H.G. (2010). Siderophore-independent iron uptake by iron-limited cells of the cyanobacterium *Anabaena Flos-Aquae*. J. Phycol. *46*, 947–957.

Wu, T., Malinverni, J., Ruiz, N., Kim, S., Silhavy, T.J., and Kahne, D. (2005). Identification of a multicomponent complex required for outer membrane biogenesis in *Escherichia coli*. Cell *121*, 235–245.

Wu, T., McCandlish, A.C., Gronenberg, L.S., Chng, S.S., Silhavy, T.J., and Kahne, D. (2006). Identification of a protein complex that assembles lipopolysaccharide in the outer membrane of *Escherichia coli*. Proc. Natl. Acad. Sci. U.S.A. *103*, 11754–11759.

Wunder, T., Bredemeier, R., Ruprecht, M., and Schleiff, E. (2009). The Omp85 proteins of Anabaena sp. PCC 7120 and *E. coli* are functionally distinct. Endocytobiosis Cell. Res. *19*, 20–30.

Wylie, J.L., and Worobec, E.A. (1995). The OprB porin plays a central role in carbohydrate uptake in *Pseudomonas aeruginosa*. J. Bacteriol. *177*, 3021–3026.

Xu, C., Fan, J., Froehlich, J.E., Awai, K., and Benning, C. (2005). Mutation of the TGD1 chloroplast envelope protein affects phosphatidate metabolism in Arabidopsis. Plant Cell *17*, 3094–3110.

Xu, X., Khudyakov, I., and Wolk, C.P. (1997). Lipopolysaccharide dependence of cyanophage sensitivity and aerobic nitrogen fixation in Anabaena sp. strain PCC 7120. J. Bacteriol. *179*, 2884–2891.

Yao, X., Jericho, M., Pink, D., and Beveridge, T. (1999). Thickness and elasticity of Gram-negative murein sacculi measured by atomic force microscopy. J. Bacteriol. *181*, 6865–6875.

Yoon, H.S., and Golden, J.W. (1998). Heterocyst pattern formation controlled by a diffusible peptide. Science *282*, 935–938.

Yoshihara, S., Geng, X., Okamoto, S., Yura, K., Murata, T., Go, M., Ohmori, M., and Ikeuchi, M. (2001). Mutational analysis of genes involved in pilus structure, motility and transformation competency in the unicellular motile cyanobacterium *Synechocystis* sp. PCC 6803. Plant Cell Physiol. *42*, 63–73.

Yoshimura, F., Zalman, L.S., and Nikaido, H. (1983). Purification and properties of *Pseudomonas aeruginosa* porin. J. Biol. Chem. *258*, 2308–2314.

Zaitseva, J., Jenewein, S., Oswald, C., Jumpertz, T., Holland, I.B., and Schmitt, L. (2005). A molecular understanding of the catalytic cycle of the nucleotide-binding domain of the ABC transporter HlyB. Biochem. Soc. Trans. *33*, 990–995.

Zak, E., Norling, B., Maitra, R., Huang, F., Andersson, B., and Pakrasi, H.B. (2001). The initial steps of biogenesis of cyanobacterial photosystems occur in plasma membranes. Proc. Natl. Acad. Sci. U.S.A. *98*, 13443–13448.

Zhang, L.C., Chen, Y.F., Chen, W.L., and Zhang, C.C. (2008). Existence of periplasmic barriers preventing green fluorescent protein diffusion from cell to cell in the cyanobacterium Anabaena sp. strain PCC 7120. Mol. Microbiol. *70*, 814–823.

Zhang, L.F., Yang, H.M., Cui, S.X., Hu, J., Wang, J., Kuang, T.Y., Norling, B., and Huang, F. (2009). Proteomic analysis of plasma membranes of cyanobacterium *Synechocystis* sp. Strain PCC 6803 in response to high pH stress. J. Proteome Res. *8*, 2892–2902.

Zhou, R., and Wolk, C.P. (2003). A two-component system mediates developmental regulation of biosynthesis of a heterocyst polysaccharide. J. Biol. Chem. *278*, 19939–19946.

Zhu, J., Kong, R., and Wolk, C.P. (1998). Regulation of hepA of Anabaena sp. strain PCC 7120 by elements 5′from the gene and by hepK. J. Bacteriol. *180*, 4233–4242.

Proteomics in Revealing the Composition, Acclimation and Biogenesis of Thylakoid Membranes

4

Natalia Battchikova and Eva-Mari Aro

Abstract

Oxygenic photosynthesis evolved in the thylakoid membrane of ancient cyanobacteria. Here we review the composition and biogenesis of cyanobacteria thylakoid membranes from the proteomics viewpoint, with the main emphasis on the macromolecular protein complexes involved in photosynthetic electron transfer. Response of the thylakoid membrane proteome to changes in environmental cues and to various stress conditions is also described and discussed in terms of dynamic modifications in metabolic and catabolic pathways in order to adjust cyanobacterial cells to a new environment.

Introduction

Oxygenic photosynthesis, i.e. using water as an electron source to produce high energy compounds and generating oxygen as a side product, is the most significant feature that distinguishes cyanobacteria among other prokaryotic organisms. More than 2 billion years ago, they evolved the ability to capture and utilize energy of the sunlight for splitting water molecules, with concomitant production of molecular oxygen, and have thereby changed the Earth atmosphere and created favourable conditions for evolution of aerobic organisms. A fundamental nature of photosynthesis has caused a keen interest in the details of the photosynthetic mechanisms, further enhanced by a huge potential of photosynthetic microbes for industrial applications directed to the development of a sustained energy supply. Extensive photosynthesis research has been conducted for decades in various disciplines of science including biophysics, biochemistry and molecular biology, and has resulted in comprehensive structural and functional investigations of proteins and protein complexes involved in the capture of light and in the photosynthesis-dependent electron flows from water splitting to transformation of energy into ATP and NADPH and, finally, to CO_2 fixation.

In cyanobacteria and their relatives, chloroplasts of green plants, the photosynthetic energy conversion machinery is primarily located in the thylakoid membrane. Linear electron flow that lays in the basis of photosynthetic light reactions is carried out by large membrane-embedded protein complexes, Photosystem I (PSI), Photosystem II (PSII), and Cytochrome b_6f complex (Cyt b_6f), accompanied by specific soluble proteins including plastocyanin and ferredoxin-NADP$^+$ reductase (FNR) (Fig. 4.1), the latter being responsible for generation of NADPH. Along with the linear electron flow, a pH gradient is established across the thylakoid membrane that powers up ATP synthase. The energy collected in the form of NADPH and ATP is used further for fixation of CO_2 by ribulose-1,5-bisphosphate carboxylase/oxygenase (Rubisco) and for multiple other metabolic processes. Beside linear electron flow, alternative electron transfer routes exist in thylakoid membranes and their proximity. They include (i) the cyclic electron transfer around PSI via NAD(P)H:Quinone oxidoreductases (NDH-1 complexes) that help to adjust the production of ATP and NADPH (Battchikova and Aro, 2007; Battchikova *et al.*, 2011a,b); (ii) flavodiiron proteins (Flvs) essential for photoreduction of O_2 in cyanobacteria (Flv1 and Flv3,

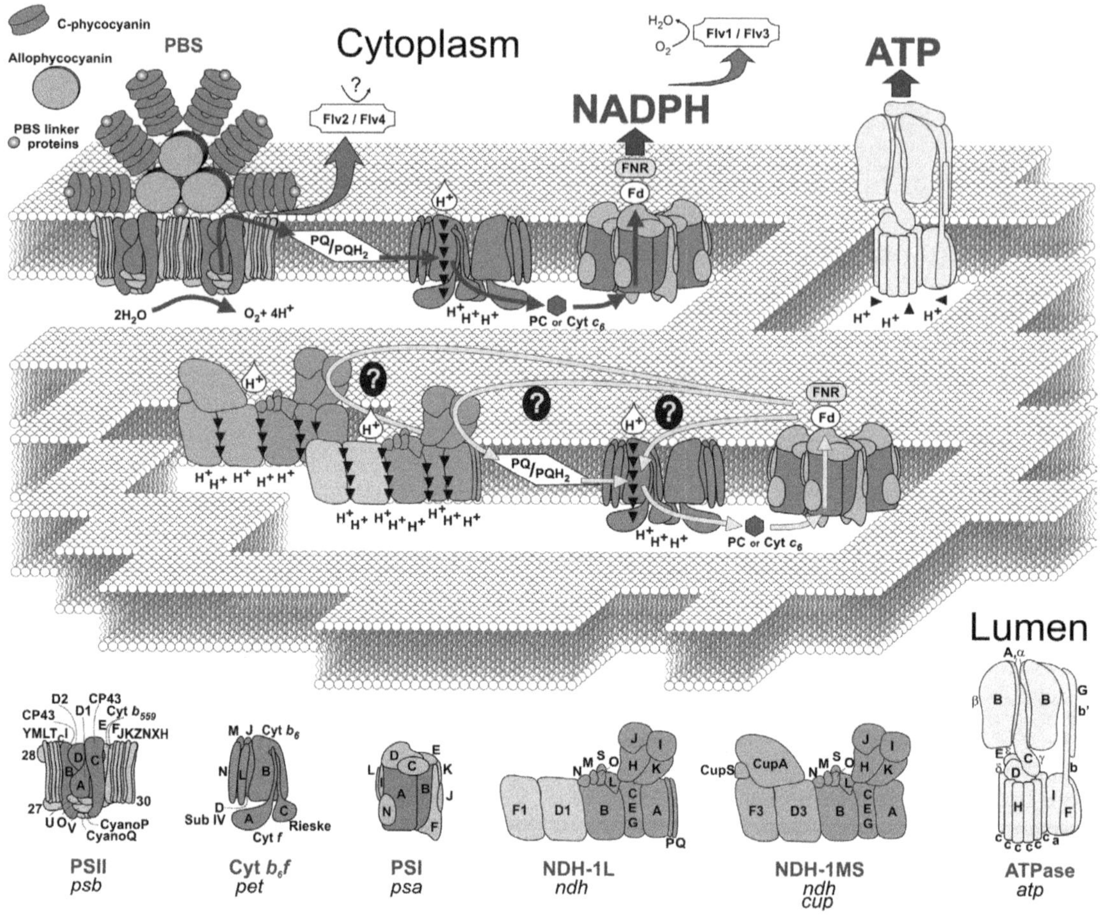

Figure 4.1 Scheme of photosynthetic protein complexes in the thylakoid membrane of *Synechocystis* 6803 and some associated proteins involved in the light reactions. Linear electron transport is shown by red arrows. Cyclic electron transport pathways shown by yellow arrows are still under debate and indicated with question marks. Subunit composition of protein complexes (in monomeric forms) is presented at the bottom of the picture. Subunits are designated by capital letters of the corresponding genes; a commonly used protein name is also indicated for some subunits. Blue triangles represent proton gradient channels. A colour version of this figure is available in the plate section at the back of the book.

Helman *et al.*, 2003) and protecting the fragile pigment-containing photosynthetic protein complexes, PSI (Flv1 and Flv3, Allahverdiyeva *et al.*, 2011, 2013) and PSII (Flv2 and Flv4, Zhang *et al.*, 2009b, 2012), from excess of light and other environment stresses; and (iii) terminal oxidases and other respiratory enzymes that provide energy for supporting the cell life in the dark (Vermaas 2001).

The wealth of data concerning the structure and functions of proteins and protein complexes participating in photosynthesis and photosynthesis-related reactions have been collected from reverse-genetic experiments and thorough investigations of mutants by biochemical and biophysical methods. The major photosynthetic complexes have been isolated in the active form and crystallized, thus clarifying the exact electron paths and molecular rearrangements making possible the energy transduction. At present, the mechanisms involved in the assembly and turnover of the photosynthetic complexes, as well as their responses to environmental challenges are under extensive research (for recent reviews, see Mulo *et al.*, 2009; Nixon *et al.*, 2010; Komenda *et al.*, 2012).

The development of proteomic techniques brought studies of the photosynthetic apparatus to a qualitatively different level. Contrary to the earlier research that has dealt mainly with a limited number of proteins or with separate protein complexes, the global investigation of the thylakoid proteome that includes both the membrane-embedded and membrane-associated peripheral proteins has a potential to unravel novel proteins with yet unknown functions in this compartment. Since the thylakoid membrane is specialized for transformation of solar energy into chemical energy, it is plausible to assume that these uncharacterized proteins are directly or indirectly involved in electron transport and/or play roles in preservation, maintenance, biogenesis, and a proper regulation of photosynthesis.

Membrane organization in cyanobacteria

The thylakoid membrane is a distinguished feature of cyanobacterial cells, and it is easily recognizable in electron microscopy pictures (Stanier, 1988). It harbours the major photosynthetic complexes participating in light reactions of photosynthesis. So far, the only known exception is *Gloeobacter violaceus*, which is devoid of thylakoids (Mangels *et al.*, 2002; Nakamura *et al.*, 2003).

In addition to the thylakoid membrane, cyanobacteria have the cytoplasmic and outer membranes, similarly to other Gram-negative bacteria (Liberton and Pakrasi, 2008; see also Chapter 3). In bacteria, the plasma membrane contributes to the cell bioenergetics by housing respiratory systems. In cyanobacteria, however, a distribution of photosynthetic and respiratory elements between thylakoid and plasma membranes is not clear at present and therefore is intensively studied; most probably, the division of tasks between two membranes is not absolutely strict and may significantly vary between cyanobacteria.

Cyanobacterial thylakoids are organized as sheets formed by bilayers of thylakoid membranes enclosing the lumenal space. Specific details of the thylakoids architecture might slightly vary among cyanobacteria. In *Synechocystis* sp. PCC 6803 (hereafter *Synechocystis* 6803), thylakoids form stacks of 3–10 parallel sheets located primarily nearby the cytoplasmic membrane (van de Meene *et al.*, 2006; Liberton *et al.*, 2006). In *Synechococcus* sp. PCC 7002 (formerly known as *Agmenellum quadruplicatum*) the thylakoid membranes are arranged inside the cell as concentric cylinders underlying the plasma membrane (Nierzwicki-Bauer *et al.*, 1983). In *Cyanothece* sp. ATCC 51142, a bunch of thylakoids that are arranged radially around a central cytoplasmic area spirals around the cell periphery (Liberton *et al.*, 2011). The peripheral band of thylakoid membranes appeared to be a dense interconnected network derived from the continuous branching and splitting of membranes. Regardless of the overall arrangement, in most cyanobacteria the space between thylakoids is sufficient to harbour a double row of phycobilisomes, large light-harvesting antenna complexes (Mullineaux, 1999). However, few cyanobacterial species have been shown to contain appressed and non-appressed thylakoid regions. These include *Prochloron* sp. (Giddings *et al.*, 1980) that contains in the thylakoid membrane chlorophyll (Chl) *a*/*b* light-harvesting complexes, similarly to chloroplasts of green plants, and *Acaryochloris marina* (Chen *et al.*, 2009), a Chl *d*-containing cyanobacterium with two discrete antenna systems, integral membrane-bound Chl-binding proteins and external phycobiliproteins.

Thylakoids converge at sites adjacent to the cytoplasmic membrane. Many places were found where thylakoid membranes and the plasma membrane are in a close proximity (Nierzwicki-Bauer *et al.*, 1983; Liberton *et al.*, 2006, 2011; van de Meene *et al.*, 2006). In some cases, thylakoid centres were found at the sites of convergence (Kunkel 1982). Thylakoid centres are distinctive cylindrical structures with thylakoid membrane sheets associated along their length (van de Meene *et al.*, 2006). However, although sites of close proximity of thylakoids and plasma membranes have been observed, no continuity between these membrane systems has been demonstrated (Liberton *et al.*, 2006). It was concluded that, most probably, cyanobacterial thylakoid membranes form a separate enclosed compartment inside the cell. A similar conclusion was made by Schneider *et al.* (2007) from experiments with fluorescent staining of live *Synechocystis* 6803 cells. Whereas van de Meene *et al.* (2006) have pinpointed an apparent continuity between thylakoid and plasma membranes at some convergence sites in *Synechocystis* 6803, such places were rare and raised doubts regarding their capacity for accomplishing the material flux necessary for thylakoid formation especially during cell growth and division. Further, the authors pointed out that the thylakoid centre core appeared to be continuous with the periplasmic space. Thus, the existence of continuity between thylakoid membranes and thylakoid centres could not be confirmed (van de Meene *et al.*, 2006). (See Chapter 5 for a further discussion on this topic.)

Detailed analysis of interactions among diverse bioenergetic pathways is a very important direction of research that aims, beside the basic science, to facilitate development of cyanobacterial biofuel production systems. Proteomic analyses of thylakoid and plasma membrane compartments play an important role in investigation of proteins involved in electron transport routes and elucidation of their localization in cells.

Challenges in proteomic analysis of thylakoids

It is difficult to draw a clear line between proteomics of thylakoids and proteomics of plasma membrane since these compartments may converge at some points causing significant problems in their separation. Moreover, their physical properties are rather similar; therefore, crude preparations of thylakoid membranes are usually contaminated with the plasma membrane (Herranen *et al.*, 2004). It is hard to isolate the entire thylakoid membrane compartment in a pure form. Earlier, sucrose density centrifugation was used to obtain a low-density fraction enriched with the plasma membrane and a high-density fraction enriched with the thylakoid membrane (Omata and Murata, 1983; Riviere *et al.*, 1990). Later, Norling *et al.* (1998) introduced the aqueous two-phase partitioning method that allowed isolation of rather pure thylakoid and plasma membrane compartments. Purity of preparations is usually demonstrated by Western blot analyses performed with antibodies to proteins whose exclusive location in only one of the membranes have been established by a variety of methods. The abundant PSII core antenna proteins, CP43 and CP47, are often used as specific markers for the thylakoid membrane. Nutrient transporters such as NrtA and SbtA might be selective markers for the plasma membrane. However, a possibility for cross contamination in membrane preparations should always be considered in analysis of proteomic data since mass spectrometry is a highly sensitive technique. On the other hand, significant efforts in purification of the thylakoid membrane vesicles increase the possibility of a concomitant loss of specific thylakoid regions.

The thylakoids contain photosynthetic macromolecular complexes composed of hydrophobic proteins embedded into the membrane, together with hydrophilic peripheral proteins exposed to the cytoplasmic space or to the lumen. Prior to analysis, it is necessary to solubilize thylakoid proteins by treatment with detergents, and a detergent is selected depending on application. Proteomic methodologies can be divided in two general categories. The first one is targeted to maximal solubilization and denaturation of proteins using strong detergents and denaturing agents in order to increase proteome coverage and/or facilitate enzymatic digestion to peptides. It can be utilized as a 'shotgun' technique (Link *et al.*, 1999) where the whole proteome is digested to peptides (by trypsin or another specific protease), and the resulting peptides are separated by 2-D or 3-D liquid chromatography and analysed by mass spectrometry (MS). Another method from the same category involves separation of a proteome to single proteins with 2-D isoelectric focusing (IEF)/SDS-PAGE followed by excision of individual protein spots, digestion to peptides and MS analysis. Both schemes are extensively used despite hydrophobic proteins tend to aggregate and precipitate during IEF electrophoresis hampering their detection. In contrast to these schemes, the alternative general proteomic methodology aims to separate protein complexes in intact form in order to study their subunit composition, assembly/disassembly stages, protein-protein or protein–cofactor interactions, etc. Here, to minimize effects on the stability of protein complexes, mildly solubilizing detergents are selected and macromolecular complexes in their oligomeric states are separated by liquid chromatography or electrophoresis in native PAGE followed by MS characterization.

Cellular proteins differ strongly in their solubility and physicochemical properties because of their various amino acid content and post-translational modifications. Membrane proteins are more difficult to isolate, digest to peptides and detect in MS experiments than soluble proteins (Santoni *et al.*, 2000). There are several technical limitations in membrane proteomic analysis. Integral proteins with many transmembrane helices are difficult to solubilize, and they tend to aggregate during enzymatic digestion to peptides. Membrane proteins are poorly digested with specific proteases routinely used in proteomics, such as trypsin that cuts after Lys or Arg, or Asp-N that cuts before Asp and Glu, because charged amino acid residues are rare in transmembrane regions. Long hydrophobic peptides are easily lost on plastic surfaces of tubes and tips as well as on reverse phase columns used for desalting of peptides and their liquid chromatography separation prior to MS analysis. Therefore, hydrophobic proteins are under-represented in lists of identified proteins in large-scale investigations of total cyanobacterial proteomes (Pisareva *et al.*, 2007, 2011; Stensjö *et al.*, 2007; Ow *et al.*, 2009; Battchikova *et al.*, 2010).

Methods used for preparation of the proteomic samples and variations in proteomic methodologies have a great impact on the results (Gan *et al.*, 2005). Wang *et al.* (2009) identified 29 integral and 95 peripheral proteins in a total membrane proteome of *Synechocystis* 6803 using 2-D IEF/SDS-PAGE separation of proteins and trypsin digestion. To improve detection of membrane proteins, a modified cleavage approach might be employed based on, for example, treatment with cyanogen bromide that cleaves after Met, or much less specific methods such as chymotrypsin digestion and acid hydrolysis. Using this strategy and a shotgun methodology, Kwon *et al.* (2010) identified 155 membrane-embedded and 36 soluble membrane-associated proteins in a total membrane proteome of *Synechocystis* 6803. Integral proteins found in this study represented about 22% of 706 predicted *Synechocystis* membrane proteins. Taking into account that a cyanobacterial proteome changes in response to various environmental conditions and, therefore, only a set of genes is expressed and might be detected at the protein level, the real percentage of identified membrane proteins might be considerably higher. As expected, the published total membrane proteomes include proteins of many functional categories from all three *Synechocystis* membranes (Wang *et al.*, 2009; Kwon *et al.*, 2010).

Proteomic investigations of thylakoid proteins

Wang *et al.* (2000) initiated proteomic studies of cyanobacterial thylakoids by analysis of peripheral thylakoid proteins using a 2-D gel-based approach. Among 78 identified proteins, many were associated with photosynthesis including allophycocyanin and phycocyanin. Proteins of several other functional categories were also found, but recovery of integral membrane proteins was poor. Later, when data about proteomes of pure outer (Huang *et al.*, 2004) and plasma (Fulda *et al.*, 2000, 2006; Huang *et al.*, 2002, 2006; Pisareva *et al.*, 2007; Zhang *et al.*, 2009a; Li *et al.*, 2012) membranes from *Synechocystis* 6803 became available, Srivastava *et al.* (2005) pointed out that proteins detected in the study of Wang *et al.* (2000) included representatives from all three membranes and the periplasm.

The proteome of highly pure *Synechocystis* 6803 thylakoids prepared by combination of sucrose gradient centrifugation and two-phase partitioning was analysed by several proteomic approaches. Srivastava *et al.* (2005) applied various 2-D IEF/SDS PAGE systems with narrow pH gradients to facilitate protein separation. Further, they enriched the preparation with integral membrane proteins by washing out the peripheral ones and, in order to avoid losses during IEF, resolved them on 1-D SDS-PAGE. Totally, 76 proteins were identified in this study, 14 of them were integral ones. Recently, Pisareva *et al.* (2011) carried out an investigation of highly pure thylakoid and plasma membranes of *Synechocystis* 6803 using the shotgun proteomic approach and a modified cleavage strategy. They succeeded in detection of 59 integral and 86 soluble peripheral proteins; from them, 15 membrane-embedded and 30 soluble proteins were found also in the plasma membrane fraction. The authors (Pisareva *et al.*, 2011) noted also that a significant overlap between three *Synechocystis* 6803 membranes was observed for predicted lipoproteins. This group included some components of periplasmic transporters, ATP binding proteins, a few subunits of photosynthetic complexes and hypothetical proteins. The authors explained that upon breakage of the cells lipoproteins can detach from the native membrane and artificially attach to any of the membranes. Therefore, localization of these proteins will need to be deduced on the basis of combined results obtained by different methods.

Identified thylakoid proteins belong to several functional categories. As expected, the largest category contains photosynthesis and respiration proteins including subunits of the major photosynthetic complexes, PSI, PSII, Cyt b_6f and ATP synthase, as well as components of NDH-1 complexes. Some of the subunits that belong to PSI, PSII and ATP synthase were found in the plasma membrane, although in smaller amounts, but representatives of NDH-1 complexes and Cyt b_6f were detected only in thylakoids. It is of note that respiratory terminal oxidases, CtaC and CtaDII, along with NdbB and NdbC proteins, which are NAD(P)H dehydrogenase type 2 (NDH-2) proteins, were found only in the plasma membrane. A number of enzymes involved in biosynthesis of cofactors, prosthetic groups and carriers, such as Chl *a* synthase (ChlG), porphobilinogen synthase (HemB), ferrochelatase (HemH), etc., were found in the thylakoid membrane. Subunits of the Sec protein translocase, SecA, D, and E, and the YidC protein were found in this compartment, too. They are involved in transfer of proteins across the membrane as well as in the correct insertion of integral proteins into the membrane (Mori and Ito, 2001). In particular, the YidC protein has been shown to be important for insertion of the D1 protein of PSII (Ossenbuhl *et al.*, 2006). It is interesting that from two signal peptidases of *Synechocystis* 6803, one (Sll0716) was found in the thylakoid membrane while the other one (Slr1377) was detected in the plasma membrane (Pisareva *et al.*, 2011).

Gloeobacter violaceus is so far the only known cyanobacterium where both the photosynthetic machinery and the respiratory apparatus are located in the plasma membrane. However, two distinct plasma membrane fraction domains were discovered in this organism, with vast differences in their proteome and pigment composition (Rexroth *et al.*, 2011). The two domains successfully separated by sucrose gradient centrifugation showed clearly different pigment and lipid composition. One domain (green fraction) contained Chl and displayed some characteristics similar to those of the thylakoid membranes in more typical cyanobacteria, whereas the other one (orange fraction) was carotenoid-rich and showed properties of a cyanobacterial plasma membrane. Proteomic analysis of these two domains demonstrated a distribution of proteins similar to the one described above for the thylakoid and plasma membrane proteins of *Synechocystis* 6803. A majority of the identified proteins, including large subunits of the PSI and PSII complexes, were present in both fractions. However, three subunits of the Cyt b_6f complex, five subunits of NDH-1 complexes and some of the small subunits that belong to PSI and PSII were found exclusively in the green fraction. In contrast, restricted location in the orange fraction was shown for enzymes involved in the metabolism of terpenoids and for a homologue of the PratA protein involved in the maturation of PSII (Klinkert *et al.*, 2004). NDH-2 (Glr3503) was predominantly located in the orange fraction. A selectivity of the membrane rupture by mechanical force at the boundaries of membrane domains suggested strong internal molecular interactions within the regions specialized to particular physiological functions, with weak spots in the membrane structure integrity between the domains (Rexroth *et al.*, 2011).

Proteomics of thylakoid protein complexes

Native electrophoresis has been an indispensable tool for analysis of many different membrane protein complexes from various organisms. Both blue native (BN) PAGE (Schägger and von Jagow, 1991) and clear native (CN) PAGE (Schägger *et al.*, 1994) use non-ionic or zwitterionic detergents in a near neutral buffer system, with or without addition of Coomassie blue for BN and CN, respectively. In general, mobility of protein complexes is proportional to molecular mass making the method an excellent tool to study an oligomeric status of isolated complexes and their subunit composition. Application of BN-PAGE for analysis of multisubunit protein complexes from the thylakoid membrane of cyanobacteria was demonstrated in the pioneering work of Barbato *et al.* (1995). They described solubilization of thylakoids from *Synechocystis* 6803 with dodecyl β-D-maltoside (DM) and separation of practically intact protein complexes. Since PSI and PSII contain Chl molecules, they moved in a gel as green bands; the identity of these bands was proven by specific antibodies.

The methodology has significantly developed from that time. Several variants of high-resolution clear native (hrCN-PAGE) and blue native (hrBN-PAGE) electrophoretic systems have become available (Wittig *et al.*, 2006, 2007; Ladig *et al.*, 2011; Järvi *et al.*, 2011). The example of BN- and CN-PAGE separation of thylakoid membrane complexes from *Synechocystis* 6803 is shown in Fig. 4.2. Optimal conditions for solubilization and native PAGE-based separation vary for different membrane protein complexes; in addition to DM,

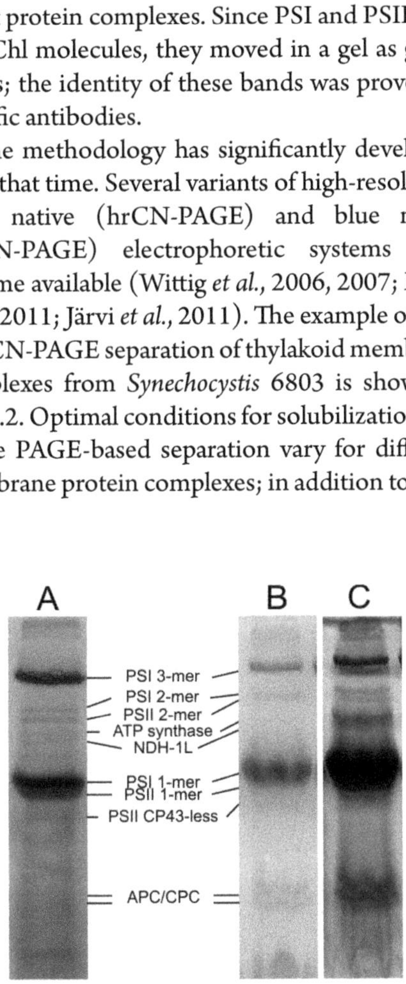

Figure 4.2 1-D native PAGE of thylakoid protein complexes. The crude thylakoid membrane fraction of *Synechocystis* 6803 was obtained as described in Zhang *et al.* (2004). Protein complexes were solubilized with 1.5% DM and separated by Blue native (BN-) PAGE (A) or Clear Native (CN-) PAGE (B) as described in Herranen *et al.* (2004) and Wittig *et al.* (2007). Gradient 4–12% PAGE was used in both (A) and (B). (C) CN-PAGE-separated complexes were stained with Coomassie Blue. A colour version of this figure is available in the plate section at the back of the book.

other detergents such as digitonin and deoxycholate have been applied for mild solubilization and separation of membrane protein complexes (Wittig *et al.*, 2006, 2007; Ladig *et al.*, 2011; Järvi *et al.*, 2011). To resolve individual components of multisubunit complexes, denaturing SDS-PAGE is usually applied in a second dimension. MS analysis significantly simplified protein identification compared to antibody detection; the details of the method are described, for example, in Mann *et al.* (2001), Aebersold and Mann (2003) and Baldwin (2003). A combination of various techniques presents a possibility to achieve even greater degree of a proteome analysis. Thus, a 3-D electrophoretic protocol was used for profound characterization of thylakoid membrane complexes from plant chloroplasts (d'Amici *et al.*, 2008). The protocol included native liquid phase IEF in the first dimension followed by BN-PAGE of multiple IEF fractions. Further, each single component resolved in the second dimension was subjected to SDS-PAGE (third dimension) to separate individual subunits and identify them by MS.

However, 1-D native PAGE and 2-D native/SDS-PAGE, in combination with MS analysis, are the preferred techniques for exploration of thylakoid membrane protein complexes from cyanobacteria. Fast and not very selective methods of thylakoid isolation are often favoured in this type of experiments in order to preserve the intact state of protein complexes; therefore, minor contaminations from plasma membrane are sometimes detected. A typical 2-D BN/SDS-PAGE map of thylakoid complexome from *Synechocystis* 6803 is shown in Fig. 4.3. The major thylakoid complexes involved in photosynthesis-driven electron transfer, which dominate the map, are described in detail below; for clarity, a proteomic point of view is accompanied by data obtained with other methods. The sequence of the discussion below follows the order of complexes involved in light harvesting and electron transfer pathways shown in Fig. 4.1.

Phycobilisomes

Phycobilisomes (PBSs) are supramolecular complexes comprising the light-harvesting pigment system in a majority of cyanobacteria except

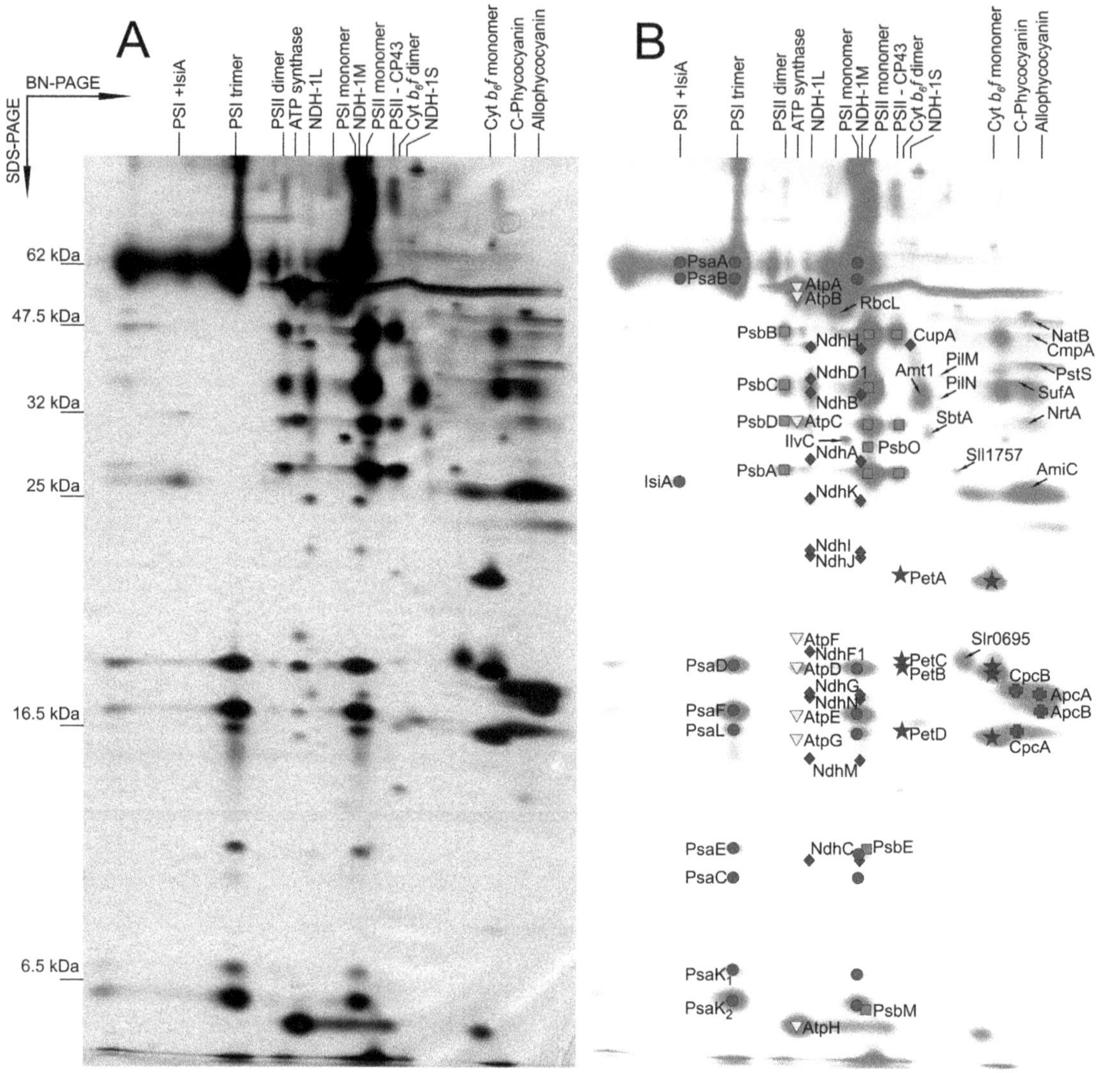

Figure 4.3 2-D BN/SDS-PAGE of thylakoid protein complexes from *Synechocystis* 6803 grown in high CO_2 conditions. (A) A silver-stained gel. (B) Schematic representation of (A), with subunits of complexes shown by following coloured symbols: phycobilisome components, blue crosses; PSII, green squares; Cyt b_6f, magenta stars; PSI, turquois circles; ATP synthase, yellow triangles; NDH-1, red diamonds. Identification of proteins is based on mass spectrometry. Proteins from cytosol or other membranes present as contaminants in the crude thylakoid membrane preparation are shown by arrows. A colour version of this figure is available in the plate section at the back of the book.

prochlorophytes (Hess *et al.*, 2001). Molecular composition of cyanobacterial phycobilisomes varies among cyanobacteria (Tandeau de Marsac and Cohen-Bazire, 1997). The phycobiliproteins allophycocyanin and phycocyanin are always present in PBS, but in some cyanobacterial species they are accompanied by other phycobiliproteins such as phycoerythrin or phycoerythrocyanin (Tandeau de Marsac and Cohen-Bazire, 1997). Phycobiliproteins contain open-chain tetrapyrroles known as phycobilins, or bilin chromophores, that are post-translationally covalently attached, via thioether bonds, to specific Cys amino acid residues (Scheer and Zhao, 2008).

Details of PBS structure can be found in specific reviews (Sidler 1994; MacCoil 1998; Liu *et al.*, 2005; see also Chapter 6). Briefly, disc-shaped phycobiliproteins are primarily composed of *α* and β polypeptides (Fig. 4.1); the additional γ subunit is present in some phycoerythrins (Liu *et al.*, 2005). Three cylinders made of allophycocyanin discs comprise a PBS core; six rods which are formed by stacked disks of phycocyanin and, if present, phycoerythrin or phycoerythrocyanin, shape hemidiscoidal PBSs. The structure is reinforced by several linker polypeptides (Liu *et al.*, 2005). Phycobiliproteins are categorized into three types: those of high energy (containing phycoerythrins or phycoerythrocyanin), intermediate energy (containing phycocyanins), and low energy (containing allophycocyanins). Energy flows from highest- to lowest-energy pigments, in accordance with the spatial structure of PBS (MacCoil, 1998).

In vivo, PBSs associate with PSII and, in some conditions, with PSI, at the cytoplasmic side of the thylakoid membrane (for review, see Bald *et al.*, 1996). However, when cyanobacterial cells are disrupted in low-ionic-strength buffers optimized for isolation and analysis of functional major photosynthetic complexes, the phycobilisomes become detached from the thylakoid membrane and dissociate into components. In *Synechocystis* 6803 and *Nostoc punctiforme*, allophycocyanin and phycocyanin could be found in hexameric forms containing three α and three β subunits, ApcA and ApcB, and CpcA and CpcB, respectively (Fig. 4.3; Herranen *et al.*, 2004; Cardona *et al.*, 2007). Additionally, a hexameric form of phycoerythrin composed of three CpeA (α) and CpeB (β) subunits was detected in a proteomic study of *Nostoc punctiforme* (Cardona *et al.*, 2007). Since phycobiliproteins contain chromophore molecules, PBS subcomplexes are easily visualized in native PAGE by a characteristic blue colour (Fig. 4.2B).

Phycobiliproteins contain other post-translation modifications beside attachment of bilin molecules. A unique modified residue, γ-*N*-methylasparagine, has been found in all phycobiliproteins isolated from cyanobacteria (Klotz and Glazer, 1987). The γ-*N*-methylasparagine residue is located at Asn-72 of CpcB (the β-72 position), in close proximity to the chromophore linked at β-82 that serves as the terminal energy acceptor in C-phycocyanin. Methylation at β-72 is highly evolutionarily conserved. No such modification is present on the homologous α position (Klotz *et al.*, 1986; Klotz and Glazer, 1987). γ-*N*-methylasparagine was also shown in the allophycocyanin subunits, ApcB and ApcF (Swanson and Glazer, 1990). Methylation of phycobiliproteins significantly contributes to the efficiency of directional energy transfer to the terminal energy acceptors in the PBS (Swanson and Glazer, 1990).

PBS components are subject to protein phosphorylation (Allen *et al.*, 1985; Sanders *et al.*, 1989). The β-phycocyanin of *Synechococcus* 6301 has been suggested to be a phosphoprotein by *in vivo* labelling with [^{32}P]orthophosphate (Mann, 1994). Using antibodies against phospho-Ser and phospho-Thr, as well as the ProQ staining specific for phosphoproteins, Piven *et al.* (2005) demonstrated that linker proteins of light harvesting antenna are phosphorylated in *Synechocystis* 6803. The authors proposed that linker dephosphorylation might act as a signal for PBS degradation in various conditions of stress such as high light and nutrient deprivation. However, Piven *et al.*, (2005) disproved the idea that allophycocyanins and phycocyanins are phosphoproteins and attributed signals obtained from ProQ staining of the corresponding protein bands to the self-fluorescence of bilin-containing proteins. Recently, Mukhopadhyay and Kennelly (2011) described the phycocyanin rod linker protein CpcD, as well as phycocyanin-α and -β subunits, as possible endogenous substrates of protein tyrosine phosphatase Slr0328 using immunodetection, MS and synthetic phosphatase substrates. Exact phosphorylation sites of PBS components remain to be elucidated.

PSII complexes

The light energy absorbed by antenna pigments excites the reaction centre Chl P680 of the PSII complex priming linear electron transfer. The lipophilic plastoquinone (PQ) molecule residing at the acceptor side of PSII, after acquiring two electrons and two protons, becomes reduced to plastoquinol (PQH_2) and diffuses into the PQ pool. Concomitantly, in the oxygen-evolving extrinsic domain at the lumenal side of PSII, water splitting occurs resulting in oxygen production. Function and structure of PSII from cyanobacteria have been extensively reviewed (see, e.g. Nelson and Ben-Shem, 2004; Nelson and Yochum, 2006; DeRuyter and Fromme, 2008).

In native and blue-native gels, PSII complexes are usually resolved as a monomer and a dimer (Fig. 4.3; Herranen *et al.*, 2004, Cardona *et al.*, 2007), and a PSII monomer lacking the CP43 protein becomes sometimes visible as a minor form. It is important to note that native electrophoresis reflects a genuine oligomeric state of PSII (and other complexes) in the preparation since the same two forms were observed in PSII complexes isolated by chromatographic methods from both cyanobacteria and green plants (Rögner *et al.*, 1987; Bald *et al.*, 1996; Watanabe *et al.*, 2009). However, the ratio of PSII dimer to monomer varies in preparations depending, for example, on DM concentrations (Watanabe *et al.*, 2009). Assignment of PSII complexes in Fig. 4.3 was based on identification of the reaction centre proteins D1 (PsbA) and D2 (PsbD), the Chl *a*-binding core antenna proteins CP47 (PsbB) and CP43 (PsbC), the α-subunit of cytochrome b_{559} (PsbE), the PsbO protein of the oxygen-evolving complex, as well as PsbM. Other subunits were apparently released from PSII upon membrane solubilization or, in the case of small ones, were not retained in the gel. Modifications of the PAGE buffer were shown to be beneficial for analysis of small subunits in characterization of the PSII complex from the thermophilic cyanobacterium *Thermosynechococcus vulcanus* (Kashino *et al.*, 2002a).

Crystallographic studies of PSII from *Thermosynechococcus elongatus* (Zouni *et al.*, 2001) revealed 17 subunits, assigned as PsbA-F, PsbH-O, PsbU, -V, and -X, -Z. The majority of them contain at least one transmembrane helix (except PsbO, PsbU and PsbV) and therefore can be regarded as integral membrane proteins (Fromme *et al.*, 2006). A high-resolution crystal structure was obtained for PSII from *Thermosynechococcus vulcanus* (Umena *et al.*, 2011) that contained 19 subunits, PsbA-F, PsbH-M, PsbO, PsbT-V, PsbX, PsbZ, and Psb30, while PsbN was missing (Fig. 4.1). According to Race and Gounaris (1993), the PsbH subunit might be phosphorylated. However, conflicting observations were described by Komenda *et al.* (2002), and the question remains to be resolved.

Introduction of a His_6 tag into the C-terminus of the CP47 protein allowed fast and effective isolation of a highly active and sufficiently pure PSII complex from *Synechocystis* 6803. A proteomic analysis of this preparation (Kashino *et al.*, 2002b) demonstrated the presence of all subunits known from crystallographic studies of *T. elongatus* (Zouni et al., 2001) except PsbN. Further, the isolated PSII complex contained other low-molecular-mass (10 to 3.6 kDa) peptides, $PsbT_c$ (Smr0001), PsbY (Sml0007), PsbZ (ycf9, Sll1281), Psb27 (Slr1645), and Psb28 (Sll1398). Some of them are genuine PSII subunits whereas others are involved in PSII biogenesis. Two different forms of the FtsH protease were also present in the isolated PSII preparation, corroborating earlier findings that the thylakoid FtsH protease is involved in the light-induced degradation of the D1 protein (Lindhal *et al.*, 2000). In addition, several novel polypeptides were found which showed homology to proteins of green plants, including Sll1638 that is homologous to PsbQ of *Arabidopsis thaliana*.

Extrinsic subunits localized at the lumenal side of PSII, together with the Mn_4-Ca cluster, constitute the oxygen-evolving complex (OEC). OEC proteins differ significantly between PSII complexes from cyanobacteria and green plants. Crystallographic and other investigations revealed that cyanobacterial OEC contains PsbU and PsbV proteins in proximity of PsbO (Roose *et al.*, 2007). Further, reverse genetics and biochemical studies showed that cyanobacterial OEC complexes comprise also a PsbP-like protein (Sll1418), a PsbQ-like protein (Sll1638) and Psb27 (Fig. 4.1). The PsbP-like protein, also designated as CyanoP (Fagerlund and Eaton-Rye, 2011), is conserved among cyanobacteria, the PsbQ-like protein, or CyanoQ, is absent in *Gloeobacter violaceus* and *Prochlorococcus* strains, while Psb27 is present also in green algae and plants (Roose *et al.*, 2007; Fagerlund and Eaton-Rye, 2011).

CyanoP, CyanoQ and Psb27 in cyanobacteria are predicted to be lipoproteins (Fagerlund and Eaton-Rye, 2011; Pisareva *et al.*, 2011). These hydrophilic proteins resist high salt washes of the thylakoid membrane indicating a rather tight membrane association that might be explained by the presence of the lipid anchor at their N-termini.

Mass spectrometry analysis corroborated lipid attachment to recombinant CyanoQ from *Synechocystis* 6803 produced in *E. coli* (Ujihara *et al.*, 2008) and to native Psb27 from *T. elongatus* (Nowaczyk *et al.*, 2006), but lipidation of CyanoP remains to be confirmed. Thornton *et al.* (2004) suggested a role in regulation of PSII activity for CyanoP and CyanoQ proteins.

Cytochrome b_6f complex

The Cyt b_6f complex mediates electron transfer between PSII and PSI by oxidizing plastoquinol (PQH_2) and reducing soluble plastocyanin or cytochrome c_6 located in the lumen (for review, see Cramer *et al.*, 2006). The complex is involved in both linear and cyclic electron flows along with respiration. A functional homodimer and a monomer of Cyt b_6f are usually seen in the 2-D maps of cyanobacterial membranes (Fig. 4.3). The two forms have been identified on the basis of four integral subunits PetA-D (Herranen *et al.*, 2004, Cardona *et al.*, 2007).

The crystal structure of the Cyt b_6f complex from the thermophilic cyanobacterium *Mastigocladus laminosus* (Kurisu *et al.*, 2003) resolved 8 subunits, cytochrome *f* (PetA), cytochrome b_6 (PetB), the iron-sulphur Rieske protein (PetC), subunit IV (PetD) and low-molecular-weight proteins (3–4kDa) PetG, -L, -M and –N (Fig. 4.1). All of them contain at least one transmembrane helix. The structure of Cyt b_6f is conserved among cyanobacteria. However, several *petC* genes have been found in some cyanobacterial genomes (Schultze *et al.*, 2009). For example, three PetC proteins have been predicted for *Synechocystis* 6803. Using PetC1–3 specific antibodies, Schultze *et al.* (2009) showed that all three proteins are produced. PetC1 and PetC2 appeared to be located in the thylakoid membrane whereas PetC3 was found in the plasma membrane. Proteomics experiments (Pisareva *et al.*, 2011) corroborated location of PetC1 and PetC3 in the corresponding membrane compartments; however, PetC2 protein was not detected. This fact might indicate lower expression of *petC2* compared to *petC1*. Further, Schultze *et al.* (2009) investigated *petC1-C3* mutants of *Synechocystis* 6803 and concluded that this organism might contain two different Cyt b_6f complexes, with PetC1 or PetC2, but that PetC3 is not a Rieske protein. The functional role of the PetC3 protein remains unclear at present. Further, the protein encoded by the *ssr2998* gene was shown to be structurally and functionally associated with the Cyt b_6f complex of *Synechocystis* 6803 (Volkmer *et al.*, 2007).

PSI complexes

The PSI complex catalyses electron transfer from plastocyanin in the lumen to ferredoxin at the stromal side of the thylakoid membrane (for review, see Chitnis 2001). Finally, FNR finishes the linear electron transfer by reducing $NADP^+$ to NADPH, which is used for carbon fixation and many other metabolic processes.

The PSI complex is usually resolved in BN-PAGE as a monomer and a trimer (Fig. 4.3; Chitnis and Chitnis, 1993; Herranen *et al.*, 2004; Kubota *et al.*, 2009; Ladig *et al.*, 2011); minute amounts of PSI dimer are sometimes detected. The same forms were observed when PSI was isolated by chromatography (Takahashi *et al.*, 1982, Rögner *et al.*, 1990). Various methods including crystallography and electron microscopy with single particle analysis corroborated a trimeric organization of the functional PSI complex both in unicellular cyanobacteria, including *Synechocystis* 6803 (Kruip *et al.*, 1997), *T. elongatus* (Jordan *et al.*, 2001), *Synechococcus* sp. PCC 7002 (Tsiotis *et al.*, 1995), *Synechococcus* sp. PCC 7942 (Boekema *et al.*, 2001), *Gloeobacter violaceus* (Mangels *et al.*, 2002) and *Acaryochloris marina* (Chen *et al.*, 2005), and in the filamentous cyanobacteria *Phormidium laminosum* (Ford and Holzenburg, 1988) and *Spirulina platensis* (Rakhimberdieva *et al.*, 2001).

Jordan *et al.* (2001) resolved the crystal structure of the PSI trimer from *T. elongatus.* A monomeric unit is composed of twelve proteins. Two large hydrophobic proteins, PsaA and PsaB, create a PSI core that is surrounded by smaller membrane-embedded subunits, PsaF, I, J, K, L, M, and X; the PsaL subunit was shown to provide most of the contacts between monomers. Three extrinsic hydrophilic proteins, PsaC, -D and -E involved in the docking of ferredoxin or flavodoxin, are situated at the stromal side of the PSI complex (Fromme *et al.*, 2001, 2006).

Mukhopadhyay and Kennelly (2011) suggested that the PsaD subunit might be phosphorylated.

In *Synechocystis* 6803, there are two *psaK* genes, and both of them are expressed (Fig. 4.3 and Naithani *et al.*, 2000). Kubota *et al.* (2010) introduced a His_6 tag into the C-terminus of PsaF and the N-terminus of PsaJ subunits and isolated *Synechocystis* PSI by affinity chromatography. The PsaK1 protein was detected in both monomer and trimer forms of the PSI complex while PsaK2 was found only in the trimer. However, the PsaK2 subunit is present in both the PSI trimer and the PSI monomer, which are resolved by 2D BN/SDS-PAGE (Fig. 4.3). Since the PSI monomer is considered to have only one binding site for the PsaK protein, Kubota *et al.* (2010) suggested that the trimers might be a mixture of heterotrimers composed of PsaK1-binding and PsaK2-binding monomers. The physiological significance of PsaK redundancy in *Synechocystis* 6803 is not clear.

Remarkably, Watanabe *et al.* (2011) discovered an unusual form of the PSI complex in the filamentous N_2-fixing cyanobacterium *Anabaena* sp. PCC 7120 (hereafter *Anabaena* 7120). Based on the mobility of PSI bands in BN-PAGE and sucrose density gradient centrifugation, the authors concluded that this organism contains the PSI monomer and the PSI tetramer but no trimer, in contrast to other cyanobacteria. In agreement with the unique character of the tetrameric form of the PSI, PsaL and PsaI subunits from *Anabaena* 7120 and some other related heterocyst-forming cyanobacteria seem to have specific attributes in structures since they cluster together in phylogenetic analysis forming unique clades distinct from the other cyanobacteria (Watanabe *et al.*, 2011). In a close relative, *Nostoc punctiforme*, several bands corresponding to the high molecular mass PSI complexes were observed in BN gels (Cardona *et al.*, 2007, 2009); however, differences between them have not been investigated. What caused appearance of a PSI tetramer in *Anabaena* 7120 (Watanabe *et al.*, 2011) during evolution is not yet known.

ATP synthase

A part of free energy gained during photosynthetic light reactions is converted into an electrochemical gradient of protons ($\Delta\mu H^+$) across the thylakoid membrane. The proton gradient serves as the energy source for F_1F_0 ATP synthase located in the same membrane. Thus, a part of sun energy is fixed in the form of ATP. Function of the ATP synthase is reviewed in von Ballmoos *et al.* (2009).

The ATP synthase complex of *Synechocystis* 6803 is easily recognized in 2-D maps since almost all subunits are well resolved (Fig. 4.3 and Herranen *et al.*, 2004; Cardona *et al.*, 2007). The ATP synthases of all species are similarly structured. The complex comprises the integral membrane F_0 domain and the hydrophilic F_1 domain exposed to the stroma (Fig. 4.1). The F_1 domain, which converts ADP and P_i to ATP, consists of five subunits, α, β, γ, δ, and ε, encoded by *atpA*, *B*, *C*, *D*, and *E*, respectively, with a stoichiometry of $\alpha_3\beta_3\gamma\delta\varepsilon$. The integral membrane F_0 domain consists of three different polypeptides, a (*atpI*), b (*atpF*), and c (*atpH*); cyanobacteria contain an additional gene, *atpG*, a duplicated and diverged form of the *atpF* gene, encoding the b′ subunit (Curtis, 1988). Multiple copies of the c subunit form a c-ring; the number of c subunits shows substantial variation in different organisms, from 10 to 15 (Vollmar *et al.*, 2009). The γ subunit constitutes a central rotating shaft that mediates the energy exchange between the proton flow at F_0 and ATP synthesis at the F_1 domain (Noji *et al.*, 1997). Subunits b and b' form a heterodimer (Dunn *et al.*, 2001) and, as can be suggested from the crystal structure of the mitochondrial enzyme (Dickson *et al.*, 2006), constitute a peripheral stalk extending from the top of the F_1 domain into the membrane where it binds the a subunit (Walker and Dickson, 2006). Crystal structures have been resolved for the F_1 domain from mitochondria (Abrahams *et al.*, 1994) as well as for the c-ring rotor of the proton-driven ATPases from chloroplasts (Vollmar *et al.*, 2009) and the cyanobacterium *Spirulina platensis* (Pogoryelov *et al.*, 2009).

NDH-1 complexes

The intracellular ATP:NADPH ratio can be dramatically increased, when cells require it, with cyclic electron flow around PSI. In this process, electrons from the acceptor side of PSI return back to the PQ pool raising $\Delta\mu H^+$ across the thylakoid membrane, and the F_1F_0ATP synthase driven by the proton gradient forms ATP without the net production of NADPH or reduced ferredoxin. Multisubunit NDH-1 complexes form one of the pathways for electron transport from PSI to the PQ pool. In addition, they participate in cell respiration and the CO_2 uptake.

Cyanobacterial NDH-1 complexes belong to a family of mitochondrial Complex I enzymes and bacterial type 1 NADH dehydrogenases that pump protons across the inner membrane of mitochondria or the plasma membrane of various bacteria. For reviews on NDH-1 of cyanobacteria, see Battchikova and Aro (2007), Ogawa and Mi (2007) and Battchikova *et al.* (2011a,b). According to reverse genetic studies, several distinctive NDH-1 complexes may coexist in cells of many cyanobacteria. They share some common subunits, NdhA–C, E, G–K, homologous to similar enzymes in other organisms, and contain specific subunits, NdhD1–D4, NdhF1–F3, CupA (ChpY) and CupB(ChpX), that may account for functional distinctions of individual complexes. Proteomic analyses corroborated this concept at the protein level and significantly advanced the understanding of the structure and subunit composition of distinct NDH-1 complexes in cyanobacteria. Based on the genomic content, four complexes, NDH-1L, NDH-1L′, NDH-1MS and NDH-1MS′, have been predicted to exist in *Synechocystis* 6803 and *T. elongatus*. The names of complexes vary in publications; for simplicity, we refer to names mentioned in the proteomic research.

The NDH-1L complex is present in all cyanobacteria (Fig. 4.1). Functional studies of several *ndh* mutants demonstrated the involvement of this complex in cyclic electron flow and respiration. NDH-1L is constitutively expressed in cyanobacterial cells. In BN-PAGE, it moves in close proximity to ATP synthase (Fig. 4.3; Herranen *et al.*, 2004; Prommeenate *et al.*, 2004). Fifteen subunits were detected in this complex by SDS-PAGE. Mass spectrometry identification and N-terminal protein sequencing showed that they include NdhA–NdhK, which are homologous to the subunits of Complex I from *E. coli*, and NdhL–NdhO, specific for cyanobacteria (Prommeenate *et al.*, 2004; Zhang *et al.*, 2005; Battchikova *et al.*, 2005). Specifically, the NDH-1L complex contains NdhD1 and NdhF1 subunits. Further, three novel small subunits, NdhP, NdhQ (Nowaczyk *et al.*, 2011) and NdhS (Battchikova *et al.*, 2011c) were discovered in proteomic analyses of NDH-1L complexes isolated from *T. elongatus* and *Synechocystis* 6803, respectively, by affinity chromatography.

It was suggested that another NDH-1 variant, the NDH-1L′ complex, would be similar to NDH-1L except that it should contain the NdhD2 subunit instead of NdhD1. Reverse genetic experiments predicted a similar function for these two complexes (Ohkawa *et al.*, 2000). According to microarray data, the *Synechocystis ndhD2* gene is up-regulated under some environmental conditions (Lee *et al.*, 2007; Summerfield and Sherman, 2008). However, the NdhD2 protein, and consequently the NDH-1L′ complex, has not been identified at the protein level. The 2-D BN/SDS PAGE protein pattern of the *ndhD2* mutant of *Synechocystis* 6803 was nearly identical to the wild-type pattern (unpublished results from our laboratory). Likewise, NdhD2 was not detected by shotgun proteomics of the *Synechocystis* 6803 membranes (Kwon *et al.*, 2010; Pisareva *et al.*, 2011). Moreover, the subunit was not found in NDH-1 complexes isolated by affinity chromatography (Prommeenate *et al.*, 2004; Zhang *et al.*, 2005; Battchikova *et al.*, 2011c). Most probably, NDH-1L′ is produced in cells in negligible amounts, thus so far avoiding detection and characterization.

The NDH-1MS complex structurally and functionally differs from NDH-1L and NDH-1L′ (Fig. 4.1). According to reverse genetics (Ohkawa *et al.*, 2000; Shibata *et al.*, 2002; Maeda *et al.*, 2002) and functional proteomic (Zhang *et al.*, 2004) studies, NDH-1MS is responsible for the low CO_2-inducible, high-affinity CO_2 acquisition. Contrary to NDH-1L, the NDH-1MS complex is strongly induced in cells grown in low-CO_2 conditions, and it is easily detectable in 2-D BN/SDS PAGE as one of the most abundant protein complexes. NDH-1MS is more fragile compared to NDH-1L, therefore it is found only at low concentrations of the detergent used for solubilization of the membranes (Zhang *et al.*, 2005). The intact NDH-1MS complex (~500 kDa) was found in the thermophilic cyanobacterium *T. elongatus* (Zhang *et al.*, 2005), whereas in

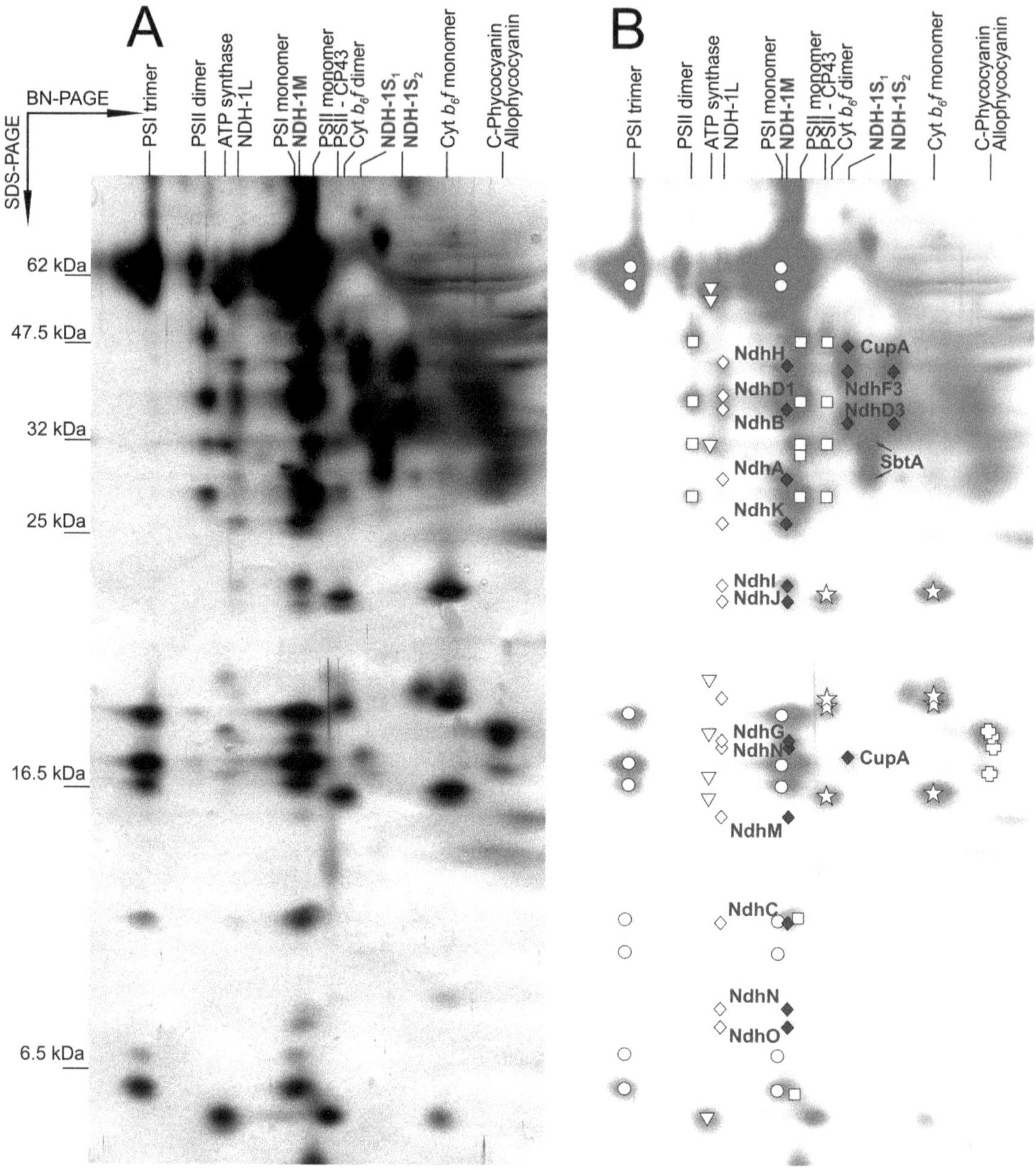

Figure 4.4 2-D BN/SDS-PAGE of thylakoid protein complexes from *Synechocystis* 6803 grown in low CO_2 conditions. (A) A silver-stained gel. (B) Schematic representation of (A), with subunits of complexes depicted by the same symbols as in Figure 4.3 but uncoloured. Identification of proteins is based on mass spectrometry. Proteins and complexes up-regulated in low CO_2 conditions (compared to high CO_2) are marked with red. A colour version of this figure is available in the plate section at the back of the book.

Synechocystis 6803 NDH-1MS dissociated in two subcomplexes, NDH-1M and NDH-1S$_1$. The latter one might degrade further forming NDH-1S$_2$ if DM concentration exceeded 1.2% (Fig. 4.4 and Herranen *et al.*, 2004). Proteomic analysis revealed that the NDH-1M subcomplex contains NdhA–C, NdhE, NdhG–O (Battchikova *et al.*, 2005) as well as NdhS (Battchikova *et al.*, 2011); however, no proteins of the NdhD and NdhF families were found in this complex. It is possible that NdhP and NdhQ discovered by Nowaczyk *et al.* (2011) are also components of NDH-1M; however, this is not experimentally proven yet. On the other hand, the NDH-1S$_1$ subcomplex comprises NdhD3, NdhF3, CupA and CupS (Zhang *et al.*, 2005). Since CupA and CupS are soluble proteins, they easily detach during sample preparation resulting in appearance of the NDH-1S$_2$ complex (NdhD3 and NdhF3, Fig. 4.4, Herranen *et al.*, 2004). It has been suggested that CupA might act as a carbonic anhydrase and catalyse CO_2 hydration (Kaplan and Reinhold, 1999; Maeda *et al.*, 2002). Tchernov *et al.* (2001) have shown that intracellular conversion of CO_2 to HCO_3^- is driven by a PSI-generated $\Delta\mu H^+$. The physiological role of CupS remains unknown since no changes in the phenotype were found in a *cupS* deletion mutant of *Synechococcus* sp. PCC 7002 (Klughammer *et al.*, 1999). Being parts of the large NDH-1MS complex, NDH-1M and NDH-1S$_1$ (NDH-1S$_1$ together with NDH-1S$_2$) show congruent changes in expression, for example, they both are strongly induced in *Synechocystis* 6803 cells grown photo-autotrophically at low CO_2 conditions, and a rise in CO_2 concentration drastically reduces their expression (Herranen *et al.*, 2004; Zhang *et al.*, 2005).

The fourth NDH-1 variant, NDH-1MS′, is supposed to contain all the subunits of NDH-1M plus NdhD4, NdhF4 and CupB. The latter subunits were shown to be involved in a constitutive, low affinity CO_2 uptake (Ohkawa *et al.*, 2000; Shibata *et al.*, 2002; Maeda *et al.*, 2002); however, the NDH-1MS' complex was difficult to detect at the protein level. The 2-D membrane proteome of the *ndhD3/D4* mutant of *Synechocystis* 6803 was indistinguishable from the one of the *ndhD3* mutant (unpublished results from our laboratory). Similarly to NdhD2 of NDH-1L′, the NdhD4, NdhF4 and CupB proteins from NDH-1MS' have not been detected by proteomic studies (Battchikova *et al.*, 2010; Kwon *et al.*, 2010; Pisareva *et al.*, 2011). Introduction of His$_6$ and cMyc tags into the C-terminus of *Synechocystis* CupB followed by affinity purification helped Xu *et al.* (2008) to demonstrate the presence of the CupB protein in a membrane-bound protein complex with a size of ~450 kDa, close to that of NDH-1MS. Nevertheless, the NDH-1MS′ complex appeared to be highly unstable during purification hampering its detailed characterization.

The specific membrane location of cyanobacterial NDH-1 complexes has for a long time been an open question. In earlier reports, Ndh subunits were found in both the plasma membrane and the thylakoid membrane of *Synechocystis* 6803 (Berger *et al.*, 1991; Pieulle *et al.*, 2000) and *Synechococcus elongatus* (strain PCC 7942, formerly known as *Anacystis nidulans*) (Dworsky *et al.*, 1995), or explicitly in the plasma membrane of *Anabaena* 7120 (Howitt *et al.*, 1993). It has been suggested that some variants of NDH-1 complexes might have a dual location in both the thylakoid membrane and the plasma membrane, or that NDH-1 complexes having dissimilar functions might reside on different membranes (Ohkawa *et al.*, 1998; Price *et al.*, 1998). However, other results (Ogawa 1992) including some obtained in proteomic studies (Zhang *et al.*, 2004; Xu *et al.*, 2008; Pisareva *et al.*, 2011) imply that all four variants of NDH-1 complexes reside largely, if not exclusively, in the thylakoid membrane of *Synechocystis* 6803. The inconsistency might be explained by differences in the purity of the membrane preparations obtained with different isolation methods. The more recent experiments that demonstrated the thylakoid localization of the NDH-1 complexes in *Synechocystis* 6803 (Zhang *et al.*, 2004; Xu *et al.*, 2008; Ohkawa *et al.*, 2001, 2002; Pisareva *et al.*, 2011) were performed with highly pure thylakoid and plasma membranes prepared by the two-phase partitioning method (Norling *et al.*, 1998). The occurrence of NDH-1 complexes in thylakoids *in vivo* has been confirmed by microscopy investigation of *Synechococcus elongatus* containing the NdhM subunit fused with GFP (Liu *et al.*, 2012). Yet, at present, a dissimilar location of low abundance complexes, NDH-1L′ and NDH-1MS′, cannot be completely excluded.

The functional mechanism of cyanobacterial NDH-1 complexes, specifically an electron donor and the protein(s) comprising the activity domain, remain an intriguing question (Battchikova *et al.*, 2011a). In chloroplasts of *Arabidopsis thaliana*, the NDH complex analogous to NDH-1L of cyanobacteria has been shown to interact with PSI, and a PSI-NDH supercomplex has been observed by BN-PAGE and 2D BN/SDS-PAGE (Peng *et al.*,

2008; Sirpio *et al.*, 2009). The ferredoxin (Fd)-binding subunit required for NDH function *in vivo* has been identified as part of the chloroplast NDH-PSI supercomplex (Yamamoto *et al.*, 2011) indicating that Fd is the electron donor to NDH from green plants. A homologous Fd-binding subunit, NdhS, has been found in NDH-1 complexes of *Synechocystis* 6803 (Battchikova *et al.*, 2011b); however, a supercomplex of NDH-1 and PSI was not detected in cyanobacteria. On the contrary, in *Synechocystis* 6803, Ma *et al.* (2006) described a large NDH-1 supercomplex (designated as Act-NDH-1Sup, Active Super-complex) that possessed NADPH dehydrogenase activity, and the activity was proportional to cyclic PSI electron flow. The proteins contributing to the NADPH activity of Act-NDH-1Sup remain unknown. Nonetheless, formation of the PSI-NDH-1 supercomplexes in cyanobacteria should not be excluded. Confocal microscopy analysis demonstrated that intracellular distribution of NDH-1 complexes in *Synechococcus elongatus* varied depending on light conditions (Liu *et al.*, 2012). When cells were exposed to moderate light, NDH-1 complexes were evenly distributed within the thylakoid membranes. Under low light, the complexes concentrated in thylakoids in discrete patches. It is possible that cyanobacterial NDH-1 complexes use different activity domains and/or electron input devices in diverse light conditions (Battchikova *et al.*, 2011a). All such putative domains, however, still remain to be discovered.

Quantitative proteomics: response of the thylakoid membrane proteome to changes in environmental conditions

The thylakoid membrane proteome, being a part of the overall cellular proteome, readily responds to changes in growth environment and to various stress conditions, reflecting dynamic modifications of metabolic and catabolic pathways in order to adjust cyanobacterial cells to a new environment. Light is the most important factor for the photosynthetic function, but other factors, such as availability of various nutrients, temperature, oxidative and heavy metal stresses etc., as well as cellular differentiation occurring in many filamentous cyanobacteria, also cause an adjustment of thylakoid proteins and protein complexes since the thylakoidal membrane is the fundamental energy producing compartment.

Alteration of gene expression in response to external or internal stimuli is often assessed at the transcriptional level. Global investigations of changes in transcript profiles become more and more affordable with development of custom microarrays and third-generation sequencing approaches that permit transcriptome profiling. These techniques that can be carried out in a high throughput manner are powerful tools to study acclimation processes since a distinct correlation is usually observed between up- or down-regulation of transcripts and protein accumulation, although the ratios of induction might not be the same (Suzuki *et al.*, 2006). However, it is important to underline that evident differences have been observed in the alteration patterns of cellular metabolism detected by transcriptomics and proteomics. For example, Suzuki *et al.* (2006) analysed the heat shock response of *Synechocystis* 6803 cells by both approaches and discovered that some proteins such as translation factors EF-G1, EF-G2 and EF-Tu were clearly up-regulated whereas their mRNA levels were not elevated. Such discrepancies are well documented in the literature (Glanemann *et al.*, 2003; Waters *et al.*, 2006; de Groot *et al.*, 2007). They, most probably, indicate participation of other mechanisms in the regulation of gene or protein expression that occur post-transcriptionally and therefore cannot be revealed by mRNA profiling.

A variety of methods is available for a quantitative proteomic analysis, both for relative and absolute quantitation (for reviews, see Ong *et al.*, 2003; Bantscheff *et al.*, 2007; Nakamura and Oda, 2007); relative quantitation techniques are mostly applied at present. 2-D IEF/SDS-PAGE-based differential proteomics was used to analyse the response of cyanobacterial cells to, for instance, heat shock (Suzuki *et al.*, 2006; Slabas *et al.*, 2006), heterotrophy (Kurian *et al.*, 2006a), acid growth conditions (Kurian *et al.*, 2006b), nitrogen-fixing conditions (Sandh *et al.*, 2011), or to compare light- and dark-cultured cells (Choi *et al.*, 2000). For quantitation, proteins were stained with fluorescent dyes (difference gel electrophoresis, DIGE; Ünlü *et al.*, 1997; van den Bergh and Arckens, 2003), or stained with SYPRO Ruby, or radioactively labelled; silver staining allows only qualitative evaluation of differential expression. It is important to note that limitations in proteomic analysis of membrane proteins described earlier for protein identification hold true also for quantitative studies. In quantitative proteomic investigations based on 2D IEF/SDS-PAGE, either only a soluble protein fraction was analysed (Kurian *et al.*, 2006a,b), or only limited numbers of membrane proteins were evaluated (Suzuki *et al.*, 2006; Slabas *et al.*, 2006; Sandh *et al.*, 2011). Sometimes, relative changes in abundance of photosynthetic complexes could be assessed by quantitation of few relatively hydrophilic subunits. Thus, Slabas *et al.* (2006) described about 2-fold up-regulation of PSI, PSII and ATP synthase during the heat shock treatment of *Synechocystis* 6803 on the basis of one or two separate subunits in each complex, and about 2-fold down-regulation of phycobilisomes based on the CpcG1 component.

Shotgun approaches are significantly more efficient for differential quantitation of membrane proteins. The iTRAQ approach (isobaric tags for relative and absolute quantification; Ross *et al.*, 2004; Aggarwal *et al.*, 2004) is a well-established technique that allows simultaneous investigation of several samples (4- and 8-plex kits are commercially available). It has been successfully applied to study responses of *Synechocystis* 6803 to CO_2 limitation (Battchikova *et al.*, 2010) and to a high salt environment (Pandhal *et al.*, 2009), and to evaluate filamentous *Nostoc* strains grown in N_2-fixing and non-N_2-fixing conditions (Stensjö *et al.*, 2007; Ow *et al.*, 2008, 2009). Studies have been targeted to total cellular proteomes but a significant bias was still observed towards the soluble proteins (Battchikova *et al.*, 2010). Moreover, only a fraction of proteins that are unambiguously identified can be quantified in shotgun experiments (Bantscheff *et al.*, 2007) since a reliable accuracy of quantitation is more difficult to achieve compared to consistent identification. To our knowledge, no shotgun experiments have been conducted to specifically investigate only the thylakoid membranes of cyanobacteria.

In contrast to 2-D IEF/SDS-PAGE, the 2-D BN (or CN)/SDS-PAGE approach is applicable for differential analysis of thylakoid membrane protein complexes. It can be combined with appropriate staining that allows reliable quantitation (DIGE, SYPRO Ruby, or radioactive labelling); however, changes could be rapidly assessed qualitatively. Comparison of Fig. 4.4 to Fig. 4.3 demonstrates an evident increase of the components of the NDH-1MS complex in cells of *Synechocystis* 6803 shifted from high-CO_2 (3%) to low-CO_2 (air) conditions as well as the induction of the SbtA protein, the plasma membrane bicarbonate transporter (Herranen *et al.*, 2004). The method is well suited for comparison of the WT thylakoid proteome with those of mutants in different components of electron transfer chains (Zhang *et al.*, 2004) and for evaluation of differences between thylakoids and other membranes (Cardona *et al.*, 2009). Since 2-D BN(or CN)/SDS-PAGE and the shotgun proteomics are complementary to each other, a combination of these approaches would provide the most comprehensive examination of changes that occur in proteins of the thylakoid membrane during acclimation to various stress factors and environmental challenges.

Acclimation of the *Synechocystis* 6803 proteome to low CO_2 conditions

The quantitative 4-plex iTRAQ proteomic approach has been used to study alterations that occur in the proteome of *Synechocystis* 6803 when cells are shifted from high CO_2 (HC, 3%) to the low CO_2 (LC, air level) condition (Battchikova *et al.*, 2010). To follow dynamic proteome changes, protein samples were collected at 0 h, 6 h, 24 h, and 72 h after the shift. The highest levels of up-regulation were observed for proteins of the NDH-1MS complex, the Na^+-dependent bicarbonate transporter SbtA, and the ABC-type transporter BCT-1, that participate in CO_2 (NDH-1MS) and bicarbonate (SbtA and BCT-1) uptake. The most dramatic changes were detected for the CO_2 uptake system located in the thylakoid membrane, i.e. the NDH-1MS complex. Three from the four specific subunits of NDH-1MS, CupA (Sll1734), CupS (Sll1735) and NdhD3 (Sll1733), were quantified in this study, and they demonstrated high up-regulation in the LC condition (4.2-, 3.6- and 2.9-fold, respectively, at 72 h). Levels of induction were slightly lower, 1.6- to 2.8-fold, for seven hydrophilic subunits, NdhH (Slr0261), NdhI (Sll0520), NdhJ (Slr1281), NdhK (Slr1280), NdhM (Slr1623), NdhN (Sll1262), and NdhO (Ssl1690), and two hydrophobic subunits, NdhA (Sll0519) and NdhG (Sll0521). Dissimilar degrees of up-regulation are well in line with occurrence of several variants of NDH-1 complexes. It is worthy to note that the hydrophilic CupB protein (Slr1302), specific for the NDH-1MS′ complex, was not detected indicating that the level of its expression in *Synechocystis* cells is rather low. Beside the CO_2 uptake system, bicarbonate transporters located in the plasma membrane were also accumulated in response to CO_2 downshift. Amounts of SbtA (Slr1512) and SbtB (Slr1513) increased in LC conditions (3.1- and 3.6-fold, respectively, at 72 h). Despite the fact that the function of the SbtB protein remains unknown, clustering of the two genes together and a coordinate manner of their expression suggest that these two proteins interact with each other and/or function together. Further, subunits of the ATP-dependent BCT-1 transporter encoded by the *cmpABCD* operon (*slr0040-slr0043*) were co-ordinately up-regulated in response to the CO_2 downshift (2.1- to 4.6-fold at 6 h).

These three carbon uptake systems exhibited markedly different dynamics of induction in the course of the LC treatment. The up-regulation of the BCT-1 transporter was fast and reached nearly the maximum level at 6 h. SbtA and SbtB also showed a significant increase at the 6-h point; however, maximum values were attained later at 24 h after CO_2 downshift. The subunits of the NDH–1MS complex exhibited progressive increase in protein accumulation up to 72 h. Despite the strong up-regulation of NDH-1MS, bicarbonate uptake mechanisms remained expressed at elevated levels at later stages. This is consistent with a readiness of cells to maximally utilize inorganic carbon, both as CO_2 and CO_3^{2-} formed upon dissolving of carbon dioxide in the growth medium. Further, up-regulation of both subunits of Rubisco, RbcL (Slr0009, 1.9-fold) and RbcS (Slr0012, 2-fold), and a moderate increase in the carboxysome shell components, CcmK1 (Slr1029), CcmK2 (Slr1029), and CcmM (Sll1031) assisted in maximizing the carbon fixation in LC conditions.

Concurrently, photoprotection and reactive oxygen species (ROS) response mechanisms became enhanced to diminish the damage by excess of excitation energy to components of the photosynthetic apparatus. Quantitation data obtained for subunits of PSI, PSII, and phycobilisomes revealed a slight down-regulation of these complexes in the course of the experiment. Enzymes involved in the biosynthesis of pigments and cofactors, such as haem, phycobilin, porphyrin, and Chl, were distinctly down-regulated. At 72 h of the LC treatment, porphobilinogen synthase (HemB, Sll1994), coproporphyrinogen III oxidase (HemF, Sll1185), geranylgeranyl hydrogenase (ChlP, Sll1091), and magnesium-protoporphyrin IX monomethyl ester cyclase (Sll1214) demonstrated marked repression (0.5- to 0.7-fold). Further, the water-soluble orange carotenoid protein (OCP, Slr1963), which is essential for phycobilisome-associated non-photochemical quenching (Wilson *et al.*, 2006), demonstrated a steady increase during acclimation of cells to LC and reached a 1.7-fold up-regulation at 72 h. The flavodiiron protein Flv2 (Sll0219) showed fast induction and reached a plateau level (1.6-fold) already at 6 h. Flv2 together with Flv4 (not detected in the study) protects the oxygen-evolving PSII complex against photo-inhibition (Zhang *et al.*, 2009b, 2012). The Flv3 (Sll0550) protein that functions in the Mehler reaction donating electrons from PSI directly to molecular oxygen (Helman *et al.*, 2003) accumulated steadily during the whole course of the low-CO_2 treatment and reached a value of 2.4 at 72 h. From enzymes protecting cells against ROS, glutathione peroxidase Gpx2 (Slr1992) demonstrated a fast response to LC conditions (1.6-fold up-regulation at 6 h) followed by a slight further increase at 72 h (1.9-fold). These mechanisms appeared to be efficient in protecting the photosynthetic complexes and other cell components against ROS, since amounts of enzymes involved in H_2O_2 dismutation, i.e. superoxide dismutase (SodB, Slr1516) and bifunctional catalase-peroxidase (KatG, Sll1987), remained unchanged. Subunits of ATP synthase and the Cyt b_6f complex, as well as soluble electron carriers plastocyanin (PetE, Sll0199) and FNR (PetH, Slr1643), did not show noticeable differences.

Harmonization of carbon and nitrogen assimilation is crucial for the optimization of cell metabolism. Results showed that the CO_2 downshift leads to a fast and significant decrease in expression of transporters and other proteins participating in the uptake of fixed nitrogen and the assimilation of this nutrient. Down-regulation was observed for components of ribosomes, chaperonins and some other proteins (Battchikova *et al.*, 2010). Since we focus here on the thylakoid proteome and thylakoid-associated proteins, they will not be discussed further.

Proteomes of filamentous strains and N_2 fixation

Extensive quantitative proteomic studies using 4- and 8-plex iTRAQ technology have been performed to examine the response of filamentous cyanobacteria to nitrogen starvation. In growth medium supplemented with ammonium, filaments of *Anabaena* 7120 and *Nostoc punctiforme* consist of vegetative cells. In nitrogen-depleted medium, extensive changes occur in cell structure and cell metabolism resulting in differentiation of some cells (5–10%) into heterocysts specialized for fixation of atmospheric N_2. Modifications in the gene expression pattern and morphogenesis that happen in the differentiation process are required for protection of O_2-sensitive nitrogenase; in detail, they are described in reviews (Flores and Herrero, 2010; Golden and Yoon, 2003; Herrero *et al.*, 2004; Xu *et al.*, 2008; Kumar *et al.*, 2010; see also Chapter 12).

In heterocysts, intracellular membranes are rearranged into 'honeycomb' structures close to the cellular poles (Lang and Fay, 1971), and the water-oxidizing activity of PSII is lacking (Thomas 1972; Thiel *et al.*, 1990), what contributes to establish a micro-oxic environment. The reducing power required for N_2 fixation comes in the form of carbohydrates imported from vegetative cells; in return, heterocysts supply fixed nitrogen to other cells in a filament. Concurrently, the lack of PSII activity results in significant alterations of energy metabolism and electron transfer pathways.

Quantitative proteomic studies of *Anabaena* 7120 and *Nostoc punctiforme* (Stensjö *et al.*, 2007; Ow *et al.*, 2008, 2009) demonstrated differential expression of many soluble proteins involved in N_2 fixation, carbon metabolism and energy production. In heterocysts and heterocyst-containing filaments, greater abundance was observed for subunits of nitrogenase and proteins related to its synthesis and assembly, for glutamine synthetase, for the oxidative part of the pentose phosphate pathway, for the TCA pathway, glycogen phosphorylase etc. Among important findings, accumulation of specific electron transfer proteins was discovered in heterocysts of *Anabaena* 7120, including ferredoxin FdxH (all1430) that may be the primary electron donor for nitrogenase, two thioredoxins (All2367 and All0513), and two flavoproteins, Flv1B (All0177) and Flv3B (All0178) (Ow *et al.*, 2008). Recently, the exclusive location of Flv1B and Flv3B in heterocysts was confirmed by fusion with the yellow fluorescence protein (YFP) and fluorescence imaging (Ermakova *et al.*, 2013). Accumulation of these flavoproteins suggests that alternative pathways based on flavin-dependent electron transporters are important in heterocysts. In contrast, the level of Flv2 (All4444) was similar in vegetative cells and heterocysts.

A significant decrease of phycobilisome proteins was observed in isolated heterocysts. Studies of *Nostoc punctiforme* (Ow *et al.*, 2009) showed that phycoerythrin, phycocyanin and allophycocyanin subunits were, on average, 7.5-, 3.5-, and 4-fold less abundant in these cells. Degradation of phycobilisomes agrees with their role as an alternative macronutrient reserve described for nitrogen-starved unicellular cyanobacteria (Schwarz and Forchhammer, 2005). It is of note that the differential expression of phycobiliproteins was insignificant when ammonium-grown whole filaments were compared with the N_2-fixing ones (Ow *et al.*, 2009). Moreover, together with a relatively minor decrease of phycoerythrin, phycocyanin and allophycocyanin subunits, in a similar study Stensjö *et al.* (2007) described a marked increase of phycobilisome linker proteins and polypeptides in filaments of *Anabaena* 7120 grown in nitrogen-limited conditions.

It has been suggested that decrease of O_2 evolving activity of PSII is accompanied by degradation of PSII proteins even though there have been some indications that PSII proteins can still be found in the mature heterocyst (Thiel *et al.*, 1990). Quantitative proteomic experiments confidently state otherwise. Levels of PSII proteins in isolated heterocysts and vegetative cells remain statistically unbiased, with only few exceptions (Stensjö *et al.*, 2007; Ow *et al.*, 2008, 2009). The conclusion derived from shotgun experiments is corroborated by studies of thylakoid protein complexes isolated from heterocysts of *Nostoc punctiforme*. 2-D BN/SDS-PAGE analysis showed the presence of PSII complexes in the monomeric form associated with PsbO (Cardona *et al.*, 2009). Moreover, the heterocyst-derived PSII complexes were capable of light-induced electron transfer but not of oxygen evolution.

A substantial increase of PSI proteins (about 2.2-fold) was detected when filaments of *Anabaena* 7120 were grown in N_2-fixing conditions (Ow *et al.*, 2008). Similarly, about 3.3-fold accumulation of PSI subunits was identified in isolated heterocysts of *Nostoc punctiforme* (Ow *et al.*, 2009). In parallel, N_2-fixing cells accumulate subunits of NDH-1 complexes (2–4-fold). The data did not allow distinguishing NDH-1 variants since only common (and mainly hydrophilic) Ndh subunits were detected. However, these observations imply that a cyclic electron flow around PSI is induced in heterocysts. Consequently, it results in increased generation of ATP that is beneficial for nitrogenase activity since nitrogen fixation is an energy demanding process. The induction of cyclic electron transport is corroborated by raised levels of ATPase subunits (about 3-fold) and components of the Cyt b_6f complex (about 2-fold) (Ow *et al.*, 2009). FNR was also upregulated (Ow *et al.*, 2008, 2009). It is of interest that plastocyanin (PetE) was strongly decreased (3.8-fold) in heterocysts of *Nostoc punctiforme* (Ow *et al.*, 2009). In *Anabaena* 7120, a significant increase (3.5-fold) of the cytochrome c_6 protein (CytA) was noted in N_2-fixing cells, in contrast to photosynthesis-related ferredoxin PetF that did not show differential regulation. Reasons for the difference in abundance of some of the soluble electron carriers is yet unknown; however, it hints to alteration of the electron transport pathway that involves PSI and Cyt b_6f.

Biogenesis of the thylakoid membrane and protein complexes

A long standing question of the physical relationship between the thylakoid membrane and the plasma membrane is intertwined with an understanding of biogenesis of the thylakoid membrane, the means by which it is formed and how the proper protein composition of thylakoid complexes is established, modified, and repaired, in order to keep the photosynthetic machinery optimally functional. Various experimental approaches, including reverse genetics, proteomics, high resolution 3-D electron microscopy, and electron tomography have been used to get insights into the biogenesis of thylakoid membranes, and conclusions were sometimes controversial. Many questions of thylakoid biogenesis including lipid and pigment biosynthesis, membrane formation, protein synthesis and assembly of photosynthetic complexes are discussed in separate reviews (Liberton and Pakrasi, 2008; Mullineaux, 2008; Nickelsen *et al.*, 2011; see also Chapter 5).

Formation of the thylakoid compartment

In general, three scenarios have been proposed for formation of the thylakoid membrane as the cell compartment. One of the hypotheses implies that during thylakoid membrane biogenesis the formation and growth of additional membrane components including pigment and protein biosynthesis occur on existing thylakoids. Liberton *et al.* (2011) studied the membrane ultrastructure of *Cyanothece* sp. ATCC 51142 and observed a number of examples of structures that could represent sites of thylakoid membrane biogenesis by a branching and splitting mechanism. Further, in several cyanobacterial species abundant lipid bodies were detected (Nierzwicki-Bauer *et al.*, 1983; Liberton *et al.*, 2006, 2011; de Meene *et al.*, 2006) which might be a source of lipids for the developing thylakoid membrane. In agreement with this hypothesis, the distribution of lipid bodies was restricted to locations among the thylakoid membrane pairs and adjacent to the plasma membrane. However, a mechanism of lipid delivery necessary for thylakoid expansion remains unknown. Likewise, this hypothesis does not explain why some subunits of the photosynthetic complexes are present also in the plasma membrane (Zak *et al.*, 2001, Srivastava *et al.*, 2006; Pisareva *et al.*, 2001), and how these complexes are incorporated into developing thylakoid membranes since thylakoid layers were found to be essentially devoid of ribosomes (van de Meene *et al.*, 2006).

Another hypothesis suggests that thylakoid membranes might arise from the plasma membrane (Zak *et al.*, 2001). In this scenario, biosynthesis of several polypeptides comprising the photosynthetic complexes, as well as an initial assembly of these complexes including incorporation of some pigment molecules, occur in the plasma membrane. Further, if membranes are contiguous, a direct transfer by diffusion might take place relocating photosynthetic precomplexes and lipids from plasma membrane to their final destination in thylakoids. Alternatively, if thylakoids and plasma membrane are completely different compartments, lipids and proteins from the plasma membrane might be delivered by a vesicle transport system similar to one existing in chloroplasts (Nevo *et al.*, 2007). In agreement with this hypothesis, some components of the photosynthetic electron transfer chain have been found in the plasma membrane, although in much lower amounts than in thylakoids. Using an immunological approach, Zak *et al.* (2001) demonstrated that some amounts of PsaA and PsaB, the core proteins of PSI, as well as D1, D2, and Cyt b_{559}, the core proteins of PSII, are present in a highly purified plasma membrane fraction. Studies of PSII subunits in right-side-out and inside-out vesicles formed from this compartment showed a heterogeneous organization of the plasma membrane (Srivastava *et al.*, 2006). Using proteomic analysis of pure membrane preparations, Pisareva *et al.* (2011) confirmed the dual location of the PSI components and, in addition, demonstrated a similar distribution for several subunits of ATP synthase. They noted, nevertheless, an extremely low content of PSII subunits in the plasma membrane. However, in disagreement with this hypothesis, high resolution studies of the membrane ultrastructure (Nierzwicki-Bauer *et al.*, 1983; Liberton *et al.*, 2006, 2011; de Meene *et al.*, 2006) could not demonstrate a clear continuity between thylakoid and plasma membrane systems despite a close proximity of these membranes at multiple sites in cells. It is important to note that no structures resembling transport vesicles were observed, either.

The third hypothesis proposes that the components are synthesized and assembled in specialized thylakoid regions, like the thylakoid centres discussed earlier (Hinterstoisser *et al.*, 1993; van de Meene *et al.*, 2006). An intermediate membrane subfraction different from both thylakoid and plasma membrane has been described (Schottkowski *et al.*, 2009a). This subfraction (marked by the presence of the PratA factor that interacts with the D1 protein, see text later) might represent a specific sub-compartment connecting plasma and thylakoid membranes where biogenesis of photosynthetic complexes takes place (Schottkowski *et al.*, 2009b, Nickelsen *et al.*, 2011).

Recently, the second and third hypotheses have been developed further (Pisareva *et al.*, 2011; Nickelsen *et al.*, 2011) by suggesting that the plasma and thylakoid membranes are not fused but dynamically connected, and that specific thylakoid structures at least temporally make contacts with the plasma membrane situated nearby. At present, although biogenesis and maintaining of thylakoid membrane organization remain open issues, many experimental results are consistent with the hypothesis of a dual location of components and assembly factors of the photosynthetic complexes.

The VIPP1 protein

The VIPP1 protein plays a crucial role in the biogenesis of thylakoid membranes. In *Arabidopsis thaliana*, deletion of *vipp1* resulted in a complete loss of thylakoid membranes in chloroplasts (Kroll *et al.*, 2001). It was suggested that VIPP1 could be involved in vesicle trafficking between the inner envelope and the thylakoid membrane of chloroplasts. Deletion of the corresponding gene (*sll0617*) in *Synechocystis* 6803, carried out in two independent laboratories (Westphal *et al.*, 2001; Fuhrman *et al.*, 2009a), demonstrated that the gene is essential for cell viability. Although full segregation of the mutant chromosomes was not achieved, the amount of the Vipp1 protein was strongly diminished in merodiploid mutant cells. In both cases, the mutation resulted in reduction of thylakoid membranes in a cell and in disturbed morphology of thylakoids (Westphal *et al.*, 2001; Fuhrman *et al.*, 2009a). In the mutant cells obtained by Westphal *et al.* (2001), no organized thylakoids could be observed, with membrane-like structures dispersed in low numbers throughout the cytoplasm, and no light-dependent oxygen evolution was detected. In contrast, Fuhrman *et al.* (2009a) observed a significant amount of remaining internal thylakoid membranes coupled with photosynthetic activity in their *sll0617* mutant. They noted that depletion of the Vipp1 protein caused a rather specific reduction and destabilization of the PSI trimers in *Synechocystis* 6803, while the amount of active PSII was not significantly changed. The results undoubtedly indicated that the cyanobacterial Vipp1 is involved in biogenesis of thylakoid membranes. However, it is not clear whether Vipp1 directly influences the thylakoid membrane formation or affects only the biogenesis and/or stability of the PSI complex. Westphal *et al.* (2001) determined that in wild-type cells, Vipp1 is localized exclusively at the plasma membrane. Fuhrman *et al.* (2009b) have found Vipp1 of *Synechocystis* 6803 at different places within the entire cell, mainly associated with membranes but also in the soluble protein fraction. Intriguingly, they have shown that Vipp1 of *Synechocystis* 6803 forms oligomeric rings of various diameters, with different monomer stoichiometries. Mostly, Vipp1 oligomers were tightly attached to both the thylakoid and the plasma membrane, though some of these rings existed in the cytoplasm in a soluble form (Fuhrman *et al.*, 2009b). The authors presented an interesting hypothesis that Vipp1 might assemble into a ring around the neck of a forming budding vesicle, similar to dynamin in eukaryotic cells. Then, stepwise release of individual Vipp1 units would form rings with smaller diameters, bringing the membranes into a close contact, which would result in membrane fusion and pitching off a vesicle. Thus, formation of Vipp1 ring structures may play a significant role in thylakoid membrane biogenesis.

Heterogeneous distribution of assembly factors of the PSII complex

Discoveries in the plasma membrane of several subunits that belong to photosynthetic complexes, as discussed above, raised the question of whether they exist in this compartment as individual (pre)proteins or whether they form any kind of assemblies. Zak *et al.* (2001) analysed the plasma membrane complexes by BN-PAGE and showed that precomplexes of both photosystems are indeed assembled there. The plasma membrane PSII core complex contained D1, D2 and both α and β subunits of Cyt b_{559}, while the PSI core complex comprised PsaA, PsaB, PsaC, and PsaD proteins.

In light, PSII complexes undergo constant turnover due to the oxidative chemistry of water splitting. The D1 protein of PSII becomes degraded during photosynthesis, especially under high light conditions, and the D1 degradation initiates a complex PSII repair cycle (for recent reviews, see Mulo *et al.*, 2009; Nixon *et al.*, 2010; Komenda *et al.*, 2012). Several PSII assembly factors have been shown to be essential for this process.

The *psbA* gene encodes a precursor of the D1 protein, pre-D1, containing a carboxyl-terminal extension (reviewed in Satoh and Yamamoto, 2007). The Ycf48 protein (Slr2034 in *Synechocystis* 6803), which is homologous to the lumenal HCF136 of *Arabidopsis thaliana*, is important for stabilization of newly synthesized pre-D1 before its binding to a D2-Cyt b_{559} pre-complex (Komenda *et al.*, 2008). In addition, Ycf48 also functions in the selective replacement of photodamaged D1 during PSII repair (Komenda *et al.*, 2008). Consistently, proteomic studies demonstrated the presence of Ycf48 in both the thylakoid and plasma membrane fractions (Pisareva *et al.*, 2011). In the maturation process, the pre-D1 extension is cleaved by CtpA, a C-terminal peptidase (Shestakov *et al.*, 2004). In *Synechocystis* 6803, the enzyme was exclusively localized in the plasma membrane (Zak *et al.*, 2001). The function of CtpA is facilitated by the PratA factor, Slr2048 (Klinkert *et al.*, 2004). PratA contains tetratricopeptide repeats and is a soluble periplasmic protein, but a part of it (10–20%) was found to be associated with a membrane (Schottkowski *et al.*, 2009a). The protein directly interacts with the C-terminus of the mature D1 protein, and the *slr2048* deletion affects the cleavage of the pre-D1 C-terminal extension (Schottkowski *et al.*, 2009a). Importantly, cellular fractionation experiments showed that the membrane-bound PratA co-sediments with an intermediate membrane subfraction but not with thylakoid or plasma membranes (Schottkowski *et al.*, 2009a). The PratA-defined membrane subfraction, or PDM, was postulated to represent the compartment where early stages of PSII biogenesis occur, and it was suggested that the PratA factor is required for efficient membrane flow from PDM sites to the thylakoid membrane (Nickelsen *et al.*, 2011). Inactivation of PratA resulted in accumulation of the pre-D1 in PDM (Schottkowski *et al.*, 2009a). Another protein that directly interacts with pre-D1 is the Slr1471 protein that is homologous to the Oxa1/Alb3 protein of *Arabidopsis thaliana*. The *slr1471* gene is vital for the viability of *Synechocystis* 6803 cells since only partial segregation of the *slr1471* deletion mutant was achieved (Spence *et al.*, 2004). Under photoautotrophic conditions, the mutation caused the disorganization of thylakoids and a decrease in cell pigmentation, but the thylakoid membranes remained reasonably functional in terms of the efficiency of PSII electron transport and the tolerance to high light stress. Further studies elucidated that the Slr1471 protein is a membrane-integral chaperone essential for the correct integration, folding, and assembly of D1 into PSII (Ossenbühl *et al.*, 2006). In proteomic experiments, Slr1471 was localized in the thylakoid membrane (Pisareva *et al.*, 2011). These proteins are involved in early stages of PSII biogenesis.

At the later stages, the complete assembly of the PSII complexes depends on some other proteins. It was mentioned above that proteomic analysis of PSII isolated from *Synechocystis* 6803 demonstrated the presence of several proteins such as Psb27 and Psb28 (Kashino *et al.*, 2002b). The Psb27 (Slr1645) protein was shown to be important for assembly of the Mn_4 cluster of PSII during *de novo* synthesis and the repair cycle (Nowaczyk *et al.*, 2006, Grasse *et al.*, 2011). Rooze and Pakrasi (2008) suggested that Psb27 prevents binding of extrinsic lumenal PSII subunits until manganese, calcium and chlorine atoms achieve an appropriate cluster formation. After complete assembly of the Mn_4 cluster, Psb27 dissociates from the PSII complex and allows the assembly of the subunits of the oxygen evolving complex. For the Psb28 (Sll1398) protein, it was shown that it is associated with the PSII complex lacking CP43 (Dobáková *et al.*, 2009). It was suggested that Psb28 is involved in the biogenesis of the PSII inner antenna CP47. Further, the Psb29 (Sll1414) protein was found to be important for resistance of *Synechocystis* 6803 cells to elevated levels of light (Keren *et al.*, 2005); however, the exact function of Psb29 remains unclear.

Location of assembly factors of the PSI complex

The correct assembly of cyanobacterial PSI complexes also requires several factors. They include Ycf3, Ycf4 and Ycf37 proteins. The role of these factors is well established in green plants and *Chlamydomonas reinhardii* (for review, see Rochaix 2011). Ycf3 contains tetratricopeptide repeats and is an essential protein that interacts with PsaA and PsaD (Naver *et al.*, 2001). Ycf4 and Ycf37 are involved at early stages of PSI assembly, and Ycf4 may act as a scaffold for PSI assembly (Rochaix 2011).

In cyanobacteria the role of these factors might slightly vary since PSI complexes are functional and seemed to be correctly assembled in the *ycf4 (sll0226)* mutant of *Synechocystis* 6803 (Wilde *et al.*, 1995) and in the *ycf37 (slr0171)* mutant of *Synechocystis* 6803 (Dühring *et al.*, 2006). However, PSI complexes were observed in smaller amounts compared to the wild-type. According to immunological studies (Zak *et al.*, 2001), Ycf3 (Slr0823) and Ycf4 are localized primarily or exclusively in the plasma membrane, whereas proteomic analysis (Pisareva *et al.*, 2011) showed about equal distribution of Ycf4 between thylakoids and the plasma membrane. These results support the suggestion of Zak *et al.* (2001) that cyanobacterial Ycf3 and Ycf4 participate in the assembly of the PSI core that occurs in the plasma membrane. In contrast, Ycf37 of *Synechocystis* 6803 was found to be a thylakoid protein (Pisareva *et al.*, 2011). Dühring *et al.* (2006, 2007) discovered that Ycf37 is involved at the latest steps of the PSI biogenesis, probably when PsaK is assembled, and might be important for preservation of PSI complexes under high light conditions. Further, the BtpA (Sll0634) protein, which is localized in thylakoids (Zak *et al.*, 1999), was found to play a role in PSI stability (Zak and Pakrasi, 2000; Zak *et al.*, 2001) since deletion of the *sll0634* gene resulted in a high degradation rate of the PsaA core protein, especially at a low temperature (Zak and Pakrasi, 2000).

Localisation of enzymes of chlorophyll biosynthesis

Incorporation of pigments is required for the proper assembly of photosynthetic complexes and their functional performance. Chlorophylls are the major pigments in PSI and PSII. Importantly, both PSI and PSII core complexes discovered by Zak *et al.* (2001) in the plasma membrane formed green bands in the BN-PAGE indicating that some Chl molecules were incorporated already at the early stages of the assembly. Moreover, it appeared that the PSI core complex from the plasma membrane could undergo charge separation (Zak *et al.*, 2001).

In photosynthetic organisms the synthesis of Chl is performed by a tetrapyrrole biosynthetic pathway (for reviews, see Vavilin and Vermaas, 2002; Chew and Bryant, 2007). Two enzymes involved in the late steps of this pathway were found in both the thylakoid and the plasma membranes by proteomic analysis of pure membrane preparations (Pisareva *et al.*, 2011). One of them is a light-dependent NADPH-protochlorophyllide oxidoreductase (POR) that catalyses the reduction of protochlorophyllide to chlorophyllide. The second one is geranylgeranyl hydrogenase (ChlP) that participates in transformation of chlorophyllide into Chl. Two enzymes are involved in this final step. Chlorophyll *a* synthase (ChlG) adds a geranylgeranyl-pyrophosphate to the chlorophyllide molecule, and ChlP reduces this carbohydrate moiety to the phytyl tail (Chew and Bryant, 2007). The order of these two enzymatic reactions is not yet clear. It has been proposed that attachment of the geranylgeranyl tail occurs prior to its reduction; however, some evidence suggested that ChlP can saturate geranylgeranyl-pyrophosphate before interaction with chlorophyllide *a* (Chew and Bryant, 2007). Pisareva *et al.* (2011) discovered ChlG, contrary to POR and ChlP, exclusively in the thylakoid membrane fraction, implying that the latest or penultimate step of Chl biosynthesis occurs in this compartment. Thus, proteomic data do not fully support the concept that Chl molecules are incorporated into photosynthetic complexes already in the plasma membrane. However, it cannot be excluded that some chlorophyll is delivered back to plasma membrane to initiate the earliest stages of the PSI and PSII assembly, or that plasma membrane-located photosynthetic complexes contain not fully mature chlorophyll molecules.

Future perspectives

Many various approaches have been developed for characterization of membrane proteins. Proteomics has proven to be a powerful tool and the essential component in systems biology research. However, the proteomic analysis of thylakoid membrane proteins still remains a challenge. Novel strategies should be designed and tested to improve separation of a pure thylakoid membrane from cyanobacterial cells and to increase proteome coverage for integral membrane proteins. To investigate acclimation of cyanobacterial cells to changes in environmental conditions and to evaluate the corresponding alterations in thylakoid proteomes, several novel quantitative proteomic methods might be exploited, including label-free quantification (Old *et al.*, 2005) and multiple reaction monitoring (MRM) (Kirkpatrick *et al.*, 2005). Furthermore, proteomics has a tremendous potential for investigation of post-translational modifications in thylakoid membrane proteins. Future advances in this field will play a crucial role in studies of protein–protein interactions and will reveal novel regulatory pathways important for the function of the photosynthetic machinery and biogenesis of cyanobacterial thylakoid membranes.

Acknowledgements

Research in the laboratory of the authors has been supported by the Academy of Finland Centre of Excellence project 118637.

References

Abrahams, J.P., Leslie, A.G.W., Lutter, R., and Walker, J.E. (1994). Structure at 2.8 Å resolution of F_1-ATPase from bovine heart mitochondria. Nature *370*, 621–628.

Aebersol., R., and Mann, M. (2003). Mass spectrometry-based proteomics. Nature *422*, 198–207.

Aggarwal, K., Choe, L.H., and Lee, K.H. (2005). Shotgun proteomics using the iTRAQ isobaric tags. Brief. Funct. Genom. Proteom. *5*, 112–120.

Allahverdiyeva, Y., Ermakova, M., Eisenhut, M., Zhang, P., Richaud, P., Hagemann, M., Cournac, L., and Aro, E.M. (2011). Interplay between flavodiiron proteins and photorespiration in *Synechocystis* sp. PCC 6803. J. Biol.Chem. *286*, 24007–24014.

Allahverdiyeva, Y., Mustila, H., Ermakova, M., Bersanini, L., Richaud, P., Ajlani, G., Battchikova, N., Cournac, L., and Aro, E.M. (2013). Flavodiiron proteins Flv1 and Flv3 enable cyanobacterial growth and photosynthesis under fluctuating light in aquatic environments. Proc. Natl. Acad. Sci. U.S.A. *110*, 4111–4116.

Allen, J.F., Sanders, C.E., and Holmes, N.G. (1985). Correlation of membrane protein phosphorylation with excitation energy distribution in the cyanobacterium *Synechococcus* 6301. FEBS Lett. *193*, 271–275.

Bald, D., Kruip, J., and Rögner, M. (1996). Supramolecular architecture of cyanobacterial thylakoid membranes: How is the phycobilisome connected with the photosystems? Photosynth. Res. *49*, 103–118.

Baldwin, M.A. (2004). Protein identification by mass spectrometry. Mol. Cell. Proteomics 3, 1–9.

von Ballmoos, C., Wiedenmann, A., and Dimroth, P. (2009). Essentials for ATP synthesis by F_1F_0 ATP synthases. Annu. Rev. Biochem. *78*, 649–672.

Barbato, R., Polverino de Laureto, P., Riconi, F., de Martini, E., and Giacometti, G.M. (1995). Pigment–protein complexes from the photosynthetic membrane of the cyanobacterium *Synechocystis* sp. PCC 6803. Eur. J. Biochem. *234*, 459–465.

Bantscheff, M., Schirle, M., Sweetman, G., Rick, J., and Kuster, B. (2007). Quantitative mass spectrometry in proteomics: a critical review. Anal. Bioanal. Chem. *389*, 1017–1031.

Battchikova, N., and Aro, E.M. (2007). Cyanobacterial NDH-1 complexes: multiplicity in function and subunit composition. Physiol. Plantarum *131*, 22–32.

Battchikova, N., Zhang, P., Rudd, S., Ogawa, T., and Aro, E.M. (2005). Identification of NdhL and Ssl1690 (NdhO) in NDH-1L and NDH-1M complexes of *Synechocystis* sp PCC 6803. J. Biol. Chem. *280*, 2587–2595.

Battchikova, N., Vainonen, J.P., Vorontsova, N., Keranen, M., Carmel, D., and Aro, E.M. (2010). Dynamic changes in the proteome of *Synechocystis* 6803 in response to CO_2 limitation revealed by quantitative proteomics. J. Proteome Res. *9*, 5896–5912.

Battchikova, N., Eisenhut, M., and Aro, E.M. (2011a). Cyanobacterial NDH-1 complexes: Novel insights and remaining puzzles. Biochim. Biophys. Acta *1807*, 935–944.

Battchikova, N., Aro, E.M., and Nixon, P.J. (2011b). Structure and physiological function of NDH-1 complexes in cyanobacteria. In Bioenergetic Processes of Cyanobacteria – From Evolutionary Singularity to Ecological Diversity, Peschek, G.A., Obinger, C., and Renger, G., eds. (Springer Verlag, Berlin, Heidelberg, New York), pp. 445–467.

Battchikova, N., Wei, L., Du, L., Bersanini, L., Aro, E.-M., and Ma, W. (2011c). Identification of novel Ssl0352 protein (NdhS), essential for efficient operation of cyclic electron transport around photosystem I, in NADPH:plastoquinone oxidoreductase (NDH-1) complexes of *Synechocystis* sp. PCC 6803. J. Biol. Chem. *286*, 36992–37001.

Berger, S., Ellersiek, U., and Steinmuller, K. (1991). Cyanobacteria contain a mitochondrial complex-I-homologous NADH-dehydrogenase. FEBS Lett. *286*, 129–132.

van den Bergh, G., and Arckens, L. (2003). Fluorescent two-dimensional difference gel electrophoresis unveils the potential of gel-based proteomics. Curr. Opin. Biotechnol. *15*, 38–43.

Boekema, E.J., Hifney, A., Yakushevska, A.E., Piotrowski, M., Keegstra, W., Berry, S., Michel, K.-P., Pistorius, E.K., and Kruip, J. (2001). A giant chlorophyll–protein complex induced by iron deficiency in cyanobacteria. Nature *412*, 745–748.

Cardona, T., Battchikova, N., Agervald, A., Zhang, P., Nagel, E., Aro, E.M., Styring, S., Lindblad, P., and Magnuson, A. (2007). Isolation and characterization of thylakoid membranes from the filamentous cyanobacterium *Nostoc punctiforme*. Physiol. Plant. *131*, 622–634.

Cardona, T., Battchikova, N., Zhang, P., Stensjö, K., Aro, E.M., Lindblad, P., and Magnuson, A. (2009). Electron transfer protein complexes in the thylakoid membranes of heterocysts from the cyanobacterium *Nostoc punctiforme*. Biochim. Biophys. Acta *1787*, 252–263.

Chen, M., Bibby, T.S., Nield, J., Larkum, A., and Barber, J. (2005). Iron deficiency induces a chlorophyll D-binding Pcb antenna system around photosystem I in *Acaryochloris marina*. Biochim. Biophys. Acta *1708*, 367–374.

Chen, M., Floetenmeyer, M., and Bibby, T.S. (2009). Supramolecular organization of phycobiliproteins in the chlorophyll d-containing cyanobacterium *Acaryochloris marina*. FEBS Lett. *583*, 2535–2539.

Chew, A.G.M., and Bryant, D.A. (2007). Chlorophyll biosynthesis in bacteria: the origins of structural and functional diversities. Annu. Rev. Microbiol. *61*, 113–129.

Chitnis, P.R. (2001). Photosystem I: Function and physiology. Annu. Rev. Plant Physiol. Plant Mol. Biol. *52*, 593–626.

Chitnis, V.P., and Chitnis, P.R. (1993). PsaL subunit is required for the formation of photosystem I trimers in the cyanobacterium *Synechocystis* sp. PCC 6803. FEBS Lett. *336*, 330–334.

Choi, J.S., Kim, D.S., Lee, J.L., Kim, S.J., Kim, S.I., Kim, Y.H., Hong, J.H., Yoo, J.S., Suh, K.H., and Park, Y.M.

(2000). Proteome analysis of light-induced proteins in *Synechocystis* sp. PCC 6803: identification of proteins separated by 2D-PAGE using N-terminal sequencing and MALDI-TOF MS. Mol. Cells *10*, 705–711.

Cramer, W.A., Zhang, H., Yan, J., Kurisu, G., and Smith, J.L. (2006). Transmembrane traffic in the cytochrome b6f complex. Annu. Rev. Biochem. *75*, 769–790.

Curtis, S.E. (1988). Structure, organization and expression of cyanobacterial ATP synthase genes. Photosynth. Res. *18*, 223–244.

D'Amici, G.M., Timperio, A.M., and Zolla, L. (2008). Coupling of native liquid phase isoelectrofocusing and blue native polyacrylamide gel electrophoresis: a potent tool for native membrane multiprotein complex separation. J. Prot. Res. 7, 1326–1340.

DeRuyter, Y.S., and Fromme, P. (2008). Molecular structure of the photosynthetic apparatus. In Molecular Biology, Genomics and Evolution of Cyanobacteria, Herrero, A., and Flores, E., eds. (Horizon Scientific Press, Norwich, UK), pp. 217–269.

Dickson, V.K., Silvester, J.A., Fearnley, I.M., Leslie, A.G.W., and Walker, J.E. (2006). On the structure of the stator of the mitochondrial ATP synthase. EMBO J. *25*, 2911–2918.

Dobáková, M., Sobotka, R., Tichý, M., and Komenda, J. (2009). Psb28 is involved in the biogenesis of the photosystem II inner antenna CP47 (PsbB) in the cyanobacterium *Synechocystis* sp. PCC6803. Plant Physiol. *149*, 1076–1086.

Dühring, U., Irrgang, K.D., Lunser, K., Kehr, J., and Wilde, A. (2006). Analysis of photosynthetic complexes from a cyanobacterial ycf37 mutant. Biochim. Biophys. Acta *1757*, 3–11.

Dühring, U., Ossenbüehl, F., and Wilde, A. (2007). Late assembly steps and dynamics of.the cyanobacterial photosystem I. J. Biol. Chem. *282*, 10915–10921.

Dunn, S.D., Kellner, E., and Lill, H. (2001). Specific heterodimer formation by the cytoplasmic domains of the b and b′ subunits of cyanobacterial ATP synthase. Biochemistry *40*, 187–192.

Dworsky, A., Mayer, B., Regelsberger, G., Fromwald, S., and Peschek, G.A. (1995). Functional and immunological characterization of both mitochondria-like and chloroplast-like electron-proton transport proteins in isolated and purified cyanobacterial membranes. Bioelectrochem. Bioenerg. *38*, 35–43.

Ermakova, M., Battchikova, N., Allahverdiyeva, Y., and Aro, E.M. (2013). Novel heterocyst-specific flavodiiron proteins in *Anabaena* sp. PCC 7120. FEBS Lett. *587*, 82–87.

Fagerlund, R.D., and Eaton-Rye, J.J. (2011). The lipoproteins of cyanobacterial photosystem II. J. Photochem. Photobiol. *104*, 191–203.

Flores, E., and Herrero, A. (2010). Compartmentalized function through cell differentiation in filamentous cyanobacteria. Nat. Rev. Microbiol. *8*, 39–50.

Ford, R.C., and Holzenburg, A. (1988). Investigation of the structure of trimeric and monomeric photosystem I reaction centre complexes. EMBO J. 7, 2287–2293.

Fromme, P., Jordan, P., and Krauss, N. (2001). Structure of photosystem I. Biochim. Biophys. Acta *1507*, 5–31.

Fromme, P., Yu, H., DeRuyter, Y.S., Jolley, C., Chauhan, D.K., Melkozernov, A., and Grotjohann, I. (2006). Structure of photosystems I and II. C. R. Chimie *9*, 188–200.

Fuhrmann, E., Gathmann, S., Rupprecht, E., Golecki, J., and Schneider, D. (2009a). Thylakoid membrane reduction affects the photosystem stoichiometry in the cyanobacterium *Synechocystis* sp. PCC 6803. Plant Phys. *149*, 735–744.

Fuhrmann, E., Bultema, J.B., Kahmann, U., Rupprecht, E., Boekema, E.J., and Schneider, D. (2009b). The vesicle-inducing protein 1 from *Synechocystis* sp. PCC 6803 organizes into diverse higher-ordered ring structures. Mol. Biol. Cell *20*, 4620–4628.

Fulda, S., Huang, F., Nilsson, F., Hagemann, M., and Norling, B. (2000). Proteomics of *Synechocystis* sp. strain PCC 6803. Identification of periplasmic proteins in cells grown at low and high salt concentrations. Eur. J. Biochem. *267*, 5900–5907.

Fulda, S., Mikkat, S., Huang, F., Huckauf, J., Marin, K., Norling, B., and Hagemann, M. (2006). Proteome analysis of salt stress response in the cyanobacterium *Synechocystis* sp. strain PCC 6803. Proteomics *6*, 2733–2745.

Gan, C.S., Reardon, K.F., and Wright, P.C. (2005). Comparison of protein and peptide prefractionation methods for the shotgun proteomic analysis of *Synechocystis* sp. PCC 6803. Proteomics *5*, 2468–2478.

Giddings, T.H., Withers, N.W., and Staehelin, L.A. (1980). Supramolecular structure of stacked and unstacked regions of the photosynthetic membranes of *Prochloron* sp., a prokaryote. Proc. Natl. Acad. Sci. U.S.A. *77*, 352–356.

Glanemann, C., Loos, A., Gorret, N., Willis, L.B., O'Brien, X.M., Lessard, P.A., and Sinskey, A.J. (2003). Disparity between changes in mRNA abundance and enzyme activity in *Corynebacterium glutamicum*: implications for DNA microarray analysis. Appl. Microbiol. Biotechnol. *61*, 61–68.

Golden, J.W., and Yoon, H.-S. (2003). Heterocyst development in *Anabaena*. Curr. Opin. Microbiol. *6*, 557–563.

Grasse, N., Mamedov, F., Becker, K., Styring, S., Rögner, M., and Nowaczyk, M.M. (2011). Role of novel dimeric photosystem II (PSII)-Psb27 protein complex in PSII repair. J. Biol. Chem. *286*, 29548–29555.

de Groot, M.J., Daran-Lapujade, P., van Breukelen, B., Knijnenburg, T.A., de Hulster, E.A., Reinders, M.J., Pronk, J.T., Heck, A.J., and Slijper, M. (2007). Quantitative proteomics and transcriptomics of anaerobic and aerobic yeast cultures reveals post-transcriptional regulation of key cellular processes. Microbiology *153*, 3864–3878.

Helman, Y., Tchernov, D., Reinhold, L., Shibata, M., Ogawa, T., Schwarz, R., Ohad, I., and Kaplan, A. (2003). Genes encoding A-type flavodiiron proteins are essential for photoreduction of O_2 in cyanobacteria. Curr. Biol. *13*, 230–235.

Herranen, M., Battchikova, N., Zhang, P., Graf, A., Sirpio, S., Paakkarinen, V., and Aro, E.-M. (2004). Towards functional proteomics of membrane protein complexes in *Synechocystis* sp. PCC 6803. Plant Physiol. *134*, 470–481.

Herrero, A., Muro-Pastor, A.M., Valladares, A., and Flores, E. (2004). Cellular differentiation and the NtcA transcription factor in filamentous cyanobacteria. FEMS Microbiol. Rev. *28*, 469–487.

Hess, W.R., Rocap, G., Ting, C.S., Larimer, F., Stilwagen, S., Jane Lamerdin, J., and Chisholm, S.W. (2001). The photosynthetic apparatus of *Prochlorococcus*: Insights through comparative genomics. Photosynth. Res. *70*, 53–71.

Hinterstoisser, B., Cichna, M., Kuntner, O., and Peschek, G.A. (1993). Cooperation of plasma and thylakoid membranes for the biosynthesis of chlorophyll in cyanobacteria: the role of thylakoid centers. J. Plant Physiol. *142*, 407–413.

Howitt, C.A., Smith, G.D., and Day, D.A. (1993). Cyanide-insensitive oxygen-uptake and pyridine-nucleotide dehydrogenases in the cyanobacterium *Anabaena* PCC7120. Biochim. Biophys. Acta *1141*, 313–320.

Huang, F., Parmryd, I., Nilsson, F., Persson, A.L., Pakrasi, H.B., Andersson, B., and Norling, B. (2002). Proteomics of *Synechocystis* sp. Strain PCC 6803: Identification of Plasma Membrane Proteins. Mol. Cell. Proteomics *1*, 956–966.

Huang, F., Hedman, E., Funk, C., Kieselbach, T., Schroder, W.P., and Norling, B. (2004). Isolation of outer membrane of *Synechocystis* sp. PCC 6803 and its proteomic characterization. Mol. Cell Proteom. *3*, 586–595.

Huang, F., Fulda, S., Hagemann, M., and Norling, B. (2006). Proteomic screening of salt-stress-induced changes in plasma membranes of *Synechocystis* sp. strain PCC 6803. Proteomics *6*, 910–920.

Järvi, S., Suorsa, M., Paakkarinen, V., and Aro, E.-M. (2011). Optimized native gel systems for separation of thylakoid protein complexes: novel super- and mega-complexes. Biochem. J. *439*, 207–214.

Jordan, P., Fromme, P., Witt, H.T., Klukas, O., Saenger, W., and Krauss, N. (2001). Three-dimensional structure of cyanobacterial photosystem I at 2.5 Å resolution. Nature *411*, 909–917.

Kaplan, A., and Reinhold, L. (1999). CO_2 concentrating mechanisms in photosynthetic microorganisms. Annu. Rev. Plant Physiol. Plant Mol. Biol. *50*, 539–570.

Kashino, Y., Koike, H., Yoshio, M., Egashira, H., Ikeuchi, M., Pakrasi, H.B., and Satoh, K. (2002a). Low-molecular-mass polypeptide components of a Photosystem II preparation from the thermophilic cyanobacterium *Thermosynechococcus vulcanus*. Plant Cell Physiol. *43*, 1366–1373.

Kashino, Y., Lauber, W., M., Carroll, J.A., Wang, Q., Whitmarsh, J., Satoh, K., and Pakrasi, H.B. (2002b) Proteomic analysis of a highly active photosystem II preparation from the cyanobacterium *Synechocystis* sp. PCC 6803 reveals the presence of novel polypeptides. Biochemistry *41*, 8004–8012.

Keren, N., Ohkawa, H., Welsh, E.A., Liberton, M., and Pakrasi, H.B. (2005). Psb29, a conserved 22 kD protein, functions in the biogenesis of photosystem II complexes in *Synechocystis* and *Arabidopsis*. Plant Cell *17*, 2768–2781.

Kirkpatrick, D.S., Gerber, S.A., and Gygi, S.P. (2005). The absolute quantification strategy: a general procedure for the quantification of proteins and post-translational modifications. Methods 35, 265–273.

Klinkert, B., Ossenbühl, F., Sikorski, M., Berry, S., Eichacker, L., and Nickelsen, J. (2004). PratA, a periplasmic tetratricopeptide repeat protein involved in biogenesis of photosystem II in *Synechocystis* sp. PCC 6803. J. Biol. Chem. *279*, 44639–44644.

Klotz, A.V., Leary, J.A., and Glazer, A.N. (1986). Post-translational methylation of asparaginyl residues. J. Biol. Chem. *261*, 15891–15894.

Klotz, A.V., and Glazer, A.N. (1987). γ-*N*-Methylasparagine in phycobiliproteins. J. Biol. Chem. *262*, 17350–17355.

Klughammer, B., Sültemeyer, D., Badger, M.R., and Price, G.D. (1999). The involvement of NAD(P)H dehydrogenase subunits, NdhD3 and NdhF3, in high-affinity CO_2 uptake in *Synechococcus* sp. PCC 7002 gives evidence for multiple NDH-1 complexes with specific roles in cyanobacteria. Mol. Microbiol. *32*, 1305–1315.

Komenda, J., Lupínková, L., and Kopecký, J. (2002). Absence of the *psbH* gene product destabilizes photosystem II complex and bicarbonate binding on its acceptor side in *Synechocystis* PCC 6803. Eur. J. Biochem. *269*, 610–619.

Komenda, J., Nickelsen, J., Tichý, M., Prášil, O., Eichacker, L.A., and Nixon, P.J. (2008). The cyanobacterial homologue of the HCF136/YCF48 is a component of an early assembly complex and is important for both the efficient assembly and repair of photosystem II in *Synechocystis* sp. PCC6803. J. Biol. Chem. *283*, 22390–22399.

Komenda, J., Sobotka, R., and Nixon, P.J. (2012). Assembling and maintaining the Photosystem II complex in chloroplasts and cyanobacteria. Curr.Opin. Plant Biol. *15*, 245–251.

Kroll, D., Meierhoff, K., Bechtold, N., Kinoshita, M., Westphal, S., Vothknecht, U.C., Soll, J., and Westhoff, P. (2001). VIPP1, a nuclear gene of *Arabidopsis thaliana* essential for thylakoid membrane formation. Proc. Natl. Acad. Sci. U.S.A. *98*, 4238–4242.

Kruip, J., Chitnis, P.R., Lagoutte, B., Rögner, M., and Boekema, E.J. (1997). Structural organization of the major subunits in cyanobacterial photosystem 1. Localization of subunits PsaC, -D, -E, -F, and -J. J. Biol. Chem. *272*, 17061–17069.

Kubota, H., Sakurai, I., Katayama, K., Mizusawa, N., Ohashi, S., Kobayashi, M., Zhang, P., Aro, E.-M., and Wada, H. (2010). Purification and characterization of photosystem I complex from *Synechocystis* sp. PCC 6803 by expressing histidine-tagged subunits. Biochim. Biophys. Acta *1797*, 98–105.

Kunkel, D.D. (1982). Thylakoid centers: structures associated with the cyanobacterial photosynthetic membrane system. Arch. Microbiol. *133*, 97–99.

Kumar, K., Mella-Herrera, R.A., and Golden, J.W. (2010). Cyanobacterial heterocysts. Cold Spring Harb. Perspect. Biol. 2, a000315.

Kurian, D., Jansén, T., and Mäenpää, P. (2006a) Proteomic analysis of heterotrophy in *Synechocystis* sp. PCC 6803. Proteomics *6*, 1483–1494.

Kurian, D., Phadwai, K., and Mäenpää, P. (2006b). Proteomic characterization of acid response in *Synechocystis* sp. PCC 6803. Proteomics *6*, 3614–3624.

Kurisu, G., Zhang, H., Smith, J.L., and Cramer, W.A. (2003). Structure of the cytochrome b_6f complex of oxygenic photosynthesis: tuning the cavity. Science *302*, 1009–1014.

Kwon, J., Oh, J., Park, C., Cho, K., Kim, S.I., Kim, S., Lee, S., Bhak, J., Norling, B., and Choi, J.S. (2010). Systematic cyanobacterial membrane proteome analysis by combining acid hydrolysis and digestive enzymes with nano-liquid chromatography-Fourier transform mass spectrometry. J. Chromatogr. A *1217*, 285–293.

Ladig, R., Sommer, M.S., Hahn, A., Leisegang, M.S., Papasotiriou, D.G., Ibrahim, M., Elkehal, R., Karas, M., Zickermann, V., Gutensohn, M., *et al.* (2011). A high-definition native polyacrylamide gel electrophoresis system for the analysis of membrane complexes. Plant J. *67*, 181–194.

Lang, N., and Fay, P. (1971). The heterocysts of blue-green algae II. Details of ultrastructure. Proc. Roy. Soc. Lond. B *178*, 193–203.

Lee, S., Ryu, J.Y., Kim, S.Y., Jeon, J.H., Song, J.Y., Cho, H.T., Choi, S.B., Choi, D., de Marsac, N.T., and Park, Y.I. (2007). Transcriptional regulation of the respiratory genes in the cyanobacterium *Synechocystis* sp. PCC 6803 during the early response to glucose feeding. Plant Physiol. *145*, 1018–1030.

Li, T., Yang, H.M., Cui, S.X., Suzuki, I., Zhang, L.F., Li, L., Bo, T.T., Wang, J., Murata, N., and Huang, F. (2012). Proteomic study of the impact of Hik33 mutation in *Synechocystis* sp. PCC 6803 under normal and salt stress conditions. J. Proteome Res. *11*, 502–514.

Liberton, M., and Pakrasi, H.B. (2008). Membrane systems in cyanobacteria. In The Cyanobacteria. Molecular Biology, Genomics and Evolution, Herrero, A., Flores, E., eds. (Caister Academic Press, Norfolk, UK), pp. 271–288.

Liberton, M., Berg, R.H., Heuser, J., Roth, R., and Pakrasi, H.B. (2006). Ultrastructure of the membrane systems in the unicellular cyanobacterium *Synechocystis* sp. strain PCC 6803. Protoplasma *227*, 129–138.

Liberton, M., Austin, J.R., Berg, R.H., and Pakrasi, H.B. (2011). Unique thylakoid membrane architecture of a unicellular N_2-fixing cyanobacterium revealed by electron tomography. Plant. Phys. *155*, 1656–1666.

Lindhal, M., Spetea, C., Hundal, T., Oppenheim, A., Adam, Z., and Andersson, B. (2000). The thylakoid FtsH protease plays a role in the light-induced turnover of the photosystem II D1 protein. Plant Cell *12*, 419–431.

Link, A.J., Eng, J., Schieltz, D.M., Carmack, E., Mize, G.J., Morris, D.R., Garvik, B.M., and Yates, J.R. III. (1999). Direct analysis of protein complexes using mass spectrometry. Nat. Biotechnol. *17*, 676–682.

Liu, L.N., Chen, X.L., Zhang, Y.Z., and Zhou, B.C. (2005). Characterization, structure and function of linker polypeptides in phycobilisomes of cyanobacteria and red algae: an overview. Biochim. Biophys. Acta *1708*, 133–142.

Liu, L.N., Bryan, S.J., Huang, F., Yu, J., Nixon, P.J., Rich, P.R., and Mullineaux, C.W. (2012). Control of electron transport routes through redox-regulated redistribution of respiratory complexes. Proc. Natl. Acad. Sci. U.S.A. *109*, 11431–11436.

Ma, W., and Mi, H. (2008). Effect of exogenous glucose on the expression and activity of NADPH dehydrogenase complexes in the cyanobacterium *Synechocystis* sp strain PCC 6803. Plant Physiol. Biochem. *46*, 775–779.

Ma, W., Deng, Y., Ogawa, T., and Mi, H. (2006). Active NDH-1 complexes from the cyanobacterium *Synechocystis* sp strain PCC 6803. Plant Cell Physiol. *47*, 1432–1436.

Maeda, S., Badger, M.R., and Price, G.D. (2002). Novel gene products associated with NdhD3/D4-containing NDH-1 complexes are involved in photosynthetic CO_2 hydration in the cyanobacterium *Synechococcus* sp. PCC7942. Mol. Microbiol. *43*, 425–435.

MacCoil, R. (1998). Cyanobacterial phycobilisomes. J. Struct. Biol. *124*, 311–334.

Mangels, D., Kruip, J., Berry, S., Rögner, M., Boekema, E.J., and Koenig, F. (2002). Photosystem I from the unusual cyanobacterium *Gloeobacter violaceus*. Photosynth. Res. *72*, 307–319.

Mann, N.H. (1994). Protein phosphorylation in cyanobacteria. Microbiology *140*, 3207–3215.

Mann, M., Hendrickson, R.C., and Pandey, A. (2001). Analysis of proteins and proteomes by mass spectrometry. Annu. Rev. Biochem. *70*, 437–473.

van de Meene, A.M.L., Hohmann, M.F., Vermass, W.F.J., and Robertson, R.W. (2006). The three-dimensional structure of the cyanobacterium *Synechocystis* sp. PCC 6803. Arch. Microbiol. *184*, 259–270.

Mori, H., and Ito, K. (2001). The Sec protein-translocation pathway. Trends Microbiol. *9*, 494–500.

Mukhopadhyay, A., and Kennelly, P.J. (2011). A low molecular weight protein tyrosine phosphatase from *Synechocystis* sp. strain PCC 6803: enzymatic characterization and identification of its potential substrates. J. Biochem. *149*, 551–562.

Mullineaux, C.W. (1999). The thylakoid membranes of cyanobacteria: structure, dynamics and function. Aust. J. Plant Physiol. *26*, 671–677.

Mullineaux, C.W. (2008). Biogenesis and dynamics of thylakoid membranes and the photosynthetic apparatus. In The Cyanobacteria. Molecular Biology, Genomics and Evolution, Herrero, A., Flores, E., eds. (Caister Academic Press, Norfolk, UK), pp. 289–304.

Mulo, P., Sicora, C., and Aro, E.-M. (2009). Cyanobacterial *psbA* gene family: optimization of oxygenic photosynthesis. Cell. Mol. Life Sci. *66*, 3697–3710.

Nakamura, T., and Oda, Y. (2007). Mass spectrometry-based quantitative proteomics. Biotechnol. Gen. Eng. Rev. *24*, 147–164.

Nakamura, Y., Kaneko, T., Sato, S., Mimuro, M., Miyashita, H., Tsuchiya, T., Sasamoto, S., Watanabe, A., Kawashima, K., Kishida, Y., *et al.* (2003). Complete genome structure of *Gloeobacter violaceus* PCC 7421, a cyanobacterium that lacks thylakoids. DNA Res. *10*, 137–145.

Naithani, S., Hou, J.-M., and Chitnis, P.R. (2000). Targeted inactivation of the *psaK1*, *psaK2* and *psaM* genes encoding subunits of Photosystem I in the cyanobacterium *Synechocystis* sp. PCC 6803. Photosynth. Res. *63*, 225–236.

Naver, H., Boudreau, E., and Rochaix, J.D. (2001). Functional studies of Ycf3: its role in assembly of photosystem I and interactions with some of its subunits. Plant Cell *13*, 2731–2745.

Nelson, N., and Ben-Shem, A. (2004). The complex architecture of oxygenic photosynthesis. Nat. Rev. Mol. Cell. Biol. *5*, 971–982.

Nelson, N., and Yochum, C.F. (2006). Structure and function of photosystems I and II. Annu. Rev. Plant Biol. *57*, 521–565.

Nevo, R., Charuvi, D., Shimoni, E., Schwarz, R., Kaplan, A., Ohad, I., and Reich, Z. (2007). Thylakoid membrane perforations and connectivity enable intracellular traffic in cyanobacteria. EMBO J. *26*, 1467–1473.

Nickelsen, J., Rengstl, B., Stengel, A., Schottkowski, M., Soll, J., and Ankele, E. (2011). Biogenesis of the cyanobacterial thylakoid membrane system – an update. FEMS Microbiol. Lett. *315*, 1–5.

Nierzwicki-Bauer, S.A., Balkwill, D.L., and Stevens, S.E. Jr. (1983). Three-dimensional ultrastructure of a unicellular cyanobacterium. J. Cell. Biol. *97*, 713–722.

Nixon, P.J., Michoux, F., Yu, J., Boehm, M., and Komenda, J. (2010). Recent advances in understanding the assembly and repair of Photosystem II. Ann. Bot. *106*, 1–16.

Noji, H., Yasuda, R., Yoshida, M., and Kinosita, K. Jr. (1997). Direct observation of the rotation of F1-ATPase. Nature *386*, 299–302.

Norling, B., Zak, E., Andersson, B., and Pakrasi, H. (1998). 2D-isolation of pure plasma and thylakoid membranes from the cyanobacterium *Synechocystis* sp. PCC 6803. FEBS Lett. *436*, 189–192.

Nowaczyk, M.M., Hebeler, R., Schlodder, E., Meyer, H.E., Warscheid, B., and Rögner, M. (2006). Psb27, a cyanobacterial lipoprotein, is involved in the repair cycle of photosystem II. Plant Cell *18*, 3121–3131.

Nowaczyk, M.M., Wulfhorst, H., Ryan, C.M., Souda, P., Zhang, H., Cramer, W.A., and Whitelegge, J.P. (2011). NdhP and NdhQ: Two novel small subunits of the cyanobacterial NDH-1 complex. Biochemistry *50*, 1121–1124.

Ogawa, T. (1992). Identification and characterization of the ictA/ndhL gene product essential to inorganic carbon transport of *Synechocystis* sp strain PCC6803. Plant Physiol. *99*, 1604–1608.

Ogawa, T., and Mi, H. (2007). Cyanobacterial NADPH dehydrogenase complexes, Photosynth. Res. *93*, 69–77.

Ohkawa, H., Sonoda, M., Katoh, H., and Ogawa, T. (1998). The use of mutants in the analysis of the CO_2-concentrating mechanism in cyanobacteria. Can. J. Bot. *76*, 1035–1042.

Ohkawa, H., Sonoda, M., Shibata, M., and Ogawa, T. (2001). Localization of NAD(P)H dehydrogenase in the cyanobacterium *Synechocystis* sp strain PCC 6803. J. Bacteriol. *183*, 4938–4939.

Ohkawa, H., Sonoda, M., Hagino, N., Shibata, M., Pakrasi, H.B., and Ogawa, T. (2002). Functionally distinct NAD(P)H dehydrogenases and their membrane localization in *Synechocystis* sp PCC6803. Funct. Plant Biol. *29*, 195–200.

Old, W.M., Meyer-Arendt, K., Aveline-Wolf, L., Pierce, K.G., Mendoza, A., Sevinsky, J.R., Resing, K.A., and Ahn, N.G. (2005). Comparison of label-free methods for quantifying human proteins by shotgun proteomics. Mol. Cell Proteom. *4*, 1487–1502.

Omata, T., and Murata, N. (1983). Isolation and characterization of the cytoplasmic membranes from the blue-green alga (cyanobacterium) *Anacystis nidulans*. Plant Cell Physiol. *24*, 1101–1112.

Ong, S.-E., Foster, L.J., and Mann, M. (2003). Methods *29*, 124–130.

Ossenbühl, F., Inaba-Sulpice, M., Meurer, J., Soll, J., and Eichacker, L.A. (2006). The *Synechocystis* sp PCC 6803 Oxa1 homolog is essential for membrane integration of reaction center precursor protein pD1. Plant Cell *18*, 2236–2246.

Ow, S.Y., Cardona, T., Taton, A., Magnuson, A., Lindblad, P., Stensjö, K., and Wright, P.C. (2008). Quantitative shotgun proteomics of enriched heterocysts from *Nostoc* sp. PCC 7120 using 8-plex isobaric peptide tags. J. Proteome Res. *7*, 1615–1628.

Ow, S.Y., Noirel, J., Cardona, T., Taton, A., Lindblad, P., Stensjö, K., and Wright, P.C. (2009). Quantitative overview of N_2 fixation in *Nostoc punctiforme* ATCC 29133 through cellular enrichments and iTRAQ shotgun proteomics. J. Proteome Res. *8*, 187–198.

Pandhal, J., Noirel, J., Wright, P.C., and Biggs, C.A. (2009). A system biology approach to investigate the response of *Synechocystis* sp. PCC6803 to a high salt environment. Saline Syst. *5*, 8.

Peng, L., Shimizu, H., and Shikanai, T. (2008). The chloroplast NAD(P)H dehydrogenase complex interacts with photosystem I in *Arabidopsis*. J. Biol. Chem. *283*, 34873–34879.

Pieulle, L., Guedeney, G., Cassier-Chauvat, C., Jeanjean, R., Chauvat, F., and Peltier, G. (2000). The gene encoding the NdhH subunit of type 1 NAD(P)H dehydrogenase is essential to survival of *Synechocystis* PCC6803. FEBS Lett. *487*, 272–276.

Pisareva, T., Shumskaya, M., Maddalo, G., Ilag, L., and Norling, B. (2007). Proteomics of *Synechocystis* sp. PCC 6803. Identification of novel integral plasma membrane proteins. FEBS J. *274*, 791–804.

Pisareva, T., Kwon, J., Oh, J., Kim, S., Ge, C., Wieslander, Å., Choi, J.-S., and Norling, B. (2011). Model for membrane organization and protein sorting in the cyanobacterium *Synechocystis* sp. PCC 6803 inferred from proteomics and multivariate sequence analyses. J. Prot. Res. *10*, 3617–3631.

Piven, I., Ajlani, G., and Sokolenko, A. (2005). Phycobilisome linker proteins are phosphorylated in *Synechocystis* sp. PCC 6803. J. Biol. Chem. *280*, 21667–21672.

Pogoryelov, D., Yildiz, O., Faraldo-Gómez, J.D., and Meier, T. (2009). High-resolution structure of the rotor ring of a proton-dependent ATP synthase. Nat. Struct. Mol. Biol. *16*, 1068–1073.

Price, G.D., Sultemeyer, D., Klughammer, B., Ludwig, M., and Badger, M.R. (1998). The functioning of the CO_2 concentrating mechanism in several cyanobacterial strains: a review of general physiological characteristics, genes, proteins, and recent advances. Can. J. Bot. *76*, 973–1002.

Prommeenate, P., Lennon, A.M., Markert, C., Hippler, M., and Nixon, P.J. (2004). Subunit composition of NDH-1 complexes of *Synechocystis* sp PCC 6803: identification of two new *ndh* gene products with nuclear-encoded homologues in the chloroplast Ndh complex. J. Biol. Chem. *279*, 28165–28173.

Race, H.L., and Gounaris, K. (1993). Identification of of the psbH gene product as a 6 kDa phosphoprotein in the cyanobacterium *Synechocystis* 6803. FEBS Lett. *323*, 35–39.

Rakhimberdieva, M.G., Boichenko, V.A., Karapetyan, N.V., and Stadnichuk, I.N. (2001). Interaction of phycobilisomes with photosystem II dimers and photosystem I monomers and trimers in the cyanobacterium *Spirulina platensis*. Biochemistry *40*, 15780–15788.

Rexroth, S., Mullineaux, C.W., Ellinger, D., Sendtko, E., Rögner, M., and Koenig, F. (2011). The plasma membrane of the cyanobacterium *Gloeobacter violaceus* contains segregated bioenergetic domains. Plant Cell *23*, 2379–2390.

Riviere, M.E., Arrio, B., Steffan, I., Molitor, V., Kuntner, O., and Peschek, G.A. (1990). Changes of some physical properties of isolated and purified plasma and thylakoid membrane vesicles from the freshwater cyanobacterium *Synechococcus* 6301 (*Anacystis nidulans*) during adaptation to salinity. Arch. Biochem. Biophys. *280*, 159–166.

Rochaix, J.-D. (2011). Assembly of the photosynthetic apparatus. Plant Physiol. *155*, 1493–1500.

Roose, J.L., and Pakrasi, H.B. (2008). The Psb27 protein facilitates manganese cluster assembly in photosystem II. J. Biol. Chem. *283*, 4044–4050.

Roose, J.L., Wegener, K.M., and Pakrasi, H.B. (2007). The extrinsic proteins of Photosystem II. Photosynth. Res. *92*, 369–387.

Ross, P.L., Huang, Y.N., Marchese, J.N., Williamson, B., Parker, K., Hattan, S., Khainovski, N., Pillai, S., Dey, S., Daniels, S., *et al.* (2004). Multiplexed protein quantitation in *Saccharomyces cerevisiae* using amine-reactive isobaric tagging reagents. Mol. Cell. Proteomics 3, 1154–1169.

Rögner, M., Dekker, J.P., Boekema, E.J., and Witt, H.T. (1987). Size, shape and mass of the oxygen-evolving photosystem II complex from the thermophilic cyanobacterium *Synechococcus* sp. FEBS Lett. *219*, 207–211.

Rögner, M., Muhlenhoff, U., Boekema, E.J., and Witt, H.T. (1990). Mono-, di- and trimeric PSI reaction center complexes isolated from the thermophilic cyanobacterium *Synechococcus* sp. Size, shape and activity. Biochim. Biophys. Acta *1015*, 415–424.

Sanders, C.E., Melis, A., and Allen, J.F. (1989). In vivo phosphorylation of proteins in the cyanobacterium *Synechococcus* 6301 after chromatic acclimation to Photosystem I or Photosystem II light. Biochim. Biophys. Acta *976*, 168–172.

Santoni, V., Molloy, M., and Rabilloud, T. (2000). Membrane proteins and proteomics: un amour impossible? Electrophoresis *21*, 1054–1070.

Sandh, G., Ran, L., Xu, L., Sundqvist, G., Bulone, V., and Bergman, B. (2011). Comparative proteomic profiles of the marine cyanobacterium *Trichodesmium erythraeum* IMS101 under different nitrogen regimes. Proteomics *11*, 406–419.

Satoh, K., and Yamamoto, Y. (2007). The carboxyl-terminal processing of precursor D1 protein of the photosystem II reaction center. Photosynth. Res. *94*, 203–215.

Schägger, H., and von Jagow, G. (1991). Blue native electrophoresis for isolation of membrane protein complexes in enzymatically active form. Anal. Biochem. *199*, 223–231.

Schägger, H., Cramer, W.A., and von Jagow, G. (1994). Analysis of molecular masses and oligomeric states of protein complexes by blue native electrophoresis and isolation of membrane protein complexes by two-dimensional native electrophoresis. Anal. Biochem. *217*, 220–230.

Scheer, H., and Zhao, K.-H. (2008). Biliprotein maturation: the chromophore attachment. Mol. Microbiol. *68*, 263–276.

Schneider, D., Fuhrmann, E., Scholz, I., Hess, W.R., and Graumann, P.L. (2007). Fluorescence staining of live cyanobacterial cells suggest non-stringent chromosome segregation and absence of a connection between cytoplasmic and thylakoid membranes. BMC Cell Biol. *8*, 39.

Schottkowski, M., Gkalympoudis, S., Tzekova, N., Stelljes, C., Schuenemann, D., Ankele, E., and Nickelsen, J. (2009a). Interaction of the periplasmic PratA factor and the PsbA (D1) protein during biogenesis of photosystem II in *Synechocystis* sp. PCC 6803. J. Biol. Chem. *284*, 1813–1819.

Schottkowski, M., Ratke, J., Oster, U., Nowaczyk, M., and Nickelsen, J. (2009b). Pitt, a novel tetratricopeptide repeat protein involved in light-dependent chlorophyll biosynthesis and thylakoid membrane biogenesis in *Synechocystis* sp. PCC6803. Mol. Plant 2, 1289–1297.

Schultze, M., Forberich, B., Rexroth, S., Dyczmons, N.G., Roegner., M., and Appel, J. (2009). Localization of cytochrome b_6f complexes implies an incomplete respiratory chain in cytoplasmic membranes of the cyanobacterium *Synechocystis* sp. PCC 6803. Biochim. Biophys. Acta *1787*, 1479–1485.

Schwarz, R., and Forchhammer, K. (2005). Acclimation of unicellular cyanobacteria to macronutrient deficiency: emergence of a complex network of cellular responses. Microbiology *151*, 2503–2514.

Shestakov, S.V., Anbudurai, P.R., Stanbekova, G.E., Gadzhiev, A., Lind, L.K., and Pakrasi, H.B. (2004). Molecular cloning and characterization of the *ctpA* gene encoding a carboxyl-terminal processing protease. Analysis of a spontaneous photosystem II-deficient mutant strain of the cyanobacterium *Synechocystis* sp. PCC 6803. J. Biol. Chem. *269*, 19354–19359.

Shibata, M., Ohkawa, H., Katoh, H., Shimoyama, M., and Ogawa, T. (2002). Two CO_2 uptake systems in cyanobacteria: four systems for inorganic carbon acquisition in *Synechocystis* sp strain PCC6803. Funct. Plant Biol. *29*, 123–129.

Sidler, W.A. (1994). Phycobilisome and phycobiliprotein structures. In Molecular Biology of the Cyanobacteria, Bryant, D.A., ed. (Kluwer Academic Publishing, Dordrecht, The Netherlands), pp. 139–216.

Sirpio, S., Allahverdiyeva, Y., Holmstrom, M., Khrouchtchova, A., Haldrup, A., Battchikova, N., and Aro, E.M. (2009). Novel nuclear-encoded subunits of the chloroplast NAD(P)H dehydrogenase complex. J. Biol. Chem. *284*, 905–912.

Slabas, A.R., Suzuki, I., Murata, N., Simon, W.J., and Hall, J.J. (2006). Proteomic analysis of the heat shock response in *Synechocystis* PCC6803 and a thermally tolerant knockout strain lacking the histidine kinase 34 gene. Proteomics *6*, 845–864.

Spence, E., Bailey, S., Nenninger, A., Moller, S.G., and Robinson, C. (2004). A homolog of Albino3/OxaI is essential for thylakoid biogenesis in the cyanobacterium *Synechocystis* sp. PCC6803. J. Biol. Chem. *279*, 55792–55800.

Srivastava, R., Pisareva, T., and Norling, B. (2005). Proteomic studies of the thylakoid membrane of *Synechocystis* sp. PCC 6803. Proteomics *5*, 4905–4916.

Srivastava, R., Battchikova, N., Norling, B., and Aro, E.M. (2006). Plasma membrane of *Synechocystis* PCC 6803: a heterogeneous distribution of membrane proteins. Arch Microbiol. *185*, 238–243.

Stanier (Cohen-Bazire), G. (1988). Fine structure of cyanobacteria. Meth. Enzymol. *167*, 157–172.

Stensjö, K., Ow, S.Y., Barrios-Llerena, M.E., Lindblad, P., and Wright, P.C. (2007). An iTRAQ-based quantitative analysis to elaborate the proteomic response of *Nostoc* sp. PCC 7120 under N_2 fixing conditions. J. Proteome Res. *6*, 621–635.

Summerfield, T.C., and Sherman, L.A. (2008). Global transcriptional response of the alkali-tolerant cyanobacterium *Synechocystis* sp. Strain PCC 6803 to a pH 10 environment. Appl. Environ. Microbiol. *74*, 5276–5284.

Suzuki, I., Simon, W., and Slabas, A.R. (2006). The heat shock response of *Synechocystis* sp. PCC 6803 analysed by transcriptomics and proteomics. J. Exp. Bot. *57*, 1573–1578.

Swanson, R., and Glazer, A.N. (1990). Phycobiliprotein methylation: effect of the γ-N-methylasparagine residue on energy transfer in phycocyanin and the phycobilisome. J. Mol. Biol. *214*, 787–796.

Takahashi, Y., Koike, H., and Katoh, S. (1982). Multiple forms of chlorophyll–protein complexes from a thermophilic cyanobacterium *Synechococcus* sp. Arch. Biochem. Biophys. *219*, 209–218.

Tandeau de Marsac, N., and Cohen-Bazire, G. (1977). Molecular composition of cyanobacterial phycobilisomes. Proc. Natl. Acad. Sci. U.S.A. *74*, 1635–1639.

Tchernov, D., Helman, Y., Keren, N., Luz, B., Ohad, I., Reinhold, L., Ogawa, T., and Kaplan, A. (2001). Passive entry of CO_2 and its energy-dependent intracellular conversion to HCO_3^- in cyanobacteria are driven by a photosystem I-generated ΔμH+. J. Biol. Chem. *276*, 23450–23455.

Thiel, T., Hartnett, T., and Pakrasi, H.B. (1990). Examination of photosystem II in heterocysts of the cyanobacterium *Nostoc* sp. ATCC 29150. In Current research in photosynthesis, volume 1, Baltscheffsky, M. ed. (Kluwer academic publishers, The Netherlands), pp. 291–294.

Thomas, J. (1972). Relationship between age of culture and occurrence of pigments of Photosystem II of photosynthesis in heterocysts of a blue-green alga. J. Bacteriol. *110*, 92–95.

Thornton, L.E., Phkawa, H., Roose, J.L., Kashino, Y., Keren, N., and Pakrasi, H.B. (2004). Homologs of plant PsbP and PsbQ proteins are necessary for regulation of Photosystem II activity in the cyanobacterium *Synechocystis* 6803. Plant Cell *16*, 2164–2175.

Tsiotis, G., Haase, W., Engel, A., and Michel, H. (1995). Isolation and structural characterization of trimeric cyanobacterial photosystem I complex with the help of recombinant antibody fragments. Eur. J. Biochem. *231*, 823–830.

Ujihara, T., Sakurai, I., Mizusawa, N., and Wada, H. (2008). A method for analyzing lipid-modified proteins with mass spectrometry. Anal. Biochem. *374*, 429–431.

Umena, Y., Kawakami, K., Shen, J.-R., and Kamiya, N. (2011). Crystal structure of oxygen-evolving photosystem II at a resolution of 1.9Å. Nature *473*, 55–60.

Ünlü, M., Morgan, M.E., and Minden, J.S. (1997). Difference gel electrophoresis: a single method for detecting changes in protein extracts. Electrophoresis *18*, 2071–2077.

Vavilin, D.V., and Vermaas, W.F.J. (2002). Regulation of the tetrapyrolle biosynthetic pathway leading to heme and chlorophyll in plants and cyanobacteria. Physiol. Plant. *115*, 9–24.

Vermaas, W.F.J. (2001). Photosynthesis and respiration in cyanobacteria. In Encyclopedia of Life Sciences, (John Wiley & Sons Ltd, Chichester, UK), pp. 245–251.

Vollmar, M., Schlieper, D., Winn, M., Büchner, C., and Groth, G. (2009). Structure of the c_{14} rotor ring of the proton translocating chloroplast ATP synthase. J. Biol. Chem., *284*, 18228–18235.

Vilkmer, T., Schneider, D., Bernát, G., Kirchhoff, H., Wenk, S.O., and Rögner, M. (2007). Ssr2998 of *Synechocystis* sp. PCC 6803 is involved in regulation of cyanobacterial electron transport and associated with the cytochrome b_6f complex. J. Biol. Chem. *282*, 3730–3737.

Walker, J.E., and Dickson, V.K. (2006). The peripheral stalk of the mitochondrial ATP synthase. Biochim. Biophys. Acta *1757*, 286–296.

Wang, Y.C., Sun, J., and Chitnis, P.R. (2000). Proteomic study of the peripheral proteins from thylakoid membranes of the cyanobacterium *Synechocystis* sp. PCC 6803. Electrophoresis *21*, 1746–1754.

Wang, Y., Xu, W., and Chitnis, P.R. (2009). Identification and bioinformatic analysis of the membrane proteins of *Synechocystis* sp. PCC 6803. Prot. Sci. 7, 11.

Watanabe, M., Iwai, M., Narikawa, R., and Ikeuchi, M. (2009). Is the photosystem II complex a monomer or a dimer? Plant Cell Physiol. *50*, 1674–1680.

Watanabe, M., Kubota, H., Wada, H., Narikawa, R., and Ikeuchi, M. (2011). Novel supercomplex organization of Photosystem I in *Anabaena* and *Cyanophora paradoxa*. Plant Cell Physiol. *52*, 162–168.

Waters, K.M., Pounds, J.G., and Thrall, B.D. (2006). Data merging for integrated microarray and proteomic analysis. Brief. Funct. Genom. Proteom. *5*, 261–272.

Westphal, S., Heins, L., Soll, J., and Vothknecht, U.C. (2001). Vipp1 deletion mutant of *Synechocystis*: a connection between bacterial phage shock and thylakoid biogenesis? Proc. Natl. Acad. Sci. U.S.A. *98*, 4243–4248.

Wilde, A., Haertel, H., Huebschmann, T., Hoffmann, P., Shestakov, S.V., and Börner, T. (1995). Inactivation of a *Synechocystis* sp. strain PCC 6803 gene with homology to conserved chloroplast open reading frame 184 increases the photosystem II-to-photosystem I ratio. Plant Cell *7*, 649–658.

Wilson, A., Ajlani, G., Verbavatz, J.M., Vass, I., Kerfeld, C.A., and Kirilovsky, D. (2006). A soluble carotenoid protein involved in phycobilisome-related energy dissipation in cyanobacteria. Plant Cell *18*, 992–1007.

Wittig, I., Braun, H.P., and Schägger, H. (2006). Blue native PAGE. Nat. Protoc. *1*, 418–428.

Wittig, I., Karas, M., and Schägger, H. (2007). High resolution clear native electrophoresis for in-gel functional assays and fluorescence studies of membrane protein complexes. Mol. Cell. Proteomics *6*, 1215–1225.

Xu, M., Ogawa, T., Pakrasi, H.B., and Mi, H. (2008). Identification and localization of the CupB protein involved in constitutive CO_2 uptake in the cyanobacterium *Synechocystis* sp strain PCC 6803. Plant Cell Physiol. *49*, 994–997.

Xu, X., Elhai, J., and Wolk, C.P. (2008). Transcriptional and developmental responses by *Anabaena* to deprivation of fixed nitrogen. In The Cyanobacteria. Molecular Biology, Genomics and Evolution, Herrero, A., Flores, E., eds. (Caister Academic Press, Norfolk, UK), pp. 271–288.

Yamamoto, H., Peng, L., Fukao, Y., and Shikanai, T. (2011). An Src homology 3 domain-like fold protein forms a ferredoxin binding site for the chloroplast NADH dehydrogenase-like complex in *Arabidopsis*. Plant Cell *23*, 1480–1493.

Zak, E., and Pakrasi, H.B. (2000). The BtpA protein stabilizes the reaction center proteins of photosystem I in the cyanobacterium *Synechocystis* sp. PCC 6803 at low temperature. Plant Physiol. *123*, 215–222.

Zak, E., Norling, B., Andersson, B., and Pakrasi, H.B. (1999). Subcellular localization of the BtpA protein in the cyanobacterium *Synechocystis* sp. PCC 6803. Eur. J. Biochem. *261*, 311–316.

Zak, E., Norling, B., Maitra, R., Huang, F., Andersson, B., and Pakrasi, H.B. (2001). The initial steps of biogenesis of cyanobacterial photosystems occur in plasma membranes. Proc. Natl. Acad. Sci. U.S.A. *98*, 13443–13448.

Zhang, L.F., Yang, H.M., Cui, S.X., Hu, J., Wang, J., Kuang, T.Y., Norling, B., and Huang, F. (2009a) Proteomic analysis of plasma membranes of cyanobacterium *Synechocystis* sp. strain PCC 6803 in response to high pH stress. J. Proteome Res. *8*, 2892–2902.

Zhang, P., Battchikova, N., Jansen, T., Appel, J., Ogawa, T., and Aro, E.-M. (2004). Expression and functional roles of the two distinct NDH-1 complexes and the carbon acquisition complex NdhD3/NdhF3/CupA/Sll1735 in *Synechocystis* sp PCC 6803, Plant Cell *16*, 3326–3340.

Zhang, P., Battchikova, N., Paakkarinen, V., Katoh, H., Iwai, M., Ikeuchi, M., Pakrasi, H.B., Ogawa, T., and Aro, E.M. (2005). Isolation, subunit composition and interaction of the NDH-1 complexes from *Thermosynechococcus elongatus* BP-1, Biochem. J. *390*, 513–520.

Zhang, P., Allahverdiyeva, Y., Eisenhut, M., and Aro, E.M. (2009b) Flavodiiron proteins in oxygenic photosynthetic organisms: photoprotection of photosystem II by Flv2 and Flv4 in *Synechocystis* sp. PCC 6803. PLoS One, *4*, e5331.

Zhang, P., Eisenhut, M., Brandt, A.M., Carmel, D., Silén, H., Vass, I., Allahverdiyeva, Y., Salminen, T., and Aro, E.M. (2012). Operon *flv4-flv2* provides cyanobacterial Photosystem II with flexibility of electron transfer. Plant Cell *24*, 1952–1971.

Zouni, A., Witt, H.T., Kern, J., Fromme, P., Krauß, N., Saenger, W., and Orth, P. (2001). Crystal structure of photosystem II from *Synechococcus elongatus* at 3.8 Å resolution. Nature *409*, 739–743.

Protein Targeting, Transport and Translocation in Cyanobacteria

5

Dirk Schneider

Abstract

In contrast to other bacteria, cyanobacterial cells are composed of six different subcellular compartments, and proteins might be localized in one of the three membrane systems (outer, cytoplasmic or thylakoid membrane) or within one of the three soluble compartments, i.e. the periplasmic space, the cytoplasm or the thylakoid lumen. As cyanobacterial cells appear to have distinct sets of proteins localized only in a single subcellular compartment, these organisms eventually have evolved mechanisms to localize proteins to specific membranes for membrane integration or for translocation across these membranes. In the present article we summarize findings on the membrane structure of cyanobacterial cells as well as on heterogeneous protein distribution, and we discuss current models aiming at explaining mechanisms involved in protein targeting and sorting in cyanobacteria.

Subcellular organization of cyanobacterial cells – the sorting problem

As Gram-negative bacteria, cyanobacteria contain a cytoplasmic membrane (CM) separated from the outer membrane by a peptidoglycan layer (Liberton *et al.*, 2006). However, in contrast to most other bacteria, cyanobacteria harbour an additional internal membrane system, the thylakoid membranes (TM). TMs represent a dense and complex interconnected membrane network, derived from continuous branching and splitting of a single membrane (Liberton *et al.*, 2012a,b). Thus, cyanobacterial cells are composed of six different subcellular compartments, and proteins might be localized in one of the three membrane systems (outer membrane, CM or TM) or within one of the three soluble compartments, i.e. the periplasmic space, the cytoplasm or the thylakoid lumen. However, it is a longstanding debate whether the CM and the TM are interconnected, and thus, whether the periplasm and thylakoid lumen are continuous.

Two models for the morphology of a cyanobacterial cell have been discussed (Mullineaux, 1999). One idea is that TM and CM represent separate entities. Several studies on separated TM and CM as well as immunological studies on ultrathin sections of entire cells have clearly shown that the CM and TM contain different subsets of proteins, just as the periplasm and the thylakoid lumen do (for details see below). This implies that either both proteins and lipids are transported between the different subcellular compartments, or that the different membrane systems both harbour full sets of the proteins needed for protein translocation/integration, lipid biosynthesis, etc. Alternatively, the periplasmic space and thylakoid lumen are continuous, and the TM is directly connected to the CM, thus representing an invagination of the CM with a specialized function. Whereas in most high-resolution structural studies no clear evidence for a physical continuity of the two membranes has been found (Liberton *et al.*, 2006, 2012a,b; Nevo *et al.*, 2007; van de Meene *et al.*, 2006, 2012), a recent analysis of a cyanobacterial mutant strain suggests that the TM and the CM might be connected under certain conditions, at least transiently (van de Meene *et al.*, 2012). However, since the complexes of the respiratory and photosynthetic electron transfer chain reside and are functional within TMs (Gantt, 1994; Nowaczyk *et al.*, 2010; Vermaas, 2001), continuity of the periplasm with the thylakoid lumen would present a challenge to the cell. Generating and maintaining a proton gradient across the TM, which is needed for ATP production, would be difficult or even impossible. Localizing proteins, possessing functions distinct to the TM and/or the thylakoid lumen, in the CM and/or periplasm – or vice versa – would be an uneconomical process and would not allow controlling metabolic processes by enzyme or metabolic pathway localization in defined subcellular compartments. Nevertheless, mainly based on observations of photosystem (PS) assembly steps, in several recent reports a (dynamic) connection of the cyanobacterial membranes has been proposed. In some cyanobacteria, such as *Synechocystis* sp. PCC 6803 (thus far the best studied cyanobacterium; hereafter referred to as *Synechocystis*), TM and CM appear to merge at regions called 'thylakoid centres' (Kunkel, 1982), as further discussed below. Nonetheless, should the internal membrane systems be interconnected, any proposed connection regions would have to be tightly sealed and the passage of proteins, lipids, etc. would have to be firmly controlled. However, as ultra structural analyses have not indicated any physical connection between internal membrane systems thus far (Liberton *et al.*, 2006, 2012a,b; Nevo *et al.*, 2007; van de Meene *et al.*, 2006), it appears to be most likely that the internal membranes represent separated entities. Furthermore, some cyanobacterial cells are able to grow completely heterotrophically in the dark, which eventually makes the TM network dispensable (Rippka, 1972). Nevertheless, even when largely reduced, cyanobacterial cells always contain at least rudimentary thylakoids in the dark (Anderson and McIntosh, 1991; Mannan and Pakrasi, 1993; Smart *et al.*, 1991; Vernotte *et al.*, 1992), indicating that TMs are no simple (temporal) invaginations of CMs. Instead, TMs might develop from existing TMs or TM precursor structures.

In the following we summarize findings on the membrane structure of cyanobacterial cells as well as on heterogeneous protein distribution, and we discuss current models aiming at explaining mechanisms involved in protein sorting and organization of subcellular compartments. As several aspects of protein targeting and translocation have been well characterized in other bacterial cells as well as in chloroplasts, which share a common ancestor with cyanobacteria, we discuss the sparse results obtained in experiments on cyanobacteria with results obtained in other bacteria or chloroplasts.

How to establish protein heterogeneity in cyanobacteria?

As cyanobacterial cells appear to have distinct sets of proteins localized only in a single subcellular compartment, the cyanobacterial cells should have evolved mechanisms to localize proteins to specific membranes for membrane integration or for translocation across these membranes (Howe *et al.*, 1996). In some cases, however, proteins have been described to localize to both, the periplasm and the thylakoid lumen, or to the CM and the TM.

Asymmetric distribution of proteins might be organized by diverse mechanisms. Mainly based on studies employing chloroplasts and bacteria, such as the enterobacterium *Escherichia coli*, transport pathways across the bacterial CM as well as across the TM have been analysed in great detail and proteins involved in protein translocation and/or integration into the respective membranes have been identified. Thus, it is possible that cyanobacteria contain different sets of translocation systems in CM versus TM and thereby ensure proper translocation or membrane integration of proteins. Such a system might also imply that proteins, localizing to one specific subcellular compartment, contain specific targeting signals, and that different targeting factors might be involved in transport of pre-proteins to one specific membrane system. Having specific targeting signals and/or factors might even eliminate the need for having different translocases.

However, it is also possible that proteins are not sorted but rather integrated in only one membrane or in both membrane systems equally. The latter possibility would explain the observation that some proteins have a dual localization.

While some proteins, including the subunits of the cytochrome b_6f complex, the terminal oxidase or electron carriers, such as cytochrome c_6 or plastocyanin, may function in both membranes or in both the thylakoid lumen and the periplasm, other proteins might be degraded after membrane integration or translocation, if not required in the respective compartment. As this would be a rather uneconomical principle, it appears to be more likely that, if all proteins are integrated into the same membrane, proteins are sorted after translocation or membrane integration, either by lateral sorting or by a vesicular transport system. While a lateral transfer within a membrane implies that the internal membrane systems are interconnected, at least transiently, for a vesicular transport, proteins would have to be packed in sorting vesicles and transferred to a defined compartment, as has been analysed in great detail for the exocytotic pathway in eukaryotes. Since it has been suggested that components of the PSs, which are active solely in the TM, are pre-assembled in the CM, sorting, either by lateral transfer or by vesicular transport, appears to take place at least from the CM to the TM, as well as from the periplasm to the thylakoid lumen.

Protein translocation and membrane integration in bacteria and chloroplasts – a brief overview

Transport of proteins across and into cellular membranes is well characterized in several systems, including the endoplasmic reticulum (ER) membrane, the bacterial inner membrane as well as the chloroplast TM. Most likely, due to their endosymbiotic origin, principles guiding protein transport across or into the chloroplast TM and across or into the inner membrane of bacteria are broadly similar (Celedon and Cline, 2012). Nevertheless, as most chloroplast proteins are encoded in the nucleus and are post-translationally imported into the organelle, new import machineries bridging the outer and inner chloroplast envelope membranes have evolved, termed the translocase of the outer chloroplast membrane (TOC) and the translocase of the inner chloroplast membrane (TIC), respectively (Bölter and Soll, 2011). While not understood in detail yet, posttranslational import and insertion of proteins localized in the inner envelope membrane most likely involves the TIC machinery, although other mechanisms have also been suggested (Bölter and Soll, 2011; Celedon and Cline, 2012).

Protein targeting and translocation in Gram-negative bacteria

Protein transport across a bacterial inner membrane (CM) and integration of membrane integral proteins has probably been best studied in the eubacterium *E. coli*. Protein subunits of protein complexes involved in protein translocation are summarized in Fig. 5.1 and Table 5.1.

For protein transport across the bacterial inner membrane, after synthesis, proteins are kept in an unfolded conformation by chaperones and are targeted to the CM by the SecA/SecB pathway (Müller *et al.*, 2001). In *E. coli*, SecB is the primary cytoplasmic chaperone involved in secretion of a subset of periplasmic and outer membrane proteins in an unfolded state (Zhou and Xu, 2005). SecB keeps precursor proteins in an unfolded state and directs pre-proteins to the SecA ATPase component of the Sec translocase prior to protein translocation into the periplasm. Besides SecB, several other soluble chaperones might also be involved in targeting pre-proteins to membrane integral translocation complexes (Müller *et al.*, 2001), and the Hsp70 chaperone DnaK as well as the *trigger factor* are also involved in keeping pre-proteins in an unfolded and translocation-competent state. The observation that a SecB protein is only encoded in α-, β- or γ-proteobacteria indicates a non-essential role of this chaperone in protein targeting, transport and/or translocation (van der Sluis and Driessen, 2006). In the absence of SecB, export of SecB-dependent pre-proteins proceeds at a slower rate in *E. coli* (Randall and Hardy, 2002) and the Hsp70 and Hsp40 chaperones DnaK and DnaJ can substitute for the SecB function (Altman *et al.*, 1991), suggesting that the individual chaperones and/or targeting factors have overlapping precursor specificities. Thus, upon loss of one targeting factor, some precursors accumulate to significant amounts while other precursors efficiently utilize

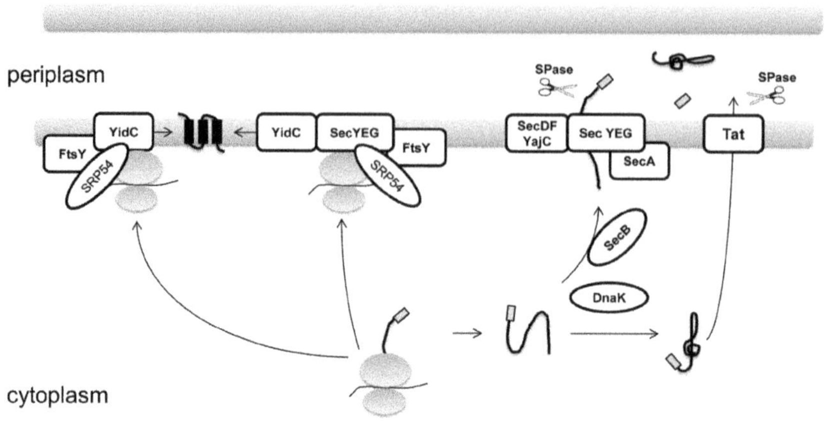

Figure 5.1 Protein trafficking pathways in *E. coli*. Individual pathways and components involved are described in the text. SPase: type I signal peptidase (LepB).

Table 5.1 Proteins implicated in protein targeting and translocation in *E. coli*, *Arabidopsis* chloroplasts and in *Synechocystis* as discussed in the text

Function	Component	*E. coli*	*Arabidopsis*	*Synechocystis*
Targeting				
	SecB	SecB	–	–
	SRP54	Ffh	cpSRP54	Slr1531
	SRP-RNA	*ffs*, 4.5 S RNA	–	*ffs*
	cpSRP43	–	cpSRP43	–
Translocation				
Sec translocase	SecA	SecA	cpSecA1, cpSecA2	Sll0616
	SecY	SecY	cpSecY1, cpSecY2	Sll1814
	SecG	SecG	–	Ssr3307
	SecE	SecE	cpSecE1, cpSecE2	Ssl3335
	SecD	SecD	–	Slr0774
	SecF	SecF	–	Slr0775
	YajC	YajC	–	–
Tat translocase	TatA	TatA	Tha4	Slr1046, Ssl2823
	TatB	TatB	Hcf106	Slr1046, Ssl2823
	TatC	TatC	cpTatC	Sll0194
	TatD	TatD	–	Sll1786
	TatE	TatE	–	Slr1046, Ssl2823
	FtsY	FtsY	cpFtsY	Slr2102
YidC/Albino3	YidC	YidC	Albino 3, Albino 4	Slr1471
Signal peptide cleavage	SPase I	LepB	At2g30440, Plsp1	Sll0716, Slr1377

the remaining targeting systems (Danese and Silhavy, 1998).

Proteins targeted to the *E. coli* CM are translocated across the membrane by the heterotrimeric Sec-translocase, composed of the membrane integral subunits SecY, SecE and SecG, which form the actual protein conducting channel (du Plessis *et al.*, 2011). A SecDF–YajC complex can associate with the SecYEG complex, and the encoding *secD*, *secF* and *yajC* genes are organized in an operon in *E. coli*. However, in contrast to SecY and SecE, the SecG protein and the additional membrane integral SecDF–YajC complex are non-essential for protein translocation in bacteria (Brundage *et al.*, 1990; du Plessis *et al.*, 2011). Proteins translocated via the Sec-translocase are typically synthesized as pre-proteins in the bacterial cytoplasm with an N-terminal signal sequence. This signal sequence has a conserved structure with a short, positively charged N-terminal region followed by a central hydrophobic region and a moderately polar C-terminal region (Von Heijne, 2002). During or directly after translocation, the N-terminal signal sequence of pre-proteins is cleaved off by a membrane-bound type I signal peptidase (LepB), resulting in the release of a mature protein into the periplasm (Auclair *et al.*, 2012).

While translocation of unfolded proteins via the Sec-pathway is the main protein export pathway across the *E. coli* CM, a limited number or proteins, which mostly bind metal cofactors, are transported in a folded conformation across the CM by the Tat (twin arginine translocation)-translocase (Müller and Klösgen, 2005). While the Sec- and Tat-signal peptides are similar in their structure, the Tat signal peptides typically contain a twin arginine motif (RR) in the positively charged N-terminal region of the signal sequence, which precedes the hydrophobic domain. At least the three proteins TatA, TatB and TatC are required for Tat translocation in *E. coli*, though the composition of the Tat-translocase varies in other bacteria. The TatA and TatB subunits have similar structures, and both are composed of a single N-terminal transmembrane helix having a cytoplasmically localized amphipathic helix and a variable region. Therefore, in some organisms it is hard to distinguish TatA from TatB homologues. Nonetheless, phylogenetic analyses have indicated that TatA homologues cluster more tightly than TatB proteins, suggesting that the sequence of TatB may have diverged more rapidly than that of TatA (Yen *et al.*, 2002). It appears that an early gene duplication event resulted in TatA and TatB, but since then only TatA has been duplicated, e.g. resulting in the TatE subunit found in *E. coli*. As *tatE* is only expressed at a low level in *E. coli* (Jack *et al.*, 2001) and *tatE* deletion appears to not affect protein export (Lee *et al.*, 2006), the functional significance of the *tatA* gene duplication, leading to the *tatE* gene, still is unclear. TatB does not appear to be essential for protein export either (Blaudeck *et al.*, 2005), and indeed the Tat machinery of Gram-positive bacteria does not have a TatB component (Barnett *et al.*, 2008) engaging only TatA and TatC proteins. Here, the TatA protein fulfils the roles of both, *E. coli* TatA and TatB (Barnett *et al.*, 2008). Furthermore, some Gram-positive bacteria as well as some archaea contain multiple *tatC* and multiple *tatA/B* genes. Noteworthy, in *Bacillus subtilis* the two *tatC* and three *tatA* genes appear to constitute separate, substrate-specific Tat systems (Jongbloed *et al.*, 2000, 2002, 2004; Pop *et al.*, 2002). For a more detailed description of the Tat-pathway see, for example, (Fröbel *et al.*, 2012; Lee *et al.*, 2006; Palmer and Berks, 2012).

Proteins that are integrated into the *E. coli* inner membrane are typically cotranslationally inserted into the membrane via the signal recognition particle (SRP)-pathway (Koch *et al.*, 2003). After synthesis of a pre-protein, the SRP binds to the signal sequences as well as to the ribosome and mediates contact with the membrane-bound SRP-receptor FtsY as well as with the SecYEG-translocase. This eventually results in release of the bound SRP and in synthesis of a protein at the Sec-translocase directly into the membrane plane. The SRP of *E. coli* and other bacteria consist of a single protein with homology to the eukaryotic SRP-subunit of 54 kDa (FFH, fifty-four homologue) and contain a small, essential RNA molecule (Koch *et al.*, 2003). While the exact function of the membrane integral protein YidC, which is involved in chaperoning membrane integration of proteins (Wang and Dalbey, 2012), still is elusive, recent data suggest that YidC alone can also act as an insertase by interaction with FtsY and SRP/ribosome/nascent chain-complexes (Welte *et al.*, 2012).

Protein targeting and translocation in chloroplasts

In chloroplasts, four different pathways have been described for integration of membrane proteins into TMs and protein translocation into the thylakoid lumen (Fig. 5.2). These mechanisms share a high degree of similarity to the mechanism described in *E. coli*. The chloroplasts' Sec-, SRP- and Tat-pathways, as well as a spontaneous insertion pathway, each handle different subsets of proteins (for recent reviews see Schüneman, 2007; Cline and Dabney-Smith, 2008).

Translocation of soluble proteins across the TM via the Sec-translocase appears to function essentially identical as in bacteria (Celedon and Cline, 2012; Wang and Dalbey, 2012). However, in contrast to bacteria, the Sec-translocon is reduced in chloroplasts to a heterodimeric complex of SecY and SecE homologues. While chloroplasts contain a SecA ATPase component associated with the SecYE-complex, a SecB chaperone is absent. Nevertheless, SecB is non-essential in bacteria and chloroplasts contain numerous other chaperones that might fulfil the SecB-function, keep pre-proteins in a transport-competent state and target pre-proteins to the Sec-translocase. Surprisingly, a second set of SecA, SecY and SecE proteins has recently been identified that appear to form a second Sec-translocase in the chloroplast inner envelope membrane (Celedon and Cline, 2012). It is likely that the assembled SecY2E2-translocase is involved in integration of pre-proteins from the chloroplast matrix into the inner envelope (Celedon and Cline, 2012). The mechanistic details of how proteins are targeted to the SecY2E2-translocase are completely enigmatic yet, and it is also unclear whether the single chloroplastic SRP54 protein interacts with both translocases. It is possible that the SecY2E2-translocase interacts with the TIC-translocase and that proteins are released into the inner envelope membrane during their import

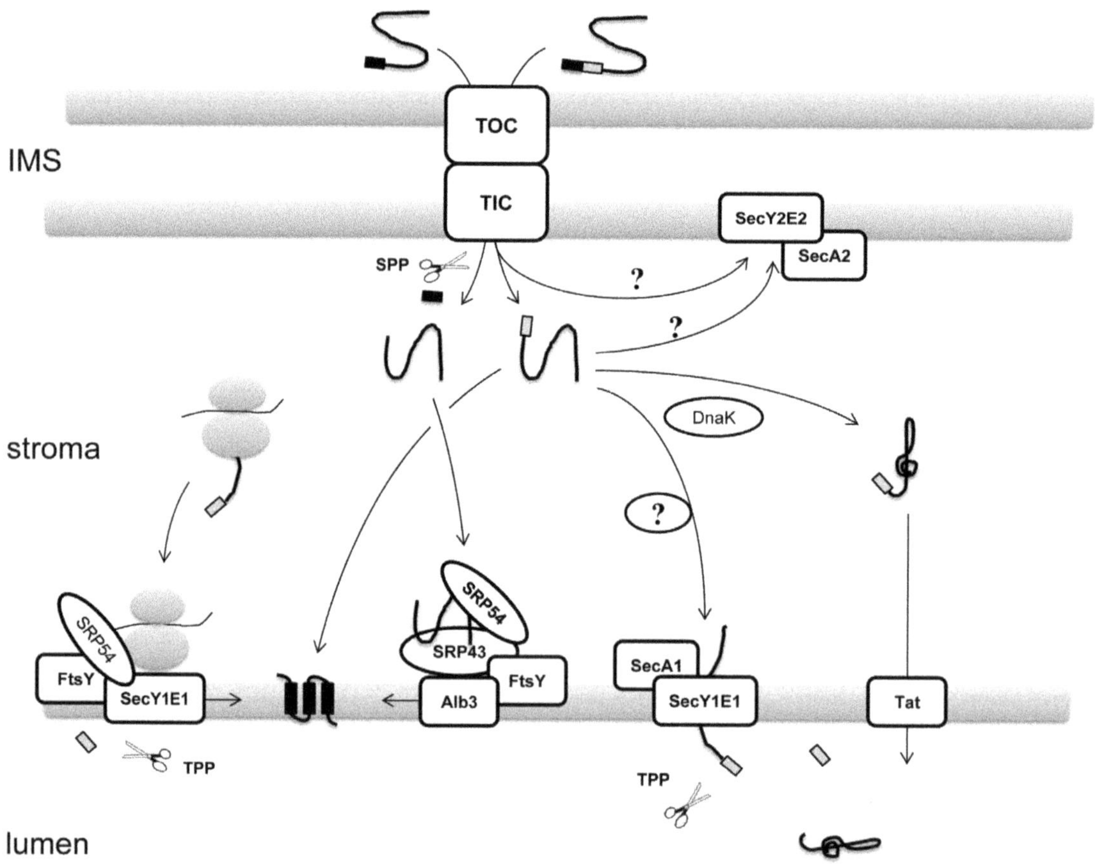

Figure 5.2 Protein trafficking pathways in chloroplasts. Individual pathways and components involved are described in the text. SPP, stromal processing protease; TPP, thylakoid processing peptidase (type I signal peptidase).

into the chloroplast via a TIC/SecY2E2-pathway (Celedon and Cline, 2012). As also two paralogues of the soluble SecA component have been identified in chloroplasts, and as SecA is involved in translocation of soluble proteins across the bacterial CM and the chloroplast TM, it is likely that also the SecY2E2-complex is involved in protein transport from the chloroplast matrix across the inner envelope membrane. Supportive of a dual localization of Sec-translocases, recent findings suggest signal peptidase activity in both, the inner envelope membrane and the TM in *Arabidopsis* chloroplasts (Hsu *et al.*, 2011; Inoue *et al.*, 2005). However, signal peptidase localization in the inner envelope membrane may be more prominent in developing chloroplasts, whereas in developed chloroplasts the signal peptidase appears to be exclusively localized in TMs (Shipman and Inoue, 2009).

The subsets of TM proteins that are encoded in the chloroplast genome are expected to co-translationally integrate into the TM, similar to the mechanisms described in *E. coli.* However, mechanistic details are less well described in chloroplasts, although studies with the PsbA protein suggest that at least a SecY protein as well as Alb3, a YidC homologue in chloroplasts, and an SRP (cpSRP54) are involved in protein integration into the TM (Celedon and Cline, 2012). This implies that also the SRP receptor FtsY as well as SecE, the second vital subunit of the Sec-translocase, are most likely involved in co-translational protein integration in chloroplasts. However, deletion of genes coding for the cpSRP54 or the FtsY protein have only limited effects, which suggests that TM proteins are targeted at least in part by different mechanisms. Furthermore, the SRP-pathway appears to additionally operate differently in chloroplasts when compared to *E. coli* (Richter *et al.*, 2010). The chloroplast cpSRP54 lacks the RNA component and a novel SRP component, the cpSRP43, is involved in a posttranslational mode of the SRP-pathway. In the posttranslational mode, the cpSRP pathway employs the Alb3 protein but not the Sec-translocase, and transported/integrated proteins do not contain a cleavable N-terminal signal sequence (Celedon and Cline, 2012). Thus far, this posttranslational SRP54/SRP43-pathway appears to integrate a single family of membrane proteins, the light-harvesting chlorophyll *a*/*b*-binding (LHC) proteins. Thus, as indicated in Fig. 5.2, LHC proteins are bound by SRP54 and SRP43 proteins after import into the chloroplast matrix and this complex interacts with cpFtsY and Alb3 for membrane integration. Noteworthy, two YidC homologues, Alb3 and Alb4, have been identified in *Arabidopsis thaliana.* They appear to have different substrate proteins, as disruption of the *alb3* and *alb4* genes, respectively, resulted in different phenotypes (Gerdes *et al.*, 2006; Sundberg *et al.*, 1997).

The Tat-pathway was originally discovered in chloroplasts, and the Tha4 (TatA) and Hcf106 (TatB) proteins, as well as the chloroplast TatC (cpTatC), have been identified (Albiniak *et al.*, 2012). The bacterial and the chloroplast Tat-pathways appear to operate similarly (see above) and both need a ΔpH across the membrane for protein translocation.

While in *E. coli* spontaneous insertion of small membrane proteins into the CM has been discussed in the past, the discovery of the YidC protein has led to the assumption that proteins do not insert spontaneously into the bacterial CM (Samuelson *et al.*, 2000). However, membrane integration of selected proteins into the chloroplast TM is not hindered by blockage of either one of the Sec, Tat or Alb3 pathways. Therefore, although insertion of proteins into plant TMs generally appears to be less sensitive to deletion of Sec- or Tat-components, when compared to *E. coli*, a pathway, which appears not to involve any proteinaceous component, is still being discussed for thylakoids (Schleiff and Klösgen, 2001). Few multi-spanning as well as a series of small, single-span membrane proteins, including subunit II of the ATP synthase (CFoII) and PS II subunits W and X (PsbW, PsbX), may use this pathway. These proteins are synthesized as pre-proteins with pre-sequences in plants, and these pre-sequences very much resemble signal sequences of proteins imported into the thylakoid lumen via the Sec-pathway (Celedon and Cline, 2012; Schünemann, 2007). Initial stages of such suggested spontaneous insertion appear to involve binding of the protein to the membrane, and thereafter both hydrophobic domains (one in the signal peptide, the other in the mature protein) insert into the membrane and form a transmembrane helical hairpin intermediate (Thompson *et al.*, 1998). Subsequently, the presequence is removed by a thylakoid processing peptidase, which leaves the mature protein inserted in the membrane with a luminal N-terminus and a stromal C-terminus. Thus, the signal peptides might simply assist in the insertion of the transmembrane section in the mature protein, as they provide an additional hydrophobic region required for formation of a TM helical hairpin structure.

Protein translocation systems in cyanobacteria: a genetic perspective

Due to the endosymbiotic origin of chloroplasts, the principles guiding translocation of proteins across the bacterial CM and the chloroplast TM are largely conserved. While cyanobacteria show many similarities to bacterial features, they are related more closely to chloroplasts in some systems (Fig. 5.3). The mechanisms involved in protein translocation or integration into the cyanobacterial CM and TM have not been widely studied yet. In Table 5.1, components of translocation systems that are encoded in the genome of *Synechocystis* sp. PCC 6803 are listed. While all essential core subunits of known translocation systems are encoded in the *Synechocystis* genome, some subunits found in *E. coli* are absent.

The Sec-translocase of *Synechocystis* is composed of the core subunits SecY, SecE and SecA, as well as of the subunits found in *E. coli* but not in chloroplasts, SecG, SecD and SecF. Thus, based on the subunit composition, the Sec-translocase of *Synechocystis* is more closely related to the *E.*

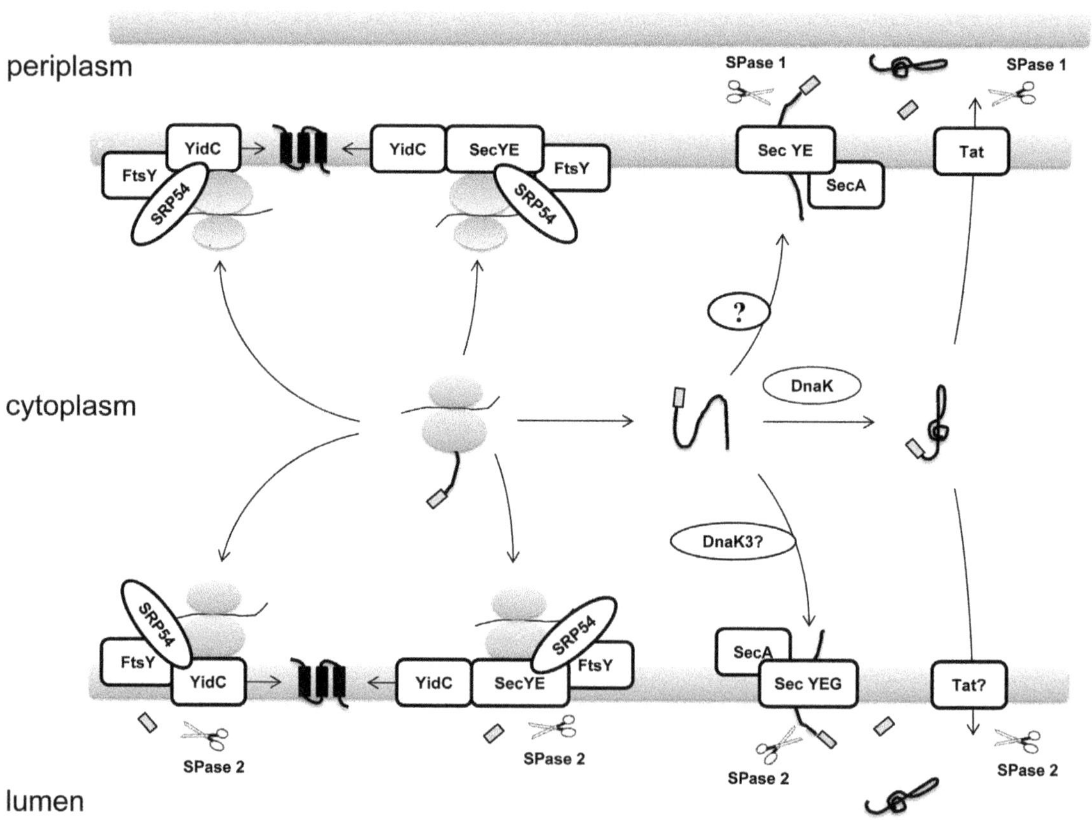

Figure 5.3 Potential protein trafficking pathways in cyanobacteria. Potential targeting pathways and components involved in protein targeting, transport and translocation in cyanobacteria are described in the text and in Table 5.1. SPase, type I signal peptidase (LepB).

coli translocase than to the chloroplast system. Nevertheless, in phylogenetic analyses, subunits of the cyanobacterial Sec-translocase cluster together with the proteins from chloroplasts (Saier, 2006), highlighting the evolutionary origin of chloroplasts. However, in contrast to (at least) *Arabidopsis*, only a single set of genes encoding subunits of the Sec-translocase have been identified to date in cyanobacteria. Thus, the existence of two Sec-complexes, each localized in one of the two different internal membranes, CM and TM – similar to the situation in the chloroplast – is unlikely. Interestingly, while also in cyanobacteria *secD* and *secF* are organized in a gene cluster, a gene encoding a homologue of the YajC protein, which is also absent in chloroplasts, has not been identified in any cyanobacterium yet. This might indicate that this non-essential protein has evolved after the emergence of cyanobacteria. While cyanobacterial genomes do not encode a SecB protein, most components of the bacterial Sec-system are present in cyanobacteria and a protein with homology to SRP54 has been identified. The cyanobacterial genomes also encode a small SRP54 RNA molecule (94 nucleotides long in *Synechocystis)*, similar to the situation in bacteria but different from chloroplasts, where the RNA component is missing. It is worth mentioning that red algae, but not green algae, mosses or higher plants' chloroplasts, possess a cpSRP54 protein with an RNA molecule (Rosenblad and Samuelsson, 2004). Since no cpSRP43 has yet been identified in cyanobacteria or red algae, the absence of the SRP-RNA and the presence of cpSRP43 might be specific to green algae, mosses and higher plants. As cyanobacteria and red algae do not contain membrane integral light–harvesting complexes, these data indicate that cpSRP43 could have appeared during the evolution of membrane integral LHC proteins, which are present in green algae and higher plants (Falk and Sinning, 2010). Recent results suggest that cpSRP43 has developed additional features and can also function independently of the cpSRP54/cpFtsY-pathway (Tzvetkova-Chevolleau *et al.*, 2007). Likely candidate genes, from which an SRP43 protein-encoding gene could have evolved, are present in cyanobacteria, although no protein fulfilling a cpSRP43-related function has yet been identified in these organisms.

The YidC/Alb3-pathway differs in bacteria versus chloroplasts. As outlined above, in bacteria, YidC is mostly associated with the Sec-translocase during integration of transmembrane proteins into the CM. Furthermore, a subset of proteins appears to integrate into the membrane co-translationally, directly via a YidC-only pathway. In chloroplasts, SRP54 as well as SRP43 are involved in chaperoning imported proteins to the TM and in posttranslational membrane integration via Alb3 (Wang and Dalbey, 2012). While the exact mechanisms are still uncertain, it is likely that YidC in cyanobacteria acts similar to the *E. coli* system, and thus the protein likely interacts with the Sec-translocase, as indicated e.g. in the case of membrane integration of the *Synechocystis* PsbA precursor protein (Ossenbühl *et al.*, 2006). The *Synechocystis* YidC protein has a conserved structure and can functionally replace the *E. coli* homologue (Gathmann *et al.*, 2008). While the *Synechocystis* YidC protein is essential for cell viability, and thus deletion of the *Synechocystis yidC* gene was impossible, depletion of the protein has demonstrated that it is crucial for TM biogenesis (Spence *et al.*, 2004).

The Tat-system in *Synechocystis* contains only homologues of the proteins TatA and TatB as well as a TatC homologue. As TatA and TatB are similar in terms of structure and sequence, it is difficult to clearly distinguish TatA from TatB homologues in cyanobacteria. A deletional analysis has shown that the genes *slr1046* and *ssl2823*, which both encode proteins homologous to TatA/B/E proteins, can both complement *E. coli* Δ*tatAE* and Δ*tatB* strains (Aldridge *et al.*, 2008). Thus, the encoded proteins have attributes of both, TatA and TatB. TatC is the most conserved of the Tat proteins and sequence conservation is particularly strong within its transmembrane domain (Bogsch *et al.*, 1998; Buchanan *et al.*, 2002). In a phylogenetic tree, the cyanobacterial proteins cluster with the chloroplast proteins (Yen *et al.*, 2002), again highlighting the likely evolutionary origin of chloroplasts. Additionally, in contrast to chloroplasts, a TatD component is encoded in cyanobacteria.

In both, protein transport via the Sec- as well

as via the Tat-system, type I signal peptidases are involved in pre-protein processing and removal of the signal sequence. While in *E. coli* only a single type I signal peptidase has been identified, a minimum of two genes coding for type I signal peptidases appear to be conserved in cyanobacterial genomes (Nakao *et al.*, 2010) (Table 5.1). As different subcellular localizations have been determined for these two peptidases, these proteins have been implicated in protein sorting in cyanobacterial cells, as discussed in greater detail below.

Whereas the SecA-, SRP- and Tat-pathways all appear to have been inherited from a cyanobacterium-like progenitor in chloroplasts, the 'spontaneous' insertion pathway used by some single-span membrane proteins appears to operate differently. While genes encoding PsbX and CFoII are present in cyanobacteria, the encoded proteins are synthesized without any presequences. However, the structure of the mature proteins appears to be very similar, and thus it is unclear why chloroplast-imported proteins need cleavable presequences whereas the cyanobacterial proteins do not. Possibly, the latter variants are inserted co-translationally and are able to integrate into the cyanobacterial TM more easily.

Protein translocation systems in cyanobacteria: subcellular localization of translocases and integrases

In recent ultra structural analyses of the cyanobacteria *Synechocystis*, *Synechococcus* sp. PCC 7942 (also known as *Synechococcus elongatus*), *Cyanothece* sp. ATCC 51142 and *Microcoleus* sp. ribosomes were mostly identified in the centre of the cells (Liberton *et al.*, 2006, 2012a,b; Nevo *et al.*, 2007; van de Meene *et al.*, 2006, 2012). Nevertheless, ribosomes have been identified in connection with both internal membranes (CM and TM) as well as in between individual thylakoid layers. However, as the space between adjacent thylakoid layers is typically occupied by densely packed phycobilisomes (Tandeau de Marsac, 2003; van de Meene *et al.*, 2006), only at places where thylakoids are in direct contact with the cytosol have ribosomes frequently been observed in close proximity to the TM (Liberton *et al.*, 2006, 2012a,b; Nevo *et al.*, 2007; van de Meene *et al.*, 2006, 2012). These structural analyses strongly suggest that proteins are synthesized at both, the CM as well as the TM, and thus, both membrane systems should contain proteins needed for protein translocation and/or integration (Fig. 5.3). Furthermore, the structural analyses might indicate a heterogenic distribution of translocase subunits within the TM, and solely where a TM plane faces the cytoplasm, defined regions exist where proteins are integrated into the TM. Such defined regions might be called biogenesis centres or biogenesis platforms.

In the early 1990s, localization of subunits of the Sec-pathway was studied by immunological analyses using purified internal membranes. The SecY and SecA proteins of the Sec-translocase were identified in both, the CM and TM of the cyanobacterium *Synechococcus* sp. PCC 7942 (Nakai *et al.*, 1993, 1994). This pioneering observation suggested that the Sec-pathway is localized in both membrane systems. The assumption of a dual localization of translocases has been further supported by the observation that, when purified from the cyanobacterium *Phormidium laminosum*, both membranes have signal peptidase (LepB) activity (Barbrook *et al.*, 1993). In addition, the observation that components of the Sec-translocase have been identified in the inner envelope membrane as well as in the TM of cyanelles, the 'primitive' plastids of glaucocystophyte algae, which resemble free-living cyanobacteria in many respects, further supports the assumption that the Sec-translocase is active in both, the cyanobacterial CM and TM (Steiner and Löffelhardt, 2005; Yusa *et al.*, 2008). In line with this, the observation that Sec-components have also been identified in the inner envelope membrane of higher plant chloroplasts also supports the idea that a dual localization of the Sec-translocase is conserved. Thus, individual Sec-subunits are most likely transported to internal membranes and mediate protein translocation and/or integration across/into the inner envelope membrane or CM as well as across/into the TM in chloroplasts of higher plants and algae, in primitive phototrophic eukaryotes and in cyanobacteria.

It has been observed that chimeric proteins, which contain a green fluorescent protein fused to a Tat signal sequence, are selectively translocated into the periplasm of *Synechocystis* (Spence *et al.*, 2003) and *Anabaena* sp. PCC 7120 (Mariscal *et al.*, 2007). This observation implies that the Tat-pathway is active at least in the cyanobacterial CM. Later on it was shown that the Tat-translocase is active in both, the TM as well as in the CM, and in line with this the TatA/B/E subunit encoded by the *Synechocystis orf slr1046* has been localized in both membranes after genetically fusing the protein to a fluorescent protein (Aldridge *et al.*, 2008). Based on these observations, it is likely that both membrane systems contain a complete set of translocase subunits and thus have active Sec- and Tat-translocases. Thus, proteins are either sorted prior to or during targeting to one specific internal membrane system or are randomly inserted and subsequently sorted or degraded.

However, the notion of an equal distribution of translocases has been challenged. After separation of membranes from *Synechocystis*, it has been proposed that the YidC homologue can only be found in purified CMs (Fulgosi *et al.*, 2002). This view has been supported by an immunogold analysis of ultra-thin *Synechocystis* sections, indicating that the YidC protein is localized predominantly in the CM (Gathmann *et al.*, 2008). In contrast, in a later proteomic study using purified *Synechocystis* membranes, the protein has been identified exclusively in TMs (Pisareva *et al.*, 2011). Furthermore, in this proteomic study, the SecA, SecE and SecD subunits of the Sec-translocase have also been identified exclusively in purified *Synechocystis* TMs (Pisareva *et al.*, 2011), indicating that solely the membrane fraction denoted as TM contains the translocases/integrases (at least the Sec-translocase). It is noteworthy that in proteomic analyses the two LepB proteins of *Synechocystis* have been identified in two different membranes, as discussed in detail below (Huang *et al.*, 2002; Pisareva *et al.*, 2011; Srivastava *et al.*, 2005).

Together, the results obtained by membrane separation are conflicting and do not allow to derive any straightforward conclusion regarding the subcellular localization of translocases/integrases. Currently it appears likely that both, CM and TM of *Synechocystis* contain an active Tat-translocase as well as a type I signal peptidase. The observed differences in subcellular localization of Sec-subunits might be a result of the experimental conditions, i.e. the exact methods used for preparation of pure membranes. However, these observed differences in the determined subcellular localization might be combined in a model for protein translocation/integration by proposing the existence of defined biogenesis centres in between CM and TM, as further discussed below (Pisareva *et al.*, 2011).

Targeting signals

Based on statistical analyses, the amino acid distribution and surface characteristics of proteins localized in different bacterial compartments might differ (Andrade *et al.*, 1998; Cedano *et al.*, 1997). Thus, such physico-chemical properties may be used to target proteins to one specific cyanobacterial compartment. However, a sorting mechanism based on surface characteristics would require synthesis and folding of the full-length protein within the cytosol prior to sorting, which is only the case when proteins are translocated by the Tat-pathway, since Sec-dependent proteins are translocated in an unfolded conformation. In line with this, a statistical analysis of a few selected *Synechocystis* proteins has indicated that the properties of the mature proteins lack any compartment-specific patterns (Rajalahti *et al.*, 2007). Already in the early 1990s it was suggested that the N-terminal signal sequences of pre-proteins are crucial for targeting proteins to distinct internal membranes in cyanobacteria (Howe *et al.*, 1996; Mackle and Zilinskas, 1994). Based on a proteomic analysis of CM and periplasmic proteins versus TM and thylakoid lumen proteins, it has been suggested that pre-proteins contain distinct physico-chemical properties in their signal sequence as well as in their mature N-terminal segments (Rajalahti *et al.*, 2007). This was an important finding as it has been shown for signal sequences of e.g. the Gram-positive bacterium *Streptococcus pyogenes* that they can contain information to target a nascent pre-protein to (structurally identical) Sec-translocases localized in different regions of the CM (Carlsson *et al.*, 2006). Thus, features of the signal sequences might be recognized by soluble factors, which direct the ribosome nascent–chain complex to defined sub-membrane regions, or signal sequences interact differently with defined membrane regions. Together, protein sorting in cyanobacteria might be controlled by features of the signal sequences as well as by the first 15 amino acids of the mature protein, which control targeting and translocation. However, such targeting mechanisms must involve other targeting factors, which are yet unknown.

Based on studies with *Arabidopsis* proteins it has been suggested that membrane integral protein domains determine sub-chloroplast targeting of transmembrane proteins to either the inner envelope membrane or the TM (Froehlich and Keegstra, 2011). The differences observed between transmembrane proteins localized in TM or the inner envelope membrane, respectively, appear to be significant for directed targeting. In a statistical analysis of membrane integral proteins with known localization in *Synechocystis*, the distribution of defined amino acid characteristics, such as the hydrophobicity/hydrophilicity, volume and polarity/charges of the side-chains has been examined (Pisareva *et al.*, 2011). Based on this analysis, neither the N-terminal transmembrane segments nor the soluble domains of TM versus CM proteins displayed statistically significant differences. However, the authors suggested that the most C-terminal located 40 amino acids allow discriminating between TM and/or CM integral proteins, since the TM proteins have more hydrophobic amino acids next to amino acids with small volume side-chains, and also the distribution of charged amino acids differ between the two subsets of proteins (Pisareva *et al.*, 2011). However, membrane integral proteins are typically co-translationally inserted into a membrane, and thus, the very C-terminal amino acids are not recognizable (and mostly not even synthesized), when the nascent chain is released into the membrane via the Sec-translocase. Therefore, this analysis implies a post-translational, post translocon mechanism for sorting transmembrane proteins between CM and TM.

Interactions with soluble factors and targeting proteins

Already early on at the ribosome, the fate of a nascent polypeptide chain is determined as the N-terminal part of the nascent chain may interact with the various cellular chaperones (Müller *et al.*, 2001). Specific interactions with chaperones determine whether the ribosome/nascent chain-complex is targeted to the Sec-translocase for co-translational protein translocation/insertion or whether protein synthesis will be completed in the cytoplasm. The two proteins SRP54 and *trigger factor* already interact with the nascent chain when the signal sequences leave the ribosomal exit tunnel, whereas other chaperones/targeting factors subsequently take over the nascent chain and prevent protein folding prior to translocation. SRP54 is involved in targeting pre-proteins to internal membranes (CM and/or TM), and only a single SRP54 as well as only a single SRP-receptor protein FtsY is encoded in cyanobacterial genomes. Thus, it appears unlikely that SRP and the SRP-receptor (alone) mediate specific targeting of proteins to defined membrane regions in cyanobacteria.

In *E. coli*, some soluble proteins, which are supposed to be post-translationally translocated across the CM, interact with the soluble SecB chaperone and are subsequently transferred to SecA and the Sec-translocon. However, while SecB is non essential, pre-proteins have to be kept in an unfolded conformation competent for membrane transport, and thus other chaperones should be involved in protein transport to the CM. Immediately after emerging from the ribosomal exit tunnel, the nascent proteins interact with the ribosome-associated chaperone *trigger factor*, and the Hsp70 chaperone DnaK might also be involved in preventing protein folding in the cytosol (Hoffmann *et al.*, 2010). This assumption is supported by the observation that deletion of the *trigger factor* gene (*tig*) or the *dnaK* gene alone is possible, whereas simultaneous deletion of the *dnaK* and *tig* genes in *E. coli* essentially causes synthetic lethality (Deuerling *et al.*, 1999; Teter *et al.*, 1999). *E. coli* can merely tolerate deletion of both chaperone genes at temperatures $< 30°C$ (Genevaux *et al.*, 2004).

In contrast to chloroplasts, in all cyanobacterial genomes completely sequenced so far, a gene coding for the ribosome-bound chaperone *trigger factor* has been identified (Nakao *et al.*, 2010). The physiological function of the gene product and a potential cooperativity with the DnaK-system, as demonstrated in *E. coli*, has, however, not been analysed in any cyanobacterium yet. DnaK is a member of the Hsp70 family of proteins that are major ATP-dependent molecular chaperones facilitating protein folding. These proteins participate in diverse cellular functions, such as DNA replication, protein degradation, protein translocation across membranes or signal transduction (Bukau *et al.*, 2006l Mayer and Bukau, 2005; Saibil, 2008). The co-chaperones DnaJ (a Hsp40) and GrpE assist DnaK to carry out these essential functions. While GrpE facilitates nucleotide exchange in DnaK, DnaJ-type co-chaperones stimulate ATP hydrolysis in DnaK and thereby facilitate the binding of polypeptide substrates in the substrate-binding pocket of DnaK (Genevaux *et al.*, 2007). Importantly, for some proteins translocated via the Tat-pathway, a direct interaction of DnaK with signal sequences has been shown, indicating an involvement of DnaK in membrane targeting (Graubner *et al.*, 2007; Oresnik *et al.*, 2001). In line with this, a DnaK protein has also been suggested to be critically involved in specific targeting of proteins to the cyanobacterial TM (Katano *et al.*, 2006; Nimura *et al.*, 1996). In all cyanobacteria, a family of multiple DnaK proteins is encoded, which has been studied to some extent in *Synechocystis* and *Synechococcus* sp. PCC 7942 (Katano *et al.*, 2006; Nimura *et al.*, 2001; Rupprecht *et al.*, 2007, 2010; Sato *et al.*, 2007a). Only the gene encoding the DnaK1 protein of *Synechocystis* has been deleted successfully, indicating a dispensable function (Rupprecht *et al.*, 2007). A dispensable function is supported by the observation that DnaK1 is not strictly conserved in cyanobacteria (Rupprecht *et al.*, 2007). However, the remaining two DnaK proteins, DnaK2 and DnaK3, appear to be essential in both *Synechococcus* and *Synechocystis.* Based on various analyses, DnaK2 has been classified as the major DnaK protein, behaving like a prototypical heat-shock protein (Rupprecht *et al.*, 2007, 2008, 2010). Noteworthy, the second essential DnaK protein, DnaK3, appears to be mainly attached to TMs in *Synechococcus* sp. PCC 7942 and *Synechocystis* (Nimura *et al.*, 1996; Rupprecht *et al.*, 2007). Based on this observation it has been proposed that this chaperone is involved in the specific targeting of proteins to the TM. The mechanism, by which DnaK3 might mediate such a specific transport is, however, unknown.

A conserved motif localized in the C-terminus of all DnaK3 proteins has been shown to be essential for the physiological function of DnaK3 and might be involved in specific protein–protein interactions or in protein–lipid interactions (Rupprecht *et al.*, 2010). Whether the various cyanobacterial DnaK proteins have an intrinsic propensity to interact with (a subset) of signal sequences, and are thus involved in targeting of some pre-proteins to translocases, as has been shown for the *E. coli* DnaK protein and some Tat-substrates, is yet unknown. Also, whether DnaK3 interacts predominantly with signal-sequences of pre-proteins that are selectively targeted to the TM has yet to be analysed. Defining the substrate specificity of the TM-bound DnaK3 protein could be crucial to dissect the proposed involvement of this specific chaperone in subcellular protein targeting in cyanobacteria. The assumption that DnaK3 is crucial for sorting proteins to thylakoids and thus for TM maintenance and/or biogenesis is supported not only by the observation that the *dnaK3* gene cannot be deleted in *Synechocystis* (Rupprecht *et al.*, 2007). While DnaK3 is highly conserved in all cyanobacteria, the cyanobacterium *Gloeobacter violaceaus*, which lacks internal TMs, is the only cyanobacterium that lacks DnaK3 as well as the DnaJ protein, which is typically encoded together with DnaK3 in a gene cluster (genes *sll1932* and *sll1933* in *Synechocystis*) (Nakamura *et al.*, 2003; Rippka *et al.*, 1974). It is possible e.g. that DnaK3 interacts with one (or more) specific DnaJ proteins that reside exclusively within the TM and that DnaK3 is thereby tethered to this membrane. Noteworthy, DnaJ proteins can also be involved in initial substrate binding and in delivering substrates to DnaK (Genevaux *et al.*, 2007; Kelley, 1998).

In all cyanobacteria sequenced so far, multiple genes encoding a family of DnaJ proteins have been identified (Dueppre *et al.*, 2011; Sato *et al.*, 2007b). In *Synechocystis*, one DnaJ protein is predicted to have membrane-spanning α-helices, and the homologue of the DnaJ protein encoded by *sll1933* has been shown to be tightly associated with TMs in *Synechococcus* sp. PCC 7942

(Dueppre *et al.*, 2011; Oguchi *et al.*, 1997). Therefore, it is possible for DnaK3 to act together with a TM-localized J-domain protein in a concerted mechanism and/or for DnaJ proteins to be involved in membrane-specific targeting. A deletion analysis has indicated that most *Synechocystis* DnaJ proteins are non-essential, indicating overlapping functions for most of them (Dueppre *et al.*, 2011). Nevertheless, as a simultaneous deletion of two *dnaJ* genes was impossible, functional overlap is limited. Noteworthy, not only the gene coding for DnaK3, but also the DnaJ protein encoded in a gene cluster together with DnaK3, is essential in *Synechocystis*. In summary, the 'classical' proteins involved in protein targeting in chloroplasts and bacteria, SecB and SRP54, may not be involved in targeting pre-proteins to defined subcellular membrane regions in cyanobacteria, and thus far only the DnaK3 protein has emerged as being one likely candidate for a targeted protein transport within these organisms.

Type I signal peptidases

During or directly after translocation, the N-terminal signal sequences of pre-proteins are cleaved off by a membrane-bound type I signal peptidase, which results in the release of a mature protein into the periplasm or the thylakoid lumen, respectively (Auclair *et al.*, 2012). Interestingly, in all cyanobacterial genomes completely sequenced so far, at least two genes encoding type I signal peptidases have been identified (Nakao *et al.*, 2010). Both proteins appear to have similar size and fold, and all amino acids crucial for signal peptide cleavage are conserved. However, regions with different charge distribution have been identified in the two proteins, and these regions include the cytoplasmic N-terminus and an extra-cytoplasmic extruding ribbon, as well as the slightly prolonged C-terminus of the *Synechocystis* peptidase Slr1377 (Rajalahti *et al.*, 2007). As the two encoded proteins (Sll0716 and Slr1377 in *Synechocystis*) have been identified to localize in different internal membranes (Huang *et al.*, 2002; Srivastava *et al.*, 2005), an involvement of these type I signal peptidases in protein sorting has been suggested (Pisareva *et al.*, 2011).

Multiple type I signal peptidases are present in chloroplasts and Gram-positive bacteria. The first *Arabidopsis* signal peptidase was discovered in 1998 (*At2g30440*) (Chaal *et al.*, 1998) and localization in the thylakoid lumen was confirmed by immunological analyses. However, a second type I signal peptidase has recently been identified in *Arabidopsis* (Plsp1) (Hsu *et al.*, 2011; Inoue *et al.*, 2005), and together with the observation of a second Sec-translocase in the chloroplast inner envelope membrane, this observation might suggest that two distinct Sec-complexes (including specific signal peptidases) exist in *Arabidopsis*. In contrast to Gram-negative bacteria, occurrence of multiple type I signal peptidases is frequently observed in Gram-positive bacteria (van Roosmalen *et al.*, 2004). In *Bacillus subtilis*, which contains only the CM, a family of seven type I signal peptidases is present and individual representatives have partly overlapping and partly exclusive substrate specificities (van Roosmalen *et al.*, 2004). Nevertheless, when multiple type I signal peptidases are present in Gram-positive bacteria, none of them appear to be essential, and thus the respective enzymes are able to complement (at least in part) each other (Tjalsma *et al.*, 1997, 1998). Based on their catalytic activity, the various *Bacillus* peptidases have been classified into groups of *major* or *minor* signal peptidases (van Roosmalen *et al.*, 2001, 2004). Differences in the catalytic activity are, however, mostly not reflected in the sequences of the respective enzymes, as the catalytic domains are highly similar (van Roosmalen *et al.*, 2004). Nevertheless, while some experimental results have indicated that the transmembrane helices of type I signal peptidases are not important for the catalytic activity (Carlos *et al.*, 2000), computational analyses have identified defined sequence features in the N-terminal domain of the proteins, which contains the membrane spanning regions (van Roosmalen *et al.*, 2001).

Thus, the observed differences in the cyanobacterial type I signal peptidases could be crucial for defined interactions of the protein with translocases. Alternatively, those differences could be important for targeting the respective enzyme to a distinct cyanobacterial membrane system. However, it appears unlikely that the two type I signal peptidases determine subcellular localization of distinct proteins in cyanobacteria, as cleavage of the signal peptide occurs after translocation of preproteins through the Sec-translocon has started. Thus, proteins have to be targeted already to a Sec-translocase before a signal peptidase activity is operative. Furthermore, as the subunits of the Sec-translocase have been identified only in the TM but one of the two signal peptidases has been identified in the CM, the question of the function of the signal peptidase in the CM arises immediately. Based on the currently available information on the subcellular distribution of translocase subunits, the only function left for the CM-localized signal peptidase would be the processing of the Tat-substrates. Consequently, the signal peptidase localized in the TM would process both, Sec and Tat-substrates, whereas the CM-localized protein would be specific for Tat-substrates. Importantly, in several cyanobacteria more than two type I signal peptidases are encoded. Even in the unique species *Gloeobacter violaceaus*, the only cyanobacterium identified to date that lacks thylakoids, two type I signal peptidases (encoded by the genes *glr2023* and *glr2481*) and one protein having high similarity to type I signal peptidases (*glr3215*) are encoded (Nakamura *et al.*, 2003). This observation suggests that the cyanobacterial type I signal peptidases might be classified as *major* or *minor* signal peptidases, respectively, similar to the situation in many Gram-positive bacteria. However, while the various paralogues in Gram-negative bacteria can substitute each other, inactivation of the *Synechocystis* LepB protein Sll0716 was possible but inactivation of the LepB-coding gene *slr1377* was not. This at least indicates that Slr1377 has a (partly) overlapping substrate specificity with Sll0716 (Zhbanko *et al.*, 2005). In the strain carrying an inactivated *sll0716* gene, the photosynthetic activity was affected and the strain became light-sensitive, suggesting that the type I signal peptide encoded by *slr0716* is involved in processing of photosynthetic preproteins. In line with this observation, processing of the PsbO precursor protein, a subunit of PS II, was affected, though not completely abolished, in the deletion strain. Based on the available data, it is likely that the two identified cyanobacterial signal peptidases process different substrates. Differential expression of the respective genes under defined growth conditions might allow the cells to posttranslationally control expression, maturation and activity of certain proteins. Detailed analyses of the substrate specificities are crucial to finally answer the question of the function of two LepB signal peptidases in cyanobacteria in the future. However, based on the presently available information, the type I signal peptidase encoded by *sll1377* might be denoted as the *major* signal peptidase in *Synechocystis*.

Proteins involved in membrane formation

A model, which has been discussed for approximately 10 years, is that proteins are sorted after translocation and/or membrane integration by an intracellular vesicle transfer system. However, thus far transport vesicles have never been unambiguously identified in cyanobacteria. In 2007 the internal organization of the cyanobacterium *Microcoleus* sp. was analysed by electron microscope tomography, and in this organism large structures resembling vesicles were described (Nevo *et al.*, 2007). These structures had diameters of 150–300 nm, and single cells contained only very few of such structures. Their rather limited number appears to be problematic for a vesicular transfer system being involved in a continuous sorting of proteins and lipids in cyanobacteria. Whereas the exact nature of the identified structures is currently enigmatic, similar structures have never been reported for any other cyanobacterium yet and they might have a different origin. As e.g. viruses infecting different cyanobacterial species have been identified in recent years, it might be possible that the structures represent viral particles within *Microcoleus* cells.

A second observation of vesicular structure has been reported (Schneider *et al.*, 2007). When a specific fluorescent dye was added to living cyanobacterial cells, the dye integrated and merely stained the cyanobacterial CM. This observation has supported the assumption that the internal membrane systems are separated or that molecular transport via a potential connection is highly controlled. About 30 min after the dye had been added, small vesicular structures became visible in the cyanobacterial cells. While this could indicate that indeed vesicles have pitched off from the CM, the exact nature of the structures has not been identified yet.

In chloroplasts, formation of vesicular structures at the inner envelope has been observed under certain experimental conditions (Vothknecht and Soll, 2005). Deletion of the *AAC27135* gene in *Arabidopsis* has resulted in a phenotype where TM biogenesis was disturbed and such vesicular structures were not observed any more (Kroll *et al.*, 2001). Based on this observation, the encoded protein was implicated in vesicle formation and in a vesicular transfer system and has been named 'vesicle inducing protein in plasmids 1' (or Vipp1). The *vipp1* gene is also conserved in cyanobacteria, and whereas deletion of *vipp1* was not possible, depletion of the gene product resulted in a severe reduction of internal TMs together with an impairment of the PS biogenesis (Fuhrmann *et al.*, 2009b; Gao and Xu, 2009; Westphal *et al.*, 2001). These observations clearly indicate that Vipp1 is essential and crucially involved in TM biogenesis and/or organization (Vothknecht *et al.*, 2012). However, an involvement in vesicle formation or a vesicular transfer system has never been shown yet. As the structure of Vipp1 is highly similar to the bacterial PspA protein (Bultema *et al.*, 2010; Fuhrmann *et al.*, 2009a), it can be suggested that Vipp1 interacts with membranes and might assist in stabilizing bilayer structures (Vothknecht *et al.*, 2012). While the observation that a Vipp1 homologue is not encoded in the thylakoid-free cyanobacterium *Gloeobacter violaceaus* could further link Vipp1 to TM biogenesis, it has to be noted that also other cyanobacteria, including several *Prochlorococcus* species, do not contain *vipp1* genes (Nakao *et al.*, 2010). Whether in these cyanobacteria the (otherwise essential) function of Vipp1 is for an unknown reason dispensable or taken over by other proteins, such as PspA, remains to be shown.

Notably, homologues of proteins involved in vesicle formation in eukaryotes have been identified in cyanobacterial genomes as well (Andersson and Sandelius, 2004). The genome of *Synechocystis* encodes homologues of eukaryotic SNARE proteins, which are involved in vesicle fusion (Kaneko *et al.*, 1996). Additionally, small GTPases with homology to eukaryotic small G-proteins, which are crucial for vesicle formation and targeting, are encoded in cyanobacteria. Unfortunately, none of these proteins have an annotated function yet. In summary, the information on potential transport vesicles in cyanobacteria is scarce and a vesicular transfer system has not yet been unequivocally shown in cyanobacteria.

Transient and/or permanent membrane connections: thylakoid centre and PratA-defined membranes

De novo assembly of PS complexes has been analysed in some detail in *Synechocystis*. In cellular membrane subfractionation experiments, membrane subfractions denoted as CM have been identified to contain several PS subunits (Smith and Howe, 1993; Zak *et al.*, 2001). Other membrane fractionation experiments have identified membrane subfractions possessing features more characteristic of TMs that have been described to represent biogenesis centres (Rengstl *et al.*, 2011). In several recent models, regions at which the TM and CM are interconnected have been proposed, and some of the separated membrane subfractions have been suggested to represent such connection zones (Nickelsen *et al.*, 2010). In some cyanobacterial species, multiple thylakoids converge at a peripheral location, and a prototypical *Synechocystis* cell contains about three to ten of such convergence sites, with an average of five (van de Meene *et al.*, 2006). A physical connection between the CM and the TM at these sites has, however, never been shown unambiguously (Liberton *et al.*, 2006, 2012b; Nevo *et al.*, 2007; van de Meene *et al.*, 2006). Frequently, the thylakoids are connected at the convergence sites by so-called 'thylakoid centres' (Kunkel, 1982), a cylindrical structure with a diameter of 40–50 nm and a length of 50–1000 nm (van de Meene *et al.*, 2006). Thylakoid centres are typically located close to the CM, and in some EM micrographs the thylakoid centres seem to be continuous with the periplasmic space (Fig. 5.4). Thylakoid centres are probably involved in establishing and/or maintaining the integrity and structural organization of the TM network. An involvement in thylakoid biogenesis as well as a crucial function in the

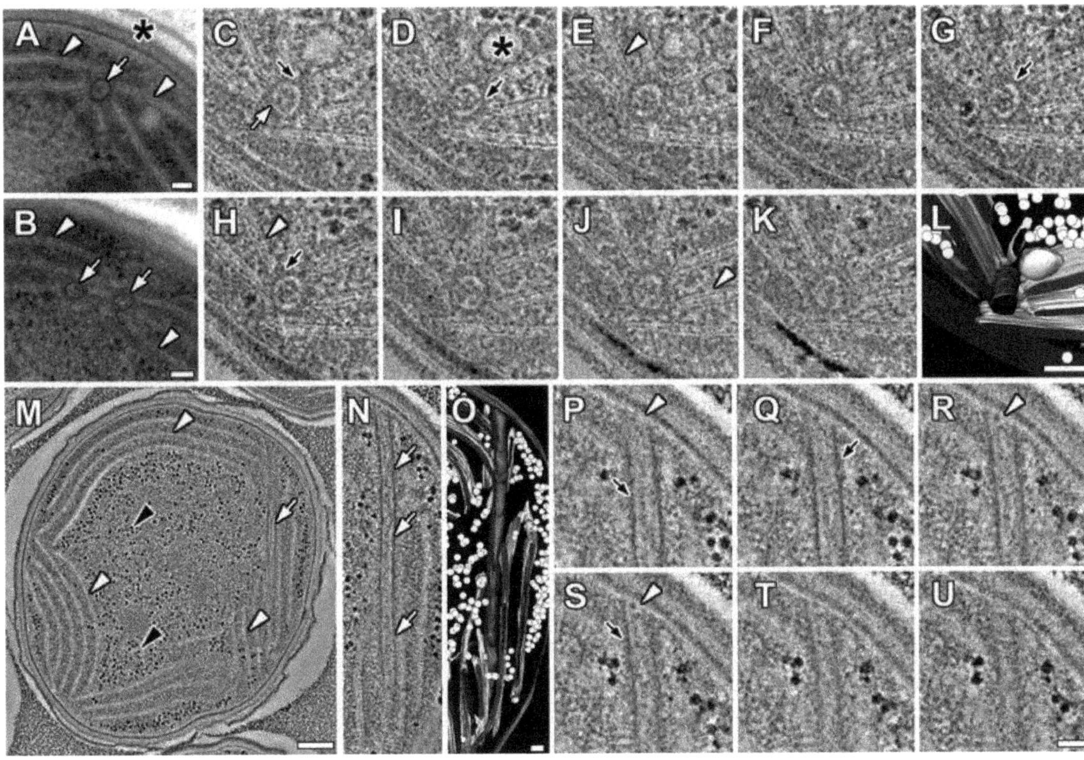

Figure 5.4 The thylakoid centre of *Synechocystis*. (A, B) Cross-sections of thylakoid membranes (white arrowheads) and thylakoid centres (white arrows). Asterisk: cell wall. Bars = 50 nm. (C–K) 1.4 nm slices taken through a selected cross-section of a thylakoid centre (white arrow), surrounding thylakoid membranes (white arrowhead) and the cytoplasm. Dense material (black arrows) coats the thylakoid centre, which appears to be associated with thylakoid membranes. Black asterisk: lipid body. Bar = 50 nm. (L) Model of a thylakoid centre reconstructed from the images shown in (C-K). Thylakoid membranes (green) surround the thylakoid centre (blue). Pink: lipid body; white: ribosomes; brown: cytoplasmic membrane. Bar = 50 nm. (M) Electron micrograph of a *Synechocystis* cell with the thylakoid centre (white arrow) extending from one side to the other of the cell. White arrowheads: thylakoid membranes; black arrowheads: ribosomes. Bar = 200 nm. (N) Longitudinal view of a thylakoid centre (arrows) and a derived model (O) with associated thylakoid membrane pairs (green), ribosomes (white) and the cytoplasmic membrane (brown). Bar = 50 nm. (P-U) 1 nm sections of the thylakoid centre (black arrow) shown in (M) and (N) and its association (white arrowhead) with the cytoplasmic membrane and periplasmic space. Bar = 50 nm. From (van de Meene *et al.*, 2006), with kind permission of Springer Science and Business Media. A colour version of this figure is available in the plate section at the back of the book.

transport of PS I and PS II subunits from the CM to the TM has been proposed (Hinterstoisser *et al.*, 1993). Such assumptions have recently been supported by biochemical analyses of PS II biogenesis in *Synechocystis*. Separation of TM membranes in a sucrose density gradient resulted in two TM subfractions, one of which has been described as 'PratA-defined membranes' (PDMs) (Rengstl *et al.*, 2011; Stengel *et al.*, 2012). Several assembly factors are enriched in these membranes, one of which is the tetratricopeptide repeat protein PratA. In a mutational study, PratA has been shown to be crucial for maturation of PS II, as inactivation of the *pratA* gene affects processing of the PsbA protein precursor, which is an early step in PS II maturation (Klinkert *et al.*, 2004).

Defined membrane fractions, having a special role in biogenesis, have also been identified in the chloroplast of the green algae *Chlamydomonas reinhardii* and have been annotated as translation (T-) zones (Uniacke and Zerges, 2007, 2009). These T-zones appear to have prominent functions in PS II assembly and most likely they are identical to the low-density membranes identified in *Chlamydomonas* before, where biogenesis of TM complexes likely occurs (Zerges and Rochaix, 1998). These low-density membranes are similar in lipid composition and their respective floating density to chloroplast inner envelope membranes, but they appear to be physically associated with TMs (Zerges and Rochaix, 1998). Furthermore, several RNA-binding proteins appear to be associated with this membrane subfraction, indicating a localized translation of defined mRNAs (Uniacke and Zerges, 2009). The mechanism for protein targeting to such membrane regions involved in biogenesis of defined TM proteins has yet to be established, however current data imply that, in addition to the chloroplast signal recognition particle cpSRP54, membrane-associated RNA-binding proteins are involved (Weis *et al.*, 2012; Zerges and Rochaix, 1998).

The PDMs identified in cyanobacteria might represent the cyanobacterial counterpart of the low-density, T-zone membranes identified in green algae, as several characteristics of the described T-zones resemble features of PDMs. Noteworthy, the T-zones have been described as a defined membrane system localized around the *Chlamydomonas* pyrenoid (Uniacke and Zerges, 2007). Cellular fractionation experiments have indicated that in cyanobacteria PratA, as well as other biogenesis factors, are associated with a membrane fraction that has a density different from that of the fractions denoted as CM or TM (Rengstl *et al.*, 2011). These PDMs have been postulated to represent the structurally defined thylakoid centres, and thus to be crucial for biogenesis of at least PS II, similar to the function annotated for T-zones (Nickelsen *et al.*, 2010). However, while the T-zones appear to be localized in the chloroplast stroma, PratA has been described to have a periplasmic localization (Klinkert *et al.*, 2004). The observation that PS II biogenesis factors are localized in the periplasm supported earlier observations that some PS I and PS II proteins might accumulate in the CM, leading to the suggestion that early steps in PS II biogenesis take place in the CM (Zak *et al.*, 2001).

Based on a recently proposed model, biogenesis centres are located at sites where CM and TM converge at the thylakoid centres (Nickelsen *et al.*, 2010). While such a model would explain the observation that PS II assembly factors have a periplasmic localization, no physical connection of TMs with CMs has ever been observed at thylakoid centres in ultra structural analyses, and such a proposed connection would have to be tightly controlled to eliminate an uncontrolled flux of protons and proteins. Furthermore, while integration of certain pre-proteins into the CM or translocation into the periplasmic space and subsequent transfer to thylakoids has been proposed previously, the subunits of the Sec-translocase have, as discussed above, been exclusively identified in TM fractions (Pisareva *et al.*, 2011). Thus, the question of how any TM (pre-)protein could ever first reach the periplasmic space or CM remains unanswered. One possible explanation is that such proteins in fact never reach the periplasm but only regions having some type of connection to the periplasm and thylakoid lumen, i.e. biogenesis centres, thylakoid centres or PDMs. But why should proteins localize to the CM or periplasmic space after insertion at biogenesis centres when they actually function in the TM? Furthermore, the two type I signal peptidases have been identified in different membrane systems in the same proteomic analysis, in which the Sec-subunits were identified only in the TM, and the analysis of Tat-subunit localization has strongly indicated a dual localization of this translocase within the CM as well as in the TM. A concentration of the Tat-subunits in localized regions, such as thylakoid centres, has not been observed.

Models of protein targeting and translocation in cyanobacteria

Post-translocon targeting

The observation that individual translocases might be localized in only one internal membrane system implies that soluble as well as membrane integral proteins are sorted after translocation. As the Sec-subunits have solely been identified in an isolated membrane fraction annotated as TM (Pisareva *et al.*, 2011), proteins are then translocated into a single membrane and sorted thereafter, if necessary. Sorting is based on internal sorting signals, as e.g. identified in membrane integral proteins. Sorting could either involve permanent or transient (spatially and temporally restricted) connections of the internal membranes, potentially including thylakoid centres and/or PDMs (Fig. 5.5Ia). Alternatively, after synthesis proteins are packed into transport vesicles, which mediate transport to the respective membrane system (Fig. 5.5Ib). Sequence determinants, which would allow discrimination of thylakoid lumen and/or periplasmic proteins as well as TM and/or CM proteins, have been identified in the sequences of mature cyanobacterial proteins having a defined subcellular localization. A final possibility is that all soluble and membrane integral proteins are randomly translocated into the periplasm and thylakoid lumen or TM and CM, respectively, and thereafter mis-targeted proteins either simply reside in the wrong compartment (without any function) or are rapidly degraded by the cellular protein degradation system (Fig. 5.5Ic). While this would explain the observed dual localization of some proteins, several, even highly abundant proteins, have not been detected in both compartments rendering such a highly uneconomically process of random translocation rather unlikely.

Pre-translocon, post-translational targeting

As some observations have indicated that the translocases are located in both membranes, protein sorting of soluble proteins might be post-translational but prior to translocation. In this case, a targeting sequence is recognized by a targeting factor that directs the pre-protein to a specific membrane for translocation (Fig. 5.5II). Differences in the pre-protein sequences, which have already been identified, could be recognized e.g. by the Hsp70-type chaperone DnaK3, which subsequently targets pre-proteins to the TM. Dual localization of some proteins could be caused by cryptic or missing targeting information in the corresponding pre-proteins resulting in targeting to both, TM and CM.

Co-translational targeting

In the case of membrane proteins a co-translational targeting mechanism is likely (Fig. 5.5III). Nevertheless, while in a statistical analysis signature sequences have been identified in mature membrane proteins, such signature sequences are located in the C-terminal part of the proteins. Thus, these sequences could be identified by targeting factors only after complete synthesis of the corresponding protein, suggesting that targeting

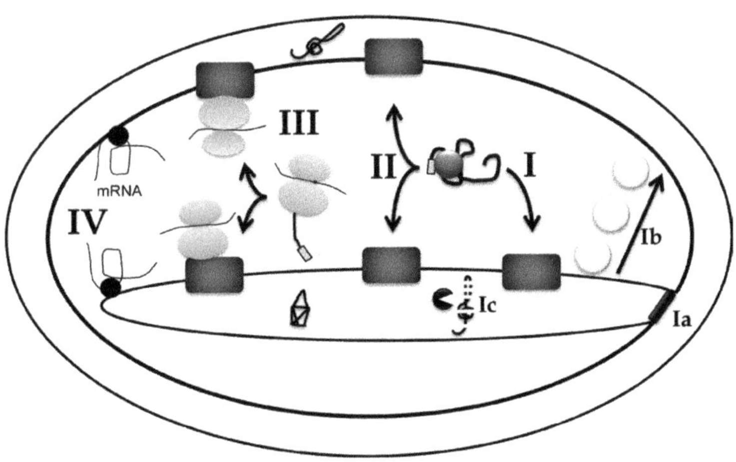

Figure 5.5 Protein targeting pathways in cyanobacteria. Proteins can be targeted after translocation and/or membrane insertion. I, Post-translocon targeting, after synthesis but prior to translocation; II, pre-translocon, post-translational targeting, directly after synthesis has started; III, co-translational targeting or the mRNA is targeted and proteins are synthesized at defined subcellular regions; IV, pre-translational targeting. For details, see text.

takes place after translation and membrane integration. As membrane integral proteins are co-translationally integrated into the membrane, other targeting mechanisms must apply, if the proteins do not follow one of the routes mentioned above. While only a single SRP54 protein and only a single SRP-receptor protein FtsY are present in cyanobacteria, it still is possible that other proteins having distinct localization interact with nascent chains and are involved in membrane-specific targeting. Furthermore, while the lipid composition of the TM and CM are highly similar, minor differences have been observed. Thus, it is possible that, for example, the signal sequences of pre-proteins interact differently with TM or CM, and that such interactions determine directionality.

Pre-translational targeting

Several RNAs have been identified in recent years that appear to contain signals for mRNA targeting to specific organelles in eukaryotes (Weis *et al.*, 2012). Also within the chloroplast, membranous subcompartments, such as the so-called 'low density membranes' or 'T-zones' have been identified at locations where mRNA appears to be concentrated. Hence, these subcompartments are supposed to be sites of translation, protein translocation or membrane integration. Even though not analysed in as much detail as in chloroplasts, the identified PratA-defined membranes could have a similar function in cyanobacteria. If assembly of the 70S ribosome takes place directly at the CM or TM, and thus synthesis of pre-proteins starts already at the correct membrane in cyanobacteria, it is not the pre-protein but the mRNA which has to be targeted (Fig. 5.5IV). In chloroplasts, it has already been observed that the RNA is connected to defined membrane regions and RNA-binding proteins have been identified. Thus, it is possible that protein sorting within cyanobacteria already starts prior to the onset of translation, since the protein-coding mRNA could be sorted to a specific membrane prior to translation. Translation initiation of cyanobacterial mRNAs should not be discussed without considering the stability of the RNAs. Recently, the average half-life of mRNAs in a cyanobacterium has been determined to be below three minutes (Steglich *et al.*, 2010). Therefore, mRNAs might be degraded rapidly in the cyanobacterial cytoplasm, even prior to translation, and only in cases where mRNAs are protected by RNA-binding proteins, degradation is slowed down. Thus, it is possible that membrane-specific RNA-binding proteins protect mRNAs from degradation and additionally localize the translation to specific subcellular compartments. If, however, synthesis of encoded pre-proteins starts in the cytoplasm prior to RNA-binding at membrane-specific RNA-binding proteins, proteins might be co-translationally targeted to internal membranes without preference for CM or TM. Thus, if the targeting information in the mRNAs is not decrypted before protein synthesis starts, proteins might be mistargeted to the wrong internal membrane system. Thus, the observation that some proteins have been identified in the 'wrong' compartments might simply reflect the limited pool of mRNAs translated prior to degradation or prior to binding via a membrane-specific RNA-binding protein. The assumption that mRNA contains targeting information would explain the observation that – at least in chloroplasts – the cpSRP54 protein is dispensable, and thus, a translation-independent mechanism of protein targeting would exist.

Epilogue: a heterogenic protein distribution in cyanobacterial subcompartments?

It can be stated that the various subcellular cyanobacterial compartments have different physiological functions and harbour defined subsets of proteins. Distribution of defined protein complexes in internal cyanobacterial membrane systems has been studied in some detail for decades. Initial studies have focused on the distribution of proteins involved in respiratory and/or photosynthetic electron transfer reactions. Essentially, all these studies relied on separation of individual membranes after cell breakage, performed either by sucrose density gradient centrifugation or by two-phase partitioning. Very rarely, such results were confirmed by immunogold analyses or by analyses of proteins fused to fluorescent proteins. In several detailed analyses of *Synechocystis* membranes separated by two-phase partitioning, Norling and co-workers have characterized the protein composition of isolated CM and TM in detail (Huang *et al.*, 2002, 2004, 2006; Pisareva *et al.*, 2007, 2011; Srivastava *et al.*, 2005, 2006; Zak *et al.*, 2001). In these studies several proteins were identified in both, the CM as well as the TM. Notably, some of the results obtained after separation of internal cyanobacterial membranes were contradictory, depending on the research group, the technique used for membrane separation and/or the analysed cyanobacterium. Although desirable, a clear picture of the subcellular localization of individual proteins in cyanobacteria has not emerged yet.

It was observed early that some subunits of PS I and PS II have defined localizations in *Anacystis nidulans.* Although PS II is only active in TMs, some of its subunits are present in both CM and TM (such as the PsbA protein, PsbO, PsaC, PsaD and PsaE), whereas others appear to be strictly confined to TMs (CP43 and CP47) (Smith *et al.*, 1992; Smith and Howe, 1993). These initial observations were essentially confirmed later in studies aiming at analysing the distribution of polypeptides in *Synechocystis*, and based on these observations a model was proposed in which the PSs are pre-assembled in the cyanobacterial CM and subsequently transferred to the TM, however by an unknown mechanism (Zak *et al.*, 2001). The observation that some factors, which appear to be involved in biogenesis of at least PS II, are localized in the cyanobacterial periplasm, further supported the idea of a PS pre-assembly in the CM and subsequent relocation to thylakoids (Nickelsen *et al.*, 2010). Nevertheless, this conclusion leaves open the question of how the assembly intermediates are exactly transported from the CM to the TM. A solution to the differently observed localization of PS subunits and assembly factors was the proposal of defined biogenesis centres in between TM and CM. However, there is no strong evidence for such connections based on structural analyses carried out to date, and finally all models do rely on membrane separation techniques.

As photosynthetic complexes are easy to detect (due to their high pigment content), a significant amount of research has focused on dissecting possible PS assembly pathways in cyanobacteria. Whereas several components of complexes of photosynthetic and respiratory electron transfer chains have solely been identified in one membrane, some subunits of the same complexes have been identified in both membranes (Pisareva *et al.*, 2011). For instance, while both of the cytochrome b_6f complex subunits PetA and PetC have been localized in both membranes, the other core subunits PetB and PetD appear to be only present in TMs. Several subunits of the NADH dehydrogenase and of the ATP synthase have been detected in both the CM and TM, whereas some subunits appear to be solely present in the TM. As these observations might imply pre-assembly of most (if not all) membrane protein complexes in the CM and subsequent complete assembly in the TM, the CM could be defined as the 'pre-assembly platform', and the main function of this membrane would be formation of such pre-assemblies. The necessary regular transfer of pre-assemblies from CM to TM has not been described yet, and it should be noted that the presence of too many pre-assemblies would probably result in a CM completely crowded with pre–assembled complexes. The available results possibly indicate mis-targeting of several integral membrane proteins, with subsequent re-targeting or degradation, but the obtained localization studies have to be analysed and discussed with great caution. As several published observations are contradictory, the exact experimental conditions used for separation of membranes and the cyanobacterial species studied appear to significantly influence the results of separation techniques. Furthermore, recent observations indicate that even the cyanobacterial TM is heterogeneous and regions with defined protein compositions appear to be present (Agarwal *et al.*, 2010, 2012; Sherman *et al.*, 1994; Vermaas *et al.*, 2008). This raises the question of which membranes or membrane patches are finally separated by any of the currently applied techniques.

The unusual cyanobacterium *Gloeobacter violaceus* contains no TMs and all photosynthetic complexes are localized in the CM. Nevertheless, after cell breakage two membranes with defined protein compositions can be separated by sucrose density centrifugation (Rexroth *et al.*, 2011). This implies that membrane separation from other cyanobacteria has not resulted in separation of CM and TM but in separation of membrane fractions, having different protein (and maybe even lipid) compositions, which would imply differences in the physico-chemical properties that are used for separation by density or surface charge. Separation of membrane systems in cyanobacteria can indeed result in isolation of more than two membrane fractions (Agarwal *et al.*, 2010; Hinterstoisser *et al.*, 1993). In chloroplasts, the specific membrane fractions termed 'low density membranes' or T-zones' (for details see above) have the same density in sucrose gradients as the inner envelope membrane (Uniacke and Zerges, 2007, 2009), although they do not appear to be connected in any way to the inner envelope membrane but either represent a separated membrane system or are connected to the TM. Furthermore, in early studies it was already shown that also in non-photosynthetic bacteria, such as *E. coli*, different membrane fractions can be isolated by sucrose density centrifugation, and analyses of EM micrographs indicated that regions with defined protein compositions might form during membrane preparation and separation (Kleemann and McConnell, 1974). Thus, in addition to a pre-existing membrane heterogeneity, which might translate in the isolation of different membrane subfractions, defined domains concentrating certain proteins are possibly present, either permanently or transiently, within the TM. Alternatively, they may be formed artificially during the process of cell harvesting, breakage and membrane preparation.

Success of membrane separation is typically judged based on two observations: firstly, it is generally assumed that the CM contains less chlorophyll than the TM; and secondly, defined marker proteins are used to ensure proper membrane separation. The chlorophyll binding proteins CP43 and CP47 are typically used as marker proteins for TMs, and thus to confirm a successful membrane separation. Noteworthy, while the reference by Bassi *et al.* (1987) is often used to indicate an exclusive TM localization of CP43 and/or CP47, subcellular localization of CP43 and CP47 has not been investigated in this study, which addressed characteristics of the PS II antenna system using purified thylakoid membranes from maize. The localization of the CP43 and CP47 markers exclusively in cyanobacterial TMs has, however, been concluded based on results obtained after (putative) separation of TM and CM (Smith and Howe, 1993). On the other hand, in a recent report the CP43 and CP47 proteins have also been identified in cyanobacterial CM fractions (Pisareva *et al.*, 2011). Thus, while it is likely that the cyanobacterial CM and TM, as well as the thylakoid lumen and periplasm, harbour different subsets of proteins, it is uncertain what is actually being separated by the currently applied methods, and the use of marker proteins is not straightforward. Nevertheless, a great deal of the currently discussed models of protein sorting depends on the results obtained by membrane separation techniques.

Acknowledgements

Work in the author's laboratory is supported by the Deutsche Forschungsgemeinschaft. I thank all my co-workers and collaborators for years of fruitful work and discussion.

References

Agarwal, R., Matros, A., Melzer, M., Mock, H.P., and Sainis, J.K. (2010). Heterogeneity in thylakoid membrane proteome of *Synechocystis* 6803. J. Proteom. *73*, 976–991.

Agarwal, R., Maralihalli, G., Sudarsan, V., Choudhury, S.D., Vatsa, R.K., Pal, H., Melzer, M., and Sainis, J.K. (2012). Differential distribution of pigment–protein complexes in the Thylakoid membranes of *Synechocystis* 6803. J. Bioenerg. Biomembr. *44*, 399–409.

Albiniak, A.M., Baglieri, J., and Robinson, C. (2012). Targeting of lumenal proteins across the thylakoid membrane. J. Exp. Bot. *63*, 1689–1698.

Aldridge, C., Spence, E., Kirkilionis, M.A., Frigerio, L., and Robinson, C. (2008). Tat-dependent targeting of Rieske iron-sulphur proteins to both the plasma and thylakoid membranes in the cyanobacterium *Synechocystis* PCC6803. Mol. Microbiol. *70*, 140–150.

Altman, E., Kumamoto, C.A., and Emr, S.D. (1991). Heat-shock proteins can substitute for SecB function during protein export in *Escherichia coli*. EMBO J. *10*, 239–245.

Anderson, S.L., and McIntosh, L. (1991). Light-activated heterotrophic growth of the cyanobacterium *Synechocystis* sp. strain PCC 6803: a blue-light-requiring process. J. Bacteriol. *173*, 2761–2767.

Andersson, M., and Sandelius, A. (2004). A chloroplast-localized vesicular transport system: a bio-informatics approach. BMC Genom. *5*, 40.

Andrade, M.A., O'Donoghue, S., and Rost, B. (1998). Adaptation of protein surfaces to subcellular location. J. Mol. Biol. *276*, 517–525.

Auclair, S.M., Bhanu, M.K., and Kendall, D.A. (2012). Signal peptidase I: Cleaving the way to mature proteins. Protein Sci. *21*, 13–25.

Barbrook, A.C., Packer, J.C., and Howe, C.J. (1993). Components of the protein translocation machinery in the thermophilic cyanobacterium *Phormidium laminosum*. Biochem. Biophys. Res. Commun. *197*, 874–877.

Barnett, J.P., Eijlander, R.T., Kuipers, O.P., and Robinson, C. (2008). A minimal Tat system from a Gram-positive organism. J. Biol. Chem. *283*, 2534–2542.

Bassi, R., Hoyer-Hansen, G., Barbato, R., Giacometti, G.M., and Simpson, D.J. (1987). Chlorophyll-proteins of the photosystem II antenna system. J. Biol. Chem. *262*, 13333–13341.

Blaudeck, N., Kreutzenbeck, P., Müller, M., Sprenger, G.A., and Freudl, R. (2005). Isolation and characterization of bifunctional *Escherichia coli* TatA mutant proteins that allow efficient Tat-dependent protein translocation in the absence of TatB. J. Biol. Chem. *280*, 3426–3432.

Bogsch, E.G., Sargent, F., Stanley, N.R., Berks, B.C., Robinson, C., and Palmer, T. (1998). An essential component of a novel bacterial protein export system with homologues in plastids and mitochondria. J. Biol. Chem. *273*, 18003–18006.

Bölter, B., and Soll, J. (2011). Protein import into chloroplasts: dealing with the (membrane) integration problem. Chem. Biochem. *12*, 1655–1661.

Brundage, L., Hendrick, J.P., Schiebel, E., Driessen, A.J.M., and Wickner, W. (1990). The purified *E. coli* integral membrane protein SecYE is sufficient for reconstitution of SecA-dependent precursor protein translocation. Cell *62*, 649–657.

Buchanan, G., Leeuw, E.d., Stanley, N.R., Wexler, M., Berks, B.C., Sargent, F., and Palmer, T. (2002). Functional complexity of the twin-arginine translocase TatC component revealed by site-directed mutagenesis. Mol. Microbiol. *43*, 1457–1470.

Bukau, B., Weissman, J., and Horwich, A. (2006). Molecular chaperones and protein quality control. Cell *125*, 443–451.

Bultema, J.B., Fuhrmann, E., Boekema, E.J., and Schneider, D. (2010). Vipp1 and PspA: related but not twins. Commun. Integr. Biol. *3*, 162–165.

Carlos, J.L., Paetzel, M., Brubaker, G., Karla, A., Ashwell, C.M., Lively, M.O., Cao, G., Bullinger, P., and Dalbey, R.E. (2000). The role of the membrane-spanning domain of Type I signal peptidases in substrate cleavage site selection. J. Biol. Chem. *275*, 38813–38822.

Carlsson, F., Stalhammar-Carlemalm, M., Flardh, K., Sandin, C., Carlemalm, E., and Lindahl, G. (2006). Signal sequence directs localized secretion of bacterial surface proteins. Nature *442*, 943–946.

Cedano, J., Aloy, P., Perez-Pons, J.A., and Querol, E. (1997). Relation between amino acid composition and cellular location of proteins. J. Mol. Biol. *266*, 594–600.

Celedon, J.M., and Cline, K. (2012). Intra-plastid protein trafficking: how plant cells adapted prokaryotic mechanisms to the eukaryotic condition. Biochim. Biophys. Acta *1833*, 341–351.

Chaal, B.K., Mould, R.M., Barbrook, A.C., Gray, J.C., and Howe, C.J. (1998). Characterization of a cDNA encoding the thylakoidal processing peptidase from *Arabidopsis thaliana*. J. Biol. Chem. *273*, 689–692.

Cline, K., and Dabney-Smith, C. (2008). Plastid protein import and sorting: different paths to the same compartments. Curr. Opin. Plant Biol. *11*, 585–592.

Danese, P.N., and Silhavy, T.J. (1998). Targeting and assembly of periplasmic and outer-membrane proteins in *Escherichia coli*. Annu. Rev. Genet. *32*, 59–94.

Deuerling, E., Schulze-Specking, A., Tomoyasu, T., Mogk, A., and Bukau, B. (1999). Trigger factor and DnaK cooperate in folding of newly synthesized proteins. Nature *400*, 693–696.

Dueppre, E., Rupprecht, E., and Schneider, D. (2011). Specific and promiscuous functions of multiple DnaJ proteins in *Synechocystis* sp. PCC 6803. Microbiology *157*, 1269–1278.

Falk, S., and Sinning, I. (2010). cpSRP43 is a novel chaperone specific for light-harvesting chlorophyll *a,b*-binding proteins. J. Biol. Chem. *285*, 21655–21661.

Fröbel, J., Rose, P., and Müller, M. (2012). Twin-arginine-dependent translocation of folded proteins. Phil. Trans. R. Soc. B Biol. Sci. *367*, 1029–1046.

Froehlich, J.E., and Keegstra, K. (2011). The role of the transmembrane domain in determining the targeting of membrane proteins to either the inner envelope or thylakoid membrane. Plant J. *68*, 844–856.

Fuhrmann, E., Bultema, J.B., Kahmann, U., Rupprecht, E., Boekema, E.J., and Schneider, D. (2009a). The vesicle-inducing protein 1 from *Synechocystis* sp. PCC 6803 organizes into diverse higher-ordered ring structures. Mol. Biol. Cell *20*, 4620–4628.

Fuhrmann, E., Gathmann, S., Rupprecht, E., Golecki, J., and Schneider, D. (2009b). Thylakoid membrane reduction affects the photosystem stoichiometry in the cyanobacterium *Synechocystis* sp. PCC 6803. Plant Physiol. *149*, 735–744.

Fulgosi, H., Gerdes, L., Westphal, S., Glockmann, C., and Soll, J. (2002). Cell and chloroplast division requires ARTEMIS. Proc. Natl. Acad. Sci. U.S.A. *99*, 11501–11506.

Gantt, E. (1994). Supramolecular membrane organization. In The Molecular Biology of Cyanobacteria, Bryant, D.A., ed. (Dordrecht, Netherlands, Kluwer Academic Publishers), pp. 119–138.

Gao, H., and Xu, X. (2009). Depletion of Vipp1 in *Synechocystis* sp. PCC 6803 affects photosynthetic activity before the loss of thylakoid membranes. FEMS Microbiol. Lett. *292*, 63–70.

Gathmann, S., Rupprecht, E., Kahmann, U., and Schneider, D. (2008). A conserved structure and function of the YidC homologous protein Slr1471 from *Synechocystis* sp. PCC 6803. J. Microbiol. Biotechnol. *18*, 1090–1094.

Genevaux, P., Keppel, F., Schwager, F., Langendijk-Genevaux, P.S., Hartl, F.U., and Georgopoulos, C. (2004). *In vivo* analysis of the overlapping functions of DnaK and trigger factor. EMBO Rep. *5*, 195–200.

Genevaux, P., Georgopoulos, C., and Kelley, W.L. (2007). The Hsp70 chaperone machines of *Escherichia coli*: a paradigm for the repartition of chaperone functions. Mol. Microbiol. *66*, 840–857.

Gerdes, L., Bals, T., Klostermann, E., Karl, M., Philippar, K., Hunken, M., Soll, J., and Schunemann, D. (2006). A second thylakoid membrane-localized Alb3/OxaI/YidC homologue is involved in proper chloroplast biogenesis in *Arabidopsis thaliana*. J. Biol. Chem. *281*, 16632–16642.

Graubner, W., Schierhorn, A., and Brüser, T. (2007). DnaK plays a pivotal role in Tat targeting of CueO and functions beside SlyD as a general Tat signal binding chaperone. J. Biol. Chem. *282*, 7116–7124.

Hinterstoisser, B., Cichna, M., Kuntner, O., and Peschek, G.N.A. (1993). Cooperation of plasma and thylakoid membranes for the biosynthesis of chlorophyll in cyanobacteria: the role of the thylakoid centers. J. Plant Physiol. *142*, 407–413.

Hoffmann, A., Bukau, B., and Kramer, G. (2010). Structure and function of the molecular chaperone Trigger Factor. Biochim. Biophys. Acta *1803*, 650–661.

Howe, C.J., Barbrook, A.C., and Packer, J.C. (1996). Protein targeting and translocation in cyanobacterial membrane biogenesis. Biochem. Soc. Trans. *24*, 750–753.

Hsu, S.-C., Endow, J.K., Ruppel, N.J., Roston, R.L., Baldwin, A.J., and Inoue, K. (2011). Functional diversification of thylakoidal processing peptidases in *Arabidopsis thaliana*. PLoS ONE *6*, e27258.

Huang, F., Parmryd, I., Nilsson, F., Persson, A.L., Pakrasi, H.B., Andersson, B., and Norling, B. (2002). Proteomics of *Synechocystis* sp. Strain PCC 6803: Identification of Plasma Membrane Proteins. Mol. Cell Proteom. *1*, 956–966.

Huang, F., Hedman, E., Funk, C., Kieselbach, T., Schroder, W.P., and Norling, B. (2004). Isolation of outer membrane of *Synechocystis* sp. PCC 6803 and its proteomic characterization. Mol. Cell Proteom. 3, 586–595.

Huang, F., Fulda, S., Hagemann, M., and Norling, B. (2006). Proteomic screening of salt-stress-induced changes in plasma membranes of *Synechocystis* sp. strain PCC 6803. Proteomics *6*, 910–920.

Inoue, K., Baldwin, A.J., Shipman, R.L., Matsui, K., Theg, S.M., and Ohme-Takagi, M. (2005). Complete maturation of the plastid protein translocation channel requires a type I signal peptidase. J. Cell Biol. *171*, 425–430.

Jack, R.L., Sargent, F., Berks, B.C., Sawers, G., and Palmer, T. (2001). Constitutive expression of *Escherichia coli tat* genes indicates an important role for the Twin-Arginine Translocase during aerobic and anaerobic growth. J. Bacteriol. *183*, 1801–1804.

Jongbloed, J.D.H., Martin, U., Antelmann, H., Hecker, M., Tjalsma, H., Venema, G., Bron, S., van Dijl, J.M., and Müller, J. (2000). TatC Is a specificity determinant for protein secretion via the Twin-arginine Translocation pathway. J. Biol. Chem. *275*, 41350–41357.

Jongbloed, J.D., Antelmann, H., Hecker, M., Nijland, R., Bron, S., Airaksinen, U., Pries, F., Quax, W.J., Van Dijl, J.M., and Braun, P.G. (2002). Selective contribution of the twin-arginine translocation pathway to protein secretion in *Bacillus subtilis*. J. Biol. Chem. *277*, 44068–44078.

Jongbloed, J.D.H., Grieger, U., Antelmann, H., Hecker, M., Nijland, R., Bron, S., and Van Dijl, J.M. (2004). Two minimal Tat translocases in *Bacillus*. Mol. Microbiol. *54*, 1319–1325.

Kaneko, T., Sato, S., Kotani, H., Tanaka, A., Asamizu, E., Nakamura, Y., Miyajima, N., Hirosawa, M., Sugiura, M., Sasamoto, S., *et al.* (1996). Sequence analysis of the genome of the unicellular cyanobacterium *Synechocystis* sp. strain PCC6803. II. Sequence determination of the entire genome and assignment of potential protein-coding regions (supplement). DNA Res. 3, 185–209.

Katano, Y., Nimura-Matsune, K., and Yoshikawa, H. (2006). Involvement of DnaK3, one of the three DnaK proteins of cyanobacterium *Synechococcus* sp. PCC7942, in translational process on the surface of the thylakoid membrane. Biosci. Biotechnol. Biochem. *70*, 1592–1598.

Kelley, W.L. (1998). The J-domain family and the recruitment of chaperone power. Trends Biochem. Sci. *23*, 222–227.

Kleemann, W., and McConnell, H.M. (1974). Lateral phase separations in *Escherichia coli* membranes. Biochim. Biophys. Acta *345*, 220–230.

Klinkert, B., Ossenbuhl, F., Sikorski, M., Berry, S., Eichacker, L., and Nickelsen, J. (2004). PratA, a periplasmic tetratricopeptide repeat protein involved in biogenesis of photosystem II in *Synechocystis* sp. PCC 6803. J. Biol. Chem. *279*, 44639–44644.

Koch, H.-G., Moser, M., and Müller, M. (2003). Signal recognition particle-depencent protein targeting, universal to all kingdoms of life. Rev. Physiol. Biochem. Pharmacol. *146*, 55–94.

Kroll, D., Meierhoff, K., Bechtold, N., Kinoshita, M., Westphal, S., Vothknecht, U.C., Soll, J., and Westhoff, P. (2001). VIPP1, a nuclear gene of *Arabidopsis thaliana* essential for thylakoid membrane formation. Proc. Natl. Acad. Sci. U.S.A. *98*, 4238–4242.

Kunkel, D.D. (1982). Thylakoid centers: structures associated with the cyanobacterial photosynthetic membrane system. Arch. Microbiol. *133*, 97–99.

Lee, P.A., Tullman-Ercek, D., and Georgiou, G. (2006). The bacterial Twin-Arginine Translocation pathway. Annu. Rev. Microbiol. *60*, 373–395.

Liberton, M., Howard Berg, R., Heuser, J., Roth, R., and Pakrasi, H.B. (2006). Ultrastructure of the membrane systems in the unicellular cyanobacterium *Synechocystis* sp. strain PCC 6803. Protoplasma *227*, 129–138.

Liberton, M., Austin, J.R., 2nd, Berg, R.H., and Pakrasi, H.B. (2012a). Insights into the complex 3-D architecture of thylakoid membranes in unicellular cyanobacterium *Cyanothece* sp. ATCC 51142. Plant Signal Behav. *6*, 566–569.

Liberton, M., Austin, J.R., 2nd, Berg, R.H., and Pakrasi, H.B. (2012b). Unique thylakoid membrane architecture of a unicellular N_2-fixing cyanobacterium revealed by electron tomography. Plant Physiol. *155*, 1656–1666.

Mackle, M., and Zilinskas, B. (1994). Role of signal peptides in targeting of proteins in cyanobacteria. J. Bacteriol. *176*, 1857–1864.

Mannan, R.M., and Pakrasi, H.B. (1993). Dark heterotrophic growth conditions result in an increase in the content of photosystem II units in the filamentous cyanobacterium *Anabaena variabilis* ATCC 29413. Plant Physiol. *103*, 971–977.

Mariscal, V., Herrero, A., and Flores, E. (2007). Continuous periplasm in a filamentous, heterocyst-forming cyanobacterium. Mol. Microbiol. *65*, 1139–1145.

Mayer, M.P., and Bukau, B. (2005). Hsp70 chaperones: cellular functions and molecular mechanism. Cell. Mol. Life Sci. *62*, 670–684.

van de Meene, A.M., Hohmann-Marriott, M.F., Vermaas, W.F., and Roberson, R.W. (2006). The three-dimensional structure of the cyanobacterium *Synechocystis* sp. PCC 6803. Arch. Microbiol. *184*, 259–270.

van de Meene, A.M., Sharp, W.P., McDaniel, J.H., Friedrich, H., Vermaas, W.F., and Roberson, R.W. (2012). Gross morphological changes in thylakoid membrane structure are associated with photosystem I deletion in *Synechocystis* sp. PCC 6803. Biochim. Biophys. Acta *1818*, 1427–1434.

Müller, M., and Klösgen, R.B. (2005). The Tat pathway in bacteria and chloroplasts (review). Mol. Membr. Biol. *22*, 113–121.

Müller, M., Koch, H.G., Beck, K., and Schafer, U. (2001). Protein traffic in bacteria: multiple routes from the ribosome to and across the membrane. Prog. Nucleic Acid Res. Mol. Biol. *66*, 107–157.

Mullineaux, C.W. (1999). The thylakoid membranes of cyanobacteria: structure, dynamics and function. Funct. Plant Biol. *26*, 671–677.

Nakai, M., Sugita, D., Omata, T., and Endo, T. (1993). Sec-Y protein is localized in both the cytoplasmic and thylakoid membranes in the cyanobacterium *Synechococcus* PCC7942. Biochem. Biophys. Res. Commun. *193*, 228–234.

Nakai, M., Nohara, T., Sugita, D., and Endo, T. (1994). Identification and characterization of the sec-A protein homologue in the cyanobacterium *Synechococcus* PCC7942. Biochem. Biophys. Res. Commun. *200*, 844–851.

Nakamura, Y., Kaneko, T., Sato, S., Mimuro, M., Miyashita, H., Tsuchiya, T., Sasamoto, S., Watanabe, A., Kawashima, K., Kishida, Y., *et al.* (2003). Complete genome structure of *Gloeobacter violaceus* PCC 7421, a cyanobacterium that lacks thylakoids (supplement). DNA Res. *10*, 181–201.

Nakao, M., Okamoto, S., Kohara, M., Fujishiro, T., Fujisawa, T., Sato, S., Tabata, S., Kaneko, T., and Nakamura, Y. (2010). CyanoBase: the cyanobacteria genome database update 2010. Nucl. Acids Res. *38*, D379-D381.

Nevo, R., Charuvi, D., Shimoni, E., Schwarz, R., Kaplan, A., Ohad, I., and Reich, Z. (2007). Thylakoid membrane perforations and connectivity enable intracellular traffic in cyanobacteria. EMBO J. *26*, 1467–1473.

Nickelsen, J., Rengstl, B., Stengel, A., Schottkowski, M., Soll, J., and Ankele, E. (2010). Biogenesis of the cyanobacterial thylakoid membrane system – an update. FEMS Microbiol. Lett. *315*, 1–5.

Nimura, K., Yoshikawa, H., and Takahashi, H. (1996). DnaK3, one of the three DnaK proteins of cyanobacterium *Synechococcus* sp. PCC7942, is quantitatively detected in the thylakoid membrane. Biochem. Biophys. Res. Commun. *229*, 334–340.

Nimura, K., Takahashi, H., and Yoshikawa, H. (2001). Characterization of the *dnaK* multigene family in the cyanobacterium *Synechococcus* sp. strain PCC7942. J. Bacteriol. *183*, 1320–1328.

Nowaczyk, M.M., Sander, J., Grasse, N., Cormann, K.U., Rexroth, D., Bernat, G., and Rögner, M. (2010). Dynamics of the cyanobacterial photosynthetic network: Communication and modification of membrane protein complexes. Eur. J. Cell Biol. *89*, 974–982.

Oguchi, K., Nimura, K., Yoshikawa, H., and Takahashi, H. (1997). Sequence and analysis of a *dnaJ* homologue gene in cyanobacterium *Synechococcus* sp. PCC7942. Biochem. Biophys. Res. Commun. *236*, 461–466.

Oresnik, I.J., Ladner, C.L., and Turner, R.J. (2001). Identification of a twin-arginine leader-binding protein. Mol. Microbiol. *40*, 323–331.

Ossenbühl, F., Inaba-Sulpice, M., Meurer, J., Soll, J., and Eichacker, L.A. (2006). The *Synechocystis* sp. PCC 6803 Oxa1 homolog is essential for membrane integration of reaction center precursor protein pD1. Plant Cell *18*, 2236–2246.

Palmer, T., and Berks, B.C. (2012). The twin-arginine translocation (Tat) protein export pathway. Nat. Rev. Micro. *10*, 483–496.

Pisareva, T., Shumskaya, M., Maddalo, G., Ilag, L., and Norling, B. (2007). Proteomics of *Synechocystis* sp. PCC 6803. Identification of novel integral plasma membrane proteins. FEBS J. *274*, 791–804.

Pisareva, T., Kwon, J., Oh, J., Kim, S., Ge, C., Wieslander, A., Choi, J.S., and Norling, B. (2011). Model for membrane organization and protein sorting in the cyanobacterium *Synechocystis* sp. PCC 6803 inferred from proteomics and multivariate sequence analyses. J. Proteome Res. *10*, 3617–3631.

du Plessis, D.J.F., Nouwen, N., and Driessen, A.J.M. (2011). The Sec translocase. Biochim. Biophys. Acta *1808*, 851–865.

Pop, O., Martin, U., Abel, C., and Müller, J.P. (2002). The Twin-arginine signal peptide of PhoD and the TatAd/Cd proteins of *Bacillus subtilis* form an autonomous Tat translocation system. J. Biol. Chem. *277*, 3268–3273.

Rajalahti, T., Huang, F., Klement, M.R., Pisareva, T., Edman, M., Sjostrom, M., Wieslander, A., and Norling, B. (2007). Proteins in different *Synechocystis* compartments have distinguishing N-terminal features: a combined proteomics and multivariate sequence analysis. J. Proteome Res. *6*, 2420–2434.

Randall, L.L., and Hardy, S.J. (2002). SecB, one small chaperone in the complex milieu of the cell. Cell Mol. Life Sci. *59*, 1617–1623.

Rengstl, B., Oster, U., Stengel, A., and Nickelsen, J. (2011). An intermediate membrane subfraction in cyanobacteria is involved in an assembly network for Photosystem II biogenesis. J. Biol. Chem. *286*, 21944–21951.

Rexroth, S., Mullineaux, C.W., Ellinger, D., Sendtko, E., Rogner, M., and Koenig, F. (2011). The plasma membrane of the cyanobacterium *Gloeobacter violaceus* contains segregated bioenergetic domains. Plant Cell *23*, 2379–2390.

Richter, C.V., Bals, T., and Schünemann, D. (2010). Component interactions, regulation and mechanisms of chloroplast signal recognition particle-dependent protein transport. Eur. J. Cell Biol. *89*, 965–973.

Rippka, R. (1972). Photoheterotrophy and chemoheterotrophy among unicellular blue-green algae. Arch. Microbiol. *87*, 93–98.

Rippka, R., Waterbury, J., and Cohen-Bazire, G. (1974). A cyanobacterium which lacks thylakoids. Arch. Microbiol. *100*, 419–436.

van Roosmalen, M.L., Jongbloed, J.D., Dubois, J.Y., Venema, G., Bron, S., and van Dijl, J.M. (2001). Distinction between major and minor *Bacillus* signal peptidases based on phylogenetic and structural criteria. J. Biol. Chem. *276*, 25230–25235.

van Roosmalen, M.L., Geukens, N., Jongbloed, J.D., Tjalsma, H., Dubois, J.Y., Bron, S., van Dijl, J.M., and Anne, J. (2004). Type I signal peptidases of Gram-positive bacteria. Biochim. Biophys. Acta *1694*, 279–297.

Rosenblad, M.A., and Samuelsson, T. (2004). Identification of chloroplast signal recognition particle RNA genes. Plant Cell Physiol. *45*, 1633–1639.

Rupprecht, E., Gathmann, S., Fuhrmann, E., and Schneider, D. (2007). Three different DnaK proteins are functionally expressed in the cyanobacterium *Synechocystis* sp. PCC 6803. Microbiology *153*, 1828–1841.

Rupprecht, E., Fuhrmann, E., and Schneider, D. (2008). Stress regulated *dnaK* expression in *Synechocystis* sp. PCC 6803. In Energy from the Sun, Allen, J., Gantt, E., Goldbeck, J., and Osmond, B., eds. (Springer Science + Media B.V., Dordrecht, Niederlande), pp. 1327–1330.

Rupprecht, E., Duppre, E., and Schneider, D. (2010). Similarities and singularities of three DnaK proteins from the cyanobacterium *Synechocystis* sp. PCC 6803. Plant Cell Physiol. *51*, 1210–1218.

Saibil, H.R. (2008). Chaperone machines in action. Curr. Opin. Struct. Biol. *18*, 35–42.

Saier, M. (2006). Protein secretion and membrane insertion systems in Gram-negative bacteria. J. Membr. Biol. *214*, 75–90.

Samuelson, J.C., Chen, M., Jiang, F., Moller, I., Wiedmann, M., Kuhn, A., Phillips, G.J., and Dalbey, R.E. (2000). YidC mediates membrane protein insertion in bacteria. Nature *406*, 637–641.

Sato, M., Nimura-Matsune, K., Watanabe, S., Chibazakura, T., and Yoshikawa, H. (2007a). Expression analysis of multiple *dnaK* genes in the cyanobacterium *Synechococcus elongatus* PCC 7942. J. Bacteriol. *189*, 3751–3758.

Sato, M., Yamahata, H., Watanabe, S., Nimura-Matsune, K., and Yoshikawa, H. (2007b). Characterization of dnaJ multigene family in the cyanobacterium *Synechococcus elongatus* PCC 7942. Biosci. Biotechnol. Biochem. *71*, 1021–1027.

Schleiff, E., and Klösgen, R.B. (2001). Without a little help from 'my' friends: direct insertion of proteins into chloroplast membranes? Biochim. Biophys. Acta *1541*, 22–33.

Schneider, D., Fuhrmann, E., Scholz, I., Hess, W.R., and Graumann, P.L. (2007). Fluorescence staining of live cyanobacterial cells suggest non-stringent chromosome segregation and absence of a connection between cytoplasmic and thylakoid membranes. BMC Cell Biol. *8*, 39.

Schünemann, D. (2007). Mechanisms of protein import into thylakoids of chloroplasts. Biol. Chem. *388*, 873–983.

Sherman, D.M., Troyan, T.A., and Sherman, L.A. (1994). Localization of membrane proteins in the cyanobacterium *Synechococcus* sp. PCC7942 (radial asymmetry in the photosynthetic complexes). Plant Physiol. *106*, 251–262.

Shipman, R.L., and Inoue, K. (2009). Suborganellar localization of plastidic type I signal peptidase 1 depends on chloroplast development. FEBS Lett. *583*, 938–942.

van der Sluis, E.O., and Driessen, A.J.M. (2006). Stepwise evolution of the Sec machinery in Proteobacteria. Trends Microbiol. *14*, 105–108.

Smart, L.B., Anderson, S.L., and McIntosh, L. (1991). Targeted genetic inactivation of the photosystem I reaction center in the cyanobacterium *Synechocystis* sp. PCC 6803. EMBO J. *10*, 3289–3296.

Smith, D., and Howe, C.J. (1993). The distribution of Photosystem I and Photosystem II polypeptides between the cytoplasmic and thylakoid membranes of cyanobacteria. FEMS Microbiol. Lett. *110*, 341–347.

Smith, D., Bendall, D.S., and Howe, C.J. (1992). Occurrence of a Photosystem II polypeptide in non-photosynthetic membranes of cyanobacteria. Mol. Microbiol. *6*, 1821–1827.

Spence, E., Sarcina, M., Ray, N., Moller, S.G., Mullineaux, C.W., and Robinson, C. (2003). Membrane-specific targeting of green fluorescent protein by the Tat pathway in the cyanobacterium *Synechocystis* PCC6803. Mol. Microbiol. *48*, 1481–1489.

Spence, E., Bailey, S., Nenninger, A., Moller, S.G., and Robinson, C. (2004). A homolog of Albino3/OxaI is essential for thylakoid biogenesis in the cyanobacterium *Synechocystis* sp. PCC6803. J. Biol. Chem. *279*, 55792–55800.

Srivastava, R., Pisareva, T., and Norling, B. (2005). Proteomic studies of the thylakoid membrane of *Synechocystis* sp. PCC 6803. Proteomics *5*, 4905–4916.

Srivastava, R., Battchikova, N., Norling, B., and Aro, E.M. (2006). Plasma membrane of *Synechocystis* PCC 6803: a heterogeneous distribution of membrane proteins. Arch. Microbiol. *185*, 238–243.

Steglich, C., Lindell, D., Futschik, M., Rector, T., Steen, R., and Chisholm, S. (2010). Short RNA half-lives in the slow-growing marine cyanobacterium *Prochlorococcus*. Genome Biol. *11*, R54.

Steiner, J.M., and Löffelhardt, W. (2005). Protein translocation into and within cyanelles. Mol. Membr. Biol. *22*, 123–132.

Stengel, A., Gugel, I.L., Hilger, D., Rengstl, B., Jung, H., and Nickelsen, J. (2012). Initial steps of photosystem II de novo assembly and preloading with manganese take place in biogenesis centers in *Synechocystis*. Plant Cell *24*, 660–675.

Sundberg, E., Slagter, J.G., Fridborg, I., Cleary, S.P., Robinson, C., and Coupland, G. (1997). ALBINO3, an *Arabidopsis* nuclear gene essential for chloroplast differentiation, encodes a chloroplast protein that shows homology to proteins present in bacterial membranes and yeast mitochondria. Plant Cell *9*, 717–730.

Tandeau de Marsac, N. (2003). Phycobiliproteins and phycobilisomes: the early observations. Photosynth. Res. *76*, 193–205.

Teter, S.A., Houry, W.A., Ang, D., Tradler, T., Rockabrand, D., Fischer, G., Blum, P., Georgopoulos, C., and Hartl, F.U. (1999). Polypeptide flux through bacterial Hsp70: DnaK cooperates with Trigger Factor in chaperoning nascent chains. Cell *97*, 755–765.

Thompson, S.J., Kim, S.J., and Robinson, C. (1998). Sec-independent insertion of thylakoid membrane proteins. J. Biol. Chem. *273*, 18979–18983.

Tjalsma, H., Noback, M.A., Bron, S., Venema, G., Yamane, K., and van Dijl, J.M. (1997). *Bacillus subtilis* contains four closely related type I signal peptidases with overlapping substrate specificities. Constitutive and temporally controlled expression of different sip genes. J. Biol. Chem. *272*, 25983–25992.

Tjalsma, H., Bolhuis, A., van Roosmalen, M.L., Wiegert, T., Schumann, W., Broekhuizen, C.P., Quax, W.J., Venema, G., Bron, S., and van Dijl, J.M. (1998). Functional analysis of the secretory precursor processing machinery of *Bacillus subtilis*: identification of a eubacterial homolog of archaeal and eukaryotic signal peptidases. Genes Dev. *12*, 2318–2331.

Tzvetkova-Chevolleau, T., Hutin, C., Noel, L.D., Goforth, R., Carde, J.-P., Caffarri, S., Sinning, I., Groves, M., Teulon, J.-M., Hoffman, N.E., *et al.* (2007). Canonical signal recognition particle components can be bypassed for posttranslational protein targeting in chloroplasts. Plant Cell *19*, 1635–1648.

Uniacke, J., and Zerges, W. (2007). Photosystem II assembly and repair are differentially localized in Chlamydomonas. Plant Cell *19*, 3640–3654.

Uniacke, J., and Zerges, W. (2009). Chloroplast protein targeting involves localized translation in Chlamydomonas. Proc. Natl. Acad. Sci. U.S.A. *106*, 1439–1444.

Vermaas, W.F. (2001). Photosynthesis and respiration in cyanobacteria. In Encyclopedia of Life Sciences (Nature Publishing Group, London), pp. 1–7.

Vermaas, W.F., Timlin, J.A., Jones, H.D., Sinclair, M.B., Nieman, L.T., Hamad, S.W., Melgaard, D.K., and Haaland, D.M. (2008). *In vivo* hyperspectral confocal fluorescence imaging to determine pigment localization and distribution in cyanobacterial cells. Proc. Natl. Acad. Sci. U.S.A. *105*, 4050–4055.

Vernotte, C., Picaud, M., Kirilovsky, D., Olive, J., Ajlani, G., and Astier, C. (1992). Changes in the photosynthetic apparatus in the cyanobacterium *Synechocystis* sp. PCC 6714 following light-to-dark and dark-to-light transitions. Photosynth. Res. *32*, 45–57.

Von Heijne, G. (2002). Targeting sequences. In Protein Targeting, Transport and Translocation, Dalbey, R., and Von Heijne, G., eds. (Academic Press, London, San Diego), pp. 35–46.

Vothknecht, U.C., and Soll, J. (2005). Chloroplast membrane transport: interplay of prokaryotic and eukaryotic traits. Gene *354*, 99–109.

Vothknecht, U.C., Otters, S., Hennig, R., and Schneider, D. (2012). Vipp1: a very important protein in plastids?! J. Exp. Bot. *63*, 1699–1712.

Wang, P., and Dalbey, R.E. (2012). Inserting membrane proteins: the YidC/Oxa1/Alb3 machinery in bacteria, mitochondria, and chloroplasts. Biochim. Biophys. Acta *1808*, 866–875.

Weis, B.L., Schleiff, E., and Zerges, W. (2012). Protein targeting to subcellular organelles via mRNA localization. Biochem. Biophys. Acta *1833*, 260–273.

Welte, T., Kudva, R., Kuhn, P., Sturm, L., Braig, D., Müller, M., Warscheid, B., Drepper, F., and Koch, H.-G. (2012). Promiscuous targeting of polytopic membrane proteins to SecYEG or YidC by the *Escherichia coli* signal recognition particle. Mol. Biol. Cell *23*, 464–479.

Westphal, S., Heins, L., Soll, J., and Vothknecht, U.C. (2001). Vipp1 deletion mutant of *Synechocystis*: a connection between bacterial phage shock and thylakoid biogenesis? Proc. Natl. Acad. Sci. U.S.A. *98*, 4243–4248.

Yen, M.-R., Tseng, Y.-H., Nguyen, E., Wu, L.-F., and Saier, M. (2002). Sequence and phylogenetic analyses of the twin-arginine targeting (Tat) protein export system. Arch. Microbiol. *177*, 441–450.

Yusa, F., Steiner, J., and Loffelhardt, W. (2008). Evolutionary conservation of dual Sec translocases in the cyanelles of *Cyanophora paradoxa*. BMC Evol. Biol. *8*, 304.

Zak, E., Norling, B., Maitra, R., Huang, F., Andersson, B., and Pakrasi, H.B. (2001). The initial steps of biogenesis of cyanobacterial photosystems occur in plasma membranes. Proc. Natl. Acad. Sci. U.S.A. *98*, 13443–13448.

Zerges, W., and Rochaix, J.D. (1998). Low density membranes are associated with RNA-binding proteins and thylakoids in the chloroplast of *Chlamydomonas reinhardtii*. J. Cell Biol. *140*, 101–110.

Zhbanko, M., Zinchenko, V., Gutensohn, M., Schierhorn, A., and Klösgen, R.B. (2005). Inactivation of a predicted leader peptidase prevents photoautotrophic growth of *Synechocystis* sp. strain PCC 6803. J. Bacteriol. *187*, 3071–3078.

Zhou, J., and Xu, Z. (2005). The structural view of bacterial translocation-specific chaperone SecB: implications for function. Mol. Microbiol. *58*, 349–357.

Chromatic Acclimation: a Many-coloured Mechanism for Maximizing Photosynthetic Light Harvesting Efficiency

6

Adam N. Bussell and David M. Kehoe

Abstract

Our understanding of chromatic acclimation, during which the light harvesting antennae or phycobilisomes of cyanobacteria are modified to maximize photon capture for photosynthesis, has grown remarkably over the past century. Originally a curiosity of a 'chameleon cyanobacteria' capable of dramatically altering its pigmentation between red and green in response to changes in the ambient light colour, multiple forms of chromatic acclimation are now known to exist and are found in a wide range of species in most habitats on Earth. This ecologically important process gives species that possess it a fitness advantage in environments with fluctuating light colour conditions. The form of chromatic acclimation found in deeper marine environments is sensitive to blue and green light and involves the selective substitution of light-absorbing chromophores for these specific wavelengths, while other types of chromatic acclimation are maximally sensitive to the green and red regions of the visible spectrum and involve the reversible replacement of both proteins and chromophores within the phycobilisomes as well as changes in many other cellular processes. Although some progress has been made in unravelling the mechanisms by which these organisms sense and respond to changes in light colour conditions, many questions remain to be answered.

Introduction

The oxygen-evolving photosynthetic organisms that were the progenitors of modern-day cyanobacteria are estimated to have evolved approximately 3.5–2.3 billion years ago (Drews, 2011). These organisms developed the capacity to perform photosynthesis using both Photosystem I (PSI) and II (PSII) and derive electrons from the splitting of water. A by-product of this reaction is oxygen, which significantly enriched the Earth's atmosphere beginning between 2.4–2.2 billion years ago (Bekker *et al.*, 2004). Originally characterized as blue-green algae in the 19th century, cyanobacteria are actually Gram-negative prokaryotes capable of performing both carbon and nitrogen fixation under aerobic conditions, in addition to their ability to generate oxygen as a by-product of photosynthesis (Stanier and Bazire, 1977). Most cyanobacteria utilize chlorophyll *a* within their PSI and PSII reaction centres and have water-soluble structures associated with the surface of the thylakoid or photosynthetic membranes called phycobilisomes (PBS), which they use to harvest light energy for photosynthesis. Cyanobacteria have adapted to colonize almost all environments on our planet in part by perfecting their capacity to acclimate their light-harvesting systems to changes in both the amount (irradiance) and colour (spectral distribution) of ambient light. One important process developed by cyanobacteria to deal with changes in their light colour environment is called chromatic acclimation (CA). Currently, three distinct types of CA have been discovered and explored to varying extents. The demarcation of these three forms of CA has been based on the manner in which a particular cyanobacterium alters the protein and/or chromophore composition of their PBS in response to specific light colours. There is evidence, however, that within each of these three types, significant variation exists in the responses during CA and the sensing mechanisms through which they are controlled. In addition, recent findings suggest that additional forms of CA are likely to exist. This chapter will describe our current understanding of the physiological responses and the signal transduction system(s) controlling each type of CA that has been analysed to date.

Studies delineating the variation in the types of CA

The process of CA as we define it today was first characterized in the early 20th century by Engelmann and Gaidukov as a light-colour-responsive process that led to changes in the pigment composition of certain species of cyanobacteria. The type of CA that Engelmann and Gaidukov characterized was named 'complementary chromatic adaptation', since *Oscillatoria sancta* cells were phenotypically red when grown in green light (GL) and green when grown in red light (RL) (Engelmann, 1902; Gaiducov, 1902, 1903), i.e. the colours of these cells were 'complementary' to the ambient light colour. Additional cyanobacterial strains were also examined but failed to show an equivalent response. Subsequent studies have demonstrated that the capacity to undergo complementary chromatic acclimation is species and strain specific and that the form of CA discovered by these researchers is only one of the three known types of CA (Boresch, 1922; Gutu and Kehoe, 2012; Tandeau de Marsac, 1977). In addition, more recent studies indicate that no changes in DNA content or arrangement occur during any known CA response. Thus, it has been noted that the original use of the word 'adaptation' to describe this phenomenon was erroneous and should be replaced by the more accurate term 'acclimation', since all types of CA so far characterized operate through changes in gene expression (Kehoe and Gutu, 2006).

Since the first description of CA, significant advances have been made in understanding how CA-capable cyanobacteria modify their pigment composition in response to light colour and the mechanisms regulating these changes. The discovery of the molecular events that occur during CA has been guided by the careful and innovative work of many researchers over the last century. Beginning in the 1960s, the contributions of Dr. Yoshihiko Fujita and his colleagues were crucial for establishing the first action spectrum for a CA-capable species, *Tolypothrix tenuis*, a filamentous species which seems to have been isolated from a soil sample in Indonesia (Fujita and Hattori, 1960). Their analyses established that CA in these species was controlled by a photoreversible pigment that responded maximally to GL (~540 nm) and RL (~630 nm). Brief exposures to either of these two light colours was sufficient to drive the shift from one response to the other. For CA in *T. tenuis*, these responses were the accumulation of specific phycobiliproteins within the PBS, with phycoerythrin (PE) becoming abundant in response to GL treatment and phycocyanin (PC) accumulation occurring after RL treatment. The function of phycobiliproteins such as PE [absorption maximum (λ_{max}) = 525 nm] and PC (λ_{max} = 620 nm) is to capture additional light energy for photosynthesis at wavelengths that are relatively poorly absorbed by chlorophyll *a*. By characterizing the action spectra of *T. tenuis*, Fujita and colleagues increased our understanding of CA by demonstrating that specific wavelengths of light trigger complex changes in phycobiliprotein composition within the PBS.

The publication of the CA action spectrum for *T. tenuis* led to the analysis of the CA action spectrum of another CA-capable cyanobacterium, the filamentous, freshwater species *Fremyella diplosiphon*. The initial investigations of the *F. diplosiphon* CA action spectrum were performed by two groups: Vogelmann and Scheibe, and Haury and Bogorad. The resulting action spectra differed significantly in the blue region of the spectrum, with the action spectrum of Bogorad's group showing a strong blue light effect (Haury and Bogorad, 1977; Vogelmann and Scheibe, 1978). Subsequent studies by Fujita's group resolved this discrepancy by producing an action spectrum that closely matched that of Vogelmann and Scheibe (Ohki *et al.*, 1982). They suggested that the conditions used by Haury and Bogorad produced an inaccurate CA action spectrum in the near UV/blue light (BL) region, which was proposed by

Haury and Bogorad to be as effective as GL or RL in regulating the cellular levels of PE and PC. Specifically, Fujita and co-workers proposed that the effect measured by Haury and Bogorad in this region of the spectrum was likely due to the action of photosynthesis, casting doubt on the proposal by Haury and Bogorad that a second, blue light-absorbing photoreceptor also controlled CA in *F. diplosiphon*. In all subsequent genetic, molecular, and biochemical studies of CA, no evidence has been found that suggests that a blue light photoreceptor plays a role in the control of CA in *F. diplosiphon*, consistent with the findings of Fujita as well as of Vogelmann and Scheibe (Gutu and Kehoe, 2012; Ohki *et al.*, 1982; Vogelmann and Scheibe, 1978).

All of these early studies focused on species that displayed dramatic changes in coloration during CA. In the 1970s, Tandeau de Marsac significantly advanced the understanding of the diversity of CA in the natural environment when she investigated the cyanobacteria culture collection at the Pasteur Institute (known as the Pasteur Culture Collection, or PCC), analysing species and strains that contained both PC and PE for their ability to undergo CA (Tandeau de Marsac, 1977). She examined 44 cyanobacterial strains that produced both PC and PE and were representative of each genus in the PCC for their ability to perform CA by measuring their relative PE or PC content after growth in RL and GL. Her study found that 32 of the 44 strains examined, or over 70%, were capable of performing CA. Interestingly, cyanobacterial species within the same genera were found to vary significantly in their ability to carry out CA. This study also identified three groups that were functionally defined by their relative PC and PE content in RL and GL. The cyanobacterial strains that made up each of these groups were phylogenetically and morphologically diverse. Group I (CA1) strains were classified by their lack of a CA response, such that no significant changes in the ratio of PC to PE were measured in GL growth conditions versus RL growth. Group II (CA2) strains were found to maintain an equivalent amount of PC during growth in both RL and GL, but increase their PE content in response to being shifted from RL to GL growth conditions (Fig. 6.1A). Group III (CA3) strains produced more PC in RL than in GL, and more PE in GL than in RL (Fig. 6.1B). Thus, it was members of the CA3

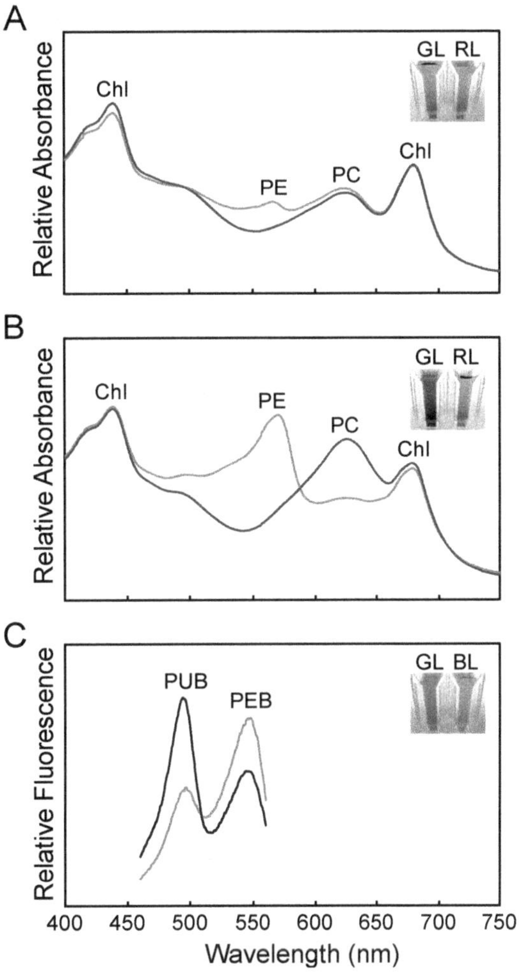

Figure 6.1 Chromatic acclimation responses and colour phenotypes. (A) Whole-cell absorption spectra of the CA2 strain *Nostoc punctiforme* after acclimation to GL and RL. (B) Whole-cell absorption spectra of the CA3 strain *F. diplosiphon* after acclimation to GL and RL. (C) Whole-cell fluorescence spectra of the CA4 strain *Synechococcus* sp. RS9916 after acclimation to GL and BL. The colour of each line denotes the light colour in which the cells were grown. Maxima for Chl, chlorophyll; PE, phycoerythrin; PC, phycocyanin; PUB, phycourobilin; and PEB, phycoerythrobilin are denoted. Insets show the colour phenotype of cells after acclimation to the specific light colour treatment indicated above each cuvette. A colour version of this figure is available in the plate section at the back of the book.

group, *O. sancta*, *T. tenuis*, and *F. diplosiphon*, which were originally described as being capable of 'complementary chromatic adaptation'. This is not surprising, since the CA3 response is much more phenotypically dramatic than the CA2 response (Fig. 6.1A and B, insets). Overall, this work was very important for its demonstration that the 'CA response' was more complex than previously imagined and that the capacity to carry out CA was not correlated with the evolutionary relatedness of cyanobacterial species. It was also noteworthy because it demonstrated that even within the CA2 and CA3 groups there is a range in the extent of the CA response itself. This very interesting finding suggests that there is likely to be variability in the signal transduction pathways controlling CA2 as well as in those regulating CA3. If so, this raises the question of whether the response variations observed in the CA2 and CA3 groups are the result of differences that have arisen randomly during the evolution of these processes in different genera and species that have not been selected upon, or if these differences are the result of different selection pressures on species growing in various light colour conditions, or both. Further analyses of the natural history of these species and the signal transduction pathways controlling CA2 and CA3 responses may provide the answer to this question.

More than 20 years after Tandeau de Marsac's first description of CA2 and CA3, a third type of CA (CA4) was described by Palenik in his analysis of marine cyanobacteria possessing PBS that contain the GL absorbing chromophore phycoerythrobilin (PEB) and the BL absorbing chromophore phycourobilin (PUB) (Palenik, 2001) (Fig. 6.1C). Instead of responding maximally to RL and GL, this form of CA appears to be BL and GL responsive, although an action spectrum has not yet been generated for this response. Phenotypically, the cells are orange in BL and pink in GL (Fig. 6.1C inset). These CA4 strains differ from the other two types of CA because during CA4, strains do not substitute one phycobiliprotein for another in response to light changes; instead, the type of open-chain tetrapyrrole (bilin) attached to PE changes.

Cyanobacterial phycobilisomes

PBS are water-soluble macromolecular light harvesting antenna structures that are associated with the thylakoid membranes of cyanobacteria, red algae, and glaucocystophytes. These structures are usually attached to the photosynthetic complexes located within these membranes and help to more effectively power photosynthesis by increasing the cross-sectional area of photon capture and energy transfer to the photosynthetic reaction centres. PBS absorb light energy within the visible region of the spectrum that is not absorbed well by chlorophyll *a* and transduce this energy to photosynthetic reaction centres (Kawamura *et al.*, 1979). PBS structures vary greatly among cyanobacterial species and red algae, existing as hemi-discoidal, hemi-ellipsoidal, bundle-shaped, block-shaped and hemi-ellidiscoidal. This section will briefly review the structure, composition, and function of the most common form among cyanobacteria and red algae, the hemi-discoidal PBS, since all known types of CA occur in organisms containing this type of light-harvesting structure. Several excellent, more comprehensive reviews are available that discuss other types of PBS structures (Adir, 2005; MacColl, 1998; Sidler, 2004; Wehrmeyer, 1990).

Hemi-discoidal PBS are composed of a series of rods that extend outwardly from either a bi-, tri-, or pentacylindrical core (Bryant *et al.*, 1979; Ducret *et al.*, 1998; Glazer *et al.*, 1979). Both rod and core PBS structures consist of approximately 12×3 nm discs, each composed of two layers of a ring-shaped trimer made up of dimers of chromophorylated alpha (α) and beta (β) subunits. These discs are then stacked to make up each cylindrical core unit, which predominantly contain α and β subunits of the phycobiliprotein allophycocyanin (AP; $\lambda_{max} = 650$ nm) (Bryant *et al.*, 1979). In freshwater and many marine species, the rods extending from the core vary in their number of discs and are composed of the α- and β-subunits of the phycobiliproteins PE and/or PC. The number of discs per rod is species specific and also dependent on the ambient environment, being particularly responsive to changes in light quantity, quality, and nutrient availability (Bryant *et al.*, 1979; Glazer *et al.*, 1979; Grossman *et al.*,

1993; Perez-Gomez *et al.*, 2012; Stowe-Evans and Kehoe, 2004).

The number of rods extending from the core also varies from six rods for bicylindrical and tricylindrical cores to eight rods for the penta-cylindrical core. The fine structure of each rod allows unidirectional energy transfer to the photosynthetic reaction centre because each disc absorbs a shorter wavelength of light than is absorbed by the adjacent, more core-proximal disc. The unidirectional energy transfer is facilitated by linker proteins, many of which are not chromophorylated, that hold adjacent discs together and connect each rod to the core. The PBS core is connected to the PSII reaction centre by the core-membrane linker (L_{cm}), which has a single bilin chromophore attached (Gao *et al.*, 2012; Lundell *et al.*, 1981). The mass of each hemi-discoidal PBS typically consists of approximately 85% phycobiliproteins and 15% linker proteins (Tandeau de Marsac and Cohen-Bazire, 1977).

AP, PC, and PE possess bilins that are covalently attached to cysteine positions within both their α and β subunits. Enzymes called lyases attach the bilins to these and other phycobiliproteins via single or double cysteinyl thioether linkages through their A and D pyrrole rings. The bilins that are attached to phycobiliproteins in species that are known to carry out CA are blue-coloured phycocyanobilin (PCB), red-coloured phycoerythrobilin (PEB), and yellow-coloured phycourobilin (PUB). In the cyanobacterium *Mastigocladus laminosus*, AP attaches PCB at positions α–84 and β–84 (Sidler *et al.*, 1981) and PC attaches three PCB, one at each position α-84, β-84, and β-155 (Frank *et al.*, 1978). In *F. diplosiphon*, PE attaches five PEB, one at each position α-84, α-143, β-50/61, β-8, and β-155 (Sidler *et al.*, 1986).

Physiology and regulation of CA3

The mechanisms and regulation of CA3 will be examined first, since this process has been studied the most extensively. The majority of these studies have examined CA3 in the freshwater filamentous cyanobacterium *F. diplosiphon*, which was isolated from a freshwater lake near New Haven, Connecticut, in 1952 and eventually deposited in the University of Texas culture collection as *Fremyella diplosiphon* UTEX 481. It was subsequently added to the PCC, where it was renamed *Calothrix* sp. PCC 7601, then *Tolypothrix* sp. PCC 7601. The majority of the genetic and molecular work on CA conducted to date, however, has been carried out on a mutant strain with shortened filaments called SF33 that was generated by Cobley's group (Cobley *et al.*, 1993). Because this was derived from the UTEX 481 strain, the name *F. diplosiphon* will continue to be used here. Other CA3-capable strains, such as *T. tenuis* and *Pseudanabaena* PCC 7409, also have been examined to some extent.

As discussed earlier, action spectra performed using two CA3-responsive species, *T. tenuis* and *F. diplosiphon*, began in the 1950s and demonstrated that a photoreversible photoreceptor likely controlled the changes in PE or PC induced by exposure to GL or RL (Diakoff and Scheibe, 1973; Fujita and Hattori, 1960; Haury and Bogorad, 1977; Ohki *et al.*, 1982; Vogelmann and Scheibe, 1978). The growth rate of *F. diplosiphon* cells kept in continuous darkness, except for a few minutes' daily exposure to GL or RL, was affected by the type of light treatment provided. Cultures given a final treatment of GL had depressed growth in the dark, and a RL treatment would reverse the effects of GL, which provided further evidence for the role of a photoreceptor in controlling CA3 (Diakoff and Scheibe, 1975). This evidence of a photoreceptor controlling CA3 and capable of sensing changes in GL and RL was novel, because it appeared to be somehow related to a class of plant photoreceptors called phytochromes, which had been shown to be responsive to changes in the RL and far-red-light regions of the spectrum (Butler *et al.*, 1959). Since the generation of the CA3 action spectra, research in this field primarily has focused on identifying the GL-RL-responsive photoreceptor(s) and the component(s) of the signal transduction pathway(s) through which it acts to control CA3.

PBS structural and gene expression studies in *F. diplosiphon*

The PBS of *F. diplosiphon* are hemi-discoidal in structure, consisting of a tricylindrical core with six outwardly extending rods. The α and β subunits of AP within the discs of the core are encoded by the *apcA1B1* genes (Conley *et al.*, 1986). There are two different types of PC, encoded by two different operons, present in *F. diplosiphon* PBS under normal growth conditions. Bryant was the first to propose the molecular basis for PC regulation by CA3 when he discovered that CA3-capable species, including *F. diplosiphon*, appeared to always have at least two different genes encoding PC, one not light regulated and the other up-regulated in red light (Bryant, 1981). In *F. diplosiphon* the core-proximal rod discs are composed of a form of PC known as PC1, which is present in PBS in both RL and GL. The α- and β-subunits of PC1 are encoded by the *cpcB1A1* genes (Conley *et al.*, 1986, 1988). The core–distal rod discs are composed of the second form of PC called PC2 in RL, and contain PE in GL (Bryant, 1981; Rosinski *et al.*, 1981) (Fig. 6.2C and D). The α- and β-subunits of PC2 are encoded by the *cpcB2A2* genes (Conley *et al.*, 1985, 1988), while the α- and β-subunits of PE are encoded by the *cpeBA* genes (Mazel *et al.*, 1986). Analysis of PE and PC in *F. diplosiphon* demonstrated that changes in PBS composition during CA3 depended on *de novo* synthesis of these phycobiliproteins and dilution of pre-existing PBS rather than being the result of turnover (Bennett and Bogorad, 1973).

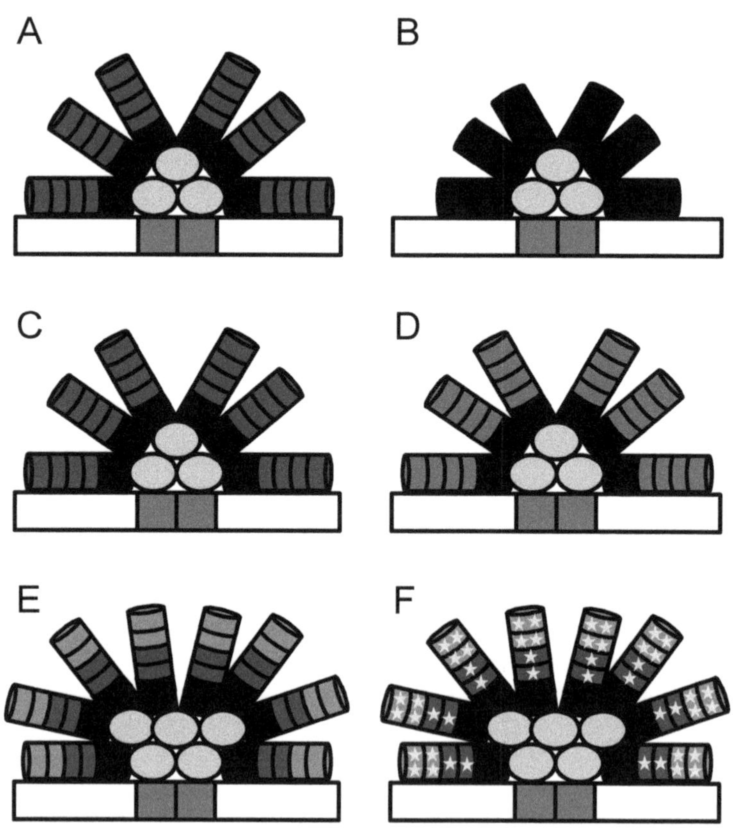

Figure 6.2 Phycobilisome structures among cyanobacterial species capable of chromatic acclimation. PBS of (A) CA2 species adapted to green light, (B) CA2 species adapted to red light, (C) CA3 species adapted to green light, (D) CA3 species adapted to red light, (E) CA4 species adapted to green light, and (F) CA4 species adapted to blue light. Yellow stars indicate the location and number of bilins changing from PEB to PUB upon a switch from green light to blue light. In the PBS models above, the thylakoid membranes are white, photosynthetic reaction centres are green, AP cores are light blue, PC1 discs are dark blue, PE discs are red, PC2 discs are medium blue, and PEII discs are orange. A colour version of this figure is available in the plate section at the back of the book.

Characterization studies of the genes encoding PC, PE, their associated linkers, and bilin synthesis and attachment enzymes have been critical for guiding studies aimed at identifying the components of the CA3 regulatory system. Identification and characterization of these genes initially demonstrated that the CA3 regulation of genes encoding PBS components occurred at the level of transcription, and later showed that CA3 also acts to regulate a subset of these genes post-transcriptionally. Analysis of the expression of the genes encoding PC1, *cpcB1A1*, revealed that these genes are transcribed as a single RNA species of approximately 1600 nucleotides, which is nearly equally abundant in both RL and GL (Conley *et al.*, 1986). The PC2 genes, found within the *cpcB2A2H2I2D2* operon (hereafter referred to as the *cpc2* operon), were transcribed as two RNA species of approximately 1600 and 3800 nucleotides that were much more abundant in RL than in GL (Conley *et al.*, 1985). The *cpc2* operon encodes the α and β subunits of PC2 as well as the genes encoding the PC2 rod linkers (Lomax *et al.*, 1987). In addition, the expression of *pcyA*, the gene encoding the ferredoxin-dependent bilin oxidoreductase that produces PCB (Frankenberg *et al.*, 2001), is also more highly expressed in RL than in GL (Alvey *et al.*, 2007). PCB is the bilin that is attached to AP, PC1, and PC2. The two subunits of PE, encoded by *cpeBA*, are transcribed in GL as a single mRNA species of approximately 1450 nucleotides (Mazel *et al.*, 1986). The absence of a larger RNA species, as in the case of *cpc2*, indicated that the PE rod linker proteins were transcribed independently of *cpeBA*. This inference proved to be correct, and the first three genes of the *cpeCDESTR* operon (designated hereafter as the *cpeC* operon) were found to encode the *F. diplosiphon* PE rod linker proteins (Federspiel and Grossman, 1990; Federspiel and Scott, 1992). The *cpeC* operon is transcribed approximately 10 times more strongly in GL than RL and in the initial and subsequent studies was found to give rise to RNA species approximately 2200, 3200, 4100 and 5200 nucleotides long (Bezy *et al.*, 2011; Federspiel and Grossman, 1990; Federspiel and Scott, 1992). Additionally, the *pebAB* genes, which encode the ferredoxin-dependent bilin oxidoreductases that produce the PE-specific chromophore PEB (Frankenberg *et al.*, 2001), are expressed more highly in GL than in RL in *F. diplosiphon* (Alvey et al., 2003).

Molecular physiology and photobiology of CA3 in *F. diplosiphon*

A series of photobiology and physiology experiments carried out by the Grossman and Briggs groups in the 1980s provided important and revealing information about the *F. diplosiphon* CA3 response at the molecular level. The results of these experiments allowed a series of hypotheses to be generated that addressed the nature of the photoreceptor(s) and signalling pathways controlling CA3. First, an examination of the expression of *cpc2* and *cpeBA* under different light qualities demonstrated that the expression of these genes increased rapidly in response to RL and GL treatments, respectively, although complete *cpeBA* RNA accumulation took longer (6 h) than that of *cpc2* RNA (2 h). Inhibiting transcription with rifampicin and measuring the half-lives of the transcripts from these genes demonstrated that these two operons were primarily transcriptionally controlled by CA3. Analysis of *cpc2* and *cpeBA* RNA loss after switching to a non-inductive light condition revealed dramatic differences for these two operons. Upon switching cells from RL to GL, the *cpc2* RNA half-life was approximately 15 min, while the *cpeBA* RNA half-life after switching cells from GL to RL was approximately 13 h (Oelmuller *et al.*, 1988a). The authors suggested that the prolonged expression of *cpeBA* mRNA after switching from GL to RL could indicate that a stable, positively acting regulatory factor was involved in the regulation of these genes in GL. They also suggested that the rapid decline of *cpc2* RNA after transfer to GL meant that either a negative factor was rapidly activated or produced or that a positively acting factor was rapidly inactivated or degraded.

The photobiology of the CA3 response in *F. diplosiphon* was examined by supplying RL and GL pulses of varying intensity and duration (Oelmuller *et al.*, 1988b). The results showed that *cpc2* and *cpeBA* RNA abundance changes are so rapid that a subsequent 10 min light treatment of the opposite light colour immediately following the first 10 min light treatment was unable to stop the effects of this previous light treatment. This differed from known phytochrome systems, in which a reversing pulse given rapidly enough after the first pulse could completely reverse the effect of the initial pulse. There was also no evidence for degradation or reversion of the photoreceptor after one light treatment or the other, as measured in early phytochrome studies. Fluence response curves for CA3 demonstrated that *cpc2* and *cpeBA* RNA levels changed from threshold to saturation levels over two orders of magnitude of photon fluence, indicating that a photoreceptor was mediating the CA3 response. However, approximately twice as much light was required to affect *cpeBA* RNA levels as *cpc2* RNA levels. This result suggested either that two different photoreceptors were involved in the CA3 regulation of these two operons, or that the signal transduction pathways leading to these two responses varied in their efficiency (Oelmuller *et al.*, 1988b).

The role of protein synthesis in the CA3 response was examined by measuring the effects of protein synthesis inhibitors on light-colour-regulated accumulation of *cpc2* and *cpeBA* RNA. These studies indicated that protein synthesis was required for *de novo cpc2* and *cpeBA* RNA synthesis, but was not required for the very rapid cessation of *cpc2* RNA production in GL or the very slow decline in *cpeBA* RNA production in RL. These latter data supported the earlier proposal that a stable, positive regulatory element is required for *cpeBA* synthesis in GL and is lost only very slowly after transfer to RL conditions (Oelmuller *et al.*, 1988a, 1989).

The regulation of CA3 in *F. diplosiphon*

A number of studies have further explored the CA3 regulation of the genes required for PBS biogenesis in *F. diplosiphon* by conducting functional analyses of the promoter regions of CA3-controlled genes. The goal of such studies was to identify CA3-responsive elements and, subsequently, the transcription factor(s) that bound such elements. Analysis of the *cpc2* promoter by the Grossman group revealed that a region from −75 to +25, relative to the transcription start site, was required for CA3-responsive transcriptional regulation. In addition, this region of DNA was bound by a protein from extracts found in cells grown in RL, but not in GL (Casey and Grossman, 1994). The 13 base pair direct repeat sequence 5′AAATTTGCACAAA3′, overlapping by two base pairs, was noted from −36 to −11 within this promoter, but the role of this sequence in CA3 regulation could not be tested because its replacement led to the loss of promoter activity. The resolution of the role of this sequence in CA3 regulation was achieved when *pcyA* was found to also be RL responsive in *F. diplosiphon*, but to retain some transcriptional activity in GL (Alvey *et al.*, 2007). The direct repeat identified by the Grossman group was also present in the *pcyA* CA3-regulated promoter, in close to the same position relative to the transcription start site, and its substitution led to the loss of CA3 regulation but retention of some transcriptional activity. This direct repeat, named the 'L box' and marked in bold in Fig. 6.3A, is thus a CA3 responsive element, and its role in CA3 will be discussed further below. In an electrophoretic mobility shift assay (EMSA) study of *cpc2* promoter regions, an additional protein, named RcaD, was found to bind to two separate regions of the *cpc2* promoter in RL and proposed to affect RL expression of some PBS components during shifts between GL and RL growth conditions (Noubir *et al.*, 2002; Sobczyk *et al.*, 1994). However, deletion of the two RcaD binding sites within the *cpc2* promoter did not affect *cpc2* expression in cells that were completely acclimated to RL (Casey and Grossman, 1994), suggesting that RcaD may play a role in regulating CA3 only while acclimation is actually occurring.

EMSA studies examining DNA-binding activity of *F. diplosiphon* protein extracts to the *cpeBA* promoter region were also conducted by several groups. Tandeau de Marsac's group identified a DNA-binding protein they named RcaA that was active only in GL-grown cell extracts and was capable of interacting with the direct repeat 5′TTGTTA3′, separated by four base pairs, present from −66 to −45 of this promoter (Sobczyk *et al.*, 1993). RcaA binding activity was found to be dependent on the phosphorylation state of the protein (Sobczyk *et al.*, 1993). The Federspiel group also identified a protein that they named PepB (phycoerythrin promoter-binding protein) that

A

cpeC -64 GAGAATTTTGCACTCAATTTGCACAAACTATT -95
pcyA -46 CTGCAAATTGCACAGATATTGCACAGGCTATA -15
cpc2 -48 TTT**AAATTTGCACAAA**ATTTGCACAAACGAAT -17

B

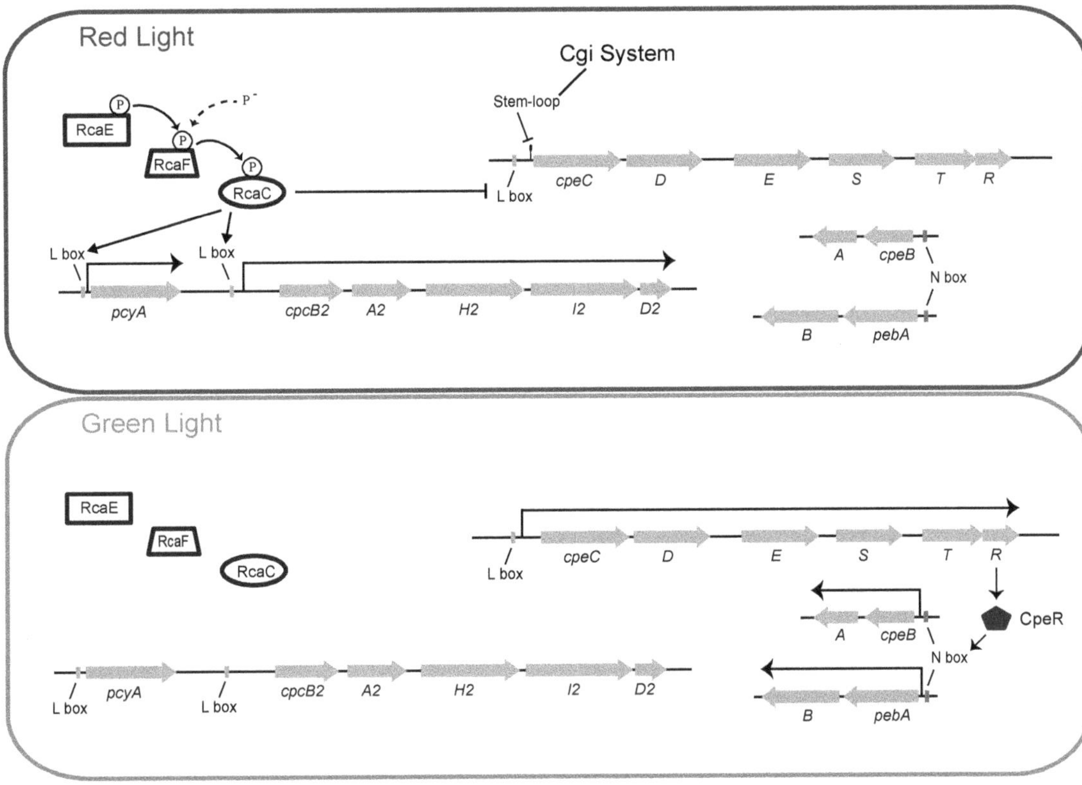

Figure 6.3 Model for CA3 regulation in *F. diplosiphon*. (A) Alignment of L box sequences upstream of *cpeC*, *pcyA*, and *cpc2* in *F. diplosiphon*. Note that the *cpeC* L box sequence is on the opposite strand, relative to transcription start, compared to *pcyA* and *cpc2*. Shaded letters indicate shared sequence identity with the consensus L box repeat. Bold letters indicate the direct repeat sequence, over lapping by two base pairs (Casey and Grossman, 1994). Numbers are for known transcription start sites in *F. diplosiphon*; for *cpc2*, numbering is based on the most recent determination (Alvey *et al.*, 2007). The line demarks the sequence that makes up the L box repeats and flanking sequence, as most recently defined (Bezy *et al.*, 2011). (B) The Rca system appears to be more phosphorylated in RL than GL, leading to the activation of *pcyA* and *cpc2* and the repression of *cpeC* through binding of RcaC to L boxes (indicated). In GL, the Rca system effect is lost and *cpeC* expression increases, leading to expression of CpeR and activation of other GL-expressed genes, possibly through N boxes (indicated). In RL, the Cgi system represses *cpeC* expression post-transcriptionally, requiring a stem–loop within the 5′ leader for this regulation, and this repression is relieved in GL. A colour version of this figure is available in the plate section at the back of the book.

bound to the same region of the *cpeBA* promoter as RcaA, although no difference in binding to this region by protein extracts from RL- or GL-grown cells was observed by this group (Schmidtgoff and Federspiel, 1993). This direct repeat has been termed the 'N box' after the shared initial of the first names of Drs. Nicole Tandeau de Marsac and Nancy Federspiel, whose groups identified this binding site (Kehoe and Gutu, 2006). An N box has also been identified upstream of the

GL-activated *pebAB* operon, encoding the PEB biosynthetic enzymes, in *F. diplosiphon* (Alvey *et al.*, 2003). It is likely that RcaA and PepB are the same protein, although this has not yet been demonstrated, and the discrepancy between RcaA and PepB binding activity in extracts from RL-grown cells has not yet been resolved. In addition, the Tandeau de Marsac group identified RcaB, a possible alternative sigma factor, which bound to the *cpeBA* promoter between the binding sites of RcaA and RNA polymerase, although the precise location was unclear (Sobczyk *et al.*, 1993). Because functional studies of the *cpeBA* promoter have not yet been performed, the roles of RcaA/PepB and RcaB in CA3 remain undefined.

In conjunction with the promoter analyses described above, mutant screens were performed in an effort to identify CA3 signal transduction mutants. Because PE and PC levels were found to be regulated at the level of transcription by CA3, efforts focused on identifying mutants with aberrant accumulation of these proteins. Such mutants displayed colour phenotypes that were very different from wild-type, and these were partitioned into a number of different classes, including red, green, black, and blue (Beguin *et al.*, 1985; Bruns *et al.*, 1989; Cobley and Miranda, 1983; Kehoe and Grossman, 1996: Tandeau de Marsac, 1983). The initial mutants, which were isolated from *F. diplosiphon*, arose spontaneously as a result of the mobilization of endogenous transposable DNA elements within the genome. Subsequently, mutants have been generated via electroporation (Chiang *et al.*, 1992), allelic exchange (Noubir *et al.*, 2002), chemicals (Kehoe and Grossman, 1997), and Tn5 mutagenesis (Cobley *et al.*, 1999; 2002). These mutants were analysed to determine any changes in PC and/or PE levels, usually by measuring whole-cell absorbance spectra. Red mutant cells showed production of high levels of PE in both GL and RL and no production of PC2 in RL. Green mutants had normal CA3 regulation of PC2 but had lost the ability to produce PE in GL. Black mutants had intermediate levels of PC2 and PE in both RL and GL, often with a slightly higher PE to PC ratio in GL and vice versa in RL. Blue mutants produced high levels of PC2 in both RL and GL, while the regulation of PE production by CA3 was normal.

While many of the genes disrupted in these mutants have now been determined, at the time of the early mutant screens the molecular genetic tools available were limited. Genes encoding specific phycobiliprotein components were unknown until the mid to late 1980s, so all studies prior to this were dependent on the characterization of PBS at the protein level. From these initial mutant screens, several different hypotheses were proposed regarding the possible mechanisms regulating CA3, and although some of these were partially accurate, they uniformly underestimated the complexity of this regulatory system. Based on their mutant studies, Cobley and Miranda suggested that PC synthesis was not repressed directly by PE and that a specific repressor of PE was co-synthesized along with PC2. The proposed model was that the genes encoding PC2 and the PE repressor were co-localized within the *F. diplosiphon* genome and that the expression of these genes was attenuated by GL (Cobley and Miranda, 1983). An alternative hypothesis was that the photoreceptor controlling CA3 was always in a biologically active state and had specific domains for interacting directly with the promoters of the PE or PC2 genes (Tandeau de Marsac, 1983).

The ability to use plasmid libraries and electroporation to routinely complement *F. diplosiphon* pigment mutants in the 1990s finally revealed some of the components of the central CA3 regulatory pathway, which had been named the Rca system (regulator for complementary chromatic adaptation) (Sobczyk *et al.*, 1993). These appeared to encode a complex two-component phosphorelay system that consisted of a cyanobacteriochrome named RcaE and two response regulators named RcaF and RcaC. Complementation of black mutants revealed the insertion of a mobile genetic element within *rcaE*, and transformation of these mutants with plasmids carrying the *rcaE* gene complemented the mutation and restored normal CA3 function (Kehoe and Grossman, 1996). RcaE is a 74 kDa protein that consists of a phytochrome-class GAF (cGMP phosphodiesterase/adenylate cyclase/FhlA) domain in its amino-terminal half, followed by a partial PAS (Per-ARNT-Sim) domain, and a sensor histidine kinase domain in its carboxy-terminal region (Kehoe and Grossman, 1996; Terauchi *et al.*, 2004). Lesions within the region of *rcaE* encoding the amino terminus of RcaE resulted in the black mutant phenotype, while lesions in *rcaE* encoding the histidine kinase region of RcaE generated red pigment mutants (Kehoe and Grossman, 1996, 1997). *rcaF*, encoding the response regulator RcaF, which was found to be located immediately downstream of *rcaE*, was identified by complementing red mutants and, based on genetic evidence, proposed to be the carrier of the phosphoryl group from RcaE to RcaC (Kehoe and Grossman, 1997). RcaF consists of a single receiver domain of 124 amino acids and contains the aspartate residue that typically acts as the site of phosphorylation (Kehoe and Grossman, 1997). RcaC was also identified by complementation of red mutants and is a 73 kDa protein that consists of a receiver domain at each end of the protein, and an OmpR/PhoB class DNA binding domain with an adjacent histidine phosphotransferase domain (Chiang *et al.*, 1992; Kehoe and Grossman, 1997).

RcaE was the first phytochrome superfamily member discovered in prokaryotes. Also a member of the sensor histidine kinase family, *in vivo* studies showed that it was equally abundant in RL and GL (Kehoe and Grossman, 1996; Terauchi *et al.*, 2004). An alanine substitution of C198 within its amino-terminal GAF domain led to the loss of bilin attachment *in vivo*, demonstrating that this cysteine was required for covalent chromophore attachment and was likely to be the site of attachment (Terauchi *et al.*, 2004). Experiments carried out to attach a bilin to C198 of RcaE *in vitro* were unsuccessful, and it is still unknown which bilin is attached to RcaE *in vivo* (Terauchi *et al.*, 2004). Genetic studies of phenotypically black and red *rcaE* mutants suggested that RcaE functions as a RL-activated kinase and a GL-activated phosphatase, since the black (null) *rcaE* mutant contained intermediate levels of *cpc2* and *cpeBA* RNA, while red *rcaE* mutants, which were proposed to have a form of RcaE that had lost its kinase activity but retained its phosphatase activity, had high *cpeBA* RNA levels and low *cpc2* RNA levels (Kehoe and Grossman, 1997). Phosphoryl groups were proposed to be transferred from RcaE to the first response regulator RcaF, then to the second response regulator in this pathway, RcaC, through its histidine phosphotransferase domain and finally to its amino-terminal receiver domain. More recent *in vitro* work has shown that purified RcaE can attach a chromophore and undergo a red-green photocycle that involves protonation and deprotonation, and that in RL, RcaE autophosphorylates to a greater extent than GL (Hirose *et al.*, 2013).

RcaC appears to function primarily as a transcription factor that is activated by a RL-dependent phosphorylation event that affects the expression of CA3-regulated genes (Alvey *et al.*, 2007; Li *et al.*, 2008). Site directed mutagenesis studies that investigated the proposed phosphorelay model demonstrated that replacement of specific histidine and aspartate residues within all three Rca components results in a loss of RL CA3 phenotypes (Li and Kehoe, 2005, 2008). The apparent phosphorylation sites within RcaC were aspartate 51, aspartate 576, and histidine 316. Substitutions of these aspartate and histidine residues in RcaC demonstrated that aspartate 51 and histidine 316 were critical for the majority of the CA3 response, while aspartate 576 was the only phosphorylation site residue not required, by itself, for CA3-regulated accumulation of *cpc2* and *cpeBA* RNAs (Li and Kehoe, 2005, 2008). RcaC abundance was also found to be CA3 regulated, with a six-fold increase in RcaC abundance in RL that apparently depended in part on the phosphorylation state of the protein (Li and Kehoe, 2008). It was proposed that the increase in RcaC abundance in RL helps to further increase the range of control of RcaC-activated promoters, particularly *cpc2*, which is expressed at least 100 times more in RL than in GL.

The current model for RcaC function is that during RL growth, RcaC is phosphorylated, increasing its binding affinity for L box regions within the promoters of Rca-regulated genes (Alvey *et al.*, 2007; Li *et al.*, 2008) (Fig. 6.3B). As described earlier, the L box sequence was originally identified in the *cpc2* promoter (Casey and Grossman, 1994). In that study, it was described as part of a larger 13 base pair direct repeat sequence (Fig. 6.3A), but through subsequent DNA binding footprint analyses and promoter studies, the L box has been shown to be a much smaller direct repeat consensus sequence and has been found near *cpeC* as well as within the *cpc2* and *pcyA* promoter regions of *F. diplosiphon* (Fig. 6.3A) (Alvey *et al.*, 2007; Casey and Grossman, 1994; Li *et al.*, 2008). Insertion of an L box sequence within the promoter of a non-light-regulated gene conferred CA3 regulation, while changing the L box within the *pcyA* promoter abolished CA3 regulation of this gene (Alvey *et al.*, 2007). Thus, the L box was shown to be both necessary and sufficient for CA3 transcriptional regulation. The L box is centred at −32 and −31, respectively, relative to the transcription start sites of *pcyA* and *cpc2*. In these promoters, it is in the 'forward' (5′-TTGCACA-3′) orientation, while for *cpeC* it is the 'reverse' (5′-TGTGCAA-3′) orientation and centred at −78 (Alvey *et al.*, 2007). The locations of L boxes relative to the transcription start sites of *pcyA*, *cpc2*, and *cpeC*, as well as their directionality and their flanking sequences, have been examined for their effects on CA3 regulation. Each of these parameters affects the form and degree of CA3 regulation of these genes (Bezy and Kehoe, 2010). RcaC binding to the L boxes of *pcyA* and *cpc2* during growth in RL leads to the activation of these genes, while binding to the L box of *cpeC* in RL represses the transcriptional activity of this operon (Alvey *et al.*, 2007; Li *et al.*, 2008). The use of L boxes for the regulation of CA3 responses appears to occur in other cyanobacterial species, as these DNA sequences have been found within the promoter regions of some PBS-encoding genes in the CA3-capable strains *Pseudanabaena* PCC 7409 and *Synechococcus* PCC 7335 (Bezy and Kehoe, 2010).

In GL, a decrease in RcaC binding activity has been proposed to occur through the phosphatase activity of RcaE (Kehoe and Grossman, 1997). This leads to the loss of the RL activated transcription of *pcyA* and *cpc2* and the RL repression of *cpeC* expression. The de-repression of the *cpeC* operon leads to the expression of *cpeBA* and the accumulation of PE within the PBS core-distal rod discs. Expression of *cpeBA* in GL requires the small activator protein CpeR, which is encoded by the final gene in the *cpeC* operon (Cobley *et al.*, 2002; Seib and Kehoe, 2002). The function of CpeR, a 14 kDa protein, is unknown at present but has limited sequence homology to serine/threonine phosphatases of the PP2C family of proteins (Seib and Kehoe, 2002). Mutations affecting the expression and function of CpeR have been characterized in both green and turquoise mutants (Alvey *et al.*, 2003; Cobley *et al.*, 2002; Seib and Kehoe, 2002). Several different studies described above identified the N box within the *cpeBA* promoter (Schmidtgoff and Federspiel, 1993; Sobczyk *et al.*, 1993), and it is possible that CpeR regulates GL-dependent gene expression through its binding to the N boxes within the promoters of *cpeBA*, *pebAB*, and other GL-induced genes (Alvey *et al.*, 2003; Kehoe and Gutu, 2006; Stowe-Evans *et al.*, 2004). Thus, CpeR, which has only been shown to be an activator of GL-induced genes, may be the same protein(s) that have been previously described as RcaA/PepB, which have only been described based on their DNA binding activity. Activation of *cpeBA* expression by CpeR would support the results of earlier CA3 research that demonstrated the need for protein synthesis during the GL activation of *cpeBA* expression (Cobley *et al.*, 2002; Oelmuller *et al.*, 1988a, 1989; Seib and Kehoe, 2002). The GL regulation of *cpeBA* by CpeR has been proposed to be a hierarchical mechanism that allows PE linker proteins to accumulate first, before expression of *cpeBA*, during the assembly of PE (Cobley *et al.*, 2002). A gene encoding CpeR is present in all cyanobacterial genomes with the capacity to encode PE, including species incapable of CA, suggesting that CpeR is universally used to regulate the expression of genes required for biogenesis of PE-containing PBS (Cobley *et al.*, 2002). If true, the subjugation of the CpeR, and perhaps the N box, regulatory system by the placement of *cpeR* under the control of the CA3-regulated *cpeC* promoter is an efficient mechanism by which CA3 regulation has evolved in *F. diplosiphon* to further increase the number of genes that are GL-activated.

The Rca system is not the only regulatory pathway affecting the RL-GL regulation of gene expression in *F. diplosiphon*, however. Evidence for a second system was first provided in studies examining genes encoding putative PEB attachment enzymes (Kahn *et al.*, 1997). The system was identified as affecting only GL-activated genes in *rcaE* and *rcaC* null mutant strains and named the Cgi system (for control of green light induction) (Alvey *et al.*, 2003; Bezy *et al.*, 2011; Kehoe and Gutu, 2006; Li and Kehoe, 2005; Seib and Kehoe, 2002) (Fig. 6.3). Several genes, including *cpeC*, *cpeBA* and *pebAB*, are all regulated through the Cgi system (the latter two may be regulated indirectly through CpeR), while many additional genes are also likely under Cgi control (Alvey *et al.*, 2003; Bezy *et al.*, 2011; Li and Kehoe, 2005; Seib and Kehoe, 2002). Because of the central role played by the *cpeC* operon in the regulation of GL-expressed gene expression, studies of the Cgi system focused on its regulation of this operon. *cpeC* RNA levels are normally 10-fold higher during growth in GL than RL in wild-type cells, while analyses of an *rcaC* null mutant strain has revealed that approximately one-third of this response is the result of the Cgi system (Bezy *et al.*, 2011; Li and Kehoe, 2005). For *cpeC*, the Cgi system regulates GL-dependent gene expression through a post-transcription initiation mechanism, likely via transcription attenuation, that requires the formation of a putative stem–loop near the CpeC translation start site within the 5′ leader sequence of *cpeC* (Bezy *et al.*, 2011; Gutu and Kehoe, 2012). The Cgi system acts by repressing *cpeC* expression in RL and does not appear to have any effect on RL-dependent gene expression. No components of the Cgi system have yet been reported.

Other CA3-regulated cellular and developmental responses in *F. diplosiphon*

Studies investigating CA3 have uncovered numerous cellular responses in addition to the regulation of genes encoding proteins required for the production of PBS that depend on growth in RL or GL. In *F. diplosiphon*, these have been observed to affect both cellular morphology and physiology. Changes in the cell shape and length of the filament occur between RL and GL, with individual cells appearing more rounded and smaller in size and average filament length being shorter in RL (Bennett and Bogorad, 1973; Bordowitz and Montgomery, 2008). The average length of RL-grown filaments has been proposed to decrease as a result of an increase in the number of necridia, or specialized sites of programmed cell death, which form within the filamentous cellular chain (Bennett and Bogorad, 1973). It has been proposed that necridia formation occurs upon introduction of cultures to RL, causing filament fragmentation, and that this process ceases after cultures become fully acclimate to RL (Bennett and Bogorad, 1973). Cellular morphology of an *rcaE* null mutant strain revealed that these cells exhibited RL phenotypic morphology independent of growth in GL or RL, suggesting the involvement of RcaE in the photoregulation of cellular morphology and filament length in *F. diplosiphon* (Bordowitz and Montgomery, 2008). The loss of TonB, a member of the cell membrane biogenesis family (see Chapter 3), has been shown to affect *F. diplosiphon* morphology in GL but not RL, and although the expression of *tonB* was higher in GL than RL, it was not controlled by RcaE (Pattanaik and Montgomery, 2010). What selective advantage is bestowed upon *F. diplosiphon* by this response in the natural environment, if any, is currently unknown.

Like many filamentous cyanobacteria, *F. diplosiphon* can produce hormogonia. This developmental pathway is characterized by the differentiation of shortened motile filaments made of cells that contain gas vesicles and pili, and is typically a response to environmental stress such as nutrient availability dependent on nutrient availability (Damerval *et al.*, 1991; Rippka *et al.*, 1979; Tandeau de Marsac, 1983; see also Chapter 12). In *F. diplosiphon*, the differentiation of vegetative cells into hormogonia has also been described as CA3 regulated (Damerval *et al.*, 1991). Examination of the effect of various light conditions on the formation of hormogonia in *F. diplosiphon* revealed that RL and replete growth medium are both necessary for 100% hormogonia differentiation, while growth of cells in replete medium under GL will inhibit cells from differentiating into hormogonia but will allow cells to form heterocysts in the absence of combined nitrogen (Campbell *et al.*, 1993; Damerval *et al.*, 1991). Hormogonia formation occurs within the first 12 to 24 h of RL exposure, and they revert back to vegetative filaments within 72 to 96 h (Damerval *et al.*, 1991). Transcript abundance of the gas vesicle genes *gvpABC* are higher in RL than GL and accumulate during the first nine hours of hormogonia differentiation, while expression of the PBS genes appears to be reduced (Csiszar *et al.*, 1987; Damerval *et al.*, 1987, 1991). The use of electron transport inhibitors in conjunction with RL and GL treatments suggested that the simultaneous increase in transcript abundance of genes expressed during hormogonia development and the repression of PBS gene expression is caused by the combined effects of CA3 and the differential excitation of PSI and PSII (Campbell *et al.*, 1993). These studies showed that RL and GL responses, especially the transient phase just after light shifts, is regulated in a complex manner by the combined effects of photoreceptor- and redox-controlled systems.

Analyses of protein abundance levels in *F. diplosiphon* cells grown in RL or GL have revealed that CA3-dependent responses extend beyond the control of PBS component proteins (Gendel *et al.*, 1979; Stowe-Evans *et al.*, 2004). Approximately 80 proteins that are not known to be directly involved in PBS biogenesis have been identified as CA3 regulated (Gendel *et al.*, 1979; Stowe-Evans *et al.*, 2004). Examination of the transcriptome in *F. diplosiphon* using DNA microarrays covering a portion of the genome revealed 17 previously uncharacterized CA3 components, including the gene *chlL* (Stowe-Evans *et al.*, 2004). The GL-mediated increase in the expression of *chlL*, encoding a subunit of protochlorophyllide oxidoreductase, is interesting because the activity of this enzyme is light-independent (Shui *et al.*, 2009; Stowe-Evans *et al.*, 2004). Along with the other enzymatic subunits encoded by *chlN* and *chlB*, the protein encoded by *chlL* carries out the penultimate step in chlorophyll *a* synthesis, and these genes are expressed at a two-fold higher level in GL than RL in *F. diplosiphon* (Shui *et al.*, 2009). This regulation might have been predicted, because species such as *F. diplosiphon* possess a second, independently evolved form of protochlorophyllide oxidoreductase that is encoded by *porA*. PorA is an unusual enzyme because its enzymatic activity requires RL. Thus, as *F. diplosiphon* is exposed to a higher ratio of GL to RL, its PorA activity would be expected to decrease, and for chlorophyll *a* production to be maintained at a constant level under such conditions, an increase in the expression of the *chlN*, *chlB*, and *chlL* genes, encoding the light-independent form of protochlorophyllide oxidoreductase, might be expected (Stowe-Evans and Kehoe, 2004). Examination of the expression of these genes in an *rcaE* null mutant showed a 50% reduction of *chlL* and *chlN* mRNA abundance in GL compared to a wild-type strain, suggesting that CA3 regulatory components affect the regulation of chlorophyll biosynthesis in *F. diplosiphon* (Shui *et al.*, 2009).

Physiology and regulation of CA2

Characterized by their ability to alter only PE accumulation within their PBS structures, CA2-capable strains are known to control this physiological process by responding to changes in the ratio of RL to GL, although an action spectrum has not been generated for a member of this group (Tandeau de Marsac, 1977; Tandeau de Marsac *et al.*, 1980). Studies of CA2-capable strains, specifically *Synechocystis* spp., *Nostoc* spp., and *Gloeotrichia* sp. UTEX 583, have revealed minor distinctions in PBS composition and regulation compared with CA3-capable strains. In the CA2-capable strain *Synechocystis* sp. PCC 6701, PBS are hemi-discoidal in shape and their rod length varies according to the colour of light, such that in GL there is one more core-distal rod disc than in RL (Tandeau de Marsac, 1983; Williams *et al.*, 1980) (Fig. 6.2A and B). The PE–PC–AP molar ratio varies from approximately 2:2:1 in GL to the near absence of PE and a 2:1 ratio of PC:AP in RL (Williams *et al.*, 1980). *Gloeotrichia* sp. UTEX 583, originally classified as a CA3 species, appears to actually belong to the CA2 group and has no detectable second operon of PC-encoding genes (Stowe *et al.*, 2011; Tandeau de Marsac, 1977). Studies investigating the expression of *cpeBA* have shown the presence of an approximately 1450 nucleotide transcript at levels almost 20-fold higher under GL than in RL (Anderson and Grossman, 1990; Neuschaefer-Rube *et al.*, 2002). Within the *cpeBA* upstream regions of the CA2-capable strains *Synechocystis* sp. PCC 6701 and *Synechocystis* sp. BO 8402, there are N box consensus sequences similar to those found upstream of *cpeBA* in *F. diplosiphon*, which have been shown to be bound by RcaA/PepB in the latter species (Cobley *et al.*, 2002; Kehoe and Gutu, 2006; Neuschaefer-Rube *et al.*, 2002; Schmidtgoff and Federspiel, 1993; Sobczyk *et al.*, 1993). Therefore, although the involvement of the N box in light colour regulation of *cpeBA* gene expression has not yet been demonstrated in either a CA2- or CA3-capable species, its widespread presence upstream of these operons strongly suggests that it may have a role in CA regulation.

Studies examining the effects of GL and RL in the CA2-capable strain *Nostoc punctiforme* have demonstrated that PE accumulation patterns vary during CA2 and have uncovered a two-component system controlling the expression of three genes encoding components of the PBS, *cpeC*, *cpcG2*, and *cpeR1* (Hirose *et al.*, 2010; Wolf and Schussler, 2005). This regulatory system is composed of the cyanobacteriochrome sensor histidine kinase that was named CcaS (for complementary chromatic adaptation) and a response regulator named CcaR. These denominations were taken from two proteins that were originally identified in *Synechocystis* sp. PCC 6803, a species that does not produce PE and was defined as not capable of performing CA2 or CA3 (Hirose *et al.*, 2008; Tandeau de Marsac, 1977). In *Synechocystis* sp. PCC 6803, CcaS/CcaR regulated the expression of *cpcG2*, which encodes an alternative rod–core linker protein that enables energy transfer to PSI (Hirose *et al.*, 2008; Kondo *et al.*, 2005). In *N. punctiforme*, the cyanobacteriochrome CcaS, an 85 kDa protein, consists of a transmembrane helix domain in its amino-terminus, followed by a phytochrome-class GAF domain, two PAS domains, and a sensor histidine kinase domain. CcaS attaches a PCB chromophore to a cysteine (C147) residue within its single GAF domain and senses changes between GL and RL (Hirose *et al.*, 2010). CcaS autophosphorylates a histidine (H537) residue within its histidine kinase domain and is proposed to transfer this phosphoryl group in GL to CcaR, based on previous *in vitro* phosphorylation studies performed with orthologues from *Synechocystis* sp. PCC 6803 (Hirose *et al.*, 2008, 2010). Thus, upon GL activation, CcaR, a 26 kDa protein consisting of a receiver domain and an OmpR-class DNA binding domain, has been proposed to bind to a region of DNA containing a direct repeat of the sequence 5′-CTTTNCN-ATTT-3′, called the G box, located within the region of DNA upstream of the *cpeC-cpcG2-cpeR1* operon (Hirose *et al.*, 2010). Although specific binding of CcaR to the G box has not yet been demonstrated and the transcription start site of this operon is not yet known, it was proposed that CcaR binding leads to the expression of the *cpeC-cpcG2-cpeR1* operon. This appears to be similar to RcaC binding to L boxes in the CA3-capable species *F. diplosiphon*, except that the transcriptional activation and repression of PBS genes by RcaC is RL mediated and has been proposed to result from the combined effects of increased RcaC levels and increased phosphorylation of RcaC in RL (Alvey *et al.*, 2007; Hirose *et al.*, 2010; Li *et al.*, 2008). *In vivo* levels of CcaR are not known.

Deletion of both *ccaS* and *ccaR* in *N. punctiforme* affected the overall expression of the *cpeC-cpcG2-cpeR1* operon. In a *ccaS* null mutant, in both GL and RL the abundance of *cpeC-cpcG2-cpeR1* RNAs were not light regulated and were reduced to levels that were between GL and RL in wild-type cells, while in a *ccaR* null mutant, these RNAs were absent in both GL and RL (Hirose *et al.*, 2010). A similar phenotype was observed previously for the CA3-dependent Rca pathway in *F. diplosiphon*, where the deletion of *rcaE* resulted in the intermediate expression of genes encoding both PC and PE components, indicating that CcaS, like RcaE, may function as both a kinase and a phosphatase, while loss of *rcaC* led to the complete loss of expression of cpc2, suggesting that the loss of CcaR, like that of RcaC, leads to the complete loss of function of the pathway (Alvey *et al.*, 2003; Kehoe and Grossman, 1997; Seib and Kehoe, 2002). If CcaS operates in the same manner as RcaE, then the intermediate levels of *cpeC-cpcG2-cpeR1* RNAs may be the result of cross-phosphorylation of CcaR by other phosphodonors in the *ccaS* null mutant strain. *N. punctiforme* also appears to contain two copies of *cpeR*, one of which is activated by GL and one that is not. The loss of *ccaS* also led to increased expression of *cpeB-cpeA* in RL, suggesting that this operon is negatively regulated by this system (Hirose *et al.*, 2010). In *Gloeotrichia* sp. UTEX 583, which appears to belong to the CA2 group, a gene with similarity to *ccaS* has been identified, suggesting that the use of CcaS in CA2 strains may be widespread (Stowe *et al.*, 2011).

Physiology and regulation of CA4

Distinctly different from both CA2- and CA3-capable species in terms of their responses to various light colours and the resulting modifications to their PBS structures, CA4-capable cyanobacteria were more recently discovered among species found in marine environments (Palenik, 2001). The capacity to carry out CA4 has thus far been found in some strains of *Synechococcus*, which is a group of cyanobacteria in the marine environment that contribute significantly to global primary productivity (Olson *et al.*, 1990; Zwirglmaier *et al.*, 2008). Specifically, CA4 can occur in strains of *Synechococcus* that contain two forms of PE, PEI and PEII (encoded by the *mpeBA* genes) (Ong and Glazer, 1991), and are capable of producing both GL-absorbing PEB and blue light (BL)-absorbing PUB. Such strains are widespread in marine environments (Toledo *et al.*, 1999), and approximately half of such strains examined in one study were capable of undergoing CA4 (Six *et al.*, 2007), making it likely to be an important process globally for optimizing photosynthetic activity in the world's oceans. During CA4 cells respond to GL and BL, although an action spectrum for this process has not been generated, acclimating to these two light colours by altering the bilins that are contained within their PBS. In these strains, PBS structures are typically hemi-discoidal and have tricylindrical or pentacylindrical cores with six to eight outwardly extending rods composed of core-proximal rod discs of PC and core-distal rod discs of PEI and PEII (Everroad *et al.*, 2006; Six *et al.*, 2007). Unlike CA3, during CA4 there apparently are no significant changes in the protein composition within the PBS (Everroad *et al.*, 2006). PCB chromophores are attached to the PC subunits, while PEB and/or PUB chromophores are attached to the PEI and PEII subunits, depending on the light colour conditions (Fig. 6.2E and F). Sequence and biochemical analyses of *Synechococcus* spp. have indicated that PEI and PEII have five and six bilin attachment sites, respectively, and that the ratio of PUB:PEB is strain specific (Everroad *et al.*, 2006; Ong and Glazer, 1991; Shukla *et al.*, 2012; Six *et al.*, 2007).

PEII is composed of the α subunit MpeA and the β-subunit MpeB, and analyses of PEII in CA4-capable strains indicate that the bilin species attached to MpeA is CA4-dependent (Everroad *et al.*, 2006; Shukla *et al.*, 2012). Mass spectrometry analysis of PEII isolated from *Synechococcus* sp. strain M16.17 and strain M11.1 identified the ratio of PUB to PEB attached to both subunits under BL and GL growth conditions. MpeA was found to attach PUB to all three chromophore attachment site cysteine residues in BL, while the ratio of PUB:PEB attached to this protein in GL was determined to be 1:2. The chromophorylation of MpeB was not CA4 dependent, as the ratio of PUB to PEB was 1:2 under both light conditions (Everroad *et al.*, 2006). CA4-mediated bilin attachment to MpeA was proposed to occur through the actions of a bilin lyase(s) and/or lyase-isomerase(s) and several candidate proteins were identified through genome analyses of CA4 strains (Dufresne *et al.*, 2008; Six *et al.*, 2007). Three putative lyase proteins, MpeV, MpeU, and MpeZ, were identified among CA4 strains and were predicted as the most likely candidates responsible for CA4-mediated MpeA bilin attachment (Six *et al.*, 2007).

The CA4-capable strain *Synechococcus* sp. RS9916 recently has been developed as a model organism for the study of this process, particularly through the achievement of single recombination disruption of specific genes via homologous recombination (Shukla *et al.*, 2012). Analyses of *mpeZ* transcript levels have demonstrated that this gene is expressed at a 6-fold higher level in BL than GL and that it encodes a functional lyase-isomerase that is required for PUB attachment in BL to MpeA (Shukla *et al.*, 2012). *In vitro* examination of MpeZ function in *Escherichia coli* and mass spectrometry analysis of PEII subunits isolated from *Synechococcus* sp. RS9916 wild-type and *mpeZ* mutant lines demonstrated that MpeZ specifically attaches PEB to cysteine-83 of MpeA and isomerizes it into PUB, but is not responsible for PUB attachment to cysteine-140 of MpeA or cysteine-139 of CpeA in BL, supporting previous data suggesting that multiple bilin lyases and/or lyase-isomerases are used during

CA4 (Everroad *et al.*, 2006; Shukla *et al.*, 2012; Six *et al.*, 2007). In an *mpeZ* disruption mutant, cell growth was slower under low irradiance BL and the fluorescence excitation ratio of PUB to PEB ($Ex_{495\,nm}$–$Ex_{550\,nm}$) was approximately 75% lower in the mutant than in the wild-type during growth in BL but not GL (Shukla *et al.*, 2012). The authors suggested that the inordinately large decrease in the $Ex_{495\,nm}$–$Ex_{550\,nm}$ ratio in the *mpeZ* disruption mutant may implicate cysteine-83 of MpeA as a significant bilin attachment site for energy transfer within the PEII subunit during growth in BL.

Conclusions and future studies

There have been significant advances in our understanding of both the process and regulation of CA since its first characterization over 100 years ago. We have learned that rather than simply a dramatic, reversible change in the colour phenotype of a single species of cyanobacteria, CA comes in many forms, is carried out in many species, and is a globally widespread process. The ecological value of CA in providing added fitness in changing light conditions has been demonstrated (Stomp *et al.*, 2008), and there is now strong evidence that different forms of CA have evolved independently in different cyanobacterial species. A relatively slow form of environmental acclimation compared to many photosynthetic processes, CA typically takes approximately a week for complete shifts in PBS composition from one state to the other. It is likely to be useful both under conditions where light colours are relatively pure as well as in mixed light colour conditions, since all available data show that the CA response functions as a rheostat rather than a switch between two isolated states.

Yet much remains to be learned about the various forms of CA, and given the tremendous range of light environments faced by cyanobacteria and the billions of years that these organisms have had to adapt to their niches, additional types of CA are very likely to be discovered, as has already been suggested by recent studies of the marine cyanobacterium *Acaryochloris marina* (Duxbury *et al.*, 2009; Gloag *et al.*, 2007). Future work will focus not only on further defining established and novel CA systems and their regulation, but will also seek to define the ubiquity of these processes and their regulatory systems across cyanobacterial groups and, ultimately, to understand the relationships between variations in CA processes, regulatory systems and the environmental conditions that have contributed to their creation.

Acknowledgements

We would like to dedicate this review to the many scientists whose outstanding research in the field of chromatic acclimation has led to our current understanding of this subject. We wish to thank the NSF, USDA, NIH and Indiana University for the financial support that has allowed us to contribute to the progress that has been made in unravelling the regulation of CA. This work was supported by the NSF under Grant No. MCB-1029414 to D.M.K.

References

Adir, N. (2005). Elucidation of the molecular structures of components of the phycobilisome: reconstructing a giant. Photosynth. Res. *85*, 15–32.

Alvey, R.M., Karty, J.A., Roos, E., Reilly, J.P., and Kehoe, D.M. (2003). Lesions in phycoerythrin chromophore biosynthesis in *Fremyella diplosiphon* reveal coordinated light regulation of apoprotein and pigment biosynthetic enzyme gene expression. Plant Cell *15*, 2448–2463.

Alvey, R.M., Bezy, R.P., Frankenberg-Dinkel, N., and Kehoe, D.M. (2007). A light regulated OmpR-class promoter element co-ordinates light-harvesting protein and chromophore biosynthetic enzyme gene expression. Mol. Microbiol. *64*, 319–332.

Anderson, L.K., and Grossman, A.R. (1990). Structure and light-regulated expression of phycoerythrin genes in wild-type and phycobilisome assembly mutants of *Synechocystis* sp. strain PCC 6701. J. Bacteriol. *172*, 1297–1305.

Beguin, S., Guglielmi, G., Rippka, R., and Cohen-Bazire, G. (1985). Chromatic adaptation in a mutant of *Fremyella diplosiphon* incapable of phycoerythrin synthesis. Biochimie *67*, 109–117.

Bekker, A., Holland, H.D., Wang, P.L., Rumble, D., Stein, H.J., Hannah, J.L., Coetzee, L.L., and Beukes, N.J. (2004). Dating the rise of atmospheric oxygen. Nature *427*, 117–120.

Bennett, A., and Bogorad, L. (1973). Complementary chromatic adaptation in a filamentous blue-green alga. J. Cell Biol. *58*, 419–435.

Bezy, R.P., and Kehoe, D.M. (2010). Functional characterization of a cyanobacterial OmpR/PhoB class transcription factor binding site controlling light color responses. J. Bacteriol. *192*, 5923–5933.

Bezy, R.P., Wiltbank, L., and Kehoe, D.M. (2011). Light-dependent attenuation of phycoerythrin gene expression reveals convergent evolution of green light sensing in cyanobacteria. Proc. Natl. Acad. Sci. U.S.A. *108*, 18542–18547.

Bordowitz, J.R., and Montgomery, B.L. (2008). Photoregulation of cellular morphology during complementary chromatic adaptation requires sensor-kinase-class protein RcaE in *Fremyella diplosiphon*. J. Bacteriol. *190*, 4069–4074.

Boresch, K. (1922). Die komplementiire chromatische adaptation. Arch. Protistenkd. *44*, 1–70.

Bruns, B.U., Briggs, W.R., and Grossman, A.R. (1989). Molecular characterization of phycobilisome regulatory mutants of *Fremyella diplosiphon*. J. Bacteriol. *171*, 901–908.

Bryant, D.A. (1981). The photoregulated expression of multiple phycocyanin species – A general mechanism for the control of phycocyanin synthesis in chromatically adapting cyanobacteria. Eur. J. Biochem. *119*, 425–429.

Bryant, D.A., Guglielmi, G., Tandeaudemarsac, N., Castets, A.M., and Cohen-Bazire, G. (1979). Structure of cyanobacterial phycobilisomes – Model. Arch. Microbiol. *123*, 113–127.

Butler, W.L., Norris, K.H., Siegelman, H.W., and Hendricks, S.B. (1959). Detection, assay, and preliminary purification of the pigment controlling photoresponsive development of plants. Proc. Natl. Acad. Sci. U.S.A. *45*, 1703–1708.

Campbell, D., Houmard, J., and Demarsac, N.T. (1993). Electron-transport regulates cellular-differentiation in the filamentous cyanobacterium *Calothrix*. Plant Cell *5*, 451–463.

Casey, E.S., and Grossman, A. (1994). *In vivo* and *in vitro* characterization of the light-regulated *cpcB2A2* promoter of *Fremyella diplosiphon*. J. Bacteriol. *176*, 6362–6374.

Chiang, G.G., Schaefer, M.R., and Grossman, A.R. (1992). Complementation of a red-light-indifferent cyanobacterial mutant. Proc. Natl. Acad. Sci. U.S.A. *89*, 9415–9419.

Cobley, J.G., and Miranda, R.D. (1983). Mutations affecting chromatic adaptation in the cyanobacterium *Fremyella diplosiphon*. J. Bacteriol. *153*, 1486–1492.

Cobley, J.G., Zerweck, E., Reyes, R., Mody, A., Seludo-Unson, J.R., Jaeger, H., Weerasuriya, S., and Navankasattusas, S. (1993). Construction of shuttle plasmids which can be efficiently mobilized from *Escherichia coli* into the chromatically adapting cyanobacterium, *Fremyella diplosiphon*. Plasmid *30*, 90–105.

Cobley, J.G., Seneviratne, L., Drong, L., Thounaojam, M., Oda, J.F., and Carroll, J. (1999). Transposition of Tn5 derivatives in the chromatically adapting cyanobacterium, *Fremyella diplosiphon*. In The Phototrophic Prokaryotes, Peschek, G.A., Loffelhardt, W., and Schmetterer, G., eds. (New-York, Kluwer Academic/Plenum Publisher), pp. 443–451.

Cobley, J.G., Clark, A.C., Weerasurya, S., Queseda, F.A., Xiao, J.Y., Bandrapali, N., D'Silva, I., Thounaojam, M., Oda, J.F., Sumiyoshi, T., *et al.* (2002). CpeR is an activator required for expression of the phycoerythrin operon (*cpeBA*) in the cyanobacterium *Fremyella diplosiphon* and is encoded in the phycoerythrin linker-polypeptide operon (*cpeCDESTR*). Mol. Microbiol. *44*, 1517–1531.

Conley, P.B., Lemaux, P.G., and Grossman, A.R. (1985). Cyanobacterial light-harvesting complex subunits encoded in two red-light induced transcripts. Science *230*, 550–553.

Conley, P.B., Lemaux, P.G., Lomax, T.L., and Grossman, A.R. (1986). Genes encoding major light-harvesting polypeptides are clustered on the genome of the cyanobacterium *Fremyella diplosiphon*. Proc. Natl. Acad. Sci. U.S.A. *83*, 3924–3928.

Conley, P.B., Lemaux, P.G., and Grossman, A. (1988). Molecular characterization and evolution of sequences encoding light-harvesting components in the chromatically adapting cyanobacterium *Fremyella diplosiphon*. J. Mol. Biol. *199*, 447–465.

Csiszar, K., Houmard, J., Damerval, T., and Demarsac, N.T. (1987). Transcriptional analysis of the cyanobacterial *gvpABC* operon in differentiated cells – occurrence of an antisense RNA complementary to three overlapping transcripts. Gene *60*, 29–37.

Damerval, T., Houmard, J., Guglielmi, G., Csiszar, K., and Demarsac, N.T. (1987). A developmentally regulated *gvpABC* operon is involved in the formation of gas vesicles in the cyanobacterium *Calothrix* 7601. Gene *54*, 83–92.

Damerval, T., Guglielmi, G., Houmard, J., and Marsac, N.T.D. (1991). Hormogonium differentiation in the cyanobacterium *Calothrix*: a photoregulated developmental process. Plant Cell 3, 191–201.

Diakoff, S., and Scheibe, J. (1973). Action spectra for chromatic adaptation in *Tolypothrix tenuis*. Plant Physiol. *51*, 382–385.

Diakoff, S., and Scheibe, J. (1975). Cultivation in dark of blue-green-alga *Fremyella diplosiphon* – photoreversible effect of green and red light on growth rate. Physiol. Plant. *34*, 125–128.

Drews, G. (2011). The evolution of cyanobacteria and photosynthesis. In Bioenergetic Processes Of Cyanobacteria: From Evolutionary Singularity to Ecological Diversity, Peschek, G.A., Obinger, C., and Renger, G., eds. (Springer-Verlag, Berlin), pp. 265–284.

Ducret, A., Muller, S.A., Goldie, K.N., Hefti, A., Sidler, W.A., Zuber, H., and Engel, A. (1998). Reconstitution, characterisation and mass analysis of the pentacylindrical allophycocyanin core complex from the cyanobacterium *Anabaena* sp. PCC 7120. J. Mol. Biol. *278*, 369–388.

Dufresne, A., Ostrowski, M., Scanlan, D., Garczarek, L., Mazard, S., Palenik, B., Paulsen, I., de Marsac, N., Wincker, P., Dossat, C., *et al.* (2008). Unraveling the genomic mosaic of a ubiquitous genus of marine cyanobacteria. Genome Biol. *9*, R90.

Duxbury, Z., Schliep, M., Ritchie, R.J., Larkum, A.W.D., and Chen, M. (2009). Chromatic photoacclimation extends utilisable photosynthetically active radiation in the chlorophyll *d*-containing cyanobacterium, *Acaryochloris marina*. Photosynth. Res. *101*, 69–75.

Engelmann, T.W. (1902). Untersuchungen uber die qualitativen Beziehungen zwieschen Absorbtion des Lichtes und Assimilation in Pflanzenzellen.

I. Das Mikrospectraphotometer, ein Apparat zur qualitativen Mikrospectralanalyse. II. Experimentelle Grundlangen zur Ermittelung der quantitativen Beziehungen zwieschen Assimilationsenergie und Absorptiongrosse. III. Bestimmung der Vertheilung der Energie im Spectrum von Sonnenlicht mittels Bacterien-Methode und quantitativen Mikrospectralanalyse. Bot. Z. *42*, 81–105.

Everroad, C., Six, C., Partensky, F., Thomas, J.-C., Holtzendorff, J., and Wood, A.M. (2006). Biochemical bases of Type IV chromatic adaptation in marine *Synechococcus* spp. J. Bacteriol. *188*, 3345–3356.

Federspiel, N.A., and Grossman, A.R. (1990). Characterization of the light-regulated operon encoding the phycoerythrin-associated linker proteins from the cyanobacterium *Fremyella diplosiphon*. J. Bacteriol. *172*, 4072–4081.

Federspiel, N.A., and Scott, L. (1992). Characterization of a light-regulated gene encoding a new phycoerythrin-associated linker protein from the cyanobacterium *Fremyella diplosiphon*. J. Bacteriol. *174*, 5994–5998.

Frank, G., Sidler, W., Widmer, H., and Zuber, H. (1978). Complete amino-acid sequence of both subunits of C-Phycocyanin from cyanobacterium *Mastigocladus laminosus*. Hoppe-Seylers Zeit. Physiol. Chem. *359*, 1491–1507.

Frankenberg, N., Mukougawa, K., Kohchi, T., and Lagarias, J.C. (2001). Functional genomic analysis of the HY2 family of ferredoxin-dependent bilin reductases from oxygenic photosynthetic organisms. Plant Cell *13*, 965–978.

Fujita, Y., and Hattori, A. (1960). Effect of chromatic lights on phycobilin formation in a blue-green alga, *Tolypothrix tenuis*. Plant Cell Physiol. *1*, 293–303.

Gaiducov, N. (1902). Uber den Einfluss farbigen Lichtes auf die Farbung lebender Oscillatorien. Abh. Preuss. Akad. Wiss. *V*, 1–36.

Gaidukov, N. (1903). Die Farbervanderung bei den Prozessen der Komplementaren Chromatischen Adaptation. Ber. Deutsch. Bot. Ges. *21*, 517–522.

Gao, X., Wei, T.-D., Zhang, N., Xie, B.-B., Su, H.-N., Zhang, X.-Y., Chen, X.-L., Zhou, B.-C., Wang, Z.-X., Wu, J.-W., *et al.* (2012). Molecular insights into the terminal energy acceptor in cyanobacterial phycobilisome. Mol. Microbiol. *85*, 907–915.

Gendel, S., Ohad, I., and Bogorad, L. (1979). Control of phycoerythrin synthesis during chromatic adaptation. Plant Physiol. *64*, 786–790.

Glazer, A.N., Williams, R.C., Yamanaka, G., and Schachman, H.K. (1979). Characterization of cyanobacterial phycobilisomes in zwitterionic detergents. Proc. Natl. Acad. Sci. U.S.A. *76*, 6162–6166.

Gloag, R.S., Ritchie, R.J., Chen, M., Larkum, A.W.D., and Quinnell, R.G. (2007). Chromatic photoacclimation, photosynthetic electron transport and oxygen evolution in the chlorophyll *d*-containing oxyphotobacterium *Acaryochloris marina*. BBA-Bioenergetics *1767*, 127–135.

Grossman, A.R., Schaefer, M.R., Chiang, G.G., and Collier, J.L. (1993). The phycobilisome, a light-harvesting complex responsive to environmental conditions. Microbiol. Mol. Biol. Rev. *57*, 725–749.

Gutu, A., and Kehoe, D.M. (2012). Emerging perspectives on the mechanisms, regulation, and distribution of light color acclimation in cyanobacteria. Mol. Plant *5*, 1–13.

Haury, J.F., and Bogorad, L. (1977). Action spectra for phycobiliprotein synthesis in a chromatically adapting cyanophyte, *Fremyella diplosiphon*. Plant Physiol. *60*, 835–839.

Hirose, Y., Shimada, T., Narikawa, R., Katayama, M., and Ikeuchi, M. (2008). Cyanobacteriochrome CcaS is the green light receptor that induces the expression of phycobilisome linker protein. Proc Nat Acad Sci U.S.A. *105*, 9528–9533.

Hirose, Y., Narikawa, R., Katayama, M., and Ikeuchi, M. (2010). Cyanobacteriochrome CcaS regulates phycoerythrin accumulation in *Nostoc punctiforme*, a group II chromatic adapter. Proc. Natl. Acad. Sci. U.S.A. *107*, 8854–8859.

Hirose, Y., Rockwell, N.C., Nishiyama, K., Narikawa, R., Ukaji, Y., Inomata, K., Lagarias, J.C., and Ikeuchi, M. (2013). Green/red cyanobacteriochromes regulate complementary chromatic acclimation via a protochromic photocycle. Proc. Natl. Acad. Sci. U.S.A. *110*, 4974–4979.

Kahn, K., Mazel, D., Houmard, J., Tandeau de Marsac, N., and Schaefer, M.R. (1997). A role for *cpeYZ* in cyanobacterial phycoerythrin biosynthesis. J. Bacteriol. *179*, 998–1006.

Kawamura, M., Mimuro, M., and Fujita, Y. (1979). Quantitative relationship between two reaction centers in the photosynthetic system of blue-green-algae. Plant Cell Physiol. *20*, 697–705.

Kehoe, D.M., and Grossman, A.R. (1996). Similarity of a chromatic adaptation sensor to phytochrome and ethylene receptors. Science *273*, 1409–1412.

Kehoe, D.M., and Grossman, A.R. (1997). New classes of mutants in complementary chromatic adaptation provide evidence for a novel four-step phosphorelay system. J. Bacteriol. *179*, 3914–3921.

Kehoe, D.M., and Gutu, A. (2006). Responding to color: the regulation of complementary chromatic adaptation. Annu. Rev. Plant Biol. *57*, 127–150.

Kondo, K., Geng, X.X., Katayama, M., and Ikeuchi, M. (2005). Distinct roles of CpcG1 and CpcG2 in phycobilisome assembly in the cyanobacterium *Synechocystis* sp PCC 6803. Photosynth. Res. *84*, 269–273.

Li, L., and Kehoe, D.M. (2005). *In vivo* analysis of the roles of conserved aspartate and histidine residues within a complex response regulator. Mol. Microbiol. *55*, 1538–1552.

Li, L., and Kehoe, D.M. (2008). Abundance changes of the response regulator RcaC require specific aspartate and histidine residues and are necessary for normal light color responsiveness. J. Bacteriol. *190*, 7241–7250.

Li, L., Alvey, R.M., Bezy, R.P., and Kehoe, D.M. (2008). Inverse transcriptional activities during complementary chromatic adaptation are controlled by the response regulator RcaC binding to red and green light-responsive promoters. Mol. Microbiol. *68*, 286–297.

Lomax, T.L., Conley, P.B., Schilling, J., and Grossman, A.R. (1987). Isolation and characterization of light-regulated phycobilisome linker polypeptide genes and their transcription as a polycistronic messenger-RNA. J. Bacteriol. *169*, 2675–2684.

Lundell, D.J., Yamanaka, G., and Glazer, A.N. (1981). A terminal energy acceptor of the phycobilisome – the 75,000-dalton polypeptide of *Synechococcus* 6301 phycobilisomes – a new biliprotein. J. Cell Biol. *91*, 315–319.

MacColl, R. (1998). Cyanobacterial Phycobilisomes. J. Struct. Biol. *124*, 311–334.

Mazel, D., Guglielmi, G., Houmard, J., Sidler, W., Bryant, D.A., and Demarsac, N.T. (1986). Green light induces transcription of the phycoerythrin operon in the cyanobacterium *Calothrix* 7601. Nucleic Acids Res. *14*, 8279–8290.

Neuschaefer-Rube, O., Boger, P., and Ernst, A. (2002). Interference of an *apcA* insertion with complementary chromatic adaptation in the diazotrophic *Synechocystis sp* strain BO 8402. Biochim. Biophys. Acta-Bioenerg. *1553*, 279–295.

Noubir, S., Luque, I., de Alda, J., Perewoska, I., de Marsac, N.T., Cobley, J.G., and Houmard, J. (2002). Co-ordinated expression of phycobiliprotein operons in the chromatically adapting cyanobacterium *Calothrix* PCC 7601: a role for RcaD and RcaG. Mol. Microbiol. *43*, 749–762.

Oelmuller, R., Conley, P.B., Federspiel, N., Briggs, W.R., and Grossman, A.R. (1988a). Changes in accumulation and synthesis of transcripts encoding phycobilisome components during acclimation of *Fremyella diplosiphon* to different light qualities. Plant Physiol. *88*, 1077–1083.

Oelmuller, R., Grossman, A.R., and Briggs, W.R. (1988b). Photoreversibility of the effect of red and green light-pulses on the accumulation in darkness of messenger RNAs coding for phycocyanin and phycoerythrin in *Fremyella diplosiphon*. Plant Physiol. *88*, 1084–1091.

Oelmuller, R., Grossman, A.R., and Briggs, W.R. (1989). Role of protein-synthesis in regulation of phycobiliprotein messenger RNA abundance by light quality in *Fremyella diplosiphon*. Plant Physiol. *90*, 1486–1491.

Ohki, K., Watanabe, M., and Fujita, Y. (1982). Action of near UV and blue-light on the photocontrol of phycobiliprotein formation – A complementary chromatic adaptation. Plant Cell Physiol. *23*, 651–656.

Olson, R.J., Chisholm, S.W., Zettler, E.R., and Armbrust, E.V. (1990). Pigments, size, and distribution of *Synechococcus* in the North Atlantic and Pacific Oceans. Limnol. Oceanogr. *35*, 45–58.

Ong, L.J., and Glazer, A.N. (1991). Phycoerythrins of marine unicellular cyanobacteria. 1. Bilin types and locations and energy-transfer pathways in *Synechococcus* spp phycoerythrins. J. Biol. Chem. *266*, 9515–9527.

Palenik, B. (2001). Chromatic adaptation in marine *Synechococcus* strains. Appl. Environ. Microbiol. *67*, 991–994.

Pattanaik, B., and Montgomery, B.L. (2010). FdTonB is involved in the photoregulation of cellular morphology during complementary chromatic adaptation in *Fremyella diplosiphon*. Microbiology *156*, 731–741.

Perez-Gomez, B., Mendoza-Hernandez, G., Cabellos-Avelar, T., Leyva-Castillo, L.E., Gutierrez-Cirlos, E.B., and Gomez-Lojero, C. (2012). A proteomic approach to the analysis of the components of the phycobilisomes from two cyanobacteria with complementary chromatic adaptation: *Fremyella diplosiphon* UTEX B590 and *Tolypothrix* PCC 7601. Photosynth. Res. *114*, 43–58.

Rippka, R., Deruelles, J., Waterbury, J.B., Herdman, M., and Stanier, R.Y. (1979). Generic assignments, strain histories and properties of pure cultures of cyanobacteria. J. Gen. Microbiol. *111*, 1–61.

Rosinski, J., Hainfeld, J.F., Rigbi, M., and Siegelman, H.W. (1981). Phycobilisome ultrastructure and chromatic adaptation in *Fremyella diplosiphon*. Ann. Bot. *47*, 1–12.

Schmidtgoff, C.M., and Federspiel, N.A. (1993). *In vivo* and *in vitro* footprinting of a light-regulated promoter in the cyanobacterium *Fremyella diplosiphon*. J. Bacteriol. *175*, 1806–1813.

Seib, L.O., and Kehoe, D.M. (2002). A turquoise mutant genetically separates expression of genes encoding phycoerythrin and its associated linker peptides. J. Bacteriol. *184*, 962–970.

Shui, J., Saunders, E., Needleman, R., Nappi, M., Cooper, J., Hall, L., Kehoe, D., and Stowe-Evans, E. (2009). Light-dependent and light-independent protochlorophyllide oxidoreductases in the chromatically adapting cyanobacterium *Fremyella diplosiphon* UTEX 481. Plant Cell Physiol. *50*, 1507–1521.

Shukla, A., Biswas, A., Blot, N., Partensky, F., Karty, J.A., Hammad, L.A., Garczarek, L., Gutu, A., Schluchter, W.M., and Kehoe, D.M. (2012). Phycoerythrin-specific bilin lyase–isomerase controls blue-green chromatic acclimation in marine *Synechococcus*. Proc. Natl. Acad. Sci. U.S.A. *109*, 20136–20141.

Sidler, W.A. (2004). Phycobilisome and Phycobiliprotein Structures. In The Molecular Biology of Cyanobacteria, Bryant, D.A., ed. (Springer, Netherlands), pp. 139–216.

Sidler, W., Gysi, J., Isker, E., and Zuber, H. (1981). The complete amino-acid-sequence of both subunits of allophycocyanin, a light harvesting protein–pigment complex from the cyanobacterium *Mastigocladus laminosus*. Hoppe-Seylers Zeit. Physiol. Chem. *362*, 611–628.

Sidler, W., Kumpf, B., Rudiger, W., and Zuber, H. (1986). The complete amino-acid-sequence of C-Phycoerythrin from the cyanobacterium *Fremyella diplosiphon*. Biol. Chem. Hoppe-Seyler *367*, 627–642.

Six, C., Thomas, J.-C., Garczarek, L., Ostrowski, M., Dufresne, A., Blot, N., Scanlan, D., and Partensky, F. (2007). Diversity and evolution of phycobilisomes in marine *Synechococcus* spp.: a comparative genomics study. Genome Biol. *8*, R259.

Sobczyk, A., Schyns, G., Tandeau de Marsac, N., and Houmard, J. (1993). Transduction of the light signal during complementary chromatic adaptation in the cyanobacterium *Calothrix sp* pcc-7601 – DNA-binding proteins and modulation by phosphorylation. EMBO J. *12*, 997–1004.

Sobczyk, A., Bely, A., Tandeau de Marsac, N., and Houmard, J. (1994). A phosphorylated DNA-binding protein is specific for the red-light signal during complementary chromatic adaptation in cyanobacteria. Mol. Microbiol. *13*, 875–885.

Stanier, R.Y., and Bazire, G.C. (1977). Phototrophic prokaryotes: the cyanobacteria. Annu. Rev. Microbiol. *31*, 225–274.

Stomp, M., van Dijk, M.A., van Overzee, H.M., Wortel, M.T., Sigon, C.A., Egas, M., Hoogveld, H., Gons, H.J., and Huisman, J. (2008). The timescale of phenotypic plasticity and its impact on competition in fluctuating environments. Am. Nat. *172*, 169–185.

Stowe-Evans, E.L., and Kehoe, D.M. (2004). Signal transduction during light-quality acclimation in cyanobacteria: a model system for understanding phytochrome-response pathways in prokaryotes. Photochem. Photobiol. Sci. *3*, 495–502.

Stowe-Evans, E.L., Ford, J., and Kehoe, D.M. (2004). Genomic DNA microarray analysis: Identification of new genes regulated by light color in the cyanobacterium *Fremyella diplosiphon*. J. Bacteriol. *186*, 4338–4349.

Stowe, W.C., Brodie-Kommit, J., and Stowe-Evans, E. (2011). Characterization of complementary chromatic adaptation in *Gloeotrichia* UTEX 583 and identification of a transposon-like insertion in the *cpeBA* operon. Plant Cell Physiol. *52*, 553–562.

Tandeau de Marsac, N. (1977). Occurrence and nature of chromatic adaptation in cyanobacteria. J. Bacteriol. *130*, 82–91.

Tandeau de Marsac, N. (1983). Phycobilisomes and complementary chromatic adaptation in cyanobacteria. Bull. Inst. Pasteur *81*, 201–254.

Tandeau de Marsac, N., and Cohen-Bazire, G. (1977). Molecular composition of cyanobacterial phycobilisomes. Proc. Natl. Acad. Sci. U.S.A. *74*, 1635–1639.

Tandeau de Marsac, N., Castets, A.M., and Cohen-Bazire, G. (1980). Wavelength modulation of phycoerythrin synthesis in *Synechocystis* sp. 6701. J. Bacteriol. *142*, 310–314.

Terauchi, K., Montgomery, B.L., Grossman, A.R., Lagarias, J.C., and Kehoe, D.M. (2004). RcaE is a complementary chromatic adaptation photoreceptor required for green and red light responsiveness. Mol. Microbiol. *51*, 567–577.

Toledo, G., Palenik, B., and Brahamsha, B. (1999). Swimming marine *Synechococcus* strains with widely different photosynthetic pigment ratios form a monophyletic group. Appl. Environ. Microb. *65*, 5247–5251.

Vogelmann, T.C., and Scheibe, J. (1978). Action spectra for chromatic adaptation in blue-green-alga *Fremyella diplosiphon*. Planta *143*, 233–239.

Wehrmeyer, W. (1990). Phycobilisomes: structures and function. In Experimental Phycology: Cell Walls and Surfaces, Reproduction, Photosynthesis, Wissner, W., Robinson, D.G., and Star, R.C., eds. (Springer Verlag, Berlin), pp. 158–172.

Williams, R., Gingrich, J., and Glazer, A. (1980). Cyanobacterial phycobilisomes. Particles from *Synechocystis* 6701 and two pigment mutants. J. Cell Biol. *85*, 558–566.

Wolf, E., and Schussler, A. (2005). Phycobiliprotein fluorescence of *Nostoc punctiforme* changes during the life cycle and chromatic adaptation: characterization by spectral confocal laser scanning microscopy and spectral unmixing. Plant Cell Environ. *28*, 480–491.

Zwirglmaier, K., Jardillier, L., Ostrowski, M., Mazard, S., Garczarek, L., Vaulot, D., Not, F., Massana, R., Ulloa, O., and Scanlan, D.J. (2008). Global phylogeography of marine *Synechococcus* and *Prochlorococcus* reveals a distinct partitioning of lineages among oceanic biomes. Environ. Microbiol. *10*, 147–161.

The Carboxysome: Function, Structure and Cellular Dynamics

7

Jeffrey C. Cameron, Markus Sutter and Cheryl A. Kerfeld

Abstract

The ability to use light energy for the accumulation and fixation of CO_2 has given cyanobacteria the ability to thrive in diverse and extreme environments. Cyanobacteria play a central role in the global carbon cycle and have changed the earth's atmosphere by generating oxygen and depleting CO_2. The CO_2 concentrating mechanism (CCM) consists of active transport systems for inorganic carbon acquisition and a distinctive protein-based organelle, the carboxysome. Recent advances in structural and systems biology and biological imaging have built upon decades of biochemical and genetic research to advance our understanding of the carboxysome. In this chapter we provide an overview of the carboxysome structure and place the carboxysome in the context of cyanobacterial metabolism and morphology.

Introduction

Carboxysomes are bacterial micro-compartments composed entirely of protein. They function in the cyanobacterial CO_2 concentrating mechanism (CCM) by encapsulating the key enzyme of carbon fixation, ribulose 1,5-bisphosphate carboxylase/oxygenase (RuBisCO), and carbonic anhydrase (CA) within a protein shell (Fig. 7.1). Co-localization of these enzymes enables cyanobacteria to generate high levels of CO_2, a substrate of RuBisCO, at the site of carbon fixation. This is advantageous because it drastically decreases the likelihood that RuBisCO will fix O_2 instead of CO_2, which generates a toxic by-product, 2-phosphoglycolate (2-PG) that has to be scavenged at the cost of cellular energy. In addition to encapsulating RuBisCO and CA, the shell also provides a barrier that prevents CO_2 from leaking into the cytosol (Cai *et al.*, 2009).

The existence of polyhedral bodies within cyanobacteria has been known for over 50 years (Jensen and Bowen, 1961). Recently the tools of synthetic biology have provided researchers the ability to probe the function of the carboxysome *in vivo* and in heterologous systems. This has enabled researchers to visualize carboxysomes in real-time within cells, revealing spatial patterns of organization and interaction with other cellular components. Structural and biochemical studies have determined the molecular arrangement of carboxysomal components, providing insight into the structure–function relationship of this micro-compartment. At the same time, new systems-wide approaches have given us a global view into the cellular physiology and function of cyanobacterial cells, offering insight into how multiple cellular pathways are interrelated and optimized for cellular productivity. This chapter foregrounds these recent findings in carboxysome research.

Carboxysome function

The CO_2 concentrating mechanism

The CO_2 concentrating mechanism (CCM) is a multicomponent system comprised of the carboxysome and a suite of inorganic carbon transporters that collectively enable cyanobacteria to proliferate under relatively low ambient atmospheric CO_2 (0.039%) and high O_2 (21%) levels (Fig. 7.1). Ultimately, the CCM

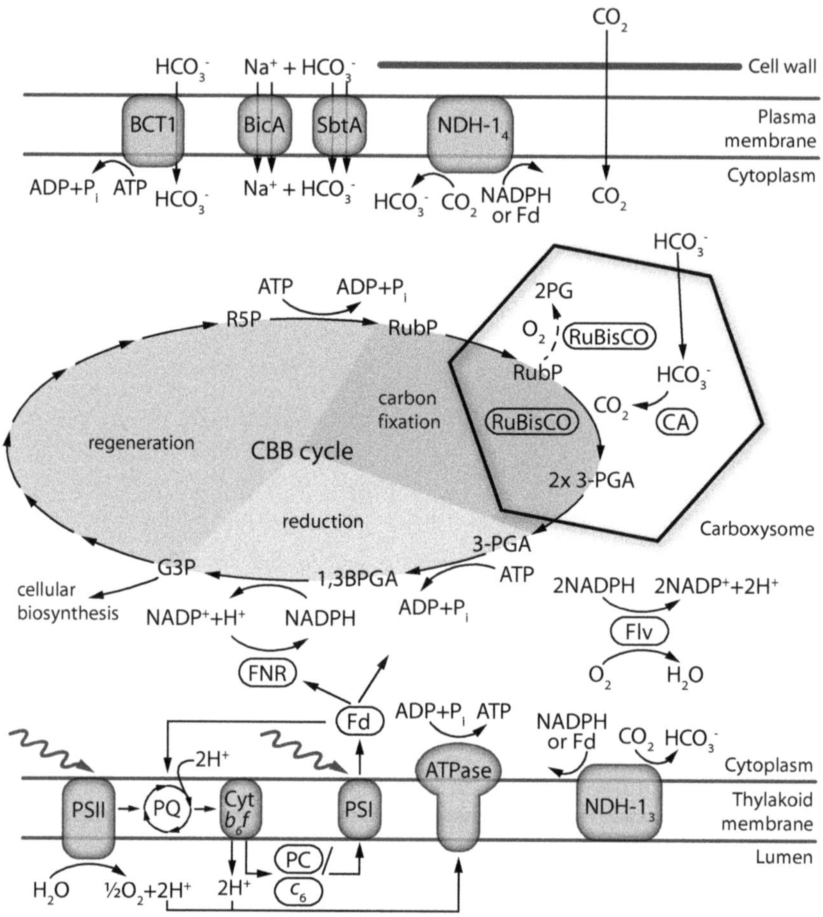

Figure 7.1 Schematic diagram showing the interrelationship of the CCM, Calvin-Benson-Bassham cycle, and the light reactions of photosynthesis. The carboxysome contains ribulose 1,5-bisphosphate carboxylase/oxygenase (RuBisCO) and carbonic anhydrase (CA). Carboxylation of RuBP generates 3-PGA, while oxygenation of RuBP results in the formation of 2-PG, a toxic metabolite that must be metabolized through photorespiration. Photosynthetically derived ATP and NADPH are used to energize the uptake of CO_2 and the Calvin-Benson-Bassham (CBB) cycle. Cyclic electron transport around photosystem I (PSI) results in the production of ATP but not NADPH. Proteins: ATPase, ATP synthase; BCT1, high-affinity bicarbonate transporter; BicA, low-affinity bicarbonate transporter; Cyt b_6f, cytochrome b_6f complex; c_6, cytochrome c_6; Fd, ferredoxin; Flv, flavodiiron protein; NDH-I_3, constitutive CO_2 uptake system; NDH-I_4, inducible CO_2 uptake system; PC, plastocyanin; PSI, photosystem I; PSII, photosystem II; SbtA, high-affinity bicarbonate transporter. Small Molecules: ATP, Adenosine-5′-triphosphate; 1,3BPGA, 1,3-bisphosphoglycerate; G3P, glyceraldehyde 3-phosphate; NADPH, reduced form of nicotinamide adenine dinucleotide phosphate; 2PG, 2-phosphoglycolate; 3PGA, 3-phosphoglycerate; PQ, plastoquinone; RubP, ribulose 1,5-bisphosphate; R5P, ribulose 5-phosphate.

results in the accumulation of CO_2 at the site of RuBisCO within the carboxysome. This leads to increased levels of carboxylation of ribulose 1,5-bisphosphate (RuBP) and formation of 3-phosphoglycerate (3-PGA). Furthermore, by saturating RuBisCO with CO_2, the CCM also functions to limit the competing oxygenation reaction that results in the production of 2-PG, a toxic metabolite. The overall effects of the CCM are achieved through the concerted functioning of multiple cellular components (Fig. 7.1).

In order to deliver sufficient substrate to the enzymes within the carboxysome, inorganic carbon (C_i) must be accumulated from the environment. The equilibrium between CO_2 and HCO_3^- ($pK_a = 6.3$ at 25°C) in solution is pH dependent, thus HCO_3^- is the predominant species under typical growth conditions (pH 8). The ionic nature of HCO_3^- results in poor diffusion across the cellular membrane. Multiple active transport systems have evolved, enabling cyanobacteria to accumulate intracellular HCO_3^- concentrations several orders of magnitude greater than the extracellular environment (Price *et al.*, 2008). In *Synechocystis* sp. PCC6803 (*Synechocystis* PCC6803) and *Synechococcus elongatus* PCC7942 (*Synechococcus* PCC7942), HCO_3^- uptake is facilitated by at least two separate transport systems located on the plasma membrane (Fig. 7.1). BCT1 is a high-affinity, low-C_i-inducible, ATP-dependent transport system encoded by the *cmpABCD* genes (Omata *et al.*, 1999). SbtA is a high-affinity, low-C_i inducible, Na^{2+}-dependent HCO_3^- transporter (Shibata *et al.*, 2002). BicA is a low-affinity, constitutively expressed, Na^{2+}-dependent HCO_3^- transporter found in *Synechocystis* PCC6803 but not *Synechococcus* PCC7942 (Price *et al.*, 2004). CO_2 is able to enter the cell by diffusing freely across the plasma membrane. However, this permeability makes it difficult to accumulate CO_2 within the cell. To prevent the loss of CO_2 from the cell and to maintain an inward gradient of CO_2 across the cellular envelope, cyanobacteria contain two NADPH-dependent systems that catalyse the hydration of CO_2 to HCO_3^-, NDH-I_4 and NDH-I_3 (Maeda *et al.*, 2002; Ohkawa *et al.*, 2000a,b; Shibata *et al.*, 2001). In *Synechocystis* PCC6803, mutants lacking four of the uptake systems (Δ*ndhD3*, Δ*ndhD4*, Δ*sbtA*, Δ*cmpA*) were able to survive in air, albeit growing more slowly than wild-type, due to the presence of the low affinity, high flux transporter, BicA; disruption of the *bicA* gene in this background results in a high-CO_2 requiring phenotype (HCR) (Xu *et al.*, 2008).

The products of the photosynthetic light reactions are required for the CCM and the CBB (Calvin–Benson–Bassham) cycle (Fig. 7.1). Electrons derived from the photolysis of water by photosystem II are used to generate the soluble reducing equivalent, NADPH, which is used during the reductive phase of the CBB cycle and for the conversion of CO_2 into HCO_3^- in the cytoplasm. The light-driven oxidation of H_2O and proton pumping across the membrane during reduction/oxidation of the quinone pool also generates a proton motive force that drives the formation of ATP. ATP is subsequently hydrolysed by the transporters of the CCM for the movement of HCO_3^- across the membrane, by enzymes of the CBB cycle, and potentially in the activation of RuBisCO. Because photosynthesis generates ATP and NADPH at a ratio of 9:7, but the CBB cycle consumes ATP and NADPH at a ratio of 9:6, multiple pathways have evolved to balance the ATP:NADPH ratio for optimal cellular productivity (Alric *et al.*, 2010; Nogales *et al.*, 2012). For instance, cyclic electron flow around photosystem I, in addition to Mehler reactions, produce ATP but not NADPH. Furthermore, polyphosphate bodies could also play a role in phosphate storage and ATP production (Nocek *et al.*, 2008), though the role of polyphosphate in cyanobacterial metabolism is not fully understood. In addition, several flavodiiron (Flv) proteins have been identified in cyanobacteria that are involved in NADPH-dependent reduction of O_2 to H_2O (Helman *et al.*, 2003) (Fig. 7.1; see also Chapter 4). In *Synechococcus* sp. PCC7002 (*Synechococcus* PCC7002), *flv1* and *flv2* were among the most responsive genes to low-CO_2 conditions in addition to components of the CCM (Ludwig and Bryant, 2012), whereas in *Synechocystis* PCC6803, *flv2* and *flv4* were highly up-regulated in response to CO_2 limitation (Hackenberg *et al.*, 2009; Hackenberg *et al.*, 2012). Ultimately, the reactions catalysed by the Flv proteins could lead to increased levels of $NADP^+$ in the cytoplasm during low-C_i conditions, where acceptors of photosynthetically derived electrons are typically limiting, to prevent the accumulation of reactive oxygen species. Moreover, $NADP^+$ and α-ketoglutarate have recently been shown to be co-repressors of CcmR, a negative regulator of many CCM components (Daley *et al.*, 2012).

Photosynthesis provides the energy to drive the CCM and the reactions of carbon fixation. However, a necessary by-product of water oxidation by photosystem II, O_2, competes with CO_2 as a substrate for RuBisCO. The toxic product of the oxygenation reaction, 2-PG, is an inhibitor of triosephosphate isomerase, an essential enzyme of the CBB cycle and the pentose phosphate pathway (Hartman *et al.*, 1975; Norman and Coleman, 1991). Photosynthetic organisms including plants, algae, and cyanobacteria have therefore evolved multiple strategies for removal and detoxification of 2-PG through photorespiration. Cyanobacteria contain both plant-like and bacterial-like pathways for the metabolism of 2-PG (Eisenhut *et al.*, 2006, 2008b). A mutant strain of *Synechocystis* PCC6803 lacking glycolate dehydrogenase exhibited a decreased growth even in the presence of high CO_2 (Eisenhut *et al.*, 2006). Despite the presence of the CCM, photorespiration still occurs; however, photorespiration was increased in a cyanobacterial strain lacking carboxysomes (Hackenberg *et al.*, 2012). Genome-scale models of photosynthetic metabolism in *Synechocystis* PCC6803 indicate that photorespiration could be essential for maximum biomass production by providing carbon skeletons for amino acid biosynthesis (Knoop *et al.*, 2010; Nogales *et al.*, 2012).

The active transport systems and CO_2 hydration components of the CCM effectively generate a high concentration (up to 40 mM) of HCO_3^- within the cell (Price *et al.*, 2008). HCO_3^- diffuses into the carboxysome, where it is converted to CO_2 by carbonic anhydrase (CA) and then combined with ribulose 1,5-bisphosphate (RuBP) by RuBisCO to produce two molecules of 3-PGA, which subsequently exit the carboxysome and are utilized by the enzymes of the CBB cycle (Fig. 7.1). In addition to the large inward flux of HCO_3^- that is maintained through the conversion of HCO_3^- to CO_2 within the carboxysome, the high concentrations of CO_2 generated within the interior of the carboxysome presumably limit the oxygenation of RuBP by RuBisCO (Kaplan *et al.*, 1991; Reinhold *et al.*, 1991) and the subsequent losses in productivity due to photorespiration (Hackenberg *et al.*, 2012).

Compared to the low-Ci inducible components of the CCM, the expression patterns of genes encoding structural (*ccmK, ccmL, ccmM, ccmN, ccmO*) components of the carboxysome exhibit only moderate differential expression under these conditions (Eisenhut *et al.*, 2007, 2008a; Hackenberg *et al.*, 2012; McGinn *et al.*, 2003; Wang *et al.*, 2004). However, it is possible that the structural genes are only activated during a transient burst upon exposure to low-C_i (Woodger *et al.*, 2003). Significant changes in the expression of carboxysomal genes were observed during exposure to different nutrient limitations and during growth on alternative nitrogen sources (Krasikov *et al.*, 2012; Ludwig and Bryant, 2012). In addition, carboxysomal genes were found to be up-regulated in response to preferential illumination of photosystem I with blue light leading to depletion of NADPH (Singh *et al.*, 2009). Carboxysomal genes were down-regulated in high light in air (Singh *et al.*, 2008), but up-regulated in high light and 1% CO_2 (Hihara *et al.*, 2001). Exposure of cells to cadmium also resulted in the down-regulation of carboxysomal genes (Houot *et al.*, 2007). Because carboxysomes are necessary for carbon fixation in the light, it is not surprising that carboxysomal genes were found to be expressed maximally during the subjective dawn in cells entrained to light/dark cycles (Ito *et al.*, 2009). In addition, carboxysomal proteins were also the most abundant during the light period in another cyanobacterium, *Cyanothece* sp. ATCC 51142 (*Cyanothece* ATCC51142) (Aryal *et al.*, 2011). While differential expression of carboxysomal genes is observed under a variety of conditions, there is not always a direct correlation between gene expression and protein translation. In fact, increased levels of RuBisCO and carboxysomal proteins were observed in a proteomic study following low-C_i exposure in contrast to the findings based on gene expression alone (Battchikova *et al.*, 2010). Regulation of the CCM at the level of transcription and translation is critical for coordination of carbon fixation with other cellular pathways. A recent analysis suggests that antisense RNA could play a role in regulation of several components of the CCM, including *ndh-I*$_3$ and *ccmK* (Mitschke *et al.*, 2011). There are still many outstanding questions regarding the regulation of the CCM and the translational coordination of the multiple structural and enzymatic components. It is clear that the CCM is a central and critical feature of cyanobacterial metabolism. As such, many CCM genes were identified as components of the core transcriptional response during multiple genetic and environmental perturbations in *Synechocystis* PCC6803 (Singh *et al.*, 2010).

Structural and catalytic elements of the carboxysome

Carboxysome subtypes: α- and β-carboxysomes

Carboxysomes are classified based on the type of RuBisCO they encapsulate. α-carboxysomes contain Form IA RuBisCO, while β-carboxysomes contain plant-like Form IB RuBisCO (Badger and Price, 2003). Form I RuBisCO is comprised of 8 large and 8 small subunits (L_8S_8) (Tabita, 1999). Structurally, Form IA RuBisCO exhibits a net negative electrostatic surface potential, while Form IB structures from β-cyanobacteria and plant sources exhibit net positive surface potentials (Zarzycki *et al.*, 2012). Typical α-type and β-type carboxysome gene clusters are shown in Fig. 7.2.

In general, marine picocyanobacteria and chemoautotrophs contain α-carboxysomes. This type of carboxysome has been extensively studied in the chemoautotroph *Halothiobacillus neapolitanus* (*H. neapolitanus*) because this organism is genetically tractable and because its carboxysomes can be readily purified to homogeneity. Recently, α-carboxysomes from the marine cyanobacterium, *Prochlorococcus marinus* MED4 (*P. marinus* MED4) were purified and their composition determined (Roberts *et al.*, 2012a).

However, most studies of cyanobacterial carboxysomes have focused on β-carboxysomes. The three well-established model organisms for cyanobacterial research, *Synechococcus* PCC7942, *Synechococcus* PCC7002, and *Synechocystis* PCC6803, all contain β-carboxysomes. However, despite the many advantages of these model organisms including sequenced genomes, genetic tractability and a large biotechnological toolkit, a full inventory of the proteins required for formation of β-carboxysomes is not available due to the difficulty in purification.

Carboxysome shell

The understanding of the carboxysome shell has been advanced significantly by structural (Cai *et al.*, 2013; Dryden *et al.*, 2009; Kerfeld *et al.*, 2005; Klein *et al.*, 2009; Sutter *et al.*, 2013; Tanaka *et al.*, 2008, 2009; Tsai *et al.*, 2007, 2009) and genetic studies (Price and Badger, 1989b; Price *et al.*, 1993; Rae *et al.*, 2012). In essence, the shell exhibits two important properties. First, the shell creates a physical barrier between the encapsulated proteins and the cytoplasm, resulting in a cellular environment that is distinct from the cytoplasmic space. Second, the shell provides selectivity for the passage of substrate and products between the cytoplasm and the interior of the compartment.

Structurally, the shell exhibits features of an icosahedral viral capsid (Yeates *et al.*, 2007) including an array of protein subunits that resemble self-assembling coat-proteins; however, carboxysome shell proteins lack primary or tertiary structural

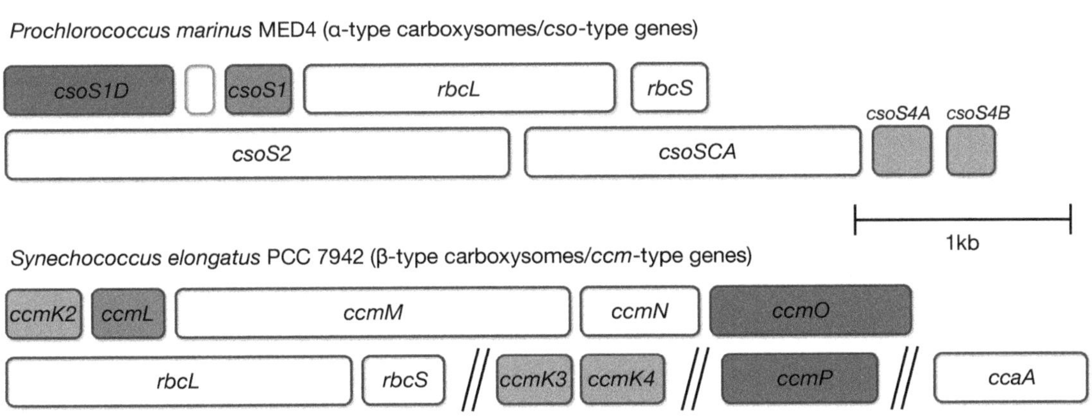

Figure 7.2 Typical gene organization for α- and β-carboxysomes. Genes coding for structural proteins are shaded from light to dark grey according to type (hexameric/Pfam00936, pentameric/Pfam03319, tandem domain). Double slashes indicate separation on the genome and the box sizes are proportional to gene size (see 1 kb scale bar).

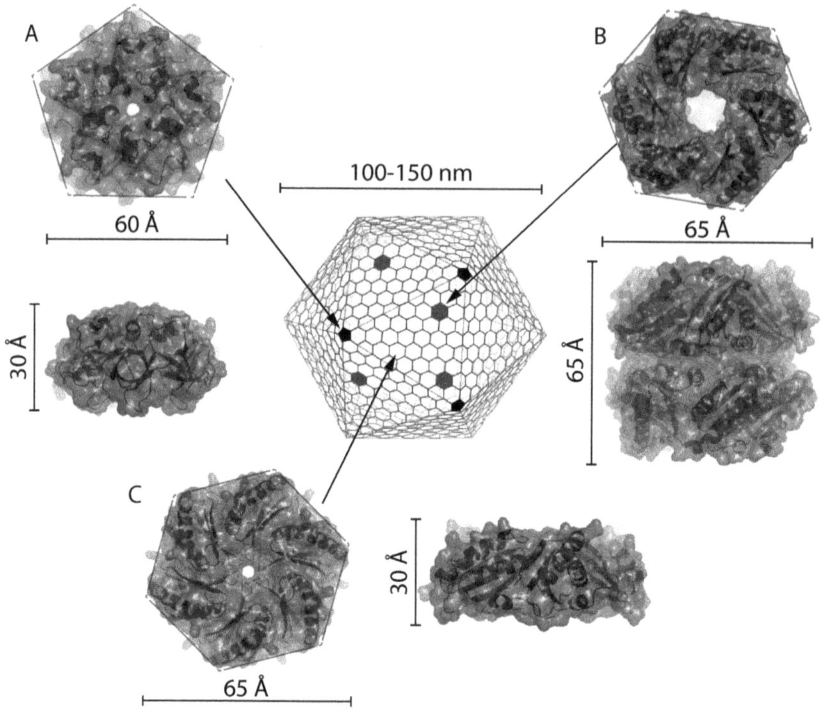

Figure 7.3 Surface and cartoon representations of selected structures of carboxysome shell proteins with their dimensions and proposed role in the carboxysome shell. (A) Pentameric CsoS4A (PDB ID 2RCF). (B) Tandem BMC-domain protein CsoS1D (3F56). (C) Hexameric CsoS1A (PDB ID 2G13).

homology to known viral capsid proteins. Geometrically, the shell is composed of at least two types of protein subunits resembling hexagons and pentagons (Fig. 7.3). The hexagonal protein oligomers are thought to assemble into sheets to form the facets (Dryden *et al.*, 2009; Kerfeld *et al.*, 2005) of an icosahedral structure. The pentagonal shaped protein oligomers provide the vertex for five adjacent facets (Tanaka *et al.*, 2008).

While only hexagonal and pentagonal shaped protein oligomers are structurally required to assemble into an icosahedron, multiple homologues of shell proteins are encoded in the genomes of carboxysome- and other micro-compartment-containing bacteria (Cai *et al.*, 2012; Kerfeld *et al.*, 2010; Kinney *et al.*, 2011; Price *et al.*, 1993; Roberts *et al.*, 2012a). Current work is focusing on the specific roles of these shell proteins in metabolite transport (Cai *et al.*, 2009; Klein *et al.*, 2009; Menon *et al.*, 2010) and structural organization (Dryden *et al.*, 2009; Kerfeld *et al.*, 2005; Rae *et al.*, 2012; Tanaka *et al.*, 2009) of the carboxysome. Two recent reviews discuss the diversity of carboxysomal shell proteins (Kinney *et al.*, 2011; Yeates *et al.*, 2011).

Single BMC-domain proteins

A conserved feature of the genomes of carboxysome-containing organisms is the presence of at least two copies of a gene encoding the bacterial micro-compartment domain (BMC; Pfam00936) (Fig. 7.2). Additionally, genes encoding other shell protein homologues are often located elsewhere in the genome. The *Synechocystis* PCC6803 genome encodes four proteins (of approximately 100 amino acids) with a single copy of the BMC-domain (*ccmK1, ccmK2, ccmK3, ccmK4*); *ccmK1* and *ccmK2* reside within the *ccm* gene cluster. In contrast, the only representative in the *cso*-operon of *P. marinus* MED4 is *csoS1*, with the tandem-BMC domain protein, CsoS1D, just upstream of the *cso* gene cluster, as in most α-cyanobacteria (Roberts *et al.*, 2011) (Fig. 7.2). Structural elucidation of CcmK2 and CcmK4 from *Synechocystis* PCC6803 revealed that these proteins crystallize as hexamers (Fig. 7.3) (Kerfeld *et al.*, 2005).

Furthermore, these and subsequent studies revealed the presence of pores (approximately 4–7 Å in diameter) at the six–fold axis of symmetry of each hexamer, suggesting that they serve as a conduit for passage of metabolites across the carboxysome shell (Dryden *et al.*, 2009; Kerfeld *et al.*, 2005; reviewed in Kinney *et al.*, 2011; Tanaka *et al.*, 2009; Tsai *et al.*, 2007, 2009). Comparison of shell protein structures also reveals highly conserved residues among the CcmK homologues including positively charged residues flanking the six-fold symmetry axes that could function in attracting the negatively charged HCO_3^- (Kinney *et al.*, 2011).

The genome of *Synechococcus* PCC7942 contains three homologues containing a single BMC domain (Fig. 7.2); a single gene (*ccmK2*) within the *ccm*-operon and a pair (*ccmK3* and *ccmK4)* located elsewhere in the genome. CcmK2 is an essential component of the carboxysome; mutants lacking *ccmK2* exhibit an HCR phenotype and cannot form carboxysomes (Price *et al.*, 1993). While fluorescently tagged CcmK4 has been demonstrated to localize to the carboxysome (Savage *et al.*, 2010), until recently, the roles of *ccmK3* and *ccmK4* have remained elusive. It is now thought that in *Synechococcus* PCC7942, CcmK3 and CcmK4 are low abundance shell proteins that play subtle roles in carboxysome function because Δ*ccmK3* and Δ*ccmK4* mutants do not display an altered phenotype, while double mutants exhibit an intermediate HCR phenotype and display abnormally clustered carboxysomes (Rae *et al.*, 2012). This is in contrast to the finding that inactivation of *ccmK4* by transposon mutagenesis in *Synechocystis* PCC6803 resulted in suboptimal photoautotrophic growth (Zhang *et al.*, 2004).

The structures of two single BMC-domain containing proteins from carboxysomes of *Halothiobacillus neapolitanus*, CsoS1A and CsoS1C, have also been solved (Tsai *et al.*, 2007, 2009); like their counterparts in β-carboxysomes, they also crystallize as hexamers and are thought to function as the major structural elements of the α-carboxysome shell (Roberts *et al.*, 2012a).

Tandem BMC-domain proteins

In addition to the single BMC-domain proteins that oligomerize into hexamers, genes encoding a fusion of two BMC domains have been identified in α- and β-cyanobacteria (Fig. 7.3). Structural analysis of CsoS1D, a tandem BMC domain protein, revealed a stacked assembly of two trimers, each appearing as pseudohexamers (Klein *et al.*, 2009). The pore at the threefold axis of symmetry is considerably larger (14 Å) than the pores in single BMC-domain shell proteins. This is presumed to facilitate transport of larger metabolites across the carboxysome shell. Interestingly, the pore in CsoS1D exists in two states, 'open' or 'closed'. The existence of multiple pore conformations could indicate an active mechanism for the regulation of metabolite flux across the carboxysome shell. Additionally, a 'closed' conformation could prevent leakage of metabolites into or out of the carboxysome. An orthologue of CsoS1D, CcmP has been identified in β-cyanobacteria, however, its function has yet to be elucidated (Cai *et al.*, 2012). Its structure has recently been determined (Cai *et al.*, 2013).

CcmO, a tandem BMC-domain protein found in β-cyanobacteria, is essential for carboxysome formation and growth in air (Marco *et al.*, 1994; Rae *et al.*, 2012). Rae *et al.* (2012) also suggest that CcmO is a major component of the carboxysome shell and showed that CcmO interacts with CcmK2 *in vivo* by analysing carboxysome enriched fractions from *Synechococcus* PCC7942. The authors also speculate that CcmO forms a 'zipper' that seals the edges of the carboxysome.

Pentameric shell proteins

Hexameric building blocks are not sufficient to form an icosahedral carboxysome structure; 12 pentamers are necessary to provide vertices (Fig. 7.3). Genes encoding a protein with a Pfam03319 domain are found in α- and β-carboxysome gene clusters (Fig. 7.2). CcmL from *Synechocystis* PCC6803 and CsoS4 from *H. neapolitanus* crystallize as pentamers (Tanaka *et al.*, 2008; Sutter *et al.*, 2013) (Fig. 7.3). However, not all proteins containing the Pfam03319 domain form a pentamer, the EutN protein from *E. coli* is a hexamer in the crystal structure (Tanaka *et al.*, 2008; Wunderlich *et al.*, 2004). The pentamer of α-carboxysomes, CsoS4, is essential for integrity of the shell (Cai *et al.*, 2009). Furthermore, disruption of *ccmL* results in an HCR phenotype and the formation of elongated carboxysomes (Price and Badger, 1989b; Price *et al.*, 1993), possibly due to the inability to seal the end of the compartment. However, elongated carboxysomes are occasionally observed in TEMs of wild-type cyanobacterial cells. Recently, it has been proposed that CcmO and CcmL interact based on the observation that expression of hexahistidine-ubiquitin tagged CcmO proteins sometimes leads to the formation of elongated carboxysomes (Rae *et al.*, 2012).

Components of the carboxysome interior – confirmed and putative

RuBisCO

RuBisCO is the major enzyme of the carboxysome and catalyses the production of 3-PGA from CO_2 and RuBP. Multiple different forms (Forms I–IV) of RuBisCO have been identified and characterized; they range widely in catalytic efficiency and relative affinity for CO_2 and O_2 (Tabita *et al.*, 2007, 2008). Functionally, the L_8S_8 RuBisCO complex is comprised of four pairs of L_2 dimers, each forming two active sites at the L–L interface. The role of the small subunits is less clear, however, recent data indicates that they increase the stability of the L_8S_8 complex in *Chlamydomonas reinhardii* and also influence catalysis (Genkov and Spreitzer, 2009). The many types of RuBisCO exhibit highly variable properties with respect to their affinity for CO_2 and rate of carboxylation (Savir *et al.*, 2010). In fact, cyanobacterial RuBisCOs exhibit high specific activities while maintaining a low affinity for CO_2. This is likely a result of the presence of the CCM that functions to dramatically increase the levels of CO_2 in the vicinity of RuBisCO to levels much higher than that achieved by higher plants.

Several studies have estimated the number of RuBisCO L_8S_8 particles encapsulated within the α-carboxysome. The numbers range from 152 in *P. marinus* MED4 (Roberts *et al.*, 2012a), 270 in *H. neapolitanus* (Heinhorst *et al.*, 2006a), and approximately 250 in *Synechococcus* WH8102 (Iancu *et al.*, 2007); this difference in number has been attributed to differences in diameter of the carboxysome. In contrast, it has been estimated that as many as 3351 RuBisCO particles are encapsulated within the β-carboxysome of *Synechococcus* PCC7942 due to a tightly packed arrangement of the L_8S_8 particles (Long *et al.*, 2007).

A recent study demonstrated that RbcL could be covalently modified with the cyanobacterial hepatotoxin, microcystin (Zilliges *et al.*, 2011). The binding of microcystin to cellular proteins was stimulated by high light and by treatment with H_2O_2. Interestingly, production of microcystin by *Microcystis aeruginosa* PCC7806 was also shown to increase cellular fitness during exposure to high light and oxidative stress. Since RbcL is thought to occur almost exclusively in the carboxysome in cyanobacteria (McKay *et al.*, 1993; Price *et al.*, 1992), it is not clear how microcystin traverses the carboxysome shell to gain access to the interior.

RuBisCO activase

Rubisco activase (Rca) is an AAA+ domain-containing protein that serves to maintain the activity of RuBisCO. Rca functions by removing inhibitory sugar phosphates, including RuBP, from un-carbamylated and catalytically incompetent RuBisCO complexes in a process that requires ATP. Subsequent carbamylation of RuBisCO and stabilization of the carbamate by Mg^{2+} results in a catalytically active complex. Rca is critical in plants; inactivation leads to an HCR phenotype (Somerville *et al.*, 1982). In cyanobacteria, Rca is found only in a subset of strains (Zarzycki *et al.*, 2012), therefore its role in cyanobacteria is less clear. However, many cyanobacterial Rcas contain an N-terminal extension with an RbcS-like domain, similar to those observed in CcmM (see below). Thus, it is conceivable that Rca could be targeted to the carboxysome in these organisms. Because Rca typically uses ATP as a substrate, this would also imply that ATP is able to traverse the carboxysomal shell in these strains. Alternatively, RuBisCO could be activated in an ATP-independent manner in cyanobacteria by inorganic phosphate; high concentrations of P_i (> 5 mM) stimulated the carbamylation of RuBisCO, while low P_i concentrations (< 5 mM) increased RuBisCO activity (Marcus and Gurevitz, 2000).

Carbonic anhydrase: CcaA, CsoSCA, CcmM

Carboxysomal carbonic anhydrase (CA) functions to convert HCO_3^- to CO_2, a substrate of

RuBisCO (Fig. 7.1). It is thought that CA activity is localized to the interior of the carboxysome based on modelling of cyanobacterial carbon flux (Reinhold *et al.*, 1991) and biochemical analysis of carboxysome-enriched fractions (Price *et al.*, 1992). Ectopic expression of human α-carbonic anhydrase in the cytosol of *Synechococcus* PCC7942 led to rapid loss of CO_2 from the cell, resulting in an HCR phenotype (Price and Badger, 1989a). Therefore, regulation of carbonic anhydrase expression, localization, and activity is crucial.

Carbonic anhydrases are grouped into different classes based on primary structure. Two classes of CA (β and γ) have been found in association with carboxysomes (reviewed in Cannon *et al.*, 2010). In α-carboxysomes, CA activity is associated with a unique member of the β-CA family, CsoSCA (Heinhorst *et al.*, 2006b; Sawaya *et al.*, 2006; So *et al.*, 2004). Bioinformatic surveys have revealed that the *csoSCA* gene is associated with the *cso*-operon in all α-cyanobacteria sequenced to date (Roberts *et al.*, 2012a). Genetic deletions of *csoSCA* in *H. neapolitanus* resulted in an HCR phenotype and impaired carbon fixation by intact, isolated carboxysomes (Dou *et al.*, 2008). Although it was noted that the isolated carboxysomes were slightly larger than those from WT, the protein composition was not significantly altered in this mutant, indicating that CsoSCA does not play a structural role in defining the carboxysome shell.

In β-cyanobacteria, two types of carboxysomal CAs have been described, CcaA (formerly IcfA) (Fukuzawa *et al.*, 1992) and CcmM (reviewed in Cannon *et al.*, 2010). The model organisms *Synechocystis* PCC6803 and *Synechococcus* PCC7942 contain both CcaA and CcmM (Cot *et al.*, 2008; Long *et al.*, 2007, 2011). CcaA is a β-CA that is found in many, but not all β-cyanobacteria (Zarzycki *et al.*, 2012). It has been shown to have CA activity (So and Espie, 1998) and appears to function as the major carboxysomal CA in *Synechocystis* PCC6803 and *Synechococcus* PCC7942 (Price *et al.*, 1992; So *et al.*, 2002a). Similar to the findings in the *H. neapolitanus* strain lacking CsoSCA, strains lacking CcaA were shown to exhibit a HCR phenotype and contain carboxysomes morphologically similar to the wild-type. In contrast to plant-type β-CAs, CcaA contains a long (~75 amino acids) C-terminal extension that is required for catalytic activity (Cannon *et al.*, 2010; So *et al.*, 2002b). Moreover, CcaA has been shown to interact with the N-terminal region of CcmM and the levels of CcaA and CcmM in the carboxysome are tightly correlated (Cot *et al.*, 2008; Long *et al.*, 2011).

CcmM is found in all β-cyanobacteria and is an essential component of the β-carboxysome; no carboxysomes were observed in a Δ*ccmM* strain of cyanobacteria (Ludwig *et al.*, 2000; Woodger *et al.*, 2005). The N-terminal domain of CcmM contains the essential features of a γ-CA including conserved residues for Zn^{2+} coordination at the active site (Cannon *et al.*, 2010; Iverson *et al.*, 2000), though its role as a CA is enigmatic. The CA activity of CcmM has, thus far, only been demonstrated in preparations from *Thermosynechococcus elongatus* BP-1, a strain lacking *ccaA* (Pena *et al.*, 2010). Interestingly, CcmM was found to be active only under oxidizing conditions. Since the cytoplasm of cyanobacteria is thought to be a reducing environment, it has been postulated that the interior of the carboxysome could maintain a redox state distinct from the cytoplasm. CcaA has been shown to be active only under oxidative conditions and inactivated in the presence of DTT (Badger and Price, 1989). However, it is known that reducing agents can inactivate various CAs (Cannon *et al.*, 2010). Redox based regulation of CA activity could prevent the counterproductive release of cytosolic CO_2 prior to incorporation into the carboxysome. In fact, RuBisCO, CcmM, and CcaA have been identified as targets of thioredoxin in cyanobacteria (Lindahl and Florencio, 2003).

While the N-terminal domain of CcmM is homologous to γ-carbonic anhydrase, its C-terminal domain contains multiple RbcS-like domains (Cannon *et al.*, 2010; Ludwig *et al.*, 2000; Price *et al.*, 1993). The number of RbcS-like repeats varies across cyanobacteria, with 3–4 being typical. The CcmM homologue in *Synechococcus* PCC7942 contains three RbcS-like domains, whereas the *Synechocystis* PCC6803 and *Acaryochloris marina* homologues contain four and five RbcS-like domains, respectively.

In *Synechococcus* PCC7942, two distinct forms of CcmM are found in cellular extracts and carboxysome-enriched fractions. The long-form, a 58 kDa polypeptide (M58), corresponds to the full-length translation product of the *ccmM* gene. The short-form is a 35 kDa C-terminal portion (M35) of the CcmM protein containing only the three RbcS-like domains. The short form is thought to arise from an internal ribosome entry site. In carboxysome-enriched fractions from *Synechococcus* PCC7942, a M35:M58 ratio of 4:1 was observed, while in preparations enriched using a His-tagged CcmM protein, the M35:M58 ratio ranged from 1:1 to 11:1 (Long *et al.*, 2007). Manipulation of these ratios by mutation of the internal start site and Shine–Dalgarno sequence suggest that a minimum ratio of M35:M58 of 1:1 is essential for carboxysome formation (Long *et al.*, 2010). In *Synechocystis* PCC6803, multiple forms of CcmM with masses of 73, 62, 52 and 36 kDa have been identified in carboxysome-enriched fractions (Cot *et al.*, 2008).

Long *et al.* (2010) presented a model for the organization of the carboxysome interior that is comprised of M35 domains forming an interlocking network of RuBisCO particles. The M58 counterpart is thought to exist as a trimer and be responsible for bridging the RuBisCO particles with the carboxysome shell and recruiting CcaA to the interior of the carboxysome. In support of this model, His-tagged versions of CcmM M35 and M58 were shown to co-purify with RbcL and RbcS in carboxysome enriched fractions of *Synechococcus* PCC7942. In addition, when co-expressed with L_8S_8 complexes in *E. coli*, the M35 form was shown to co-purify with RbcL and RbcS (Long *et al.*, 2007). Finally, yeast two-hybrid data indicate that the N-terminal portion of CcmM could directly interact with CcmK1, CcmK2, CcmL, and CcaA in *Synechocystis* PCC6803 (Cot *et al.*, 2008; Sato *et al.*, 2007).

Components of unknown function: CcmN and CsoS2

The *ccmN* gene is conserved among β-cyanobacteria and is found exclusively within the *ccm* gene cluster (Kinney *et al.*, 2012). The product of the *ccmN* gene is composed of two distinct domains separated by a poorly conserved linker region of variable length. Like CcmM, the N-terminal region of CcmN also contains bacterial hexapeptide repeat domains; thus, a structural similarity between the N-terminal domains of these proteins has been proposed (Kinney *et al.*, 2012). The C-terminus of CcmN from *Synechococcus* PCC7942 contains a conserved 18 amino acid peptide that is critical for carboxysome formation (Kinney *et al.*, 2012). Peptides with similar composition have been found in many other enzymes thought to be associated with bacterial micro-compartments (BMCs), carboxysome-like organelles of heterotrophic organisms (Kerfeld *et al.*, 2010); thus, these small peptides could represent a conserved feature required for the assembly of micro-compartments and/or targeting to them (Kinney *et al.*, 2012). Recently, several studies have demonstrated that a subset of these peptides can function as targeting signals for the encapsulation of proteins into BMCs (Choudhary *et al.*, 2012; Fan *et al.*, 2012).

Interactions between CcmN and CcmM identified in a yeast two-hybrid screen were confirmed through heterologous expression and immunoprecipitation (Cot *et al.*, 2008). Thereby, CcmN, CcmM, CcaA and RuBisCO may form a complex that is competent in the dehydration of HCO_3^- to CO_2 and carbon fixation in β-cyanobacteria. Further investigation of the CcmM:CcmN complex reveals that the N-terminal domain of CcmN interacts with CcmM, while the C-terminal 18 amino acid peptide interacts with CcmK2 (Kinney *et al.*, 2012). Moreover, removal of this peptide prevents interaction with CcmK2 and carboxysome biogenesis.

Likewise, α-carboxysome operons contain a gene encoding a protein of unknown function. CsoS2 is a large protein (~800 amino acids) conserved across α-cyanobacterial genomes. In *H. neapolitanus* it makes up about 12% of the total weight of a carboxysome (Heinhorst *et al.*, 2006a). It is present in two forms migrating at 85 and 120 kDa in SDS-PAGE, the different size being attributed to possible glycosylation (Baker *et al.*, 1999). It has an unusually high isoelectric point of above 9 and contains numerous short (three amino acid) repeats (Cannon *et al.*, 2003). The overall sequence conservation between homologues is low compared to shell proteins and to date there is no available information about its three-dimensional structure.

Cellular organization and dynamics of carboxysomes

Cyanobacteria depend on the carboxysome for growth in atmospheric CO_2 levels. Thus, there is a significant selection pressure for the distribution of a carboxysome to each daughter cell upon cell division. Recent data suggest that carboxysomes are actively organized to promote this equal distribution (Fig. 7.4). Moreover, it is becoming evident that cyanobacteria exhibit a highly organized cellular interior that is likely involved in the coordination of multiple processes.

Spatial distribution and interaction with subcellular components

The ability to express proteins with fluorescently labelled tags has greatly increased our understanding of subcellular dynamics by allowing the direct visualization of proteins in live cells. This technique has been applied to carboxysomes (Jain *et al.*, 2012; Kinney *et al.*, 2012; Savage *et al.*, 2010), resulting in a new understanding of how the enzymes involved in cyanobacterial carbon fixation are arranged and partitioned between cells during cell division. Ground-breaking work (Savage *et al.*, 2010) demonstrated that a labelled carboxysomal shell protein, CcmK4-YFP, and RbcL-CFP are co-localized in discrete fluorescent puncta that are evenly distributed along the longitudinal axis of the cells of *Synechococcus* PCC7942, a rod-shaped cyanobacterium. In log-phase cells grown in light/dark cycles, an average of four carboxysomes per cell was reported, with the spacing between individual carboxysomes being proportional to the length of the cell (Fig. 7.4).

In addition to the linear spatial arrangement, diffusion of the carboxysomes was shown to be constrained, suggesting an interaction with the cytoskeletal network (Fig. 7.4). In support of this hypothesis, deletions of the actin homologue, MreB, and a homologue of the chromosome partitioning protein, ParA, resulted in altered carboxysome arrangement. ParA-like genes have

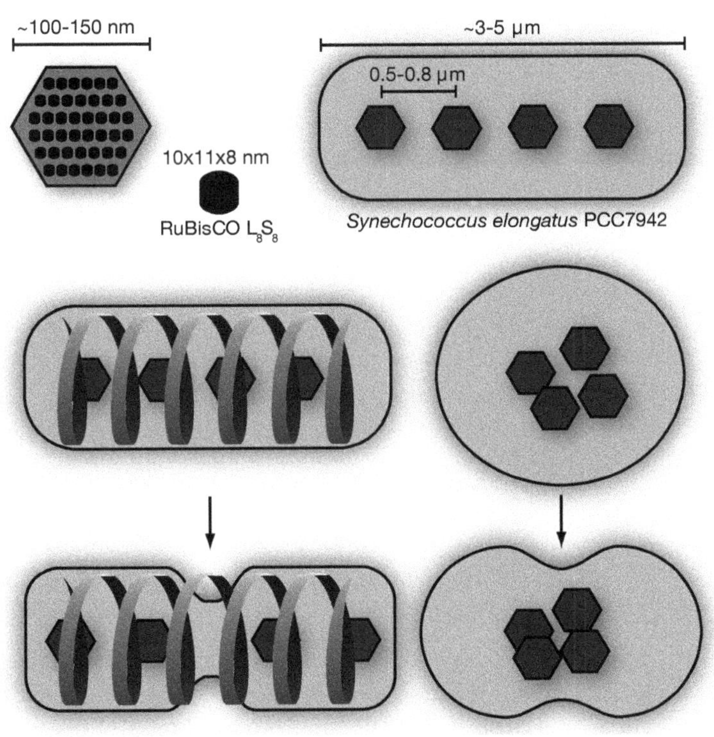

Figure 7.4 Schematic diagram illustrating the organization of carboxysomes in unicellular cyanobacterial cells. In elongated cells, carboxysomes are evenly distributed and possibly interact with the cytoskeleton. In spherical cells, carboxysomes are found in the thylakoid-free region, often in clusters.

been shown to be important in the partitioning of protein complexes and use DNA as a guide (Roberts *et al.*, 2012b). Similarly, the ParA homologue in cyanobacteria has been implicated in actively maintaining the distribution of the carboxysomes; the Δ*parA* strain exhibits random distribution of carboxysomes to daughter cells (Savage *et al.*, 2010). This results in cases where progeny cells do not inherit a carboxysome; these cells exhibit reduced fitness due to the decreased ability to fix carbon resulting in a slower growth rate. Interestingly, ParA homologues are found clustered with CcmK3 and CcmK4 in several cyanobacteria (*Fischerella* sp. PCC9605, *Fischerella* sp. PCC9339, *Calothrix* sp. PCC7103, *Nostoc punctiforme* PCC73120, *Anabaena cylindrica* PCC7122, and *Tolypothrix* sp. PCC9009), and in *Chamaesiphon* sp. PCC6605 a ParA homologue is encoded within the *ccm*-operon. Likewise, ParA homologues are found in the *cso* gene clusters in a number of chemoautotrophic organisms (Cai and Kerfeld, unpublished); however, they have not yet been observed near the *cso* operons of cyanobacteria. In *Cyanothece* ATCC51142, ParA-like proteins were shown to be most abundant in the light, coinciding with the peak in abundance of carboxysomal proteins (Aryal *et al.*, 2011). A recent proteomics study found increased levels of MreB, and decreased levels of ParA and RbcL in an *ftn6* mutant of *Synechococcus* PCC7942 that is deficient in cell division (Koksharova *et al.*, 2007). However, a second cell division mutant analysed in the study, *ftn2*, had increased levels of ParA and MreB, while a significant reduction in RbcL levels were still observed. Additionally, CcmM abundance was increased in *ftn6* and decreased in *ftn2* compared with the wild-type.

A recent study shows the subcellular arrangement of chromosomes in *Synechococcus* PCC7942 as discrete foci spatially distributed across the length of the cell (Jain *et al.*, 2012). Moreover, the position of carboxysomes and chromosomes were found to be mutually exclusive, resulting in an alternating pattern of chromosomes and carboxysomes. Surprisingly, this order was maintained in the Δ*parA* strain, indicating that multiple mechanisms are likely involved in the spacing of intracellular components.

While it is evident that spatial arrangement plays a critical role in partitioning of chromosomes and carboxysomes in *Synechococcus* PCC7942, it is not clear how these processes occur in cyanobacteria with other morphologies (Fig. 7.4). The model organism *Synechocystis* PCC6803 is spherical and contains approximately six carboxysomes per cell that are located in the thylakoid-free region of the interior (van de Meene *et al.*, 2006). A similar arrangement and number was observed in the small, spheroid, marine cyanobacterium *P. marinus* MED4 (Ting *et al.*, 2007). In *Cyanothece* ATCC51142, an ellipsoid, unicellular nitrogen-fixing cyanobacteria, the carboxysomes are often found clustered in the thylakoid-free interior (Fig. 7.5) (Liberton *et al.*, 2011). Interestingly, analysis of *Cyanothece* ATCC51142 ultrastructure revealed the presence of 21 carboxysomes ranging in size from 250–600 nm in diameter (Liberton *et al.*, 2011), the latter is three times the diameter of a typical β-carboxysome. As previously observed, carboxysomes in *Cyanothece* ATCC51142 were shown to cluster with polyphosphate granules (Liberton *et al.*, 2011); although frequently observed, a functional connection between these cellular inclusions has not been identified (Iancu *et al.*, 2010). Polyphosphate accumulation has been associated with nutrient limitation and following phosphate-repletion of phosphate-limited cultures of cyanobacteria (Grillo and Gibson, 1979). Polyphosphate could also play a role in the balance of ATP and NADPH levels; a polyphosphate kinase (PPK2) found in many bacteria, including cyanobacteria (Zhang *et al.*, 2002), was shown to use polyphosphate as a donor for the conversion of ADP to ATP (Nocek *et al.*, 2008).

Conclusions and future directions

Our understanding of carboxysome structure, function, and organization is rapidly advancing. Recent studies have shed light on the organization of the shell, the interior, and the connections between the two. Finally, a glimpse into the organization of carboxysomes within the cell and the interactions therein by using fluorescently labelled carboxysomes will lead to new insights into the process of carboxysome assembly and the dynamics of individual components.

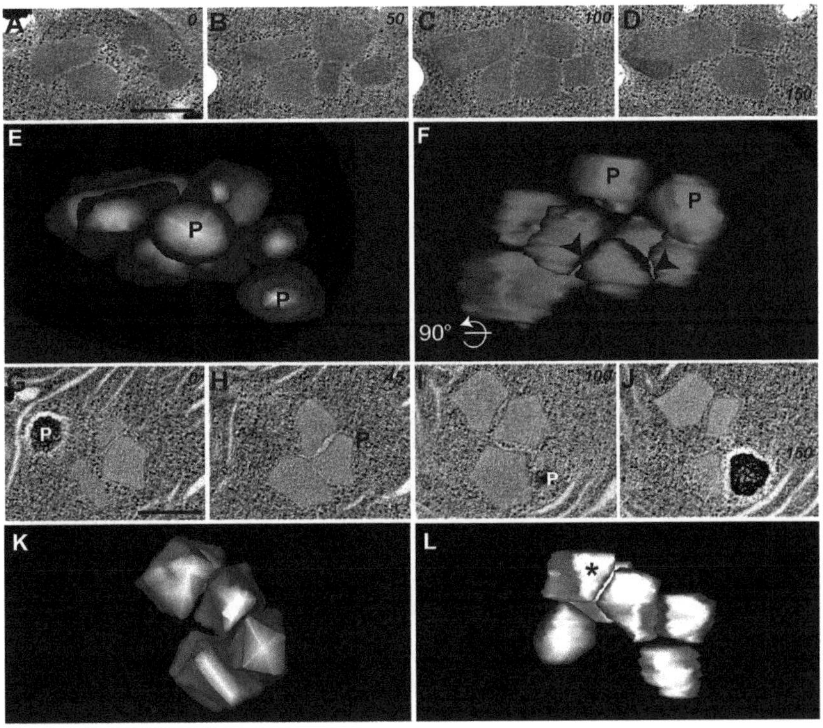

Figure 7.5 Carboxysomes cluster with polyphosphate in *Cyanothece*. Tomographic data shows carboxysomes and polyphosphate bodies (P) in the central cytoplasmic region. (A-D) and (G-H) Tomographic slice images through clusters (constructed from 3 superimposed serial 2 nm slices). (E) Model of the carboxysome cluster; (F) is a view of (E) rotated 90 degrees around the horizontal axis. Individual carboxysomes follow the shape of neighbours (arrowheads), so a tight cluster is formed. (K) Model of the cluster shown in (G-J); (L) is a rotated view. Triangular faces can be seen on some carboxysomes (asterisk) Bar = 200 nm. Figure courtesy of Michelle Liberton and Himadri Pakrasi (Washington University in St. Louis, MO, USA).

Acknowledgements

We would like to thank members of the Kerfeld laboratory for collegial discussions during the preparation of this chapter. We would especially like to thank Fei Cai and Seth Axen for helpful discussions. Funding for this work was provided by the NSF Emerging Frontiers Program (EF 1105897). M.S. was supported by a Swiss National Science Foundation Postdoctoral Fellowship.

References

Alric, J., Lavergne, J., and Rappaport, F. (2010). Redox and ATP control of photosynthetic cyclic electron flow in *Chlamydomonas reinhardtii* (I) aerobic conditions. Biochim. Biophys. Acta *1797*, 44–51.

Aryal, U.K., Stockel, J., Krovvidi, R.K., Gritsenko, M.A., Monroe, M.E., Moore, R.J., Koppenaal, D.W., Smith, R.D., Pakrasi, H.B., and Jacobs, J.M. (2011). Dynamic proteomic profiling of a unicellular cyanobacterium *Cyanothece* ATCC51142 across light-dark diurnal cycles. BMC Syst. Biol. *5*, 194.

Badger, M.R., and Price, G.D. (1989). Carbonic anhydrase activity associated with the cyanobacterium *Synechococcus* PCC7942. Plant Physiol. *89*, 51–60.

Badger, M.R., and Price, G.D. (2003). CO_2 concentrating mechanisms in cyanobacteria: molecular components, their diversity and evolution. J. Exp. Bot. *54*, 609–622.

Battchikova, N., Vainonen, J.P., Vorontsova, N., Keranen, M., Carmel, D., and Aro, E.M. (2010). Dynamic changes in the proteome of *Synechocystis* 6803 in response to CO_2 limitation revealed by quantitative proteomics. J. Proteome Res. *9*, 5896–5912.

Cai, F., Menon, B.B., Cannon, G.C., Curry, K.J., Shively, J.M., and Heinhorst, S. (2009). The pentameric vertex proteins are necessary for the icosahedral carboxysome shell to function as a CO_2 leakage barrier. PLoS One *4*, e7521.

Cai, F., Kerfeld, C.A., and Sandh, G. (2012). Bioinformatic Identification and Structural Characterization of a New Carboxysome Shell Protein. In Functional Genomics and Evolution of Photosynthetic Systems, Burnap, R., and Vermaas, W., eds. (Springer, Netherlands). pp. 345–356.

Cai, F., Sutter, M., Cameron, J., Stanley, D., and Kerfelf, C.A. (2013). Tandem BMC-domain proteins of carboxysomes form a subcompartment within a microcompartment. J. Biol. Chem. *288*, 16055–16063.

Cannon, G.C., Heinhorst, S., and Kerfeld, C.A. (2010). Carboxysomal carbonic anhydrases: Structure and role in microbial CO_2 fixation. Biochim. Biophys. Acta *1804*, 382–392.

Choudhary, S., Quin, M.B., Sanders, M.A., Johnson, E.T., and Schmidt-Dannert, C. (2012). Engineered protein nano-compartments for targeted enzyme localization. PLoS One 7, e33342.

Cot, S.S., So, A.K., and Espie, G.S. (2008). A multiprotein bicarbonate dehydration complex essential to carboxysome function in cyanobacteria. J. Bacteriol. *190*, 936–945.

Daley, S.M., Kappell, A.D., Carrick, M.J., and Burnap, R.L. (2012). Regulation of the cyanobacterial CO_2-concentrating mechanism involves internal sensing of $NADP^+$ and α-ketogutarate levels by transcription factor CcmR. PLoS One 7, e41286.

Dou, Z., Heinhorst, S., Williams, E.B., Murin, C.D., Shively, J.M., and Cannon, G.C. (2008). CO_2 fixation kinetics of Halothiobacillus neapolitanus mutant carboxysomes lacking carbonic anhydrase suggest the shell acts as a diffusional barrier for CO_2. J. Biol. Chem. *283*, 10377–10384.

Dryden, K.A., Crowley, C.S., Tanaka, S., Yeates, T.O., and Yeager, M. (2009). Two-dimensional crystals of carboxysome shell proteins recapitulate the hexagonal packing of three-dimensional crystals. Protein Sci. *18*, 2629–2635.

Eisenhut, M., Kahlon, S., Hasse, D., Ewald, R., Lieman-Hurwitz, J., Ogawa, T., Ruth, W., Bauwe, H., Kaplan, A., and Hagemann, M. (2006). The plant-like C2 glycolate cycle and the bacterial-like glycerate pathway cooperate in phosphoglycolate metabolism in cyanobacteria. Plant Physiol. *142*, 333–342.

Eisenhut, M., von Wobeser, E.A., Jonas, L., Schubert, H., Ibelings, B.W., Bauwe, H., Matthijs, H.C.P., and Hagemann, M. (2007). Long-term response toward inorganic carbon limitation in wild-type and glycolate turnover mutants of the cyanobacterium *Synechocystis* sp. strain PCC 6803. Plant Physiol. *144*, 1946–1959.

Eisenhut, M., Huege, J., Schwarz, D., Bauwe, H., Kopka, J., and Hagemann, M. (2008a). Metabolome phenotyping of inorganic carbon limitation in cells of the wild-type and photorespiratory mutants of the cyanobacterium *Synechocystis* sp. strain PCC 6803. Plant Physiol. *148*, 2109–2120.

Eisenhut, M., Ruth, W., Haimovich, M., Bauwe, H., Kaplan, A., and Hagemann, M. (2008b). The photorespiratory glycolate metabolism is essential for cyanobacteria and might have been conveyed endosymbiontically to plants. Proc. Natl. Acad. Sci. U.S.A. *105*, 17199–17204.

Fan, C., Cheng, S., Sinha, S., and Bobik, T.A. (2012). Interactions between the termini of lumen enzymes and shell proteins mediate enzyme encapsulation into bacterial microcompartments. Proc. Natl. Acad. Sci. U.S.A. *109*, 14995–15000.

Fukuzawa, H., Suzuki, E., Komukai, Y., and Miyachi, S. (1992). A gene homologous to chloroplast carbonic anhydrase (icfA) is essential to photosynthetic carbon dioxide fixation by *Synechococcus* PCC7942. Proc. Natl. Acad. Sci. U.S.A. *89*, 4437–4441.

Genkov, T., and Spreitzer, R.J. (2009). Highly conserved small subunit residues influence Rubisco large subunit catalysis. J. Biol. Chem. *284*, 30105–30112.

Grillo, J.F., and Gibson, J. (1979). Regulation of phosphate accumulation in the unicellular cyanobacterium *Synechococcus*. J. Bacteriol. *140*, 508–517.

Hackenberg, C., Engelhardt, A., Matthijs, H.C., Wittink, F., Bauwe, H., Kaplan, A., and Hagemann, M. (2009). Photorespiratory 2-phosphoglycolate metabolism and photoreduction of O_2 cooperate in high-light acclimation of *Synechocystis* sp. strain PCC 6803. Planta *230*, 625–637.

Hackenberg, C., Huege, J., Engelhardt, A., Wittink, F., Laue, M., Matthijs, H.C., Kopka, J., Bauwe, H., and Hagemann, M. (2012). Low-carbon acclimation in carboxysome-less and photorespiratory mutants of the cyanobacterium *Synechocystis* sp. strain PCC 6803. Microbiology *158*, 398–413.

Hartman, F.C., LaMuraglia, G.M., Tomozawa, Y., and Wolfenden, R. (1975). The influence of pH on the interaction of inhibitors with triosephosphate isomerase and determination of the pK_a of the active-site carboxyl group. Biochemistry *14*, 5274–5279.

Heinhorst, S., Cannon, G.C., and Shively, J.M. (2006a). Carboxysomes and carboxysome-like inclusions. In Complex intracellular structures in prokaryotes, Shively, J.M., ed. (Springer, Berlin), pp. 141–164.

Heinhorst, S., Williams, E.B., Cai, F., Murin, C.D., Shively, J.M., and Cannon, G.C. (2006b). Characterization of the carboxysomal carbonic anhydrase CsoSCA from *Halothiobacillus neapolitanus*. J. Bacteriol. *188*, 8087–8094.

Helman, Y., Tchernov, D., Reinhold, L., Shibata, M., Ogawa, T., Schwarz, R., Ohad, I., and Kaplan, A. (2003). Genes encoding A-type flavoproteins are essential for photoreduction of O_2 in cyanobacteria. Curr. Biol. *13*, 230–235.

Hihara, Y., Kamei, A., Kanehisa, M., Kaplan, A., and Ikeuchi, M. (2001). DNA microarray analysis of cyanobacterial gene expression during acclimation to high light. Plant Cell *13*, 793–806.

Houot, L., Floutier, M., Marteyn, B., Michaut, M., Picciocchi, A., Legrain, P., Aude, J.-C., Cassier-Chauvat, C., and Chauvat, F. (2007). Cadmium triggers an integrated reprogramming of the metabolism of *Synechocystis* PCC6803, under the control of the Slr1738 regulator. BMC Genom. *8*, 350.

Iancu, C.V., Ding, H.J., Morris, D.M., Dias, D.P., Gonzales, A.D., Martino, A., and Jensen, G.J. (2007). The structure of isolated *Synechococcus* strain WH8102 carboxysomes as revealed by electron cryotomography. J. Mol. Biol. *372*, 764–773.

Iancu, C.V., Morris, D.M., Dou, Z., Heinhorst, S., Cannon, G.C., and Jensen, G.J. (2010). Organization, structure, and assembly of -carboxysomes determined by electron cryotomography of intact cells. J. Mol. Biol. *396*, 105–117.

Ito, H., Mutsuda, M., Murayama, Y., Tomita, J., Hosokawa, N., Terauchi, K., Sugita, C., Sugita, M., Kondo, T., and Iwasaki, H. (2009). Cyanobacterial daily life with Kai-based circadian and diurnal genome-wide transcriptional control in *Synechococcus elongatus*. Proc. Natl. Acad. Sci. U.S.A. *106*, 14168–14173.

Iverson, T.M., Alber, B.E., Kisker, C., Ferry, J.G., and Rees, D.C. (2000). A closer look at the active site of γ-class carbonic anhydrases: high-resolution crystallographic studies of the carbonic anhydrase from *Methanosarcina thermophila*. Biochemistry *39*, 9222–9231.

Jain, I.H., Vijayan, V., and O'Shea, E.K. (2012). Spatial ordering of chromosomes enhances the fidelity of chromosome partitioning in cyanobacteria. Proc. Natl. Acad. Sci. U.S.A. *109*, 13638–13643.

Jensen, T.E., and Bowen, C.C. (1961). Organization of the centroplasm in *Nostoc pruniforme*. Proc. Iowa Acad. Sci. *68*, 89–96.

Kaplan, A., Schwarz, R., Lieman-Hurwitz, J., and Reinhold, L. (1991). Physiological and molecular aspects of the inorganic carbon-concentrating mechanism in cyanobacteria. Plant Physiol. *97*, 851–855.

Kerfeld, C.A., Sawaya, M.R., Tanaka, S., Nguyen, C.V., Phillips, M., Beeby, M., and Yeates, T.O. (2005). Protein structures forming the shell of primitive bacterial organelles. Science *309*, 936–938.

Kerfeld, C.A., Heinhorst, S., and Cannon, G.C. (2010). Bacterial microcompartments. Annu. Rev. Microbiol. *64*, 391–408.

Kinney, J.N., Axen, S.D., and Kerfeld, C.A. (2011). Comparative analysis of carboxysome shell proteins. Photosynth. Res. *109*, 21–32.

Kinney, J.N., Salmeen, A., Cai, F., and Kerfeld, C.A. (2012). Elucidating essential role of conserved carboxysomal protein CcmN reveals common feature of bacterial microcompartment assembly. J. Biol. Chem. *287*, 17729–17736.

Klein, M.G., Zwart, P., Bagby, S.C., Cai, F., Chisholm, S.W., Heinhorst, S., Cannon, G.C., and Kerfeld, C.A. (2009). Identification and structural analysis of a novel carboxysome shell protein with implications for metabolite transport. J. Mol. Biol. *392*, 319–333.

Knoop, H., Zilliges, Y., Lockau, W., and Steuer, R. (2010). The metabolic network of *Synechocystis* sp. PCC 6803: Systemic properties of autotrophic growth. Plant Physiol. doi:110.157198

Koksharova, O.A., Klint, J., and Rasmussen, U. (2007). Comparative proteomics of cell division mutants and wild-type of Synechococcus sp. strain PCC 7942. Microbiology *153*, 2505–2517.

Krasikov, V., Aguirre von Wobeser, E., Dekker, H.L., Huisman, J., and Matthijs, H.C. (2012). Time-series resolution of gradual nitrogen starvation and its impact on photosynthesis in the cyanobacterium *Synechocystis* PCC 6803. Physiol. Plant *145*, 426–439.

Liberton, M., Austin, J.R., 2nd, Berg, R.H., and Pakrasi, H.B. (2011). Unique thylakoid membrane architecture of a unicellular N_2-fixing cyanobacterium revealed by electron tomography. Plant Physiol. *155*, 1656–1666.

Lindahl, M., and Florencio, F.J. (2003). Thioredoxin-linked processes in cyanobacteria are as numerous as in chloroplasts, but targets are different. Proc. Natl. Acad. Sci. U.S.A. *100*, 16107–16112.

Long, B.M., Badger, M.R., Whitney, S.M., and Price, G.D. (2007). Analysis of carboxysomes from *Synechococcus* PCC7942 reveals multiple Rubisco complexes with carboxysomal proteins CcmM and CcaA. J. Biol. Chem. *282*, 29323–29335.

Long, B.M., Tucker, L., Badger, M.R., and Price, G.D. (2010). Functional cyanobacterial β-carboxysomes have an absolute requirement for both long and short forms of the CcmM protein. Plant Physiol. *153*, 285–293.

Long, B.M., Rae, B.D., Badger, M.R., and Price, G.D. (2011). Over-expression of the β-carboxysomal CcmM protein in *Synechococcus* PCC7942 reveals a tight co-regulation of carboxysomal carbonic anhydrase (CcaA) and M58 content. Photosynth. Res. *109*, 33–45.

Ludwig, M., and Bryant, D.A. (2012). Acclimation of the Global Transcriptome of the Cyanobacterium *Synechococcus* sp. Strain PCC 7002 to Nutrient Limitations and Different Nitrogen Sources. Front. Microbiol. *3*, 145.

Ludwig, M., Sultemeyer, D., and Price, G.D. (2000). Isolation of *ccmKLMN* Genes From the Marine Cyanobacterium, *Synechococcus* sp. PCC7002 (Cyanophyceae), and Evidence that CcmM is Essential for Carboxysome Assembly. J. Phycol. *36*, 1109–1118.

Maeda, S., Badger, M.R., and Price, G.D. (2002). Novel gene products associated with NdhD3/D4-containing NDH-1 complexes are involved in photosynthetic CO_2 hydration in the cyanobacterium, *Synechococcus* sp. PCC7942. Mol. Microbiol. *43*, 425–435.

Marco, E., Martinez, I., Ronen-Tarazi, M., Orus, M.I., and Kaplan, A. (1994). Inactivation of *ccmO* in *Synechococcus* sp. Strain PCC 7942 Results in a Mutant Requiring High Levels of CO_2. Appl. Environ. Microbiol. *60*, 1018–1020.

Marcus, Y., and Gurevitz, M. (2000). Activation of cyanobacterial RuBP-carboxylase/oxygenase is facilitated by inorganic phosphate via two independent mechanisms. Eur. J. Biochem. *267*, 5995–6003.

McGinn, P.J., Price, G.D., Maleszka, R., and Badger, M.R. (2003). Inorganic carbon limitation and light control the expression of transcripts related to the CO_2-concentrating mechanism in the cyanobacterium *Synechocystis* sp. strain PCC6803. Plant Physiol. *132*, 218–229.

McKay, R.M.L., Gibbs, S.P., and Espie, G.S. (1993). Effect of dissolved inorganic carbon on the expression of carboxysomes, localization of Rubisco and the mode of inorganic carbon transport in cells of the cyanobacterium *Synechococcus* UTEX 625. Arch. Microbiol. *159*, 21–29.

van de Meene, A.M., Hohmann-Marriott, M.F., Vermaas, W.F., and Roberson, R.W. (2006). The three-dimensional structure of the cyanobacterium

Synechocystis sp. PCC 6803. Arch. Microbiol. *184*, 259–270.

Menon, B.B., Heinhorst, S., Shively, J.M., and Cannon, G.C. (2010). The carboxysome shell is permeable to protons. J. Bacteriol. *192*, 5881–5886.

Mitschke, J., Georg, J., Scholz, I., Sharma, C.M., Dienst, D., Bantscheff, J., Voss, B., Steglich, C., Wilde, A., Vogel, J., *et al.* (2011). An experimentally anchored map of transcriptional start sites in the model cyanobacterium *Synechocystis* sp. PCC6803. Proc. Natl. Acad. Sci. U.S.A. *108*, 2124–2129.

Nocek, B., Kochinyan, S., Proudfoot, M., Brown, G., Evdokimova, E., Osipiuk, J., Edwards, A.M., Savchenko, A., Joachimiak, A., and Yakunin, A.F. (2008). Polyphosphate-dependent synthesis of ATP and ADP by the family-2 polyphosphate kinases in bacteria. Proc. Natl. Acad. Sci. U.S.A. *105*, 17730–17735.

Nogales, J., Gudmundsson, S., Knight, E.M., Palsson, B.O., and Thiele, I. (2012). Detailing the optimality of photosynthesis in cyanobacteria through systems biology analysis. Proc. Natl. Acad. Sci. U.S.A. *109*, 2678–2683.

Norman, E.G., and Colman, B. (1991). Purification and characterization of phosphoglycolate phosphatase from the cyanobacterium *Coccochloris peniocystis*. Plant Physiol. *95*, 693–698.

Ohkawa, H., Pakrasi, H.B., and Ogawa, T. (2000a). Two types of functionally distinct NAD(P)H dehydrogenases in *Synechocystis* sp. strain PCC6803. J. Biol. Chem. *275*, 31630–31634.

Ohkawa, H., Price, G., Badger, M., and Ogawa, T. (2000b). Mutation of *ndh* Genes Leads to Inhibition of CO_2 Uptake Rather than HCO_3^- Uptake in *Synechocystis* sp. Strain PCC 6803. J. Bacteriol. *182*, 2591–2596.

Omata, T., Price, G.D., Badger, M.R., Okamura, M., Gohta, S., and Ogawa, T. (1999). Identification of an ATP-binding cassette transporter involved in bicarbonate uptake in the cyanobacterium *Synechococcus* sp. strain PCC 7942. Proc. Natl. Acad. Sci. U.S.A. *96*, 13571–13576.

Pena, K.L., Castel, S.E., de Araujo, C., Espie, G.S., and Kimber, M.S. (2010). Structural basis of the oxidative activation of the carboxysomal γ-carbonic anhydrase, CcmM. Proc. Natl. Acad. Sci. U.S.A. *107*, 2455–2460.

Price, G.D., and Badger, M.R. (1989a). Expression of human carbonic anhydrase in the cyanobacterium *Synechococcus* PCC7942 Creates a high CO_2-requiring phenotype: evidence for a central role for carboxysomes in the CO_2 concentrating mechanism. Plant Physiol. *91*, 505–513.

Price, G.D., and Badger, M.R. (1989b). Isolation and characterization of high CO_2-Requiring-mutants of the cyanobacterium *Synechococcus* PCC7942: two phenotypes that accumulate inorganic carbon but are apparently unable to generate CO_2 within the carboxysome. Plant Physiol. *91*, 514–525.

Price, G.D., Coleman, J.R., and Badger, M.R. (1992). Association of Carbonic Anhydrase Activity with Carboxysomes Isolated from the Cyanobacterium *Synechococcus* PCC7942. Plant Physiol. *100*, 784–793.

Price, G.D., Howitt, S.M., Harrison, K., and Badger, M.R. (1993). Analysis of a genomic DNA region from the cyanobacterium *Synechococcus* sp. strain PCC7942 involved in carboxysome assembly and function. J. Bacteriol. *175*, 2871–2879.

Price, G.D., Woodger, F.J., Badger, M.R., Howitt, S.M., and Tucker, L. (2004). Identification of a SulP-type bicarbonate transporter in marine cyanobacteria. Proc. Natl. Acad. Sci. U.S.A. *101*, 18228–18233.

Price, G.D., Badger, M.R., Woodger, F.J., and Long, B.M. (2008). Advances in understanding the cyanobacterial CO_2-concentrating-mechanism (CCM): functional components, C_i transporters, diversity, genetic regulation and prospects for engineering into plants. J. Exp. Bot. *59*, 1441–1461.

Rae, B.D., Long, B.M., Badger, M.R., and Price, G.D. (2012). Structural determinants of the outer shell of β-carboxysomes in *Synechococcus elongatus* PCC 7942: Roles for CcmK2, K3-K4, CcmO, and CcmL. PLoS One 7, e43871.

Reinhold, L., Kosloff, R., and Kaplan, A. (1991). A model for inorganic carbon fluxes and photosynthesis in cyanobacterial carboxysomes. Can. J. Bot. *69*, 984–988.

Roberts, E.W., Cai, F., Kerfeld, C.A., Cannon, G.C., and Heinhorst, S. (2012a). Isolation and characterization of the *Prochlorococcus* carboxysome reveal the presence of the novel shell protein CsoS1D. J. Bacteriol. *194*, 787–795.

Roberts, M.A., Wadhams, G.H., Hadfield, K.A., Tickner, S., and Armitage, J.P. (2012b). ParA-like protein uses nonspecific chromosomal DNA binding to partition protein complexes. Proc. Natl. Acad. Sci. U.S.A. *109*, 6698–6703.

Sato, S., Shimoda, Y., Muraki, A., Kohara, M., Nakamura, Y., and Tabata, S. (2007). A Large-scale Protein protein Interaction Analysis in *Synechocystis* sp. PCC6803. DNA Res. *14*, 207–216.

Savage, D.F., Afonso, B., Chen, A.H., and Silver, P.A. (2010). Spatially ordered dynamics of the bacterial carbon fixation machinery. Science *327*, 1258–1261.

Savir, Y., Noor, E., Milo, R., and Tlusty, T. (2010). Cross-species analysis traces adaptation of Rubisco toward optimality in a low-dimensional landscape. Proc. Natl. Acad. Sci. U.S.A. *107*, 3475–3480.

Sawaya, M.R., Cannon, G.C., Heinhorst, S., Tanaka, S., Williams, E.B., Yeates, T.O., and Kerfeld, C.A. (2006). The structure of β-carbonic anhydrase from the carboxysomal shell reveals a distinct subclass with one active site for the price of two. J. Biol. Chem. *281*, 7546–7555.

Shibata, M., Ohkawa, H., Kaneko, T., Fukuzawa, H., Tabata, S., Kaplan, A., and Ogawa, T. (2001). Distinct constitutive and low-CO_2-induced CO_2 uptake systems in cyanobacteria: genes involved and their phylogenetic relationship with homologous genes in other organisms. Proc. Natl. Acad. Sci. U.S.A. *98*, 11789–11794.

Shibata, M., Katoh, H., Sonoda, M., Ohkawa, H., Shimoyama, M., Fukuzawa, H., Kaplan, A., and Ogawa, T. (2002). Genes essential to sodium-dependent bicarbonate transport in cyanobacteria: function and phylogenetic analysis. J. Biol. Chem. *277*, 18658–18664.

Singh, A., Elvitigala, T., Bhattacharyya-Pakrasi, M., Aurora, R., Ghosh, B., and Pakrasi, H. (2008). Integration of carbon and nitrogen metabolism with energy production is crucial to light acclimation in the cyanobacterium *Synechocystis*. Plant Physiol. *148*, 467–478.

Singh, A.K., Bhattacharyya-Pakrasi, M., Elvitigala, T., Ghosh, B., Aurora, R., and Pakrasi, H.B. (2009). A systems-level analysis of the effects of light quality on the metabolism of a cyanobacterium. Plant Physiol. *151*, 1596–1608.

Singh, A.K., Elvitigala, T., Cameron, J.C., Ghosh, B.K., Bhattacharyya-Pakrasi, M., and Pakrasi, H.B. (2010). Integrative analysis of large scale expression profiles reveals core transcriptional response and coordination between multiple cellular processes in a cyanobacterium. BMC Syst. Biol. *4*, 105.

So, A.K., and Espie, G.S. (1998). Cloning, characterization and expression of carbonic anhydrase from the cyanobacterium *Synechocystis* PCC6803. Plant Mol. Biol. *37*, 205–215.

So, A.K., John-McKay, M., and Espie, G.S. (2002a). Characterization of a mutant lacking carboxysomal carbonic anhydrase from the cyanobacterium *Synechocystis* PCC6803. Planta *214*, 456–467.

So, A.K., Cot, S.S., and Espie, G.S. (2002b). Characterization of the C-terminal extension of carboxysomal carbonic anhydrase from *Synechocystis* sp. PCC6803. Funct Plant Biol *29*, 183–194.

So, A.K., Espie, G.S., Williams, E.B., Shively, J.M., Heinhorst, S., and Cannon, G.C. (2004). A novel evolutionary lineage of carbonic anhydrase (epsilon class) is a component of the carboxysome shell. J. Bacteriol. *186*, 623–630.

Somerville, C.R., Portis, A.R., and Ogren, W.L. (1982). A Mutant of *Arabidopsis thaliana* Which Lacks Activation of RuBP Carboxylase *In Vivo*. Plant Physiol. *70*, 381–387.

Sutter, M., Wilson, S.C., Deutsch, S., and Kerfeld, C.A. (2013). Two new high-resolution crystal structures of carboxysome pentamer proteins reveal high structural conservation of CcmL orthologs among distantly related cyanobacterial species. Photosynth. Res. *in press*. doi:11120-0139909-z

Tabita, F.R. (1999). Microbial ribulose 1, 5-bisphosphate carboxylase/oxygenase: a different perspective. Photosynth. Res. *60*, 1–28.

Tabita, F.R., Hanson, T.E., Li, H., Satagopan, S., Singh, J., and Chan, S. (2007). Function, structure, and evolution of the RubisCO-like proteins and their RubisCO homologs. Microbiol. Mol. Biol. Rev. *71*, 576–599.

Tabita, F.R., Satagopan, S., Hanson, T.E., Kreel, N.E., and Scott, S.S. (2008). Distinct form I, II, III, and IV Rubisco proteins from the three kingdoms of life provide clues about Rubisco evolution and structure/function relationships. J. Exp. Bot. *59*, 1515–1524.

Tanaka, S., Kerfeld, C.A., Sawaya, M.R., Cai, F., Heinhorst, S., Cannon, G.C., and Yeates, T.O. (2008). Atomic-level models of the bacterial carboxysome shell. Science *319*, 1083–1086.

Tanaka, S., Sawaya, M.R., Phillips, M., and Yeates, T.O. (2009). Insights from multiple structures of the shell proteins from the β-carboxysome. Protein Sci. *18*, 108–120.

Ting, C.S., Hsieh, C., Sundararaman, S., Mannella, C., and Marko, M. (2007). Cryo-electron tomography reveals the comparative three-dimensional architecture of *Prochlorococcus*, a globally important marine cyanobacterium. J. Bacteriol. *189*, 4485–4493.

Tsai, Y., Sawaya, M.R., Cannon, G.C., Cai, F., Williams, E.B., Heinhorst, S., Kerfeld, C.A., and Yeates, T.O. (2007). Structural analysis of CsoS1A and the protein shell of the *Halothiobacillus neapolitanus* carboxysome. PLoS Biol. *5*, e144.

Tsai, Y., Sawaya, M.R., and Yeates, T.O. (2009). Analysis of lattice-translocation disorder in the layered hexagonal structure of carboxysome shell protein CsoS1C. Acta Crystallogr. D Biol. Crystallogr. *65*, 980–988.

Wang, H.L., Postier, B.L., and Burnap, R.L. (2004). Alterations in global patterns of gene expression in *Synechocystis* sp. PCC 6803 in response to inorganic carbon limitation and the inactivation of *ndhR*, a LysR family regulator. J. Biol. Chem. *279*, 5739–5751.

Woodger, F., Badger, M., and Price, G. (2003). Inorganic Carbon Limitation Induces Transcripts Encoding Components of the CO_2-Concentrating Mechanism in *Synechococcus* sp. PCC7942 through a Redox-Independent Pathway. Plant Physiol. *133*, 2069–2080.

Woodger, F.J., Badger, M.R., and Price, G.D. (2005). Sensing of inorganic carbon limitation in *Synechococcus* PCC7942 is correlated with the size of the internal inorganic carbon pool and involves oxygen. Plant Physiol. *139*, 1959–1969.

Wunderlich, Z., Acton, T.B., Liu, J., Kornhaber, G., Everett, J., Carter, P., Lan, N., Echols, N., Gerstein, M., Rost, B., *et al.* (2004). The protein target list of the Northeast Structural Genomics Consortium. Proteins *56*, 181–187.

Xu, M., Bernát, G., Singh, A., Mi, H., Rögner, M., Pakrasi, H., and Ogawa, T. (2008). Properties of Mutants of *Synechocystis* sp. Strain PCC 6803 Lacking Inorganic Carbon Sequestration Systems. Plant Cell Physiol. *49*, 1672–1677.

Yeates, T.O., Tsai, Y., Tanaka, S., Sawaya, M.R., and Kerfeld, C.A. (2007). Self-assembly in the carboxysome: a viral capsid-like protein shell in bacterial cells. Biochem. Soc. Trans. *35*, 508–511.

Yeates, T.O., Thompson, M.C., and Bobik, T.A. (2011). The protein shells of bacterial microcompartment organelles. Curr. Opin. Struct. Biol. *21*, 223–231.

Zarzycki, J., Axen, S.D., Kinney, J.N., and Kerfeld, C.A. (2012). Cyanobacterial-based approaches to improving photosynthesis in plants. J. Exp. Bot. *64*, 787–798.

Zhang, H., Ishige, K., and Kornberg, A. (2002). A polyphosphate kinase (PPK2) widely conserved in bacteria. Proc. Natl. Acad. Sci. U.S.A. *99*, 16678–16683.

Zhang, S., Laborde, S.M., Frankel, L.K., and Bricker, T.M. (2004). Four novel genes required for optimal photoautotrophic growth of the cyanobacterium *Synechocystis* sp. strain PCC 6803 identified by *in vitro* transposon mutagenesis. J. Bacteriol. *186*, 875–879.

Zilliges, Y., Kehr, J.C., Meissner, S., Ishida, K., Mikkat, S., Hagemann, M., Kaplan, A., Borner, T., and Dittmann, E. (2011). The cyanobacterial hepatotoxin microcystin binds to proteins and increases the fitness of microcystis under oxidative stress conditions. PLoS One *6*, e17615.

Glycogen, a Dynamic Cellular Sink and Reservoir for Carbon

8

Yvonne Zilliges

Abstract

All microorganisms accumulate carbon and energy reserves to cope with starvation conditions temporally present in the environment. In cyanobacteria, glycogen biosynthesis is the main strategy for metabolic sink and storage of photosynthetically fixed carbon. Glycogen biosynthesis is therefore tightly coupled to light and dark reactions of photosynthesis. Inversely, the process of glycogen degradation provides carbon and energy for adverse cellular conditions.

This review summarizes the current knowledge of the chemical properties and structure of glycogen and starch-like reserves in cyanobacteria, the different enzymology and regulation of glycogen biosynthesis and degradation, and the function of glycogen metabolism in cyanobacteria. A special focus is drawn on its roles in photosynthetic efficiency, during the process of nitrogen chlorosis and for the steady-state of anabolic and catabolic reactions, especially under unbalanced growth conditions.

Introduction

Cyanobacteria often live in a fluctuating environment, where essential nutrients are available in an intermittent fashion. A large pulse of nutrients may be followed by a period of nutrient deprivation. In order to cope with such inconsistent conditions, most microorganisms store excess nutrients in the form of reserve polymers that can be used as internal degradable resources in periods of nutrient deprivation. The major advantage of polymer formation is the stable, even so readily accessible storage of energy, carbon, nitrogen or phosphate and its minimal interference with other cellular processes (e.g. osmotic pressure).

A variety of cellular reserve materials of prokaryotic organisms are visible as granular cytoplasmic inclusions. Cyanobacteria produce four chemically different reserve polymers (Fig. 8.1): glycogen, poly-β-hydroxybutyric acid, cyanophycin and polyphosphate (Allen, 1984). In cyanobacteria, among these inclusions, much more information is available on glycogen, which

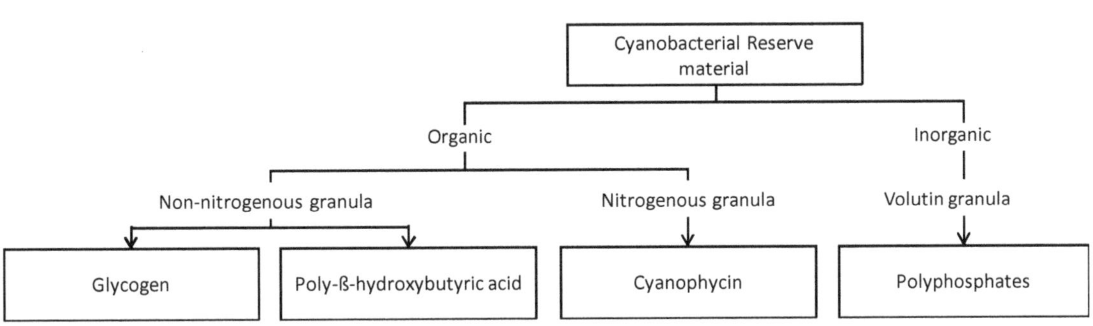

Figure 8.1 Overview of the four different kinds of cyanobacterial reserve materials and classification according to their chemical nature.

will be addressed in detail in this chapter, and cyanophycin, which will be discussed in the next chapter of this book (Chapter 9). Merely, some general information about poly-β-hydroxybutyric acid and polyphosphate inclusions will be given at the end of this chapter.

Glycogen is the prevalent glucose (carbon) polymer in eubacteria (including cyanobacteria), archaea and animals, but not in plants and algae. A micrograph in which glycogen granules can be appreciated is shown in Fig. 8.2. The figurative sense of this Greek compound word (Gr. γλυκογόνο) is 'sweet happening' (Gr. γλυκός – glykos, sweet; Gr. γίγνομαι – gignomai, happen). In the 19th century, the French scientist Claude Bernard first elucidated the chemical and physical properties of glycogen (also called dextrin in older texts) that is stored in the liver tissue (Young, 1957). In the 1950s, the Argentinean scientist Luis Federico Leloir and his co-workers intensively studied the properties of nucleoside-diphosphate sugars. These studies indicated that biosynthesis and degradation of glycogen occur by different pathways (Cabib, 1970). The presence of these non-amphibolic pathways implied a distinct enzymology and regulation of glycogen synthesis and degradation that have been well characterized by Jack Preiss and co-workers in the last decades (Preiss, 2006).

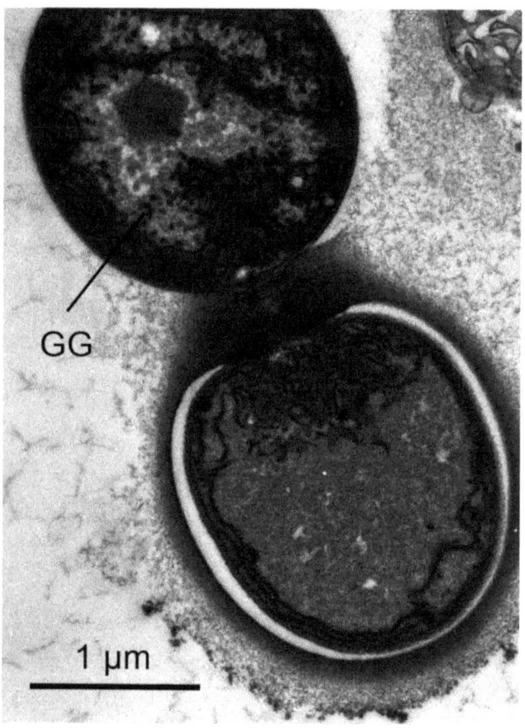

Figure 8.2 Transmission electron micrograph of part of a filament of *Anabaena* sp. PCC 7120 showing a vegetative cell and a heterocyst. Glycogen granules are clearly visible in the vegetative cell, but they are not present in the heterocyst. (Micrograph courtesy of Victoria Merino-Puerto and Iris Maldener.)

In photoautotrophic organisms (cyanobacteria, algae and plants), glucan polymer formation is closely linked to the photosynthetic light and dark reactions. The carbon assimilation reactions, also referred to as the Calvin–Benson–Bassham cycle or the reductive pentose phosphate pathway, use NADPH and ATP provided by the light reactions to reduce carbon dioxide to triose phosphate, 3-phosphoglycerate (3-PGA). Via gluconeogenesis, two molecules of triose phosphate are finally converted to glucose-6-phosphate (see Fig. 8.6). Glucose-6-phosphate is isomerized by phosphoglucomutase (PGM) to glucose-1-phosphate, the precursor of glycogen in cyanobacteria and starch in plants and algae. A multitude of enzymes and regulatory mechanisms of photoautotrophic life-style have been conserved during the endosymbiotic event. However, the formation of glycogen inclusions has been substituted by the formation of starch granules in plants and algae (Ball *et al.*, 2011).

Structures of glycogen and starch-like reserves in cyanobacteria

Glycogen is a homogeneous, water-soluble polyglucan composed of linear chains usually consisting of 9–13 (1–4)-linked α-D-glucose residues, a size within the range (10–14) indicated for bacterial glycogens (Merrick, 1979). These glucan chains are interlinked via (1–6)-α-D-glucosidic linkages (correspond to 7–10% of *O*-glucosidic linkages) that form a highly branched structure of about 10^7–10^8 Da (Wilson *et al.*, 2010). The frequent, symmetric and binary branching principle facilitates a limited spherical growth of the glycogen particle (Fig. 8.3A and C). To maintain

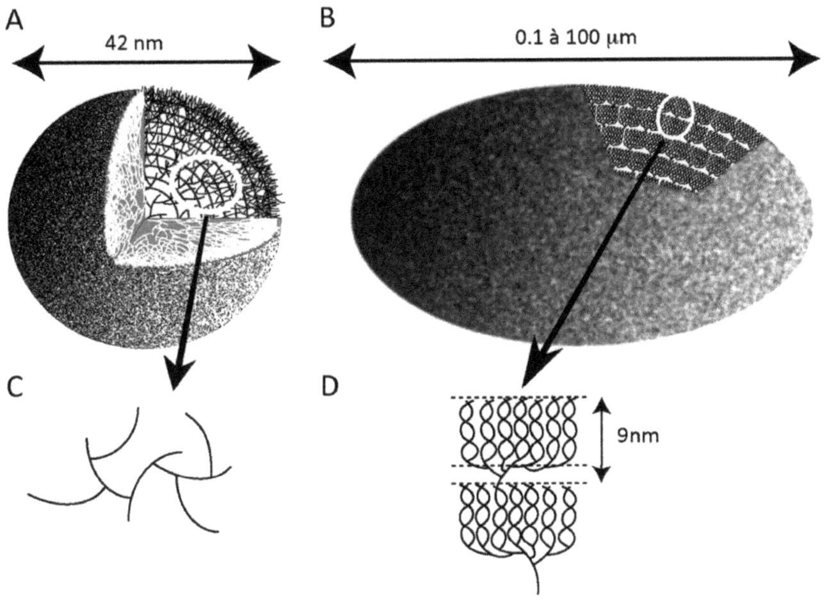

Figure 8.3 Schematic representation of glycogen (A) and starch (B) granules, from Ball *et al.*, 2011 (Oxford university press with permission). Glycogen and amylopectin (the major constituent of the native, water-insoluble starch) are polyglucans composed of linear chains of (1–4)-linked α-D-glucose residues (represented by lines) and interlinked via (1–6)-α-D-glucosidic linkages (represented by intersections). The frequent, symmetric and binary branching principle (exemplified in C) enables only a limited spherical growth of the glycogen particle with a predictable maximal diameter of 42 nm (displayed in A). By contrast, a lower frequency and asymmetric distribution of α-(1–6) branches in amylopectin (exemplified for two amylopectin clusters in D) allow the formation of intramolecular tandem clusters with a nearly unlimited spherical growth of the whole starch granule (displayed in B).

a complex granule structure a coordinated action of enzymatic activities and different structural proteins are required.

In some cyanobacteria such as *Cyanothece* sp. ATCC 51142, *Cyanobacterium* sp. MBIC10216, *Myxosarcina burmensis* and *Synechococcus* sp. BG043511, distinct α-polyglucans in terms of structure (semi-crystalline α-glucans with a tandem-cluster structure), chain length distribution, molecular size and length of the most abundant (1–4)-α-glucan chain have been found (Nakamura *et al.*, 2005; Schneegurt *et al.*, 1994, 1997; Suzuki *et al.*, 2013). These are structural intermediates between the typical glycogen and the amylopectin of the starch granule, and therefore are designated as semi-amylopectin or cyanobacterial starch (Suzuki *et al.*, 2013). Amylopectin, the major constituent (65–90%) of the native, water-insoluble starch, is also composed of (1–4)-linked and (1–6)-interlinked α-D-glucose residues (Ball *et al.*, 2011). Both a lower frequency (4–6%) and asymmetric distribution of α-(1–6) branches in amylopectin allow the formation of intramolecular tandem clusters with a nearly unlimited spherical growth of the whole starch granule (Fig. 8.3B and D).

The synthesis of semi-crystalline, starch-like polymers enables, similar as in plants, the storage of larger, osmotically inert carbon amounts (photosynthate). Therefore, it is supposed that cyanobacteria, which need to store significantly larger amounts of polyglucans to maintain cell integrity, function and viability (e.g. in diazotrophic cyanobacteria), are forced to change glycogen metabolism into the synthesis of starch-like polymers (Deschamps *et al.*, 2008). Thus, it is likely that starch biosynthesis had already originated from the cyanobacterial ancestor and the ability for glycogen synthesis, which has been kept in cyanobacteria, was lost early during endosymbiosis (Ball *et al.*, 2011; Ball and Morell, 2003; Patron and Keeling, 2005).

Enzymology of glycogen metabolism in cyanobacteria

Polymerization

The first committed step of both glycogen synthesis in eubacteria and starch synthesis in plants is the production of the glycosyl donor ADP-glucose by the ADP-glucose pyrophosphorylase (AGPase, GlgC) (EC 2.7.7.27), also referred to as glucose-1-phosphate adenylyltransferase (Ball and Morell, 2003). AGPase, which is localized in the cytoplasm and only loosely associated with the granule (as observed in *Cyanothece* sp. ATCC 51142) (Schneegurt *et al.*, 1997), catalyses the formation of the glycosyl donor ADP-glucose from glucose-1-phosphate and ATP (Fig. 8.4).

Reaction 1: ATP + α-D-glucose 1-phosphate → diphosphate + ADP-glucose

The essential activity of AGPase in cyanobacterial glycogen synthesis has been successfully verified by mutagenesis analysis, primarily performed in *Synechocystis* sp. PCC 6803 (Gründel *et al.*, 2012; Wu *et al.*, 2001) and *Synechococcus elongatus* PCC 7942 (Suzuki

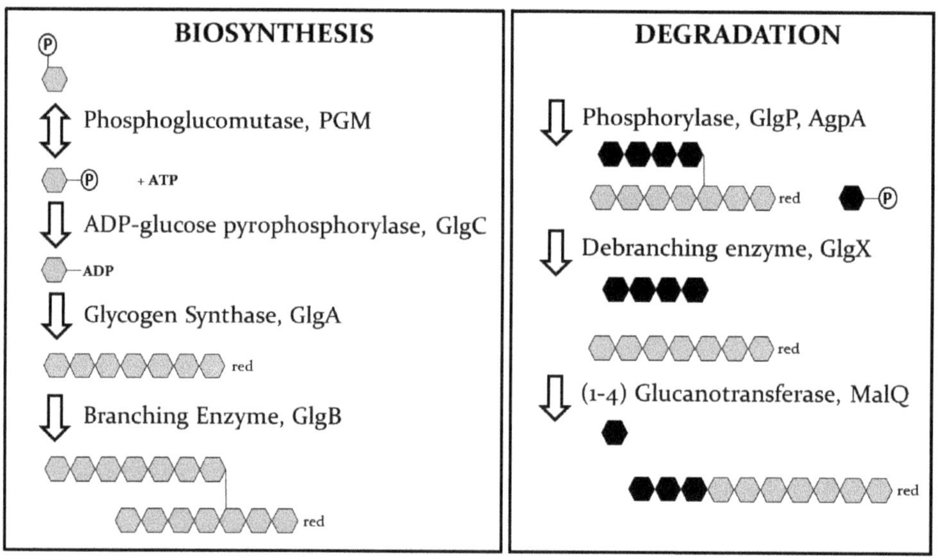

Figure 8.4 Bacterial glycogen metabolism. The pathways outlined concern glycogen metabolism in most cyanobacteria and is supported by various biochemical (cf. Text) and genetic studies (cf. Table 8.2). The different chemical properties of nucleoside-phosphate sugars imply a distinct enzymology and regulation of glycogen synthesis and degradation. The precursor of glycogen biosynthesis is glucose-1-phosphate that is isomerized from glucose-6-phosphate by the activity of phosphoglucomutase (PGM). The first committed step of glycogen biosynthesis is the synthesis of the glycosyl donor, namely ADP-glucose, from glucose-1-phosphate and ATP, catalysed by the ADP-glucose pyrophosphorylase (GlgC). The glucose moiety of ADP-glucose is transferred to the non-reducing end of a linear α-(1–4) glucan primer, a reaction catalysed by glycogen synthases (GlgA). The α-(1–6) glycosylic branches are introduced into the growing polymer by the branching enzyme (GlgB). A distinct enzymology is necessary for the degradation of glycogen granula. The glycogen phosphorylase (GlgP, AgpA) transfers orthophosphate to the non-reducing end of glucose residues, releasing glucose-1-phosphate. This process continues until the last four glucose residues are left on each branch. Outer chains of four glucose residues define the final product of phosphorylase action. The following GlgX-type of direct debranching enzymes are quite specific in their substrate requirements and will not cleave the branches of outer chains of glycogen if the chains exceed four glucose residues in length. These enzymes hydrolyse at the α-(1–6) branching point, producing a maltotetraose, which is further metabolized by the α-(1–4) glucanotransferase (MalQ) and maltodextrin phosphorylases. Abbreviation: red, reducing end.

et al., 2010). Moreover, *in vitro* assays as well as site-directed mutagenesis of the gene encoding AGPase (*glgC, agp*) from *Anabaena* sp. PCC 7120 and *Synechocystis* sp. PCC 6803 confirmed the presence, catalytic mechanism and regulation of this enzyme in cyanobacteria (Iglesias *et al.*, 1991). Additionally, it has been reported that ADP-glucose is formed by the activity of sucrose synthase (SusA) participating in a sucrose cycle (e.g. in heterocyst-forming cyanobacteria) (Cumino *et al.*, 2007; Curatti *et al.*, 2008). In a *susA* mutant of *Anabaena* sp. PCC 7119, the glycogen level is significantly reduced although the AGPase activity is twice as high as in the wild-type strain (Cumino *et al.*, 2002; Curatti *et al.*, 2008). On the other hand, ADP-glucose synthesized by AGPase could also be used as a glycosyl donor for glycosylglycerol synthesis in the osmotic stress response (Hagemann and Erdmann, 1994).

In the second step, a glycogen synthase (GlgA) (EC 2.4.1.21), also referred to as glycosyltransferase or α-(1–4)-glucan synthase, use the ADP-glucose as a sugar donor to elongate a pre-existing α-(1–4)-glucan chain. Thereby, the glucose moiety of ADP-glucose is transferred to the non-reducing end of the linear α-glucan primer forming an α-(1–4)-glycosylic linkage (Fig. 8.4). The cyanobacterial glycosyltransferases belong to the family 5 of glycosyltransferases (GT5; see CAZypedia, http://www.cazypedia.org/), also referred to as glycogen (starch) synthases (Cid *et al.*, 2002). It has been shown that the GT5 type glycogen synthase of *Agrobacterium tumefaciens* is capable of autoglycosylation, and therefore this enzyme does not need the presence of another protein (such as metazoan glycogenin) to prime glycogen synthesis (Ugalde *et al.*, 2003). No prokaryotic glycogenin homologue has been identified in any of the glycogen-synthesizing bacteria. Therefore it is proposed that the eubacterial glycogen synthase not only elongates α-(1–4)-linked glucans but also forms the primer required for the elongation process by catalysing its own glucosylation.

Reaction 2: ADP-glucose + [(1–4)-α-D-glucosyl]^{n} → ADP + [(1–4)-α-D-glucosyl]$^{n+1}$

In most cyanobacteria at least one type of glycosyltransferase (GlgA1) has been found (Beck *et al.*, 2012). Various cyanobacteria even have two isoenzymes (GlgA1, GlgA2). According to maximum likelihood phylogeny, the GlgA1 homologue is the unique glycosyltransferase in most eubacteria (Cid *et al.*, 2002; Patron and Keeling, 2005). The cyanobacterial GlgA2 isoform is closely related to plant and green algal soluble starch synthases SS3 and SS4, which are predominantly localized in the plastid stroma. Here, each enzyme has a specific role in determining granule structure (Edwards *et al.*, 2002; Fulton *et al.*, 2002). Putative distinctive roles of both cyanobacterial GlgA isoenzymes in glycogen granule formation have not yet been investigated. However, mutagenesis of each *glgA* gene in *Synechocystis* sp. PCC 6803 and *Synechococcus* sp. PCC 7002 has shown that each isoenzyme is able to synthesize an α-polyglucan (Gründel *et al.*, 2012; Xu *et al.*, 2012b). Similarly to the mutant of AGPase, inactivation of the unique glycogen synthase in *Synechococcus elongatus* PCC 7942 (Suzuki *et al.*, 2010) or of both glycogen synthases in *Synechocystis* sp. PCC 6803 and *Synechococcus* sp. PCC 7002 (Gründel *et al.*, 2012; Xu *et al.*, 2012b) leads to a complete abrogation of glycogen biosynthesis. Thus, the activity of both AGPase and glycogen synthase is essential for biosynthesis of polyglucans in cyanobacteria.

The final conformation of the glycogen (as well as of the starch) particle depends mainly on the action of the branching enzyme(s) (GBE, GlgB) (EC 2.4.1.18) and the debranching enzyme(s) (DBE, GlgX) (EC 3.2.1.) that catalyse the addition and removal of α-(1–6)-linked ramifications, respectively (Suzuki *et al.*, 2007; Yoo *et al.*, 2002). Both are glycosylhydrolases (GH13) with a transglycosylation activity. The debranching enzyme is also involved in depolymerization processes and is therefore described in detail in the next section of this chapter. The branching enzyme, also known as α-(1–4)-glucan-6-glycosyltransferases, hydrolyses an α-(1–4)-linkage within a pre-existing α-(1–4)-linked glucan and transfers six to eight glucose segments in α-(1–6) position to an α-glucan acceptor (Preiss and Romeo, 1994) (Fig.

8.4). The branching increases the number of non-reducing ends that in turn enables further glucan polymerization or depolymerization.

Reaction 3: 2 × α-(1–4)-D-glucan → α-(1–4)-D-glucan-α-(1–6)-(α-(1–4)-D-glucan)

In most cyanobacterial genomes only one gene (*glgB*) encoding α-(1–4)-glucan-6-glycosyltransferase is annotated (Beck *et al.*, 2012). Just a few cyanobacteria have multiple potential branching enzymes (Beck *et al.*, 2012) similar to those involved in starch granule formation. These starch branching isoenzymes differ in substrate specificity and expression pattern (Mutisya *et al.*, 2003; Yao *et al.*, 2004). Mutagenesis of the single *glgB* gene in *Synechocystis* sp. PCC 6803 clearly indicates that this protein has a key function in formation of branch linkages (Yoo *et al.*, 2002). The polyglucan synthesized in the mutant is significantly reduced in its molecular mass and degree of polymerization, and hence it is less water-soluble than that of the wild-type (Yoo *et al.*, 2002).

Depolymerization

The degradation of glycogen to provide readily usable energy in the form of phosphorylated sugar intermediates requires different enzymatic activities. In cyanobacteria, this process is not as well understood as glycogen synthesis. Most of the reactions have been deduced from processes in *Escherichia coli* (Deschamps *et al.*, 2008). Complete degradation of the polysaccharides is achieved via the coordinated action of phosphorylase and glycogen (starch)-debranching enzyme (phosphorolytic pathway) and a variable combination of amylases, glucosidases, and α-(1–4) glucanotransferases (hydrolytic pathway) (Deschamps *et al.*, 2008).

In the first step, glycogen phosphorylase (GlgP, AgpA) (EC 2.4.1.1) releases α-D-glucose 1-phosphate via the transfer of orthophosphate to the non-reducing end (Fig. 8.4). Glycogen phosphorylases are glycosyltransferases that belong to family 35 of glycosyltransferases (GT35; see CAZypedia, http://www.cazypedia.org/). The successive breakdown of α-(1–4)-glycosylic linkages proceeds to leave the last four glucose residues on each branch.

Reaction 1: $[(1–4)\text{-}\alpha\text{-D-glucosyl}]^{n}$ + phosphate → $[(1–4)\text{-}\alpha\text{-D-glucosyl}]^{n-1}$ + α-D-glucose 1-phosphate

Two closely related types of glycogen phosphorylases have been found in cyanobacteria, designated GlgP and AgpA (Deschamps *et al.*, 2008; Jacobsen *et al.*, 2011). Most cyanobacteria have several *glgP* homologues (e.g. *Synechocystis* sp. PCC 6803) (Beck *et al.*, 2012) or several *agpA* homologues (e.g. *Nostoc punctiforme* PCC 73102) (Deschamps *et al.*, 2008; Jacobsen *et al.*, 2011). A few cyanobacteria even have both *glgP* and *agpA* homologues (e.g. *Synechococcus* sp. PCC 7002) (Jacobsen *et al.*, 2011). The two *glgP* homologues of *Synechocystis* sp. PCC 6803 are differentially expressed in response to distinct environmental conditions and might have divergent functions in glycogen degradation (Fu and Xu, 2006).

In a second step, the external four glucose residue segments are further degraded via the isoamylase-type debranching enzymes (DBE, GlgX) (EC 3.2.1.68) (Ball and Morell, 2003). Glycosyl-hydrolases of this type directly cleave the α-(1–6) branch thereby releasing a malto-oligosaccharide of four glucose residues (maltotetraose) as it is known from studies in *Escherichia coli* (Dauvillee *et al.*, 2005). The bacterial GlgX type of debranching enzyme hydrolyses only those external branches (ECL4, exterior chain length 4) that have been digested by a glycogen phosphorylase (Fig. 8.4).

Reaction 2: [external four glucose-residue segment]-glycogen → glycogen + maltotetraose

According to a neighbour-joining tree and a maximum likelihood phylogeny, there are three different groups of isoamylase-type DBEs (GlgX) in cyanobacteria (Patron and Keeling, 2005; Suzuki *et al.*, 2007): a specific cyanobacterial-type (e.g. in *Synechococcus elongatus* PCC 7942, *Prochlorococcus marinus*, *Synechocystis* sp. PCC 6803, *Trichodesmium erythraeum* IMS101), a general eubacterial-type (e.g. in *Nostoc punctiforme*, *Trichodesmium erythraeum* IMS101), and a plant-type DBE (e.g. in *Synechocystis* sp. PCC 6803, *Trichodesmium erythraeum* IMS101). So far, there is only one successful study of mutagenesis of a glgX gene reported for *Synechococcus elongatus* PCC 7942. This study supports the function of GlgX in removal of α-(1–6)-linked ramifications in the glycogen macromolecule (Suzuki *et al*., 2007). Surprisingly, some cyanobacterial species such as *Anabaena* sp. PCC 7120, *Gloeobacter violaceus* PCC 7421 and *Thermosynechococcus elongatus* BP-1 lack the *glgX* gene (Suzuki *et al*., 2007) and thus may have other (yet unknown) DBEs.

According to the catalytic reactions known from studies in *Escherichia coli*, the maltotetraose is further degraded by α-(1–4)-glucanotransferase and the above described phosphorylase(s) (Boos and Shuman, 1998). The α-(1–4)-glucanotransferase (α-GTase, MalQ) (EC 2.4.1.25), which belongs to the α-amylase family, catalyses a reversible intermolecular glucan transferring reaction. Here, the malto-oligosaccharides are disproportionated into longer chains, thereby generating a malto-oligosaccharide series of increasing length (Fig. 8.4). These longer chains of malto-oligosaccharides will be degraded to maltotetraose by the action of a phosphorylase (Fig. 8.4).

Reaction 3: maltotetraose + maltotetraose → maltoheptaose + D-glucose

Reaction 4: $[(1\to4)\text{-}\alpha\text{-D-glucosyl}]^n$ + phosphate → $[(1\to4)\text{-}\alpha\text{-D-glucosyl}]^{n-1}$ + α-D-glucose 1-phosphate

In each of the so far annotated cyanobacterial genomes at least one *malQ* homologue, encoding α-(1–4)-glucanotransferase, has been identified (Nakao et al., 2010). An *in vitro* characterization of the α-(1–4)-glucanotransferase from *Synechocystis* sp. PCC 6803 confirms the transglycosylating activity on various maltooligosaccharides (Lee *et al*., 2009). Moreover, α-amylase-like sequences have been found in the genomes of *Nostoc* sp. PCC 7119 and *Synechococcus* sp. RF1 (Reyes-Sosa *et al*., 2010; Wing-Ming *et al*., 1994). This type of glycosylhydrolases (EC 3.2.1.1) randomly cleaves only α-(1–4)-glycosylic linkages.

Regulation of cyanobacterial glycogen metabolism

Glycogen levels

In heterotrophic microorganisms, several factors such as macronutrient availability, pH, temperature and salinity have been described to control glycogen accumulation (Preiss, 1984; Wilson *et al*., 2010). In phototrophic cyanobacteria, the intracellular glycogen level (on average 1–5% w/w dry weight) varies mainly as a function of light (photosynthetic activity) and macronutrient availability.

In the light, the photosynthate from fixed carbon is placed in glycogen granules; in the dark glycogen is mobilized and used as an endogenous respiratory substrate. Therefore, the polyglucan content and the granule formation follow a diurnal pattern in most cyanobacteria: assimilation in light periods and breakdown in dark periods (Griffiths *et al*., 1987; Schneegurt *et al*., 1994). Even in cyanobacteria with a quite constant glycogen amount per cell, a turn-over of glycogen during a diurnal cycle has been documented (Lehmann and Wober, 1976). Moreover, it has been shown that high light stress leads to a significant increase of the glycogen pool in *Spirulina maxima* (De Philippis *et al*., 1992b). No glycogen synthesis is observed in non-photoactive cells (meaning stop of photosynthetic light and dark reactions), either cells that are in the dark or dormant and specialized resting cells (Gorl *et al*., 1998; Gründel *et al*., 2012; Sarma and Kanta, 1979).

The magnitude of the glycogen pool varies in response to the carbon availability. Cells grown under high CO_2 conditions store significantly higher amounts of glycogen than those grown under low CO_2 conditions (Eisenhut *et al*., 2007). Inversely, a transient carbohydrate anaplerosis, e.g. when cells are shifted from high carbon to low carbon conditions, is compensated by the glycogen pool (Eisenhut *et al*., 2008). Some cyanobacteria are capable of both auto- and heterotrophic growth. In some cases, the organic compound and CO_2 are simultaneously assimilated by the cell (photomixotrophic growth). When such conditions of carbon excess are combined with nitrogen depletion, a rapid and large polyglucan accumulation (on average 60–70% w/w dry weight) is observed in cyanobacteria (Elmorjani and Herdman, 1987; Yoo *et al.*, 2007). However, glycogen synthesis immediately stops in *Synechocystis* sp. and *Anabaena* sp. in the dark (Gründel *et al.*, 2012; Sarma and Kanta, 1979). These observations indicate a tight link of polyglucan synthesis with both the photosynthetic light and dark reactions and the reactions of the central carbon/nitrogen metabolism.

The accumulation of glycogen reserves is a general characteristic of the microbial stringent response to deficiencies in macronutrients such as nitrogen (in non-diazotrophic organisms) or sulfur, where cells switch from a growth-related mode (active state) to a maintenance/survival mode (dormant state) (Kaprelyants *et al.*, 1993; Schwarz and Forchhammer, 2005; Tandeau de Marsac and Houmard, 1993). In agreement with this, increased glycogen levels have also been detected in the stationary phase, in which nutrients become limited (Lehmann and Wober, 1976). Inversely, addition of the corresponding macronutrient to a starved, glycogen-rich culture leads to a rapid degradation of glycogen reserves (Allen and Smith, 1969; Lehmann and Wober, 1976; Stevens *et al.*, 1981).

In diazotrophic cyanobacteria, glycogen accumulates also during active photosynthesis and is used, mainly via the oxidative pentose phosphate (OPP) pathway, to supply energy and reducing power for temporally or spatially (in heterocysts) separated N_2 fixation (Ernst and Boger, 1985; Ernst *et al.*, 1990; Fay, 1992; Griffiths *et al.*, 1987; Jensen *et al.*, 1986; Zhang *et al.*, 2006). Although glycogen granules are not evident in heterocysts (Fig. 8.2), glycogen can be found in these differentiated cells (Valladares *et al.*, 2007). This indicates that the polyglucan pool has an important role in the diazotrophic life-style.

The control mechanisms underlying glycogen synthesis and glycogen degradation in cyanobacteria have been well characterized at the enzymatic activity and transcriptional initiation levels. Putative regulatory post-transcriptional modifications have not yet been investigated in detail. However, a global transcriptional start site mapping in *Synechocystis* sp. PCC 6803 revealed a potential antisense transcript (asRNA) originating from the complementary strand of the *glgC* gene (Mitschke *et al.*, 2011). Such antisense transcripts could act in transcription termination, codegradation, control of translation, transcriptional interference or enhanced stability of their respective target

Table 8.1 Regulation of glycogen biosynthesis and degradation

Enzyme	Gene	Regulation of expression by	Regulation of enzymatic activity by	Reference
Phosphoglucomutase (PGM)	*pgm*		Thioredoxin (activation)	Lindahl and Florencio (2003)
ADP-glucose pyrophosphorylase (AGPase)	*glgC*		Thioredoxin (activation) 3-PGA (activation, ultrasensitive) P_i (repression, ultrasensitive)	Iglesias *et al.* (1991b), Lindahl and Florencio (2003)
Glycogen synthase	*glgA2*		Thioredoxin (activation)	Lindahl and Florencio (2003)
1,4-α-Glucan branching enzyme (GBE)	*glgB*		Thioredoxin (activation)	Lindahl and Florencio (2003)
Glycogen phosphorylase	*glgP*	SigE, (Hik8) OmpR		Azuma *et al.* (2011), Ehira and Ohmori (2011), Osanai *et al.* (2005), Singh and Sherman (2005)
Glycogen debranching enzyme (DBE)	*glgX*	SigE, (Hik8) OmpR		Azuma *et al.* (2011), Ehira and Ohmori (2011), Osanai *et al.* (2005), Singh and Sherman (2005)

Abbreviations: Hik8, histidine kinase 8; OmpR, OmpR-type response regulator; 3-PGA, 3-phosphoglycerate, P_i, inorganic phosphate; SigE, RNA polymerase group 2 sigma factor SigE.

transcripts (Georg and Hess, 2011). The mechanism of action of the *glgC* antisense RNA remains to be defined.

Regulation of enzyme activity

The activity of enzymes involved in glycogen synthesis is mainly regulated (i) by post-translational redox modifications and (ii) by effectors (Table 8.1). The biochemical link between light reactions and the regulation of carbon metabolism is provided by the ferredoxin/thioredoxin system (Gleason, 2004). In a proteomic approach in *Synechocystis* sp. PCC 6803, all enzymes involved in glycogen synthesis, namely AGPase (GlgC), glycogen synthase (GlgA2) and branching enzyme (GlgB), were identified as thioredoxin targets that are activated in the light (Table 8.1) (Lindahl and Florencio, 2003; Lindahl and Kieselbach, 2009). This regulatory principle explains the fact that glycogen synthesis occurs only in the light and stops immediately in the dark. Moreover, it was shown that the phosphoglucomutase (PGM, EC 5.4.2.2), which catalyses the interconversion (phosphate transfer) of glucose 1-phosphate and glucose 6-phosphate, is also redox-controlled in *Synechocystis* sp. PCC 6803 (Lindahl and Florencio, 2003). This enzyme is inactivated by oxidation and reactivated by thioredoxin (TrxA) *in vitro* (Table 8.1). The phosphoglucomutase represents a metabolic branch point linking glycolysis, the oxidative pentose phosphate pathway and glycogen metabolism and thus is directly involved in the processes of storage (glycogen synthesis) and utilization (e.g. respiration via glycolysis and OPP pathway) of polyglucans (Egli *et al.*, 2010). Starch synthesis is also activated by the thioredoxin system, but only at the level of AGPase (Schuermann and Buchanan, 2008). Here, trehalose-6-phosphate acts as a messenger to report cytosolic carbon status (e.g. sucrose level), promoting thioredoxin-mediated activation of AGPase in the plastids (Kolbe *et al.*, 2005). An indication of a thioredoxin-dependent regulation of glycogen degradation as observed for starch degradation (Geigenberger *et al.*, 2005; Schuermann and Buchanan, 2008) has only been reported for a specific amylase (Amy1) of *Nostoc* sp. PCC 7119 (Reyes-Sosa *et al.*, 2010).

The ferredoxin/thioredoxin system also acts on other light-driven processes such as pH shifts and effector concentration to modulate the enzymatic activity in response to light. Thus, glycogen synthesis is allosterically regulated by effector molecules at the level of AGPase, its first committed step (Ballicora *et al.*, 2003; Preiss, 1984). 3-Phosphoglycerate (3-PGA), an intermediate of the reductive pentose phosphate cycle that signals high carbon and energy content, and inorganic phosphate (P_i), which is indicative of a low energy level, are the most potent activator and inhibitor, respectively, of the cyanobacterial homotetrameric AGPase, as is also the case for the algal and plant heterotetrameric AGPase (Iglesias *et al.*, 1991; Levi and Preiss, 1976). Despite a different quaternary structure, both types of AGPases (bacterial-type and plant starch-type) are strictly regulated by the ratio of 3-PGA to P_i and are immunologically related to each other (Iglesias *et al.*, 1991; Preiss and Sivak, 1998). The inhibitor P_i elicits an ultrasensitive (sigmoidal) response in the enzyme towards 3-PGA activation, allowing an efficient response to minimal changes of the activator 3-PGA, even in a high P_i background (Gomez-Casati *et al.*, 1999, 2003). It has been proposed that this ultrasensitive mechanism is not operative under constant illumination, being functional during transitions from light to dark (Gomez-Casati *et al.*, 2003). Ultrasensitive regulations are widespread among prokaryotic and eukaryotic cells and are involved in different cellular processes: in *turning on or off* of metabolic pathways (Xu *et al.*, 2012a), in signal amplification in kinase cascades (Goldbeter and Koshland, 1984) and in protein translocations (Ferrell Jr, 1998). In eukaryotic glycogen synthesis, glycogen synthase (here the first committed step in glycogen synthesis) and glycogen phosphorylase are regulated in an ultrasensitive mode by a unique regulatory network to circumvent a futile cycle by simultaneous synthesis and degradation of glycogen (Mutalik and Venkatesh, 2005). In summary, thioredoxin-dependent activation and 3-PGA/P_i ultrasensitive response in glycogen synthesis effectively couples storage polysaccharide synthesis to the light and dark reactions of photosynthesis. On the other hand, the regulation of enzyme activity leading to glycogen degradation in cyanobacteria remains to be investigated. A simple adaptation of what is known from other bacterial regulatory networks seems inappropriate. Although all bacterial characterized phosphorylases appear to follow the same catalytic mechanism and show a high conservation of internal protein core, active site and cofactor-binding site, they differ in regulatory properties and substrate specificity (Schinzel and Nidetzky, 1999).

Regulation of gene expression

A minimal gene set for glycogen biosynthesis (*glgC*, *glgA* and *glgB*) and degradation (*glgP* and *glgX*) is conserved in all so far annotated cyanobacterial genomes (Beck *et al.*, 2012). Although the genes encoding enzymes of glycogen metabolism are functionally related and ought to be co-regulated and simultaneously expressed in the process of glycogen synthesis or degradation, none of them are co-localized in an operon structure in cyanobacteria (Memon *et al.*, 2012). By contrast, a number of bacteria possess a single glycogen operon comprising all *glg* genes, as in *Bacillus subtilis*, or two tandemly arranged operons (*glgBX* and *glgCAP*), as in *Escherichia coli* (Wilson *et al.*, 2010). Nevertheless, in various cyanobacterial genomes some glycogen structural genes are closely localized and obviously co-expressed with different enzymes of the central carbon metabolism. This is the case of GlgP with glycolytic NAD-specific glyceraldehyde-3-phosphate dehydrogenase (Gap1) in *Synechococcus* sp. PCC 7942 (Koksharova *et al.*, 2004) and of GlgC with the second OPP enzyme 6-phosphogluconate dehydrogenase (Gnd) in most *Prochlorococcus* strains (Nakao *et al.*, 2010), indicating a putative functional relation.

The expression of some genes involved in glycogen metabolism is modulated at the transcriptional level (Table 8.1). In *Synechocystis* sp. PCC 6803 the expression of *glgP* and *glgX* is primarily induced in the dark (Osanai *et al.*, 2005). An RNA polymerase group 2 sigma factor, namely SigE, regulates the expression of genes involved in sugar catabolism such as *glgP* and *glgX* in the dark (Osanai *et al.*, 2005, 2007). Here, SigE acts as a positive regulator of sugar catabolism (Osanai *et al.*, 2007). The SigE expression is under circadian rhythm control (Osanai *et al.*, 2008). Therefore, *glgP* transcription often shows a diurnal expression pattern as observed in *Synechococcus* sp. WH8103, *Prochlorococcus* sp. MED4, and *Cyanothece* sp. ATCC 51142 (Toepel *et al.*, 2008; Wyman and Thom, 2012; Zinser *et al.*, 2009). Starch degradation in plants is also under control of the circadian clock (Graf *et al.*, 2010; Graf and Smith, 2011). Moreover, in *Synechocystis* sp. PCC 6803, the histidine kinase Hik8 is involved in the (potential circadian) regulation of *glgP* and *glgX* transcription (Singh and Sherman, 2005). In *Synechococcus* sp. PCC 7942 the orthologue of Hik8, the SasA protein, interacts with the KaiC protein that is part of the circadian oscillator, and SasA is important for robust circadian rhythmicity (Iwasaki *et al.*, 2000; Kageyama *et al.*, 2003). Although transcription and translation is thought to be closely linked in prokaryotes, a temporal protein expression pattern has only been verified for structural, granule-associated proteins in *Cyanothece* sp. ATCC 51142 (Schneegurt *et al.*, 1997). The supposed, but as yet inconclusive, circadian control of glycogen degradation in cyanobacteria might originate from the impact of glycogen metabolism on the energetic and carbon status of the cell and vice versa. The intracellular ATP/ADP ratio affects the phosphorylation rhythm of the KaiC protein. Changes in this ratio lead to phase shifts *in vitro* as well as *in vivo* (Rust *et al.*, 2011).

The transcription levels of genes involved in glycogen degradation such as *glgP2* also rise significantly in response to low CO_2 conditions in *Synechocystis* sp. PCC 6803 incubated in the light (Eisenhut *et al.*, 2007; Wang *et al.*, 2004). Likewise, the transcription of *glgP2* and *glgX* increases in response to nitrogen starvation in non-diazotrophic as well as in diazotrophic cyanobacteria (Ehira and Ohmori, 2006; Osanai *et al.*, 2005, 2006; Summers and Meeks, 1996; Tolonen *et al.*, 2006; von Wobeser *et al.*, 2011). Simultaneously, the expression of glycogen phosphorylase GlgP2 was strongly up-regulated, as detected by a proteomic approach in *Synechocystis* sp. PCC 6803 (Wegener *et al.*, 2010). In contrast, the initial glycogen accumulation after the onset of nitrogen starvation does not correspond to an induction of *glgA* expression (von Wobeser *et al.*, 2011). It is likely that the PII protein and the global transcription factor NtcA are involved in the glycogen degradation process, since the PII-deficient

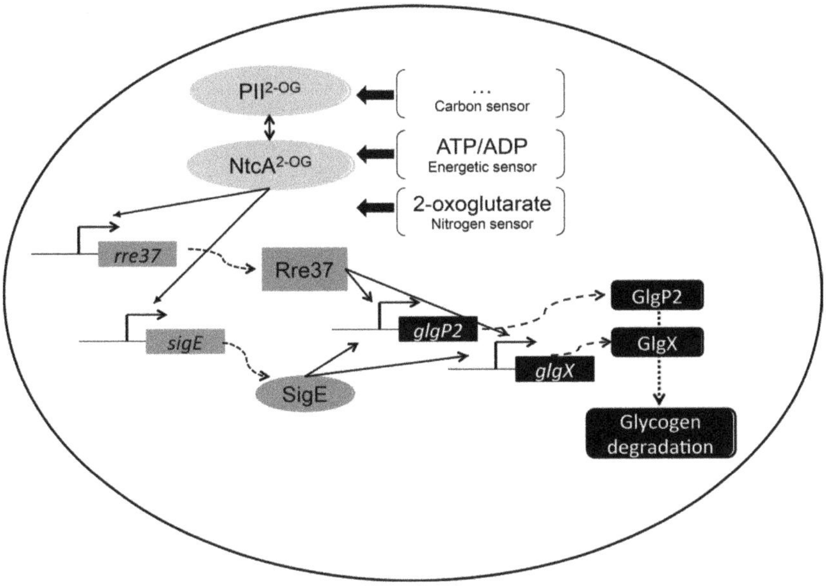

Figure 8.5 Simplified illustration of the transcriptional regulation of *glgP2* and *glgX* genes in *Synechocystis* sp. PCC 6803 in response to nitrogen depletion. Different nutritional (2-oxoglutarate as a metabolic signal for cellular nitrogen/carbon status; additional yet unknown carbon signal) and energetic (ATP/ADP ratio) signals are integrated in different conformational and covalent modifications of the PII protein that in turn activates in combination with NtcA the expression of the RNA polymerase sigma factor SigE and the OmpR-type response regulator Rre37. Both factors are directly involved in the transcription of *glgP2* and *glgX* in response to nitrogen depletion.

mutant of *Synechococcus* sp. PCC 7942 accumulates more glycogen than the wild-type strain (Forchhammer, 2004). The mechanism for this role of PII is well known and conserved in microorganisms (Forchhammer, 2004). The cellular carbon, nitrogen, energy and redox status, which are allosterically sensed through ATP and 2-oxoglutarate (Forchhammer, 2004), are integrated in different conformational and covalent modifications of the PII protein. This in turn induces, in combination with the global transcriptional regulator NtcA and other factors, the genes encoding the RNA polymerase group 2 sigma factor SigE (a conserved NtcA-binding motif, $GTAN_8TAC$, is found in the promoter region of *sigE*; Muro-Pastor *et al.*, 2001) and response regulators of the OmpR-type family such as Rre37 (Azuma *et al.*, 2011) in *Synechocystis* sp. PCC 6803, as well as NrrA in the diazotrophic cyanobacterium *Anabaena* sp. PCC 7120 (Ehira and Ohmori, 2011) (Fig. 8.5). Both the RNA polymerase group 2 sigma factor SigE and the OmpR-type transcriptional regulator are conserved in many cyanobacterial strains. They act independently as regulators in the nitrogen-dependent activation of catabolic genes including *glgP2* and *glgX* (Azuma *et al.*, 2011; Ehira and Ohmori, 2006; Osanai *et al.*, 2005, 2006, 2008; Summers and Meeks, 1996; Tolonen *et al.*, 2006; von Wobeser *et al.*, 2011).

In summary, as far as is known the pathways of glycogen synthesis and degradation are regulated at different levels in cyanobacteria: glycogen synthesis mainly at the level of enzymatic activity linked to light and dark reactions of photosynthesis, and glycogen degradation mainly at the level of transcription in response to the cellular carbon, nitrogen and energetic demand.

Role of glycogen metabolism in cyanobacteria

According to flux-balance analyses in *Synechocystis* sp. PCC 6803, the glycogen metabolism appears to be vital to produce all precursors required for biomass formation and to regenerate all cellular metabolites and cofactors (Knoop *et al.*, 2010).

Table 8.2 Phenotype of mutants inactivated in glycogen metabolism in *Synechocystis* sp. PCC 6803, *Synechococcus* sp. PCC 7002 and *Synechococcus elongatus* PCC 7942

Inactivated gene	Phenotype	Cyanobacterial strain(s)	References
glgC	Polyglucan-deficient phenotype	PCC 6803, PCC 7002, PCC 7942	Carrieri *et al.* (2012),
	Reduced viability in dark–light transitions	PCC 6803	Gründel *et al.* (2012),
	Reduced respiratory activity in the dark	PCC 6803, PCC 7002, PCC 7942	Hickman *et al.* (2013),
	Reduced photosynthetic capacity	PCC 6803, PCC 7002, PCC 7942	Miao *et al.* (2003a,b),
	Salt-sensitive phenotype	PCC 6803, PCC 7002, PCC 7942	Suzuki *et al.* (2010)
	Glucose-sensitive phenotype	PCC 6803	
	Non-bleaching phenotype	PCC 6803, PCC 7002, PCC 7942	
	Overflow phenotype	PCC 6803, PCC 7942	
	High light-sensitive phenotype	PCC 6803	
	Low inoculum-sensitive phenotype	PCC 6803	
glgA$_n$[a]	Polyglucan-deficient phenotype	PCC 6803, PCC 7002, PCC 7942	Gründel *et al.* (2012),
	Reduced viability in dark-light transitions	PCC 6803, PCC 7002	Suzuki *et al.* (2010),
	Reduced respiratory activity in dark	PCC 7002, PCC 7942	Xu *et al.* (2012b)
	Reduced photosynthetic capacity	PCC 6803, PCC 7002, PCC 7942	
	Salt-sensitive phenotype	PCC 7002, PCC 7942	
	Glucose-sensitive phenotype	PCC 6803	
	Non-bleaching phenotype	PCC 6803, PCC 7002	
	Overflow phenotype	PCC 6803	
	High light-sensitive phenotype	PCC 6803	
	Low inoculum-sensitive phenotype	PCC 6803	
glgA1	Polyglucan synthesized	PCC 6803, PCC 7002	Gründel *et al.* (2012), Xu *et al.* (2012b)
glgA2	Polyglucan synthesized	PCC 6803, PCC 7002	Gründel *et al.* (2012), Xu *et al.* (2012b)
glgB	Changed polyglucan structure	PCC 6803	Yoo *et al.* (2002)
glgP1	Temperature-sensitive phenotype	PCC 6803	Fu and Xu (2006)
glgP2	Low carbon phenotype	PCC 6803	Fu and Xu (2006)
glgX	Changed polyglucan structure Increased respiratory activity	PCC 7942	Suzuki *et al.* (2007)

[a]*n*, All glycogen synthases inactivated.

Based on this assumption, the inactivation of glycogen metabolism should result in a non-viable phenotype. However, various mutants with an inactivated glycogen metabolism have recently been reported (Table 8.2). Successful inactivation of AGPase (*glgC*) or of glycogen synthase(s) (*glgA*) has been achieved in two facultatively heterotrophic cyanobacteria, *Synechocystis* sp. PCC 6803 (Carrieri *et al.*, 2012; Gründel *et al.*, 2012; Wu *et al.*, 2001) and *Synechococcus* sp. PCC 7002 (Xu *et al.*, 2012b), as well as in one obligate photoautotrophic strain, *Synechococcus elongatus* PCC 7942 (Hickman *et al.*, 2013; Suzuki *et al.*, 2010), leading to a complete stop of glycogen synthesis. An impaired degradation of glycogen has been achieved via inactivation of glycogen phosphorylase (*glgP*) in *Synechocystis* sp. PCC 6803 (Fu and Xu, 2006) or indirectly by inactivation of the histidine kinase Hik8 (Singh and Sherman, 2005), of the RNA polymerase group 2 sigma factor SigE (Osanai *et al.*, 2005; Osanai *et al.*, 2007) or of the PII protein (Forchhammer, 2004).

Although the growth of mutants defective in glycogen metabolism most closely corresponds to that of the wild-type parental strain under steady-state conditions (normal photoautotrophic growth conditions), the consequences of impaired polyglucan synthesis or degradation are numerous and drastic (Table 8.2). The mutant cells have both a reduced photosynthetic capacity (oxygen evolution) and reduced respiratory oxygen consumption, are sensitive to low *inocula* (cell densities), to high light, to high temperature, to dark to light transitions, to low carbon conditions, to macronutrient starvation or to salt (osmotic) stress, indicating the prominent role of glycogen metabolism for cell viability rather than simply for growth (Table 8.2). The loss in viability most likely reflects the role of the glycogen pool as a dynamic reservoir and sink for carbon and electrons. This dual function as a reserve and redox buffer substance was already proposed by Carr (1988). A totally new functional aspect is indicated by the very striking phenotypes of overflow metabolism (spilling of partially oxidized carbonic acids), of inhibited chlorosis (in response to nitrogen limitation) and of photomixotrophic sensitivity (Table 8.2). The metabolism of glycogen likely is an essential cellular tool for the sensing and integration of the carbon status and for the adequate regulation of various metabolic processes in cyanobacteria, especially under unbalanced growth conditions (Gründel *et al.*, 2012; Xu *et al.*, 2012b). Three functional aspects will be considered in detail: (i) the essentiality of polyglucan metabolism for photosynthetic efficiency, (ii) the function of glycogen metabolism in the acclimation process to nitrogen starvation, and (iii) the role of glycogen metabolism in the acclimation to salt stress.

Essentiality of polyglucan metabolism for photosynthetic efficiency: acclimation to changing light and inorganic carbon availability

The significant reduction of photosynthetic efficiency (oxygen evolution rate per chlorophyll) in mutants fully defective in glycogen synthesis is a common feature of all currently published polyglucan-free mutants in photosynthetic organisms, including cyanobacteria (Eichelmann and Laisk, 1994; Gründel *et al.*, 2012; Miao *et al.*, 2003b; Sun *et al.*, 1999; Suzuki *et al.*, 2010). In what way the photosynthetic apparatus and the light reactions in the thylakoid membrane are affected by the block in glycogen synthesis that finally leads to decreased oxygen evolution rates is yet unknown. Because the synthesis of glycogen is only forced under conditions of active photosynthesis, the pathway is supposed to involve integration of a carbon signal from photosynthesis (Gründel *et al.*, 2012). The potential modified (non-integrated) *photosynthetic carbon signal* caused by the block in glycogen synthesis may result in modifications of the photosynthetic apparatus. In line with this, modifications of antenna size and/or the photosystem content were also observed during transitions from high CO_2 to low CO_2 conditions (Hihara *et al.*, 1998). By contrast, mutants affected in glycogen degradation (Δ*glgP2*) show a low carbon phenotype slightly similar to that observed for carboxysome-less or photorespiratory mutants (Fu and Xu, 2006) (Table 8.2). In plants, a starch-less mutant exhibits a low RubisCO activity (Sicher and Kremer, 1996) suggesting that starch may act as an electron acceptor in photorespiration (Eichelmann and Laisk, 1994; Sun *et al.*, 1999). In cyanobacteria, it has been proposed that the process of glycogen degradation in the light is involved in anaplerotic reactions of the Calvin–Benson–Bassham cycle during acclimation to changing CO_2 concentrations (Eisenhut *et al.*, 2008). A sink of carbon due to a transition from high CO_2 conditions to low CO_2 conditions is compensated by glycogen degradation (Eisenhut *et al.*, 2007; Wang *et al.*, 2004).

According to the general opinion, the glycogen is used as an endogenous carbon and energy reserve to maintain cell integrity, function and viability in 'photosynthetic starvation periods' such as darkness (Smith, 1982). Various turnover measurements, diurnal-determined glycogen profiles and mutagenesis studies of both glycogen synthesis (cf. Table 8.2: mutants have a lower respiratory oxygen consumption and an impaired viability) and the OPP pathway (mainly involved in the respiration of glycogen) support this opinion (Smith, 1982). Glycogen seems to be the major, although not the exclusive, oxidizable reserve for dark maintenance in cyanobacteria (Gründel *et al.*, 2012; Xu *et al.*, 2012b). Sugars, lipids or chlorophyll may partly compensate for the lack of glycogen as it is observed in plants (Huber and Hanson, 1992).

The function of glycogen metabolism in acclimation to nitrogen starvation

In photosynthetic organisms such as cyanobacteria, most stresses are integrated and responded via the activity of enzymes of carbon/

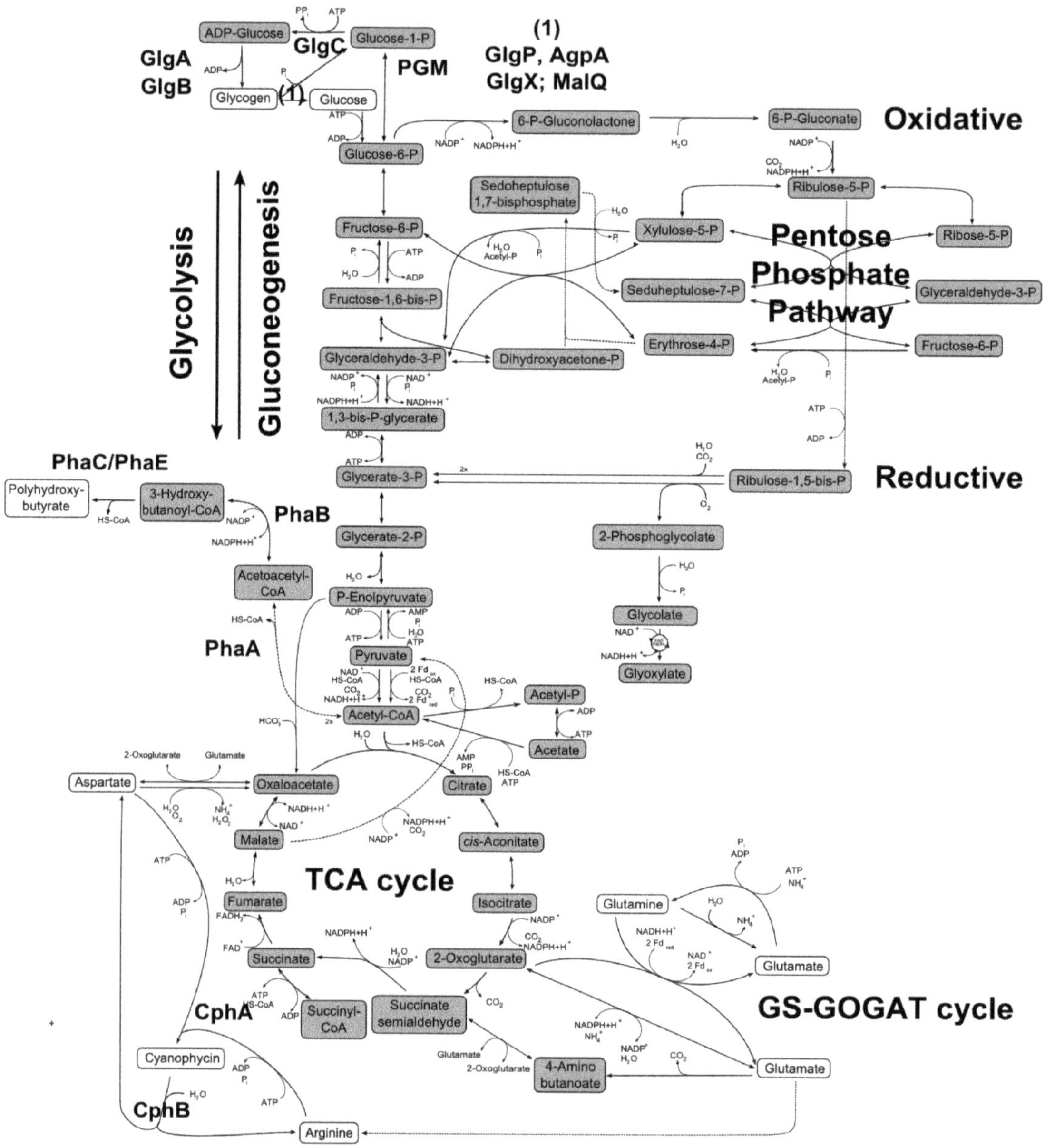

Figure 8.6 Illustration of central carbon/nitrogen metabolism of *Synechocystis* sp. PCC 6803 highlighting the different pathways of organic reserve polymer metabolism, according to Knoop *et al.* (2010). Abbreviations are: CphA, cyanophycin synthetase (EC 6.3.2.29); CphB: cyanophycinase (EC 3.4.15.6); GlgA, glycogen synthase (EC 2.4.1.21); GlgB, glycogen branching enzyme (EC 2.4.1.18); GlgC, ADP glucose pyrophosphorylase, AGPase (EC 2.7.7.27); GlgP and AgpA, glycogen phosphorylase (EC 2.4.1.1); GlgX, glycogen debranching enzyme (EC 3.2.1.68); GOGAT, glutamate synthase; GS, glutamine synthetase; MalQ, α-(1–4)-glucanotransferase, α-GTase (EC 2.4.1.25); PGM, phosphoglucomutase; PhaA, PHA-specific β-ketothiolase/Acetyl-CoA acetyltransferase (EC 2.3.1.9); PhaB, PHA-specific acetoacetyl-CoA reductase (EC 1.1.1.36); PhaC/PhaE, poly(3-hydroxyalkanoate) synthase (EC 2.3.1).

nitrogen metabolism, because the cell depends on a balanced carbon/nitrogen ratio (Wegener *et al.*, 2010). Here, glycogen polymer formation and mobilization becomes a central part of this network (cf. Fig. 8.6). After the onset to nitrogen starvation, two major processes immediately occur in parallel: the shift of photosynthetic carbon flow to glycogen reserves (Allen and

Smith, 1969; Carrieri *et al.*, 2012; Gorl *et al.*, 1998; Miller *et al.*, 2002; Sauer *et al.*, 1999; Stevens *et al.*, 1981) leads to a massive increase of the intracellular glycogen pool (up to 70% w/w per dry weight), and the lack of ammonia for operation of the GS-GOGAT cycle leads to an increase of the 2-oxoglutarate level (cf. Fig. 8.6). The metabolic signal 2-oxoglutarate interacts with the PII protein and the NtcA transcription factor, which in turn induce the expression of genes involved in the nitrogen signalling cascade such as *nblA*.

The small adaptor protein NblA binds simultaneously to the phycobiliproteins and to a Clp-protease that leads to a proteolytic degradation of the phycobilisome complex, referred to as nitrogen chlorosis (Karradt *et al.*, 2008). It is assumed that phycobilisome degradation fulfils two functions: preventing photodamage and providing amino acids for protein synthesis during the acclimation process. Inversely, as soon as a nitrogen source is available, the phycobilisomes are resynthesized and glycogen is rapidly degraded.

Both the process of glycogen synthesis and the process of nitrogen chlorosis only occur under conditions of active photosynthesis (light and CO_2 available) (Sarma and Kanta, 1979; Wood and Haselkorn, 1988). Furthermore, the degradation of the phycobilisome complex is tightly coupled to the process of glycogen synthesis. The progression of nitrogen chlorosis correlates with the level of glycogen accumulation (Gründel *et al.*, 2012; Riethman *et al.*, 1988; Xu *et al.*, 2012b). This carbon/nitrogen ratio balancing mechanism is obviously blocked in cells with an impaired glycogen biosynthesis (Table 8.2). Inversely, significant changes in glycogen accumulation are not reported from *nbl* (non-bleaching)-mutants affected in the nitrogen signalling cascade (e.g. Δ*nblA*). This indicates that glycogen biosynthesis is potentially directly involved in the metabolic control of nitrogen chlorosis in cyanobacteria (Gründel *et al.*, 2012; Xu et al., 2012b).

In heterotrophic bacteria, the block of glycogen synthesis results in an uncoupling of anabolic and catabolic reactions after the onset of nitrogen starvation that in turn leads to spilling reactions in order to get rid of energy excess (Russell, 2007; Russell and Cook, 1995; Steinbuchel and Schlegel, 1989). The loss of partially oxidized carbonic acids such as pyruvate and 2-oxoglutarate (referred to as overflow metabolism) is also observed in cyanobacterial mutants defective in glycogen synthesis under conditions of nitrogen deprivation (Table 8.2). However, such a leakage of energetic compounds only occurs under conditions of active photosynthesis (Carrieri *et al.*, 2012; Gründel *et al.*, 2012; Hickman *et al.*, 2013). It is likely that the initial massive glycogen synthesis is required to sense, balance and integrate the carbon status that in turn allows a steady-state of anabolic and catabolic reactions and thus the activation of catabolic processes which enable the switch from a growth-related mode (anabolism) to a dormant (survival) mode (catabolism) in non-diazotrophic cyanobacteria (Schwarz and Forchhammer, 2005). A similar function as an integrator of the energy and carbon signal is also proposed for animal glycogen metabolism (Greenberg *et al.*, 2006).

The presence of an intracellular chemically and osmotically inert store of glycogen enables most cyanobacteria (even those such as *Synechocystis* sp. PCC 6803 that are capable to synthesize both carbon polymers glycogen and PHB) to survive macronutrient-starvation conditions (e.g. nitrogen starvation) and to recover rapidly (Gründel *et al.*, 2012). In contrast, the viability of cells deficient in glycogen synthesis is lost with prolonged nitrogen starvation (Gründel *et al.*, 2012). It has been shown that those microorganisms that are able to closely adjust the rate of glycogen degradation to their maintenance energy requirement survive for the longest periods (Dawes, 1992). The essential role of glycogen metabolism in cellular viability and persistence under stress conditions and fluctuating environments in cyanobacteria is consistent with the observed physiological function in other free-living organisms. Only parasitic lifestyles seem to be related to reduction and eventual abolishment of glycogen metabolism (Henrissat *et al.*, 2002).

Interference of polymeric and soluble carbohydrate metabolism under conditions of salt stress

In addition to glycogen, cyanobacteria synthesize different osmolytes to balance their intracellular osmotic pressure in response to salt stress (Hagemann, 2010). The biosynthetic pathways for osmolytes such as sucrose or glucosylglycerol/glucosylglycerate and for glycogen share the same (e.g. ADP-glucose as a glycosyl donor) or closely related precursors (Hagemann and Erdmann, 1994). Moreover, glycogen is used as a precursor for transglucosidase reactions (Katoh *et al.*, 2004; MacKay and Norton, 1987; Tel-Or *et al.*, 1986). Blocks in glycogen synthesis lead to an altered osmotic stress response although viability under hypersaline growth conditions is only slightly affected (Miao *et al.*, 2003a; Suzuki *et al.*, 2010; Xu *et al.*, 2012b). Depending on the tolerance of different cyanobacteria to salt stress (stenohaline, euryhaline), the levels of different osmolytes are altered in mutants defective in glycogen synthesis. Glycogen-deficient cells of the stenohaline strains *Synechococcus elongatus* PCC 7942 and *Synechocystis* sp. PCC 6803 accumulate less sucrose (shown for both strains) and more glycosylglycerol (shown for *Synechocystis* sp. PCC 6803) than the respective parental wild-type strain in response to salt stress (Miao *et al.*, 2003a; Suzuki *et al.*, 2010). The euryhaline *Synechococcus* sp. PCC 7002 generally has a higher intracellular level of soluble sugars under both standard and hypersaline growth conditions (Xu *et al.*, 2012b). Interestingly, these substrates seem to be utilized under dark anoxic conditions – a process referred to as autofermentation (Xu *et al.*, 2012b).

Other cyanobacterial inclusion bodies

Polyhydroxybutyrate (PHB) granules

A few cyanobacterial strains of the genera *Anabaena, Cyanothece, Microcystis, Nostoc, Oscillatoria, Synechococcus* and *Synechocystis* can synthesize (in addition to glycogen) another carbon reserve, referred as polyhydroxybutyrate (PHB) (Stal, 1992) (cf. Fig. 8.1). PHB is a polyester of β-hydroxybutyric acid. The ability to synthesize both carbon reserves is shared by some purple bacteria and is exception to the principle that only one kind of carbon reserve material is formed by a given species. In *Synechocystis* sp. PCC 6803, the transcription of genes involved in PHB synthesis (*phaA, phaB, phaC* and *phaE*) appears to be controlled by the circadian clock (Kucho *et al.*, 2005). The pathway for the synthesis of PHB starts with the condensation of two molecules of acetyl-CoA to acetoacetyl-CoA by the β-ketothiolase (PhaA) (Asada *et al.*, 1999; Taroncher-Oldenberg *et al.*, 2000) (Fig. 8.6). This product is then reduced to hydroxybutyryl-CoA by the acetoacetyl-CoA reductase (PhaB). The heterodimeric PHA synthase (PhaC/PhaE) finally catalyses the polymerization of PHB monomers (Hein *et al.*, 1998). Surprisingly, neither the extracellular nor the intracellular depolymerases required for mobilization of PHB granules, as found in other bacteria, have been identified in any of the PHB-producing cyanobacterial strains. The biological function of PHB in the context of reserve metabolism remains to be established for cyanobacteria. It is assumed that in cyanobacteria and purple bacteria, which are able to synthesize both glycogen and PHB photosynthetically, the polyester acts mainly as a regulator of the intracellular reduction charge (De Philippis *et al.*, 1992a).

Polyphosphate (volutin) granules

The polyphosphate (volutin) granules serve as phosphate and energy reserves (see Fig. 8.1). Excess phosphate favours the formation of polyphosphate bodies, whereas phosphorus depletion leads to a rapid degradation of the granules (Allen, 1984). *In situ* X-ray dispersive analysis demonstrated the sequestering of metals into polyphosphate bodies, suggesting that the negatively charged bodies serve as storage sites for metal ions (Jensen *et al.*, 1982). Polyphosphate is synthesized by polyphosphate kinase (the product of the *ppk* gene) and degraded by polyphosphatase (the product of *ppx*). Both genes are widely present in sequenced cyanobacterial genomes (see http://img.jgi.doe.gov/cgi-bin/w/main.cgi).

Final remarks

Cyanobacteria can synthesize a wide range of biopolymers comprising four different chemical classes: polysaccharides (e.g. glycogen, starch-like reserves), polyesters (e.g.

poly-β-hydroxybutyrate), polyamides (e.g. cyanophycin) and inorganic polyanhydrides (e.g. polyphosphates). The enzymology of biosynthesis and mobilization as well as the role of effector molecules in those enzymatic processes is best known for glycogen. This biopolymer is produced under conditions of carbon and energy excess, mainly to balance the nutritional and redox, and thus the energetic status of the cell. An activation of biosynthesis or of degradation is (as far as known) tuned to the nutritional and energetic demands of the cell (or the cell community). A comprehensive network of different regulatory levels (transcription initiation, posttranscriptional processing, translation initiation, posttranslational modifications) is most likely involved, however yet insufficiently understood, to control the glycogen metabolism and the chemical properties of the polyglucan in response to different environmental conditions.

The glycogen pool acts as a reserve compound as well as a sink. Beside this classical definition of dual polymer function, glycogen serves also as part of a protective structure. Cells with an endowed glycogen metabolism have a substantial advantage. Here, the glycogen metabolism has an enormous impact on cellular viability and persistence under stress conditions and in fluctuating environments. Recent studies show that the glycogen metabolism itself has an eminent role for the steady-state of anabolic and catabolic reactions. It directly interferes with metabolic processes and is highly interconnected with a variety of cellular processes: the stringent response (e.g. nitrogen limitation: nitrogen chlorosis), the general stress response (e.g. heat), energy and cellular redox status and osmotic stress. A better understanding of the fundamental processes involved in glycogen metabolism and regulation will provide insights into the biological function and give clues for improved engineering approaches towards a valuable and renewable biopolymer production or as a platform for photocatalytic organic acid and soluble sugar synthesis (Carrieri *et al.*, 2012; Rehm, 2010; Xu *et al.*, 2012b).

Acknowledgements

I gratefully acknowledge Wolfgang Lockau for his cooperation and sustained support of my research. Our work was financially supported by the German Ministry of Science within the framework of FORSYS-Partner (BMBF-31101939). I also would like to express my gratitude to Antonia Herrero and Enrique Flores for their guidance and valuable and critical feedback to improve the manuscript.

References

Allen, M.M. (1984). Cyanobacterial cell inclusions. Annu. Rev. Microbiol. *38*, 1–25.

Allen, M.M., and Smith, A.J. (1969). Nitrogen chlorosis in blue-green algae. Arch. Microbiol. *69*, 114–120.

Asada, Y., Miyake, M., Miyake, J., Kurane, R., and Tokiwa, Y. (1999). Photosynthetic accumulation of poly-(hydroxybutyrate) by cyanobacteria – the metabolism and potential for CO_2 recycling. Int. J. Biol. Macromol. *25*, 37–42.

Azuma, M., Osanai, T., Hirai, M.Y., and Tanaka, K. (2011). A response regulator Rre37 and an RNA polymerase sigma factor SigE represent two parallel pathways to activate sugar catabolism in a *Cyanobacterium Synechocystis* sp PCC 6803. Plant Cell Physiol. *52*, 404–412.

Ball, S.G., and Morell, M.K. (2003). From bacterial glycogen to starch: understanding the biogenesis of the plant starch granule. Annu. Rev. Plant Biol. *54*, 207–233.

Ball, S., Colleoni, C., Cenci, U., Raj, J.N., and Tirtiaux, C. (2011). The evolution of glycogen and starch metabolism in eukaryotes gives molecular clues to understand the establishment of plastid endosymbiosis. J. Exp. Bot. *62*, 1775–1801.

Ballicora, M.A., Iglesias, A.A., and Preiss, J. (2003). ADP-glucose pyrophosphorylase, a regulatory enzyme for bacterial glycogen synthesis. Microbiol. Mol. Biol. Rev. *67*, 213–225.

Beck, C., Knoop, H., Axmann, I., and Steuer, R. (2012). The diversity of cyanobacterial metabolism: genome analysis of multiple phototrophic microorganisms. BMC Genom. *13*, 56.

Boos, W., and Shuman, H. (1998). Maltose/maltodextrin system of Escherichia coli: transport, metabolism, and regulation. Microbiol. Mol. Biol. Rev. *62*, 204.

Cabib, E. (1970). Research on Sugar Nucleotides Brings Honor to Argentinian Biochemist. Science *170*, 608–609.

Carr, N.G. (1988). Nitrogen reserves and dynamic reservoirs in cyanobacteria. In Biochemistry of the algae and cyanobacteria, Rogers, L.J., and Gallon, J.R., eds. (Clarendon Press, Oxford), pp. 13–21.

Carrieri, D., Paddock, T., Maness, P.C., Seibert, M., and Yu, J.P. (2012). Photo-catalytic conversion of carbon dioxide to organic acids by a recombinant cyanobacterium incapable of glycogen storage. Energy Environ. Sci. 5, 9457–9461.

Cid, E., Geremia, R.A., Guinovart, J.J., and Ferrer, J.C. (2002). Glycogen synthase: towards a minimum catalytic unit? FEBS Lett. *528*, 5–11.

Cumino, A., Curatti, L., Giarrocco, L., and Salerno, G.L. (2002). Sucrose metabolism: Anabaena sucrose-phosphate synthase and sucrose-phosphate phosphatase define minimal functional domains shuffled during evolution. FEBS Lett. *517*, 19–23.

Cumino, A.C., Marcozzi, C., Barreiro, R., and Salerno, G.L. (2007). Carbon cycling in *Anabaena* sp PCC 7120. Sucrose synthesis in the heterocysts and possible role in nitrogen fixation. Plant Physiol. *143*, 1385–1397.

Curatti, L., Giarrocco, L.E., Cumino, A.C., and Salerno, G.L. (2008). Sucrose synthase is involved in the conversion of sucrose to polysaccharides in filamentous nitrogen-fixing cyanobacteria. Planta *228*, 617–625.

Dauvillee, D., Kinderf, I.S., Li, Z., Kosar-Hashemi, B., Samuel, M.S., Rampling, L., Ball, S., and Morell, M.K. (2005). Role of the *Escherichia coli* glgX gene in glycogen metabolism. J. Bacteriol. *187*, 1465–1473.

Dawes, E.A. (1992). Storage polymers in Prokaryotes. In Prokaryotic Structure and Function: A New Perspective, Dow, C., and Cole, J.A., eds. (Bath, Avon, The Bath Press), pp. 82–91.

De Philippis, R., Ena, A., Guastiini, M., Sili, C., and Vincenzini, M. (1992a). Factors affecting poly-β-hydroxybutyrate accumulation in cyanobacteria and in purple non-sulfur bacteria. FEMS Microbiol. Lett. *103*, 187–194.

De Philippis, R., Sili, C., and Vincenzini, M. (1992b). Glycogen and poly-{beta}-hydroxybutyrate synthesis in Spirulina maxima. J. Gen. Microbiol. *138*, 1623–1628.

Deschamps, P., Colleoni, C., Nakamura, Y., Suzuki, E., Putaux, J.L., Buleon, A., Haebel, S., Ritte, G., Steup, M., Falcon, L.I., *et al.* (2008). Metabolic symbiosis and the birth of the plant kingdom. Mol. Biol. Evol. *25*, 536–548.

Edwards, A., Vincken, J.P., Suurs, L., Visser, R.G.F., and Zeeman, S. (2002). Discrete forms of amylose are synthesized by isoforms of GBSSI in Pea. Plant Cell *14*, 1767–1785.

Egli, B., Koeling, K., Koehler, C., Zeeman, S.C., and Streb, S. (2010). Loss of cytosolic phosphoglucomutase compromises gametophyte development in *Arabidopsis*. Plant Physiol. *154*, 1659–1671.

Ehira, S., and Ohmori, M. (2006). NrrA, a nitrogen-responsive response regulator facilitates heterocyst development in the cyanobacterium *Anabaena* sp strain PCC 7120. Mol. Microbiol. *59*, 1692–1703.

Ehira, S., and Ohmori, M. (2011). NrrA, a Nitrogen-regulated response regulator protein, controls glycogen catabolism in the nitrogen-fixing cyanobacterium *Anabaena* sp. strain PCC 7120. J. Biol. Chem. *286*, 38109–38114.

Eichelmann, H., and Laisk, A. (1994). CO_2 uptake and electron-transport rates in wild-type and a starchless mutant of *Nicotiana sylvestris* – the role and regulation of starch synthesis at saturating CO_2 concentrations. Plant Physiol. *106*, 679–687.

Eisenhut, M., von Wobeser, E.A., Jonas, L., Schubert, H., Ibelings, B.W., Bauwe, H., Matthijs, H.C.P., and Hagemann, M. (2007). Long-term response toward Inorganic carbon limitation in wild-type and glycolate turnover mutants of the cyanobacterium *Synechocystis* sp. strain PCC 6803. Plant Physiol. *144*, 1946–1959.

Eisenhut, M., Huege, J., Schwarz, D., Bauwe, H., Kopka, J., and Hagemann, M. (2008). Metabolome phenotyping of inorganic carbon limitation in cells of the wild-type and photorespiratory mutants of the cyanobacterium *Synechocystis* sp. strain PCC 6803. Plant Physiol. *148*, 2109–2120.

Elmorjani, K., and Herdman, M. (1987). Metabolic control of phycocyanin degradation in the cyanobacterium *Synechocystis* PCC 6803 – a glucose effect. J. Gen. Microbiol. *133*, 1685–1694.

Ernst, A., and Boger, P. (1985). Glycogen accumulation and the induction of nitrogenase activity in the heterocyst-forming cyanobacterium *Anabaena variabilis* J. Gen. Microbiol. *131*, 3147–3153.

Ernst, A., Reich, S., and Boger, P. (1990). Modification of dinitrogenase reductase in the cyanobacterium *Anabaena variabilis* due to C-starvation and ammonia. J. Bacteriol. *172*, 748–755.

Fay, P. (1992). Oxygen relations of nitrogen fixation in cyanobacteria. Microbiol. Rev. *56*, 340–373.

Ferrell Jr, J.E. (1998). How regulated protein translocation can produce switch-like responses. Trends Biochem. Sci. *23*, 461–465.

Forchhammer, K. (2004). Global carbon/nitrogen control by PII signal transduction in cyanobacteria: from signals to targets. FEMS Microbiol. Rev. *28*, 319–333.

Fu, J., and Xu, X.D. (2006). The functional divergence of two *glgP* homologues in *Synechocystis* sp. PCC 6803. FEMS Microbiol. Lett. *260*, 201–209.

Fulton, D.C., Edwards, A., Pilling, E., Robinson, H.L., and Fahy, B. (2002). Role of granule-bound starch synthase in determination of amylopectin structure and starch granule morphology in potato. J. Biol. Chem. *277*, 10834–10841.

Geigenberger, P., Kolbe, A., and Tiessen, A. (2005). Redox regulation of carbon storage and partitioning in response to light and sugars. J. Exp. Bot. *56*, 1469–1479.

Georg, J., and Hess, W.R. (2011). cis-Antisense RNA, another level of gene regulation in bacteria. Microbiol. Mol. Biol. Rev. *75*, 286–300.

Gleason, F. (2004). Thioredoxins in cyanobacteria: structure and redox regulation of enzyme activity. In The Molecular Biology of Cyanobacteria, Bryant, D., ed. (Springer, Netherlands), pp. 715–729.

Goldbeter, A., and Koshland, D.E. (1984). Ultrasensitivity in biochemical systems controlled by covalent modification – interplay between zero-order and multistep effects. J. Biol. Chem. *259*, 14441–14447.

Gomez-Casati, D.F., Aon, M.A., and Iglesias, A.A. (1999). Ultrasensitive glycogen synthesis in Cyanobacteria. FEBS Lett. *446*, 117–121.

Gomez-Casati, D.F., Cortassa, S., Aon, M.A., and Iglesias, A.A. (2003). Ultrasensitive behavior in the synthesis of storage polysaccharides in cyanobacteria. Planta *216*, 969–975.

Gorl, M., Sauer, J., Baier, T., and Forchhammer, K. (1998). Nitrogen Starvation-induced chlorosis in *Synechococcus* PCC 7942: adaptation to long-term survival. Microbiology *144*, 2449–2458.

Graf, A., and Smith, A.M. (2011). Starch and the clock: the dark side of plant productivity. Trends Plant Sci. *16*, 169–175.

Graf, A., Schlereth, A., Stitt, M., and Smith, A.M. (2010). Circadian control of carbohydrate availability for growth in *Arabidopsis* plants at night. Proc. Natl. Acad. Sci. U.S.A. *107*, 9458–9463.

Greenberg, C.C., Jurczak, M.J., Danos, A.M., and Brady, M.J. (2006). Glycogen branches out: new perspectives on the role of glycogen metabolism in the integration of metabolic pathways. Am. J. Physiol. *291*, E1-E8.

Griffiths, M.S.H., Gallon, J.R., and Chaplin, A.E. (1987). The diurnal pattern of dintrogen fixation by cyanobacteria *in situ*. N. Phytol. *107*, 649–657.

Gründel, M., Scheunemann, R., Lockau, W., and Zilliges, Y. (2012). Impaired glycogen synthesis causes metabolic overflow reactions and affects stress responses in the cyanobacterium *Synechocystis* sp. PCC 6803. Microbiology *158*, 3032–3043.

Hagemann, M. (2010). Molecular biology of cyanobacterial salt acclimation. FEMS Microbiol. Rev. *35*, 87–123.

Hagemann, M., and Erdmann, N. (1994). Activation and pathway of glucosylglycerol synthesis in the cyanobacterium *Synechocystis* sp. PCC 6803. Microbiology *140*, 1427–1431.

Hein, S., Tran, H., and Steinbuchel, A. (1998). *Synechocystis* sp. PCC6803 possesses a two-component polyhydroxyalkanoic acid synthase similar to that of anoxygenic purple sulfur bacteria. Arch. Microbiol. *170*, 162–170.

Henrissat, B., Deleury, E., and Coutinho, P.M. (2002). Glycogen metabolism loss: a common marker of parasitic behaviour in bacteria? Trends Genet. *18*, 437–440.

Hickman, J.W., Kotovic, K.M., Miller, C., Warrener, P., Kaiser, B., Jurista, T., Budde, M., Cross, F., Roberts, J.M., and Carleton, M. (2013). Glycogen synthesis is a required component of the nitrogen stress response in *Synechococcus elongatus* PCC 7942. Algal Res. *2*, 98–106.

Hihara, Y., Sonoike, K., and Ikeuchi, M. (1998). A novel gene, pmgA, specifically regulates photosystem stoichiometry in the cyanobacterium *Synechocystis* species PCC 6803 in response to high light. Plant Physiol. *117*, 1205–1216.

Huber, S.C., and Hanson, K.R. (1992). Carbon partitioning and growth of a starchless mutant of *Nicotiana sylvestris*. Plant Physiol. *99*, 1449–1454.

Iglesias, A.A., Kakefuda, G., and Preiss, J. (1991). Regulatory and structural properties of the cyanobacterial ADP-glucose pyrophosphorylases. Plant Physiol. *97*, 1187–1195.

Iwasaki, H., Williams, S.B., Kitayama, Y., Ishiura, M., Golden, S.S., and Kondo, T. (2000). A kaiC-interacting sensory histidine kinase, SasA, necessary to sustain robust circadian oscillation in cyanobacteria. Cell *101*, 223–233.

Jacobsen, J.H., Rosgaard, L., Sakuragi, Y., and Frigaard, N.U. (2011). One-step plasmid construction for generation of knock-out mutants in cyanobacteria: studies of glycogen metabolism in *Synechococcus* sp PCC 7002. Photosynth. Res. *107*, 215–221.

Jensen, B.B., Cox, R.P., and Burris, R.H. (1986). Isolation of cyanobacterial heterocysts with high and sustained dinitrogen-fixation capacity supported by endogenous reductants. Arch. Microbiol. *145*, 241–247.

Jensen, T.E., Baxter, M., Rachlin, J.W., and Jani, V. (1982). Uptake of heavy-metals by *Plectonema boryanum* (Cyanophyceae) into cellular components, especially polyphosphate bodies – an X-ray energy dispersive study. Environ. Poll. Series A *27*, 119–127.

Kageyama, H., Kondo, T., and Iwasaki, H. (2003). Circadian formation of clock protein complexes by KaiA, KaiB, KaiC, and SasA in cyanobacteria. J. Biol. Chem. *278*, 2388–2395.

Kaprelyants, A.S., Gottschal, J.C., and Kell, D.B. (1993). Dormancy in non-sporulating bacteria. FEMS Microbiol. Rev. *104*, 271–286.

Karradt, A., Sobanski, J., Mattow, J., Lockau, W., and Baier, K. (2008). NblA, a Key Protein of Phycobilisome Degradation, Interacts with ClpC, a HSP100 Chaperone Partner of a Cyanobacterial Clp Protease. J. Biol. Chem. *283*, 32394–32403.

Katoh, H., Asthana, R.K., and Ohmori, M. (2004). Gene expression in the *Cyanobacterium Anabaena* sp PCC7120 under desiccation. Microb. Ecol. *47*, 164–174.

Knoop, H., Zilliges, Y., Lockau, W., and Steuer, R. (2010). The metabolic network of *Synechocystis* sp. PCC 6803: systemic properties of autotrophic growth. Plant Physiol. *154*, 410–422.

Koksharova, O.A., Brandt, U., and Cerff, R. (2004). The gap1 operon of the cyanobacterium *Synechococcus* PCC 7942 carries a gene encoding glycogen phosphorylase and is induced under anaerobic conditions. Microbiology 73, 326–329.

Kolbe, A., Tiessen, A., Schluepmann, H., Paul, M., Ulrich, S., and Geigenberger, P. (2005). Trehalose 6-phosphate regulates starch synthesis via posttranslational redox activation of ADP-glucose pyrophosphorylase. Proc. Natl. Acad. Sci. U.S.A. *102*, 11118–11123.

Kucho, K., Okamoto, K., Tsuchiya, Y., Nomura, S., Nango, M., Kanehisa, M., and Ishiura, M. (2005). Global analysis of circadian expression in the cyanobacterium *Synechocystis* sp strain PCC 6803. J. Bacteriol. *187*, 2190–2199.

Lee, B.-H., Oh, D.-K., and Yoo, S.-H. (2009). Characterization of 4-alpha-glucanotransferase from *Synechocystis* sp. PCC 6803 and its application to various corn starches. N. Biotechnol. *26*, 29–36.

Lehmann, M., and Wober, G. (1976). Accumulation, mobilization and turn-over of glycogen in blue-green bacterium *Anacystis nidulans*. Arch. Microbiol. *111*, 93–97.

Levi, C., and Preiss, J. (1976). Regulatory properties of ADP-glucose pyrophosphorylase of blue-green bacterium *Synechococcus* 6301. Plant Physiol. *58*, 753–756.

Lindahl, M., and Florencio, F.J. (2003). Thioredoxin-linked processes in cyanobacteria are as numerous as in chloroplasts, but targets are different. Proc. Natl. Acad. Sci. U.S.A. *100*, 16107–16112.

Lindahl, M., and Kieselbach, T. (2009). Disulphide proteomes and interactions with thioredoxin on the track towards understanding redox regulation in chloroplasts and cyanobacteria. J. Proteom. *72*, 416–438.

MacKay, M.A., and Norton, R.S. (1987). C-13 Nuclear-magnetic-resonance study of biosynthesis of glycosylglycerol by a cyanobacterium under osmotic stress. J. Gen. Microbiol. *133*, 1535–1542.

Memon, D., Singh, A., Pakrasi, H., and Wangikar, P. (2012). A global analysis of adaptive evolution of operons in cyanobacteria. Antonie Van Leeuwenhoek, 1–16.

Merrick, J.M. (1979). Metabolism of reserve materials. In The Photosynthetic Bacteria, Clayton, R.K., and Sistrom, W.R., eds. (Plenum Press, New York), pp. 199–219.

Miao, X.L., Wu, Q.Y., Wu, G.F., and Zhao, N.M. (2003a). Sucrose accumulation in salt-stressed cells of agp gene deletion-mutant in cyanobacterium *Synechocystis* sp PCC 6803. FEMS Microbiol. Lett. *218*, 71–77.

Miao, X.L., Wu, Q.Y., Wui, G.F., and Zhao, N.M. (2003b). Changes in photosynthesis and pigmentation in an agp deletion mutant of the cyanobacterium *Synechocystis* sp. Biotechnol. Lett. *25*, 391–396.

Miller, S.R., Martin, M., Touchton, J., and Castenholz, R.W. (2002). Effects of nitrogen availability on pigmentation and carbon assimilation in the cyanobacterium *Synechococcus* sp strain SH-94-95. Arch. Microbiol. *177*, 392–400.

Mitschke, J., Georg, J., Scholz, I., Sharma, C.M., Dienst, D., Bantscheff, J., Voss, B., Steglich, C., Wilde, A., Vogel, J., *et al.* (2011). An experimentally anchored map of transcriptional start sites in the model cyanobacterium *Synechocystis* sp PCC6803. Proc. Natl. Acad. Sci. U.S.A. *108*, 2124–2129.

Muro-Pastor, A.M., Herrero, A., and Flores, E. (2001). Nitrogen-regulated group 2 sigma factor from *Synechocystis* sp. strain PCC 6803 involved in survival under nitrogen stress. J. Bacteriol. *183*, 1090–1095.

Mutalik, V.K., and Venkatesh, K.V. (2005). Quantification of the glycogen cascade system: the ultrasensitive responses of liver glycogen synthase and muscle phosphorylase are due to distinctive regulatory designs. Theoret. Biol. Med. Model. *2*, 1–18.

Mutisya, J., Sathish, P., Sun, C., Andersson, L., Ahlandsberg, S., Baguma, Y., Palmqvist, S., Odhiambo, B., Aman, P., and Jansson, C. (2003). Starch branching enzymes in sorghum (*Sorghum bicolor*) and barley (*Hordeum vulgare*): Comparative analyses of enzyme structure and gene expression. J. Plant Physiol. *160*, 921–930.

Nakamura, Y., Takahashi, J., Sakurai, A., Inaba, Y., Suzuki, E., Nihei, S., Fujiwara, S., Tsuzuki, M., Miyashita, H., Ikemoto, H., *et al.* (2005). Some Cyanobacteria synthesize semi-amylopectin type alpha-polyglucans instead of glycogen. Plant Cell Physiol. *46*, 539–545.

Nakao, M., Okamoto, S., Kohara, M., Fujishiro, T., Fujisawa, T., Sato, S., Tabata, S., Kaneko, T., and Nakamura, Y. (2010). CyanoBase: the cyanobacteria genome database update 2010. Nucleic Acids Res. *38*, D379–381.

Osanai, T., Kanesaki, Y., Nakano, T., Takahashi, H., Asayama, M., Shirai, M., Kanehisa, M., Suzuki, I., Murata, N., and Tanaka, K. (2005). Positive regulation of sugar catabolic pathways in the cyanobacterium *Synechocystis* sp PCC 6803 by the group 2 sigma factor sigE. J. Biol. Chem. *280*, 30653–30659.

Osanai, T., Imamura, S., Asayama, M., Shirai, M., Suzuki, I., Murata, N., and Tanaka, K. (2006). Nitrogen induction of sugar catabolic gene expression in *Synechocystis* sp PCC 6803. DNA Res. *13*, 185–195.

Osanai, T., Azuma, M., and Tanaka, K. (2007). Sugar catabolism regulated by light- and nitrogen-status in the cyanobacterium *Synechocystis* sp PCC 6803. Photochem. Photobiol. Sci. *6*, 508–514.

Osanai, T., Ikeuchi, M., and Tanaka, K. (2008). Group 2 sigma factors in cyanobacteria. Physiologia Plantarum *133*, 490–506.

Patron, N.J., and Keeling, P.J. (2005). Common evolutionary origin of starch biosynthetic enzymes in green and red algae. J. Phycol. *41*, 1131–1141.

Preiss, J. (1984). Bacterial glycogen synthesis and its regulation. Annu. Rev. Microbiol. *38*, 419–458.

Preiss, J. (2006). Bacterial glycogen inclusions: enzymology and regulation of synthesis. In Inclusions in Prokaryotes, Shively, J.M., ed. (Springer, Berlin/Heidelberg), pp. 71–108.

Preiss, J., and Romeo, T. (1994). Molecular biology and regulatory aspects of glycogen biosynthesis in bacteria. Prog. Nucleic Acid Res. Mol. Biol. *47*, 299–329.

Preiss, J., and Sivak, M.N. (1998). Biochemistry, molecular biology and regulation of starch synthesis. Genet. Engin. *20*, 177–223.

Rehm, B.H.A. (2010). Bacterial polymers: biosynthesis, modifications and applications. Nat. Rev. Microbiol. *8*, 578–592.

Reyes-Sosa, F., Molina-Heredia, F., and De la Rosa, M. (2010). A novel α-amylase from the cyanobacterium *Nostoc* sp. PCC 7119. Appl. Microbiol. Biotechnol. *86*, 131–141.

Riethman, H.C., Mawhinney, T.P., and Sherman, L.A. (1988). Characterization of phycobilisome glycoproteins in the cyanobacterium *Anacystis nidulans* R2. J. Bacteriol. *170*, 2433–2440.

Russell, J.B. (2007). The energy spilling reactions of bacteria and other organisms. J. Mol. Microbiol. Biotechnol. *13*, 1–11.

Russell, J.B., and Cook, G.M. (1995). Energetics of bacterial growth – Balance of anabolic and catabolic reactions. Microbiol. Rev. *59*, 48–62.

Rust, M.J., Golden, S.S., and O'Shea, E.K. (2011). Light-driven changes in energy metabolism directly entrain the cyanobacterial circadian oscillator. Science *331*, 220–223.

Sarma, T.A., and Kanta, S. (1979). Biochemical studies on sporulation in blue green algae. Zeitschrift für Allgemeine Mikrobiologie *19*, 571–575.

Sauer, J., Gorl, M., and Forchhammer, K. (1999). Nitrogen starvation in *Synechococcus* PCC 7942: involvement of glutamine synthetase and NtcA in phycobiliprotein degradation and survival. Arch. Microbiol. *172*, 247–255.

Schinzel, R., and Nidetzky, B. (1999). Bacterial alpha-glucan phosphorylases. FEMS Microbiol. Lett. *171*, 73–79.

Schneegurt, M.A., Sherman, D.M., Nayar, S., and Sherman, L.A. (1994). Oscillating behavior of carbohydrate granule formation and dinitrogen fixation in the cyanobacterium *Cyanothece* sp. strain ATCC 51142. J. Bacteriol. *176*, 1586–1597.

Schneegurt, M.A., Sherman, D.M., and Sherman, L.A. (1997). Composition of the carbohydrate granules of the cyanobacterium, *Cyanothece* sp strain ATCC 51142. Arch. Microbiol. *167*, 89–98.

Schuermann, P., and Buchanan, B.B. (2008). The ferredoxin/thioredoxin system of oxygenic photosynthesis. Antiox. Redox Signal. *10*, 1235–1273.

Schwarz, R., and Forchhammer, K. (2005). Acclimation of unicellular cyanobacteria to macronutrient deficiency: emergence of a complex network of cellular responses. Microbiology *151*, 2503–2514.

Sicher, R.C., and Kremer, D.F. (1996). Rubisco activity is altered in a starchless mutant of *Nicotiana sylvestris* grown in elevated carbon dioxide. Environ. Exp. Bot. *36*, 385–391.

Singh, A.K., and Sherman, L.A. (2005). Pleiotropic effect of a histidine kinase on carbohydrate metabolism in *Synechocystis* sp strain PCC 6803 and its requirement for heterotrophic growth. J. Bacteriol. *187*, 2368–2376.

Smith, A.J. (1982). Modes of cyanobacterial carbon metabolism. In The Biology of Cyanobacteria, Carr, N.G., and Whitton, B.A., eds. (Blackwell Scientific Publications, Oxford), pp. 47–85.

Stal, L.J. (1992). Poly(hydroxyalkanoate) in cyanobacteria: an overview. FEMS Microbiol. Lett. *103*, 169–180.

Steinbuchel, A., and Schlegel, H.G. (1989). Excretion of pyruvate by mutants of *Alcaligenes eutrophus*, which are impaired in the accumulation of the Poly(beta-hydroxybutyric acid) (PHB), under conditions permitting synthesis of PHB. Appl. Microbiol. Biotechnol. *31*, 168–175.

Stevens, S.E., Jr., Balkwill, D.L., and Paone, D.A.M. (1981). The effects of nitrogen limitation on the ultrastructure of the cyanobacterium *Agmenellum quadruplicatum*. Arch. Microbiol. *130*, 204–212.

Summers, M.L., and Meeks, J.C. (1996). Transcriptional regulation of *zwf*, encoding glucose-6-phosphate dehydrogenase, from the cyanobacterium *Nostoc punctiforme* strain ATCC 29133. Mol. Microbiol. *22*, 473–480.

Sun, J.D., Okita, T.W., and Edwards, G.E. (1999). Modification of carbon partitioning, photosynthetic capacity, and O-2 sensitivity in Arabidopsis plants with low ADP-glucose pyrophosphorylase activity. Plant Physiol. *119*, 267–276.

Suzuki, E., Umeda, K., Nihei, S., Moriya, K., Ohkawa, H., Fujiwara, S., Tsuzuki, M., and Nakamura, Y. (2007). Role of the GlgX protein in glycogen metabolism of the cyanobacterium, *Synechococcus elongatus* PCC 7942. Biochim. Biophys. Acta *1770*, 763–773.

Suzuki, E., Ohkawa, H., Moriya, K., Matsubara, T., Nagaike, Y., Iwasaki, I., Fujiwara, S., Tsuzuki, M., and Nakamura, Y. (2010). Carbohydrate metabolism in mutants of the cyanobacterium *Synechococcus elongatus* PCC 7942 defective in glycogen synthesis. Appl. Environ. Microbiol. *76*, 3153–3159.

Suzuki, E., Onoda, M., Colleoni, C., Ball, S., Fujita, N., and Nakamura, Y. (2013). Physicochemical variation of cyanobacterial starch, the insoluble alpha-glucans in cyanobacteria. Plant Cell Physiol. *54*, 465–473.

Tandeau de Marsac, N., and Houmard, J. (1993). Adaptation of cyanobacteria to environmental stimuli: new steps towards molecular mechanisms. FEMS Microbiol. Lett. *104*, 119–189.

Taroncher-Oldenberg, G., Nishina, K., and Stephanopoulos, G. (2000). Identification and analysis of the polyhydroxyalkanoate-specific beta-ketothiolase and acetoacetyl coenzyme A reductase genes in the cyanobacterium *Synechocystis* sp strain PCC6803. Appl. Environ. Microbiol. *66*, 4440–4448.

Tel-Or, E., Spath, S., Packer, L., and Mehlhorn, R.J. (1986). C-13 NMR-studies of salt shock-induced carbohydrate turnover in the marine cyanobacterium *Agmenellum quadruplicatum*. Plant Physiol. *82*, 646–652.

Toepel, J., Welsh, E., Summerfield, T.C., Pakrasi, H.B., and Sherman, L.A. (2008). Differential transcriptional analysis of the cyanobacterium *Cyanothece* sp strain ATCC 51142 during light-dark and continuous-light growth. J. Bacteriol. *190*, 3904–3913.

Tolonen, A.C., Aach, J., Lindell, D., Johnson, Z.I., Rector, T., Steen, R., Church, G.M., and Chisholm, S.W. (2006). Global gene expression of *Prochlorococcus* ecotypes in response to changes in nitrogen availability. Mol. Syst. Biol. *2*, 53.

Ugalde, J.E., Parodi, A.J., and Ugalde, R.A. (2003). De novo synthesis of bacterial glycogen: *Agrobacterium tumefaciens* glycogen synthase is involved in glucan initiation and elongation. Proc. Natl. Acad. Sci. U.S.A. *100*, 10659–10663.

Valladares, A., Maldener, I., Muro-Pastor, A.M., Flores, E., and Herrero, A. (2007). Heterocyst development and diazotrophic metabolism in terminal respiratory oxidase mutants of the cyanobacterium *Anabaena* sp. strain PCC 7120. J. Bacteriol. *189*, 4425–4430.

Wang, H.L., Postier, B.L., and Burnap, R.L. (2004). Alterations in global patterns of gene expression in *Synechocystis* sp PCC 6803 in response to inorganic carbon limitation and the inactivation of *ndhR*, a LysR family regulator. J. Biol. Chem. *279*, 5739–5751.

Wegener, K.M., Singh, A.K., Jacobs, J.M., Elvitigala, T.R., Welsh, E.A., Keren, N., Gritsenko, M.A., Ghosh, B.K., Camp, D.G., Smith, R.D., *et al.* (2010). Global proteomics reveal an atypical strategy for carbon/nitrogen assimilation by a cyanobacterium under diverse environmental perturbations. Mol. Cell. Proteom. *9*, 2678–2689.

Wilson, W.A., Roach, P.J., Montero, M., Baroja-Fernández, E., Muñoz, F.J., Eydallin, G., Viale, A.M., and Pozueta-Romero, J. (2010). Regulation of glycogen metabolism in yeast and bacteria. FEMS Microbiol. Rev. *34*, 952–985.

Wing-Ming, C., Hsueh-Mei, C., Hso-Freng, Y., Jei-Fu, S., and Tan-Chi, H. (1994). The aerobic nitrogen-fixing *Synechococcus* RF-1 containing uncommon polyglucan granules and multiple forms of α-amylase. Curr. Microbiol. *29*, 201–205.

von Wobeser, E.A., Ibelings, B.W., Bok, J., Krasikov, V., Huisman, J., and Matthijs, H.C.P. (2011). Concerted changes in gene expression and cell physiology of the cyanobacterium *Synechocystis* sp strain PCC 6803 during transitions between nitrogen and light-limited growth. Plant Physiol. *155*, 1445–1457.

Wood, N.B., and Haselkorn, R. (1988). Control of phycobiliprotein proteolysis and heterocyst differentiation in *Anabaena*. J. Bacteriol. *141*, 1375–1385.

Wu, G.F., Shen, Z.Y., Wu, Q.Y., and Zhao, N.M. (2001). Molecular cloning and construction of agp gene deletion-mutant in cyanobacterium *Synechocystis* sp PCC 6803. Acta Botanica Sinica *43*, 512–516.

Wyman, M., and Thom, C. (2012). Temporal orchestration of glycogen synthase (GlgA) Gene expression and glycogen accumulation in the oceanic picoplanktonic *Cyanobacterium Synechococcus* sp. Strain WH8103. Appl. Environ. Microbiol. *78*, 4744–4747.

Xu, Y.-F., Amador-Noguez, D., Reaves, M.L., Feng, X.-J., and Rabinowitz, J.D. (2012a). Ultrasensitive regulation of anapleurosis via allosteric activation of PEP carboxylase. Nature Chem. Biol. *8*, 562–568.

Xu, Y., Tiago Guerra, L., Li, Z., Ludwig, M., Charles Dismukes, G., and Bryant, D.A. (2012b). Altered carbohydrate metabolism in glycogen synthase mutants of *Synechococcus* sp. strain PCC 7002: Cell factories for soluble sugars. Metabol. Engin. *16C*, 56–67.

Yao, Y., Thompson, D.B., and Guiltinan, M.J. (2004). Maize starch-branching enzyme isoforms and amylopectin structure. In the absence of starch-branching enzyme IIb, the further absence of starch-branching enzyme Ia leads to increased branching. Plant Physiol. *136*, 3515–3523.

Yoo, S.H., Spalding, M.H., and Jane, J.L. (2002). Characterization of cyanobacterial glycogen isolated from the wild-type and from a mutant lacking of branching enzyme. Carbohydrate Res. *337*, 2195–2203.

Yoo, S.H., Keppel, C., Spalding, M., and Jane, J.L. (2007). Effects of growth condition on the structure of glycogen produced in cyanobacterium *Synechocystis* sp PCC6803. Int. J. Biol. Macromol. *40*, 498–504.

Young, F.G. (1957). Bernard, Claude and the discovery of glycogen – a century of retrospect. Br. Med. J. *1*, 1431–1437.

Zhang, C.C., Laurent, S., Sakr, S., Peng, L., and Bedu, S. (2006). Heterocyst differentiation and pattern formation in cyanobacteria: a chorus of signals. Mol. Microbiol. *59*, 367–375.

Zinser, E.R., Lindell, D., Johnson, Z.I., Futschik, M.E., Steglich, C., Coleman, M.L., Wright, M.A., Rector, T., Steen, R., McNulty, N., *et al.* (2009). Choreography of the transcriptome, photophysiology, and cell cycle of a minimal photoautotroph *Prochlorococcus*. PLoS One *4*, e5135.

Cyanophycin, a Cellular Nitrogen Reserve Material

9

Antonia Herrero and Mireia Burnat

Abstract

Cyanophycin is a biopolymer that serves as a nitrogen cellular reserve and occurs in most, albeit not all cyanobacteria. In this chapter, available information on the enzymes of cyanophycin metabolism and on the expression of the corresponding genes is reviewed. Regulation of cyanophycin production in response to the C to N balance of the cells and, in the case of heterocyst-forming cyanobacteria, the cellular specificity of cyanophycin metabolism are also addressed. Genes encoding cyanophycin metabolism proteins have also been found in bacteria other than cyanobacteria, and the polymer has been considered of potential biotechnological use. Some studies aiming at the production of cyanophycin in different bacteria and eukaryotes for its large-scale accumulation are also summarized.

Introduction

Cyanophycin is thought to represent a dynamic reservoir of nitrogen that is accumulated in the cells in the form of membraneless cytoplasmic granules (Fig. 9.1). As an example, the unicellular cyanobacterium *Cyanothece* sp. ATCC 51142, which temporarily separates photosynthetic O_2 evolution and N_2 fixation along daily variations, accumulates cyanophycin in the dark periods when it fixes N_2 and degrades cyanophycin in the light periods, when photosynthesis is performed (Sherman *et al.*, 1998). Daily dynamics of cyanophycin accumulation coincident with the periods of N_2 fixation has also been reported for the filamentous cyanobacterium *Trichodesmium* spp., a globally relevant marine diazotroph (Finzi-Hart *et al.*, 2009). Cyanophycin accumulates conspicuously under unbalanced growth conditions, e.g. in the stationary growth phase, under unbalanced growth including high light or CO_2 supply, or under stress conditions (phosphate or sulfur starvation) that do not involve nitrogen (N) starvation (Allen, 1984; Simon, 1987). Cyanophycin is also synthesized under conditions of N sufficiency during transitions between different N

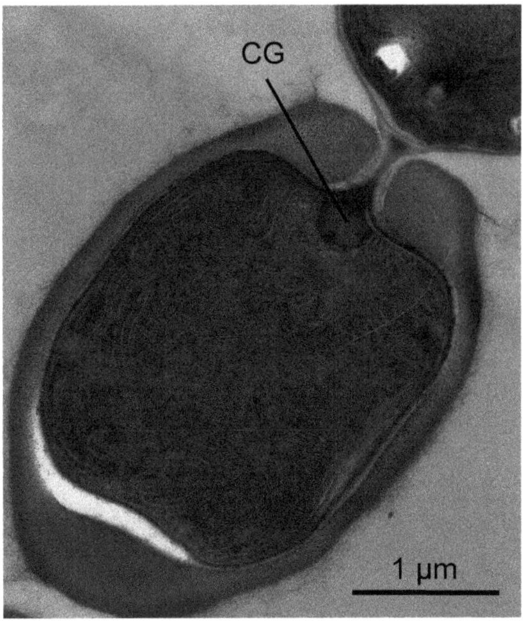

Figure 9.1 Transmission electron micrograph of a heterocyst of *Anabaena* sp. PCC 7120 showing a polar cyanophycin granule (CG). Sample prepared by high-pressure cryo-fixation as described by Merino-Puerto *et al.* (2011). Micrograph courtesy of Victoria Merino-Puerto, Heinz Schwarz and Iris Maldener.

supplies (Mackerras *et al.*, 1990) and, as shown by the use of $^{15}NO_3^-$ or $^{15}NH_4^+$ as a N source, it can be synthesized from external N sources as well as from N made available by degradation of cellular proteins (Allen *et al.*, 2005). Both the rate and the extent of nitrogen incorporation into cyanophycin depend on the source of N (ammonium or nitrate) available and whether the cells had been starved for N (Kolodny *et al.*, 2006).

In the filamentous heterocyst-forming cyanobacteria, during diazotrophic growth conditions, cyanophycin accumulates conspicuously at the heterocyst poles besides in vegetative cells (Lang *et al.*, 1972; Sherman *et al.*, 2000). In this case, how the N contained in the heterocyst granules is mobilized, and the mechanism of a putative transfer of this N to the neighbouring vegetative cells represent issues of long-standing interest in the biology of these organisms.

Although formerly thought to be a cyanobacterial characteristic, with the massive availability of bacterial genomic sequences, genes encoding cyanophycin metabolism enzymes have been detected to occur in a wide range of prokaryotic organisms with different metabolic options including phototrophy, anaerobic and aerobic respiration, fermentation and chemolitoautotrophy (Krehenbrink *et al.*, 2002).

The cyanophycin granule polypeptide

The cyanophycin granule is composed of a multi-L-arginyl-L-aspartate (arginine residues linked to a polyaspartate backbone) polypeptide (Simon and Weathers, 1976). In this polymer, the amino acids aspartate and arginine are present in an almost 1:1 ratio because nearly all the aspartate residues in the backbone are linked to arginine residues. Rather than having a uniform molecular mass, the polymer is in the range between 25 and 100 kDa (Simon, 1971, 1976). The unitary multi-Arg-Asp chains that constitute cyanophycin are water insoluble and spontaneously associate into granules that include a substantial fraction of β-pleated sheet structure (Simon *et al.*, 1980).

Although in most cases cyanophycin is composed exclusively of Asp and Arg, some other constituent amino acids have been reported occasionally. Thus, besides Arg and Asp, Lys has been found, in place of Arg, in cyanophycin isolated from *E. coli* expressing cyanophycin synthetase (see below) from *Synechocystis* sp. PCC 6803 or from *Anabaena variabilis* ATCC 29413 (Ziegler *et al.*, 1998; Berg *et al.*, 2000). Additionally, in an *in vitro* reaction catalysed by *Synechocystis* cyanophycin synthetase, Lys, citrulline and ornithine have been shown to be incorporated into cyanophycin-like polymers, although at lower rates than Arg (see Berg *et al.*, 2000). Regarding glutamic acid, it has been described to occur *in vivo*, in place of Arg, in *Synechocystis* sp. PCC 6308 incubated under N-limiting conditions (Merritt *et al.*, 1994). Glu has also been found in cyanophycin synthesized *in vitro* with cell extracts from this cyanobacterium (Aboulmagd *et al.*, 2000).

Enzymes and genetics of cyanophycin metabolism

Cyanophycin synthetase

Cyanophycin is non-ribosomically synthesized by the action of cyanophycin synthetase [multi-L-arginyl-poly(L-aspartic acid) synthetase], which catalyses the elongation of a cyanophycin primer with L-Asp and, afterwards, the addition of L-Arg involving the conversion of ATP to ADP (Simon, 1976) (1 mol of ATP per mol of amino acid incorporated [Aboulmagd *et al.*, 2001]). Thus, cyanophycin synthetase catalyses the formation of two types of amide bonds: α-peptide bonds between the aspartyl residues that form the backbone of the polymer, and isopeptide bonds between the β-carboxyl group of aspartyl residues in the backbone and the α-amino group of arginine in the branches (Fig. 9.2). The enzyme was first purified from *Anabaena variabilis* (Ziegler *et al.*, 1998), and it is a homodimer of the 96 kDa product of the *cphA* gene that was first identified in the genomic sequence of *Synechocystis* sp. PCC 6803 (Ziegler *et al.*, 1998). Interestingly, cyanophycin synthetase shows similarity to peptide synthetases involved in the biosynthesis of murein. Indeed, the N- and C-terminal parts of cyanophycin synthetase show similarity to different superfamilies of ATP-dependent ligases. This observation led to the prediction of two ATP-binding and catalytic

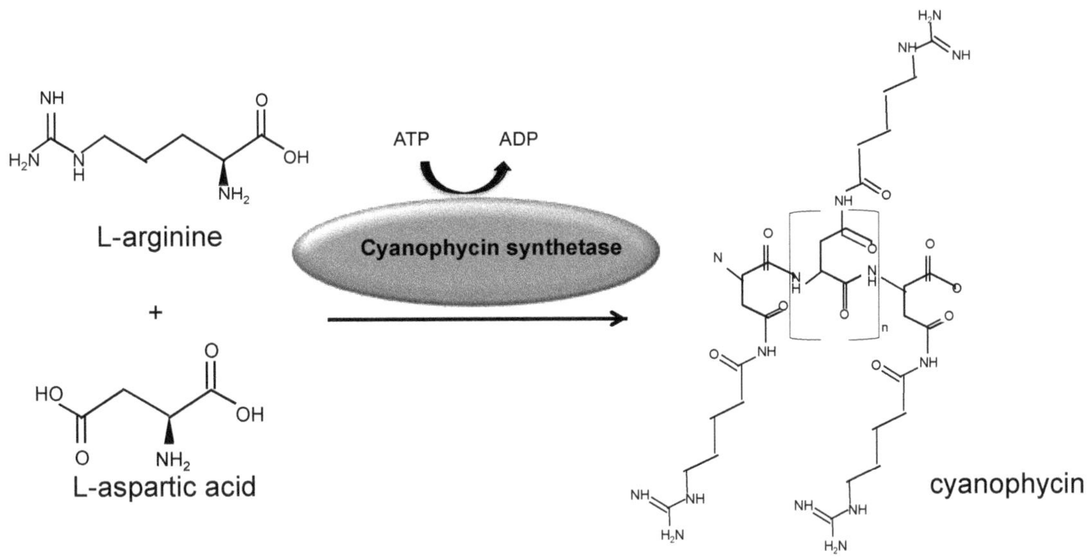

Figure 9.2 Synthesis of cyanophycin. A single enzyme, cyanophycin synthetase, is involved in cyanophycin synthesis. The enzyme requires L-arginine, L-aspartic acid, ATP, magnesium ions and, *in vitro*, a cyanophycin primer (not depicted). Cyanophycin synthetase alternately adds aspartate to the α-carboxylate and arginine to the β-carboxylate group of the terminal aspartate of the growing chain.

sites in cyanophycin synthetase, each involved in the addition of one of the amino acid substrates during the elongation reaction, bearing the C-terminal part the active site for the incorporation of aspartate and the N-terminal part the one for arginine incorporation (Berg *et al.*, 2000).

Extracts of *E. coli* expressing the *cphA* gene from *Synechocystis* show cyanophycin synthetase activity. Moreover, these cells synthesize a cyanophycin-like material, indicating that CphA catalyses *de novo* cyanophycin synthesis, in contrast to the *in vitro* activity that requires a cyanophycin primer. Interestingly, the enzyme from *Thermosynechococcus elongatus* strain BP-1 has been reported to catalyse primer-independent synthesis of cyanophycin (Arai and Kino, 2008). *In vitro*, the elongation reaction takes place at the C-terminal part of the cyanophycin primer, and the incorporation of amino acids to the cyanophycin polymer occurs step-by-step and not in the form of Asp-Arg dipeptides (Berg *et al.*, 2000). Finally, the fact that Glu can be incorporated into cyanophycin *in vivo* in *Synechocystis* sp. PCC 6308, but not *in vitro* with purified cyanophycin synthetase from this strain has led to suggest that the *in vivo* Glu incorporation could respond to an as yet unknown enzyme component (Aboulmagd *et al.*, 2001).

Cyanophycinase

Cyanophycin degradation is catalysed by a specific hydrolytic enzyme, cyanophycinase, the 29.4 kDa product of the *cphB* gene, which was first identified in the genome of *Synechocystis* sp. PCC 6803 as a gene directing the expression in *E. coli* of a product catalysing as a C-terminal exopeptidase the hydrolysis of cyanophycin to a peptide consisting of β-Asp-Arg (Richter *et al.*, 1999) (Fig. 9.3). The crystal structure of cyanophycinase from *Synechocystis* has been solved at 1.5 Å resolution, showing that the enzyme is a dimer and confirming that it is a serine-type protease. The enzyme would recognize the dipeptide up from the cleavable bond in the polypeptide chain, and details of the sterics of substrate recognition are available (Law *et al.*, 2009). CphB shows similarity, at both the protein sequence and structural levels, to an aspartyl dipeptidase from *E. coli* and the PepE dipeptidase from *Salmonella typhimurium* (Richter *et al.*, 1999; Law *et al.*, 2009). Based on sequence comparisons between the *Synechocystis* CphB and *Salmonella* PepE, residues Ser^{132},

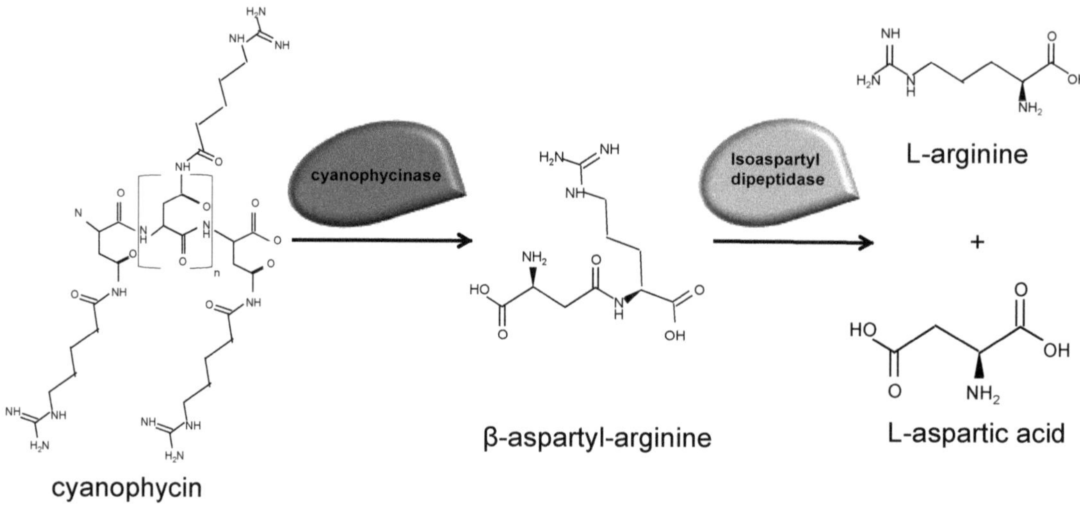

Figure 9.3 Degradation of cyanophycin. Two enzymes are involved in cyanophycin degradation: cyanophycinase, an exopeptidase that hydrolyses cyanophycin releasing β-aspartyl-arginine peptides, and an isoaspartyl dipeptidase that hydrolyses the peptide bond of β-aspartyl-arginine.

Glu[150] and His[174] (numeration of *Synechocystis*) have been proposed to constitute the catalytic triad of ChpB (Richter *et al.*, 1999).

cphA and *chpB* gene expression

In *Synechocystis* sp. PCC 6803, it has been suggested that the *cphA* (cyanophycin synthetase) and *cphB* (cyanophycinase) genes are expressed independently (Richter *et al.*, 1999). In *Anabaena* sp. PCC 7120, two gene clusters containing *cphB* and *cphA* genes have been identified (Picossi *et al.*, 2004) (Fig. 9.4). The *cph1* cluster is expressed at higher levels and constitutes an operon, although *cphA1* could additionally be expressed from an independent promoter. In the *cph2* cluster, the *cphB2* and *cphA2* genes are divergently oriented and thus expressed in separate transcripts. Expression of the *cph2* cluster is detected at levels considerably lower than the *cph1* cluster and, indeed, a mutational analysis indicated that the latter cluster contributes more than the former to cyanophycin accumulation in the *Anabaena* filaments (Picossi *et al.*, 2004).

Aspartyl-arginine dipeptidase

The β-L-aspartyl-L-arginine (β-Asp-Arg) dipeptide resulting from cleavage of cyanophycin by cyanophycinase is further hydrolysed to its amino

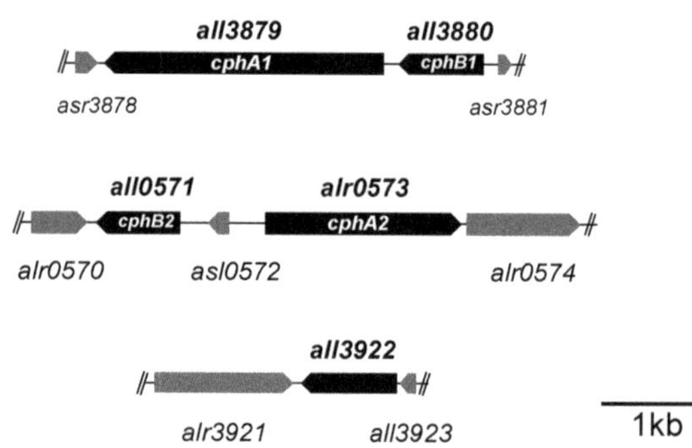

Figure 9.4 Clusters of cyanophycin metabolism related genes in *Anabaena* sp. PCC 7120. Cyanophycin synthetase (*cphA*), cyanophycinase (*cphB*) and the isoaspartyl dipeptidase (*all3922*) genes are highlighted in black.

acid constituents, Asp and Arg, by isoaspartyl dipeptidase (Fig. 9.3). β-Asp-Arg hydrolysing activity was found in extracts of *Synechocystis* sp. PCC 6803, which was attributed to the product of gene *sll0422* annotated as 'plant-type asparaginase' (Richter *et al.*, 1999). Confirming this identification, isoaspartyl-dipeptidase activity was shown in the asparaginases from *Synechocystis* sp. PCC 6803 (Sll0422) and *Anabaena* sp. PCC 7120 (All3922) expressed in *E. coli*, as well as in plant-type asparaginases in general (Hejazi et al., 2002). Besides deamidating asparagine, the enzymes hydrolyse a range of isoaspartyl dipeptides, including those that could result from proteolytic degradation of modified proteins containing isoaspartyl residues and, with high catalytic efficiency, also β-Asp-Arg from the degradation of cyanophycin.

The mature isoaspartyl dipeptidases of *Synechocystis* sp. PCC 6803 and *Anabaena* sp. PCC 7120 have molecular weights of around 70 kDa and consist of dimers of two subunits (α-subunit derived from the N-terminal part of the precursor; β-subunit derived from the C-terminal part) that arise from autoproteolytic cleavage at the Gly–Thr bond (G^{179}-T^{180}, numeration of the *Anabaena* polypeptide) within the conserved sequence GT(I/V)G of a single primary gene product (Hehazi *et al.*, 2002). Recently, a mutant of *Anabaena* devoid of All3922 has been generated, and this mutant accumulates the β-Asp-Arg peptide confirming the involvement of the enzyme in cyanophycin metabolism (Burnat *et al.*, in press).

The complete sequence of the genomes of about 150 cyanobacteria is currently available (Shih *et al.*, 2013). Genes encoding the three cyanophycin metabolism enzymes, cyanophycin synthetase, cyanophycinase and isoaspartyl dipeptidase, are widely distributed among cyanobacteria. Although these genes are not present in every cyanobacterium, they are found in phylogenetically very distant strains, from morphologically complex cyanobacteria belonging to *Rivulariaceae* to the phylogenetically deep-branching cyanobacterium *Gloeobacter violaceaus* (see http://genome.microbedb.jp/cyanobase/;http://img.jgi.doe.gov/cgi-bin/w/main.cgi).

C-to-N balance regulation

As mentioned above, cyanophycin accumulation has been shown in various cyanobacteria, both unicellular non-nitrogen fixing and filamentous or unicellular N_2 fixers, to respond to the N regime. In both *Synechocystis* sp. PCC 6803 (Maheswaran *et al.*, 2006) and *Anabaena* sp. PCC 7120 (Paz-Yepes *et al.*, 2009), it has been shown that the C-to-N sensory regulator P_{II} (the *glnB* gene product) influences cyanophycin accumulation. In *Synechocystis*, cyanophycin is absent in a P_{II}-null mutant, and in cells grown under nitrate-limiting conditions cyanophycin rapidly accumulates upon the supply of ammonium in the wild-type but not in the P_{II} mutant. *N*-acetyl-L-glutamate kinase, which catalyses the committed step of the cyclic pathway for arginine biosynthesis, is a direct target of P_{II}. This enzyme is strongly activated by complex formation with the unphosphorylated form of P_{II}, which indicates a low C-to-N ratio (Heinrich *et al.*, 2004). Thus, P_{II} affects cyanophycin accumulation through an effect on arginine synthesis under N sufficiency (Maheswaran *et al.*, 2006). Under diazotrophic conditions, a P_{II} mutant (Paz-Yepes *et al.*, 2009) or a *prpS* (encoding a putative P_{II} phosphatase) mutant (Laurent *et al.*, 2004) of *Anabaena* does not exhibit the cyanophycin polar granules characteristic of the wild-type heterocysts. However, whether in these mutants, which are impaired in diazotrophic growth, the lack of heterocyst cyanophycin responds to an effect of the mutation on arginine accumulation, on nitrogenase activity, or on other uncharacterized targets remains to be studied.

Cell specificity of cyanophycin metabolism in heterocyst-forming cyanobacteria

In both *Anabaena cylindrica* and *Anabaena* sp. PCC 7120 incubated under diazotrophic conditions, higher activities of cyanophycin synthetase and cyanophycinase have been found in cell extracts of heterocysts than of vegetative cells (Gupta and Carr, 1981). In strain PCC 7120, the cell specificity of the expression of the gene cluster responsible for the bulk of cyanophycin metabolism, *cph1*, has been studied. Globally, both *cphA1* (cyanophycin synthetase) and *cphB1* (cyanophycinase) appear to be expressed in both vegetative cells and heterocysts under diazotrophic growth conditions (Picossi *et al.*, 2004). The genes are cotranscribed in both cell types, but at increased levels in heterocysts. Additionally, *cphA1* is expressed monocistronically also preferably in the latter cells. At the molecular level, cotranscription takes place from three promoters located upstream of *cphB1* that show different cell specificities: $P_{cphB1-1}$, operating in heterocysts and dependent on the global transcriptional regulator NtcA; $P_{cphB1-2}$, a low-activity consensus-type promoter with no cell specificity; and $P_{cphB1-3}$, an NtcA-dependent promoter preferentially used in vegetative cells. On the other hand, monocistronic expression of *cphA1* would take place from at least two promoters located in the *cphB1-cphA1* intergenic region: $P_{cphA1-3}$, representing a consensus-type promoter active in vegetative cells; and $P_{cphA1-2}$, directing NtcA-dependent expression in heterocysts. This complex promoter arrangement ensures certain level of expression of *cphB1* and *cphA1* in both cell types, vegetative cells and heterocysts, and also leads to a stronger expression of *cphA1* than of *cphB1* in heterocysts, responding for the conspicuous accumulation of cyanophycin observed in these differentiated cells (Picossi *et al.*, 2004).

With regard to isoaspartyl dipeptidase, in strain PCC 7120 the *all3922* gene product is present at higher levels in vegetative cells than in heterocysts (M. Burnat *et al.*, unpublished). Thus, under diazotrophic conditions, the metabolism of cyanophycin appears to be compartmentalized in different cell types. Whereas synthesis from fixed N and accumulation into granules would take place preferentially in the heterocysts, degradation would take place stepwise in different cells: in heterocysts cyanophycin would be hydrolysed producing dipeptides that would be transferred to the vegetative cells, where they would be further broken down releasing the constituent amino acids to be used in metabolism

In *Anabaena* sp. PCC 7120 a mutational study has shown that strains lacking cyanophycin synthetase (*cphA1* and/or *cphA2*) are little affected in diazotrophic growth (Picossi *et al.*, 2004) (this is also the case for an *A. variabilis* ATCC 29413 cyanophycin synthetase mutant; Ziegler *et al.*, 2001). However, diazotrophic growth is significantly decreased in cyanophycinase (*cphB1* and *cphB1 cphB2*) mutants (Picossi *et al.*, 2004) or in isoaspartyl dipeptidase mutants (M. Burnat *et al.*, unpublished), which show conspicuous cyanophycin granules in both vegetative cells and heterocysts. Thus, although the path of the fixed nitrogen through cyanophycin seems to be dispensable, the capacity of mobilization of the nitrogen stored in this reservoir once it has been synthesized appears to significantly influence N nutrition.

Cyanophycin and cyanophycin-metabolism enzymes in other bacteria

Genes putatively encoding cyanophycin synthetase and cyanophycinase have been identified in the sequenced genomes of quite diverse bacteria. For instance, by 2002 genes homologous to the *Synechocystis* sp. PCC 6803 *cphA* were reported to occur in the genomic sequence of seven bacteria, out of 65 searched for, of the genera *Acinetobacter, Bordetella, Clostridium, Desulfobacterium* and *Nitrosomonas*, with identities in the range of 37–40% at the amino acid level with regard to the *Synechocystis* sequence (Krehenbrink *et al.*, 2002). Moreover, the aerobic bacterium *Acinetobacter* sp. strain DSM 587 accumulates cyanophycin under phosphate-limited conditions, and the *cphA* gene from *Acinetobacter* (Krehenbrink *et al.*, 2002) or from the anaerobe *Desulfitobacterium hafniense* (Ziegler *et al.*, 2002) cloned in *E. coli* leads to expression of cyanophycin synthetase activity and accumulation of cyanophycin-like polymers. Regarding enzymes of cyanophycin degradation, already by 2002 eleven eubacteria were selected as able to use cyanophycin as a carbon source for growth through the use of extracellular cyanophycinases. The purified extracellular enzyme from *Pseudomonas anguilliseptica* strain B1, CphE, exhibits high specificity for cyanophycin and, as its cyanobacterial counterparts, hydrolyses cyanophycin to β-Asp-Arg dipeptides, despite exhibiting only 27–28% identity to intracellular cyanophycinases from cyanobacteria (Obst *et al.*, 2002).

More recently, the presence of cyanophycin-metabolism genes has been searched for in 570 microbial genomes, identifying putative cyanophycin synthetases in 44 prokaryotes, of which 31 also possess cyanophycinase genes. In addition, 24 of the strains encoding cyanophycinases also bear putative isoaspartyl dipeptidases. Based on the organization of these genes in the genome, a classification of ten groups of ChpA and eight groups of ChpB enzymes has been proposed. Thus, the ability to synthesize and degrade cyanophycin appears to be widely distributed among prokaryotes, although it seems to be a eubacterial trait found neither in Eukarya nor in Archaea (Füser and Steinbüchel, 2007).

Applied aspects

Cyanophycin as well as its derived dipeptides have been considered of potential industrial use. In the chemical industry, cyanophycin has been considered as a potential source of biopolymers, e.g. the polyaspartate polymers that can be generated by dearginylation of cyanophycin represent a biodegradable alternative for polyacrilates, and the cyanophycin constituent amino acids can be used as feedstock for the production of various other chemicals, e.g. nylons. Also, the dipeptides have potential uses in medicine and pharmacy as natural food additives and therapeutic source of amino acids (Mooibroek *et al.*, 2007).

To increase the extent of cyanophycin accumulation and of source biomass production beyond the levels attained with the wild-type strains of cyanobacteria, *E. coli* strains bearing the *cphA* gene from *Synechocystis* sp. PCC 6803 (Frey *et al.*, 2002) or engineered versions of *cphA* from *Nostoc ellipsosporum* (Hai *et al.*, 2006) have been used to produce cyanophycin, which accumulates to up 24% and 34.5% (wt/wt dry matter), respectively, although requiring complex growth media. These systems could yield isolated cyanophycin at the kilogram scale. The industrially relevant host *Saccharomyces cerevisiae* has also been used for expression of the *cphA* gene from *Synechocystis* sp. PCC 6308, although the accumulation of cyanophycin was lower in this system (15.3% wt/wt), even when production of cyanophycin containing amino acids alternative to arginine was improved by the use of *Saccharomyces* strains defective in arginine synthesis (Steinle *et al.*, 2009). Interestingly, in the latter case, in addition to Asp–Arg only and Lys-containing polymers, variants including citrulline or ornithine were obtained. Also the yeast *Pichia pastoris* has been used as a host of mutant derivatives of the *Synechocystis* sp. PCC 6308 *cphA* gene, yielding up to 23.3% (wt/wt) cyanophycin (Steinle *et al.*, 2010). In addition to differences in the chemical composition, the polymer produced in some of these recombinant strains contains soluble and insoluble forms of the polypeptide, which has a smaller, albeit more homogenous, molecular mass (approximately, 25–30 kDa) than the natural one. Finally, procedures have been developed for the large-scale isolation of cyanophycin from biomass (Sallam *et al.*, 2009; Steinle and Steinbüchel, 2010) and for the further production of β-dipeptides by treatment with massively produced extracellular cyanophycinase from *Pseudomonas alcaligenes* (Sallam *et al.*, 2009, 2011).

Production of cyanophycin has also been attempted in plants, in transgenic tobacco and potato, where the cyanophycin synthetase gene from *Thermosynechococcus elongatus* BP-1 was constitutively expressed and a cyanophycin-like polymer was produced up to 1.1% dry weight (Neumann *et al.*, 2005). Improvement of large-scale production and isolation of cyanophycin has been achieved by localizing the production in potato amyloplasts and chloroplasts, which also reduced the negative phenotypic effects of cyanophicin production in the plant (Neubauer *et al.*, 2012). Thus, the industrial exploitation of cyanophycin as a biotechnological resource appears promising.

Acknowledgements

We thank Enrique Flores for critical reading of the manuscript. Work in the authors' laboratory was supported by grants BFU2010-17980 and BFU2011-22762 from the Dirección General de Investigación y Gestión del Plan Nacional de Investigación, Desarrollo e Innovación, Spain, co-financed by FEDER. M. Burnat was the recipient of an FPI Fellowship from the Spanish Government.

References

Aboulmagd, E., Oppermann-Sanio, F.B., and Steinbüchel, A. (2000). Molecular characterization of the cyanophycin synthetase from *Synechocystis* sp. strain PCC6308. Arch. Microbiol. *174*, 297–306.

Aboulmagd, E., Oppermann-Sanio, F.B., and Steinbüchel, A. (2001). Purification of *Synechocystis* sp. strain PCC6308 cyanophycin synthetase and its characterization with respect to substrate and primer specificity. Appl. Environ. Microbiol. *67*, 2176–2182.

Allen, M.M. (1984). Cyanobacterial cell inclusions. Ann. Rev. Microbiol. 38, 1–25.

Allen, M.M., Yuen, C., Medeiros, L., Zizlsperger, N., Farooq, M., and Kolodny, N.H. (2005). Effects of light and chloramphenicol stress on incorporation of nitrogen into cyanophycin in *Synechocystis* sp. PCC 6803. Biochim. Biophys. Acta *1725*, 241–246.

Arai, T., and Kino, K. (2008). A cyanophycin synthetase from *Thermosynechoccocus elongatus* BP-1 catalyzes primer-independent cyanophycin synthesis. Appl. Microbiol. Biotechnol. *81*, 69–78.

Berg, H., Ziegler, K., Piotukh, K., Baier, K., Lockau, W., and Volkmer-Engert, R. (2000). Biosynthesis of the cyanobacterial reserve polymer multi-L-arginyl-poly-L-aspartic acid (cyanophycin). Mechanism of the cyanophycin synthetase reaction studied with synthetic primers. Eur. J. Biochem. *267*, 5561–5570.

Burnat, M., Herrero, A., and Flores, E. (2014). Compartmentalized metabolism in the diazotrophic filaments of a heterocyst-forming cyanobacterium. Proc. Natl. Acad. Sci. U.S.A., in press. (doi/10.1073/pnas.1318564111).

Finzi-Hart, J.A., Pett-Ridge, J., Weber, P.K., Popa, R., Fallon, R.P., Gunderson, T., Hutcheon, I.D., Nealson, K.H., and Capone, D.G. (2009). Fixation and fate of C and N in the cyanobacterium *Trichodesmium* using nanometer-scale secondary ion mass spectrometry. Proc. Natl. Acad. Sci. U.S.A. *106*, 6345–6350.

Frey, K.M., Oppermann-Sanio, F.B., Schmidt, H., and Steinbüchel, A. (2002). Technical-scale production of cyanophycin with recombinant strains of *Escherichia coli*. Appl. Environ. Microbiol. *68*, 3377–3384.

Füser, G., and Steinbüchel, A. (2007). Analysis of genome sequences for genes of cyanophycin metabolism: Identifying putative cyanophycin metabolizing prokaryotes. Macromol. Biosci. *7*, 278–296.

Gupta, M., and Carr, N.G. (1981). Enzyme activities related to cyanophycin metabolism in heterocysts and vegetative cells of *Anabaena* spp. J. Gen. Microbiol. *125*, 17–23.

Hai, T., Frey, K.M., and Steinbüchel, A. (2006). Engineered cyanophycin synthetase (CphA) from *Nostoc ellipsosporum* confers enhanced CphA activity and cyanophycin accumulation to *Escherichia coli*. Appl. Environ. Microbiol. *72*, 7652–7660.

Heinrich, A., Maheswaran, M., Ruppert, U., and Forchhammer, K. (2004). The *Synechococcus* elongatus P_{II} signal transduction protein controls arginine synthesis by complex formation with *N*-acetyl-L-glutamate kinase. Mol. Microbiol. *52*, 1303–1314.

Hejazi, M., Piotukh, K., Mattow, J., Deutzmann, R., Volkmer-Engerts, and Lockau, W. (2002). Isoaspartyl dipeptidase activity of plant-type asparaginases. Biochem. J. *364*, 129–136.

Kolodny, N.H., Bauer, D., Bryce, K., Klucevsek, K., Lane, A., Medeiros, L., Mercer, W., Moin, S., Park, D., Petersen, J., *et al.* (2006). Effect of nitrogen source on cyanophycin synthesis in *Synechocystis* sp. strain PCC 6308. J. Bacteriol. *188*, 934–940.

Krehenbrink, M., Oppermann-Sanio, B.F., and Steinbüchel, A. (2002). Evaluation of non-cyanobacterial genome sequences for occurrence of genes encoding proteins homologous to cyanophycin synthetase and cloning of an active cyanophycin synthetase from *Acinetobacter* sp. strain DSM 587. Arch. Microbiol. *177*, 371–380.

Lang, N.J., Simon, R.D., and Wolk, C.P. (1972). Correspondence of cyanophycin granules with structured granules in *Anabaena cylindrica*. Arch Mikrobiol. *83*, 313–320.

Laurent, S., Forchhammer, K., González, L., Heulin, T., Zhang, C.-C., and Bédu, S. (2004). Cell-type specific modification of PII is involved in the regulation of nitrogen metabolism in the cyanobacterium *Anabaena* PCC 7120. FEBS Lett. *576*, 261–265.

Law, A.M., Sandy, Lai, S.W.S., Tavares, J., and Kimber, M.S. (2009). The structural basis of β-peptide-specific cleavage by the serine protease cyanophycinase. J. Mol. Biol. *392*, 393–404.

Mackerras, A.H., de Chazal, N.M., and Smith, G.D. (1990). Transient accumulation of cyanophycin in *Anabaena cylindrica* and *Synechocystis* 6308. J. Gen. Microbiol. *136*, 2057–2065.

Maheswaran, M., Ziegler, K., Lockau, W., Hagemann, M., and Forchhammer, K. (2006). P_{II}-regulated arginine synthesis controls accumulation of cyanophycin in *Synechocystis* sp. strain PCC 6803. J. Bacteriol. *188*, 2730–2734.

Merino-Puerto, V., Mariscal, V., Schwarz, H., Maldener, I., Mullineaux, C., Herrero, A., and Flores, E. (2011). FraH is required for reorganization of intracellular membranes during heterocyst differentiation in *Anabaena* sp. strain PCC 7120. J. Bacteriol. *193*, 6815–6823.

Merritt, M.V., Sid, S.S., Mesh, L., and Allen, M.M. (1994). Variations in the amino acid composition of cyanophycin in the cyanobacterium *Synechocystis* sp. PCC 6308 as a function of growth conditions. Arch. Microbiol. *162*, 158–166.

Mooibroek, H., Oosterhuis, N., Giuseppin, M., Toonen, M., Franssen, H., Scott, E., Sanders, J., and Steinbüchel, A. (2007). Assessment of technological options and economical feasibility for cyanophycin biopolymer and high-value amino acid production. Appl. Microbiol. Biotechnol. *77*, 257–267.

Neubauer, K., Hühns, M., Hausmann, T., Klemke, F., Loackau, W., Kahmann, U., Pistorius, E.K., Kragl, U., and Broer, I. (2012). Isolation of cyanophycin from tobacco and potato plants with constitutive plastidic chpATe gene expression. J. Bacteriol. *158*, 50–58.

Neumann, K., Stephan, D.P., Ziegler, K., Hühns, M., Broer, I., Lockau, W., and Pistorius, E.K. (2005). Production of cyanophycin, a suitable source for the biodegradable polymer polyaspartate, in transgenic plants. Plant Biotechnol. J. *3*, 249–258.

Obst, M., Oppermann-Sanio, F.B., Luftmann., H., and Steinbüchel, A. (2002). Isolation of cyanophycin-degrading bacteria, cloning and characterization of an extracellular cyanophycinase gene (*cphE*) from *Pseudomonas anguilliseptica* strain BI. The *cphE* gene from *P. anguilliseptica* BI encodes a cyanophycin hydrolyzing enzyme. J. Biol. Chem. *277*, 25096–25105.

Paz-Yepes, J., Flores, E., and Herrero, A. (2009). Expression and mutational analysis of the *glnB* genomic region in the heterocyst-forming cyanobacterium *Anabaena* sp. strain PCC 7120. J. Bacteriol. *191*, 2353–2361.

Picossi, S., Valladares, A., Flores, E., and Herrero, A. (2004). Nitrogen-regulated genes for the metabolism of cyanophycin, a bacterial nitrogen reserve polymer. Expression and mutational analysis of two cyanophycin synthetase and cyanophycinase gene clusters in the heterocyst-forming cyanobacterium *Anabaena* sp. PCC 7120. J. Biol. Chem. *279*, 11582–11592.

Richter, R., Hejazi, M., Kraft, R., Ziegler, K., and Lockau, W. (1999). Cyanophycinase, a peptidase degrading the cyanobacterial reserve material multi-L-arginyl-poly-L-aspartic acid (cyanophycin). Molecular cloning of the gene of *Synechocystis* sp. PCC 6803, expression in *Escherichia coli* and biochemical characterization of the purified enzyme. Eur. J. Biochem. *263*, 163–169.

Sallam, A., Kast, A., Przybilla, S., Meiswinkel, T., and Steinbüchel, A. (2009). Biotechnological process for production of β-dipeptides from cyanophycin on a technical scale and its optimization. Appl. Environ. Microbiol. *75*, 29–38.

Sallam, A., Kalkandzhiev, D., and Steinbüchel, A. (2011). Production optimization of cyanophycinase ChpEal from *Pseudomonas alcaligenes* DIP1. AMB Express *1*, 38.

Sherman, L.A., Meunier, P., and Colón-López, M.S. (1998). Diurnal rithms in metabolism: A day in the life of a unicellular, diazotrophic cyanobacterium. Photosynth. Res. *58*, 25–42.

Sherman, D.M., Tucker, D., and Sherman, L.A. (2000). Heterocyst development and localization of cyanophycin in N_2-fixing cultures of *Anabaena* sp. PCC 7120 (cyanobacteria) J. Phycol. *36*, 932–941.

Shih, P.M., Wu, D., Latifi, A., Axen, S.D., Fewer, D.P., Talla, E., Calteau, A., Cai, F., Tandeau de Marsac, N., Rippka, R., *et al.* (2013). Improving the coverage of the cyanobacterial phylum using diversity-driven genome sequencing. Proc. Natl. Acad. Sci. U.S.A. *110*, 1053–1058.

Simon, R.D. (1971). Cyanophycin granules from the blue-green alga *Anabaena cylindrica*: A reserve material consisting of copolymers of aspartic acid and arginine. Porc. Natl. Acad. Sci. U.S.A. *68*, 265–267.

Simon, R.D. (1976). The biosynthesis of multi-L-arginyl-poly(L-aspartic acid) in the filamentous cyanobacterium *Anabaena cylindrica*. Biochim. Byophys. Acta *422*, 407–418.

Simon, R.D. (1987). Inclusion bodies in the cyanobacteria: cyanophycin, polyphosphate, polyhedral bodies. In The Cyanobacteria, Fay, P., and van Baalen, C., ed. (Elsevier, Amsterdam, The Netherlands), pp. 199–225.

Simon, R.D., and Wheathers, P. (1976). Determination of the structure of the novel polypeptide containing aspartic acid and arginine which is found in cyanobacteria. Biochim. Biophys. Acta *420*, 165–176.

Simon, R.D., Lawry, N.H., and McLendon, G.L. (1980). Structural characterization of the cyanophycin granule polypeptide of *Anabaena cylindrica* by circular dichroism and Raman spectroscopy. Biochim. Biophys. Acta *626*, 277–281.

Steinle, A., and Steinbüchel, A. (2010). Establishment of a simple and effective isolation method for cyanophycin from recombinant *Saccharomyces cerevisiae*. Appl. Microbiol. Biotechnol. *85*, 1393–1399.

Steinle, A., Bergander, K., and Steinbüchel, A. (2009). Metabolic engineering of *Saccharomyces cerevisiae* for production of novel cyanophycins with an extended range of constituent amino acids. Appl. Environ. Microbiol. *75*, 3437–3446.

Steinle, A., Witthoff, S., Krause, J.P., and Steinbüchel, A. (2010). Establishment of cyanophycin biosynthesis in *Pichia pastoris* and optimization by use of engineered cyanophycin synthetases. Appl. Environ. Microbiol. *76*, 1062–1070.

Ziegler, K., Diener, A., Herpin, C., Richter, R., Deutzmann, R., and Lockau, W. (1998). Molecular characterization of cyanophycin synthetase, the enzyme catalyzing the biosynthesis of the cyanobacterial reserve material multi-L-arginyl-poly-L-aspartate (cyanophycin). Eur. J. Biochem. *254*, 154–159.

Ziegler, K., Dirk, P.S., Pistorius, E.K., Ruppel, H.G., and Lockau, W. (2001). A mutant of the cyanobacterium *Anabaena variabilis* ATCC 29413 lacking cyanophycin synthetase: growth properties and ultrastructural aspects. FEMS Microbiol. Lett. *196*, 13–18.

Ziegler, K., Deutzmann, R., and Lockau, W. (2002). Cyanophycin synthetase-like enzymes of non-cyanobacterial eubacteria: characterization of the polymer produced by a recombinant synthetase of *Desulfitobacterium hafniense*. Z. Naturforsch. *57*, 522–529.

Gas Vesicles

Judith Herzfeld

10

Abstract

Gas vesicles are hollow, organelles that provide buoyancy in aquatic prokaryotes and allow them to regulate their depth in the water column. Classic studies have shown them to be rigid and permeable to gases as large as C_4F_8. The wall of the hollow structure is formed entirely of protein. The principal component, GvpA (~7.5 kDa) is a metamorphic protein, adopting two conformations in an asymmetric dimer. These assemble into an amyloid (i.e. open-ended, cross-β) ribbon that wraps around the vesicle axis in a shallow helix and presents an aliphatic face to the interior. This aliphatic face is expected to cause evaporation of liquid water from the interior of nascent vesicles and prevent water condensation inside mature vesicles. A second protein GvpC, adheres to the GvpA shell and strengthens it. A great deal remains to be learned about how other members of the *gvp* gene cluster cooperate with the two main structural proteins in assembling and disassembling vesicles. Of particular interest is the fact that, despite the amyloid properties of the vesicles, the cells are able to dismantle them, in order to descend in the water column and apparently recycle protein from vesicles that have collapsed.

Introduction

Gas vesicles, the uniquely gas-filled buoyancy organelles of aquatic microorganisms have been objects of curiosity since they were first discovered at the end of the 19th century (Ahlborn, 1895; Klebahn, 1895; Strodtmann, 1895). However, their unusual features have also rendered them challenging to characterize. The majority of what we now know is due to decades of persistent, creative, meticulous, and broadly collaborative work by Anthony Walsby. Walsby has also written several thorough reviews (Walsby, 1972, 1975, 1994), and the most recent of these remains must reading for anyone interested in gas vesicles. This chapter will take that 1994 review as its point of departure, summarizing its major conclusions and describing more recent work in greater detail. In the latter period, progress has been enabled by advances in mass spectrometry, nuclear magnetic resonance, genetics, and molecular modelling. Particular attention will be paid to the important constraints that experimental results place on modelling. Although the focus will be on cyanobacteria, the strong homology between gas vesicles of all prokaryotes will justify including information from other species where equivalent information is not available for cyanobacteria.

Morphology

Mature gas vesicles in cyanobacteria comprise hollow cylinders capped at each end by hollow cones (Fig. 10.1). The vesicles are rigid structures that withstand moderate applications of pressure with little change in volume, but collapse irreversibly to a flattened envelope when a certain critical pressure is reached. While permeable to very large gas molecules, the vesicle wall prevents liquid water from entering or accumulating inside its cavity. A full understanding of these exceptional and intriguing properties requires knowledge of the wall structure at the molecular level, as discussed later in this chapter. However, we focus first on larger scale features.

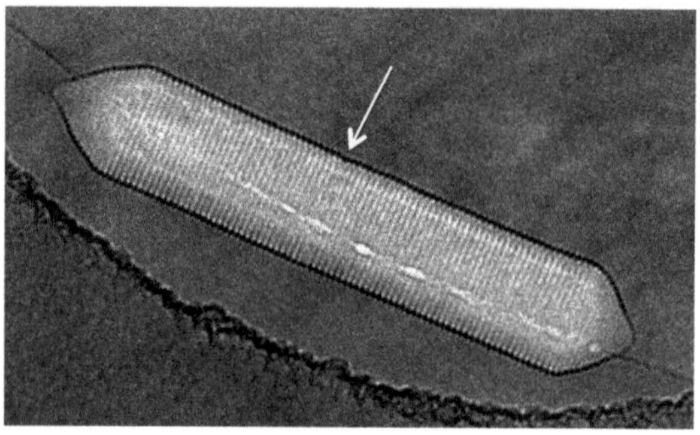

Figure 10.1 Electron micrograph of a 292 nm × 52 nm gas vesicle from *Anabaena flos-aquae*. Notice the very low electron density inside the vesicle, the ribs running across the vesicle, and the bipolar structure of the vesicle indicated by the orientations of the rib projections on either side of the seam marked by the arrow. (Courtesy of Chen Xu and Marina Belenky, Brandeis University.)

While the lengths of gas vesicles vary widely, even within a single cell, the widths for a given organism vary very little (Walsby and Bleything, 1988). On the other hand, the widths vary considerably between organisms. In different cyanobacteria the average critical collapse pressure, which varies from 3 to 12 bars, correlates inversely with the width of the gas vesicle, as expected from mechanical considerations (narrow pipes are stronger than wide ones of the same wall thickness). As expected from natural selection, narrower, stronger gas vesicles are found in cyanobacteria living in deeper water bodies (Walsby, 1972, 1975, 1994).

When produced in abundance, gas vesicles occupy large portions of the cell, but minimize interference with other constituents by collecting into hexagonally packed bundles known as gas vacuoles (Bowen and Jensen, 1965; Jost, 1965; Smith and Peat, 1967; Whitton and Peat, 1969). This bundling seems to be driven entirely by packing constraints (Onsager, 1949), as the vesicles disperse when the cells are lysed (Jones and Jost, 1970).

Closer examination of the vesicles by electron microscopy, reveals ridges running roughly perpendicular to the vesicle axis. In freeze fracture preparations, these ribs are also evident on the interior surface (Jones and Jost, 1970). As expected from symmetry arguments (Walsby, 1972), gas vesicles are bipolar. Careful scrutiny of the edges of the images, where the electron density of the vesicle wall is least obscured by the low density of the hollow interior, shows that the two ends meet at an irregular rib located in the cylindrical region, although not necessarily in the middle (see the arrow in Fig. 10.1). At high resolution, Moiré patterns show that the ribs on each side take the form of a low-pitch helix, rather than a stack of hoops (Offner *et al.*, 1998). Growth presumably occurs at one end or the other of these helices, i.e. at the conical tips or at the irregular rib.

X-ray diffraction patterns reveal crystalline order, with reflections that correspond to a wall thickness of 1.95 nm, a rib-to-rib distance of 4.57 nm, and a repeat distance along the rib of 1.15 nm (Blaurock and Walsby, 1976).

Composition

Gas vesicles are built entirely of protein. Although lipids are generally important constituents of cellular organelles, none are found in gas vesicles (Jones and Jost, 1970; Walsby and Buckland, 1969). This comports with the crystalline order of the vesicles. Also missing are phosphates (Walsby and Hayes, 1988) and carbohydrates (Jones and Jost, 1970).

Vesicles remain intact after washing with urea or detergent and these stripped bodies comprise the fundamental shell of the vesicle. Although it is generally supposed that some special molecules must be incorporated into these shells, particularly at the cone tips and at the irregular rib, a single small protein, GvpA accounts for essentially all of the shell (Walsby and Hayes, 1988) and antibodies raised against a 13 residue peptide corresponding to the N-terminus of GvpA bind very close to the cone tips (Powell *et al.*, 1991). GvpA typically comprises about 70 amino acid residues (Hayes *et al.*, 1986; Tandeau de Marsac *et al.*, 1985), consistent with the size of the crystallographic unit cell, and is not cross-linked or otherwise post-translationally modified (Belenky *et al.*, 2004). The monomers are only known to dissociate in highly protic solvents (e.g. 80% formic acid) (Walker and Walsby, 1983). Thus, while GvpA contains an unusually large proportion of aliphatic residues (Carugo, 2008; Jones and Jost, 1970), and has a net negative charge at physiological pHs, the charge balance among the protic residues appears to play a major role in stabilizing the shells.

Vesicles that have not been washed with urea are coated with another, ~3-fold larger protein, GvpC (Hayes *et al.*, 1992; Walsby and Hayes, 1988). Whereas GvpA is highly hydrophobic, GvpC is relatively hydrophilic. Whereas GvpA is unusually rich in aliphatic residues, GvpC is unusually rich in glutamine, alanine, glutamate and phenylalanine. Both the conical and cylindrical portions of the vesicles are coated with GvpC and saturation occurs at an approximate molar ratio of GvpA:GvpC = 25:1 (Buchholz *et al.*, 1993). Removing GvpC weakens the vesicles so that they collapse at a lower pressure (e.g. ~2 bars rather than ~6 bars in *Anabaena flos-aquae*) (Walsby and Hayes, 1988), and replacing it restores the original critical pressure (Hayes *et al.*, 1992).

Assembly, disassembly and control

Investigators have been unable to refold dissolved GvpA molecules, much less reassemble gas vesicles, after denaturation and dissociation in highly protic solvent; dialysis with physiological solvent yields only an amorphous precipitate. Moreover, investigators have not been able to re-erect vesicles that have been collapsed by a pressure pulse. On the other hand, the organisms themselves are apparently able to recycle subunits from collapsed vesicles to nascent vesicles (Hayes and Walsby, 1984) via a process that survives in a cell-free lysate (Lehmann and Jost, 1972). It seems likely that the cell accomplishes both assembly and disassembly by providing chaperones for the GvpA monomers.

Nascent vesicles start as small bicones that grow until they reach a characteristic diameter (Lehmann and Jost, 1971; Waaland and Branton, 1969). Growth beyond that point takes the form of a cylinder between the two cones. In view of the difficulty of changing cones to cylinders, it seems most likely that growth occurs by monomer addition at the irregular rib between the two ends of the vesicle and that this functions as an insertion seam.

The number of gas vesicle genes varies from 9 to 14 among the organisms that have been studied (Dunton and Walsby, 2005; Horne *et al.*, 1991; Kinsman and Hayes, 1997; Mlouka *et al.*, 2004). Gas vesicle production varies with the number of copies of the *gvpA* gene, from one in *Pseudanabaena*, which produces few gas vesicles (Damerval *et al.*, 1991), to two in *Calothrix* (Damerval *et al.*, 1987), to five in *Anabaena* (Hayes and Powell, 1995), to as many as seven in *Planktothrix*, which produces many gas vesicles in each cell (Beard *et al.*, 2000, 2002). Gas vesicle proteins other than GvpA and GvpC, are present in low copy number and challenging to detect. However, gas vesicle genes have been amenable to study in transformants of *Haloferax volcanii* because it has no gas vesicle genes of its own (Englert *et al.*, 1992a; Horne *et al.*, 1991). As a result, at least one role has been identified for each of 12 of the 14 proteins in the dominant *Halobacterium* gas vesicle gene cluster:

- Gene deletion studies found that the eight gas vesicle genes A, O, F, G, J, K, L and M suffice for gas vesicle formation (Offner *et al.*, 2000).
- Antibodies raised to the products of genes F, G, J, L and M were found to bind to gas vesicles (Shukla and DasSarma, 2004). Some of these proteins have also been detected in purified gas vesicles by LC-MS/MS proteomics, along with I, N and O (Chu *et al.*, 2011). Thus, including A and C, a total of 10 proteins are apparently associated in some manner with the vesicle wall.
- Regulatory proteins have also been identified. GvpE acts as a transcriptional activator and GvpD is involved in repression (Hofacker *et al.*, 2004). Deletion of D and E reduces the number of vesicles (Offner *et al.*, 2000), while deletion of D alone results in a huge excess of gas vesicles (Englert *et al.*, 1992b). (No counterparts to these genes have been found in cyanobacteria.)
- In the dominant *Halobacterium* gene cluster, only GvpH and GvpK have no known role at this time.

Functional features

In gas vesicles, nature has evolved a remarkably elegant mechanism by which cells can evade the relentless gravitational drive towards sedimentation. Better yet, by adjusting the number of vesicles, cells can regulate their depth in the water column, e.g. for access to light and oxygen. By comparison, flagella have been judged cheaper to build, but more expensive to operate, making the use of flagella more suitable for more rapidly dividing organisms and investment in gas vesicles more suitable for slower growing organisms (Walsby, 1994).

Gas vesicles are not balloon like: they are neither elastic nor actively filled, but rather rigid and passively filled (Walsby, 1969, 1982a). To be effective, they have to withstand pressure from two sources: (1) the cell turgor pressure, which varies from < 1 bar in halophiles to as much as 5 bar in freshwater habitats, and (2) the hydrostatic pressure of the water column, which increases by 1 bar for every 10 m of descent.

The crystalline walls of the vesicles have a critical yield pressure that depends on the vesicle width (as well as the GvpC reinforcement mentioned above): the narrower the vesicles, the higher the pressure at which they collapse (Hayes and Walsby, 1986; Walsby and Bleything, 1988). On the other hand, the smaller volume-to-surface ratio of narrower vesicles reduces the buoyancy effected by a given amount of protein. Thus, in economical fashion, organisms in which the gas vesicles are subjected to lower pressures, have evolved wider, weaker gas vesicles. For example gas vesicles in cyanobacteria from shallow brine pools (*Dactylococcopsis, Aphanothece*) collapse at a pressure of only 3 bars; those in deeper freshwater lakes (*Anabaena, Microcystis*) at 6 to 11 bars, and those in tropical oceans (*Trichodesmium*) at 12 bars or more (Walsby, 1978). For most of these organisms growth occurs only within the top 20 m of the water column (with pressures up to only 2 bars). However, stronger gas vesicles are required when the organisms are carried down to greater depths by vertical mixing (Walsby *et al.*, 1998).

For a vesicle to stabilize a gas–water interface, and for a gas–water interface to stabilize a gas vesicle, the interior, gas-facing surface should be hydrophobic and the exterior, water-facing surface should be hydrophilic. This is in fact what has been found. In particular, lyophilized vesicles are readily suspended in water, but not in oil (Walsby, 1971), and when layers of collapsed and dried vesicles are rehydrated, water only enters between alternate surfaces (Worcester, 1975). In addition, X-ray diffraction shows a higher electron density at the outer surface of the vesicle wall than at the inner surface (Blaurock and Walsby, 1976), consistent with relatively H-rich residues at the inner surface and relatively N- and O-rich residues at the outer surface.

Gas vesicles are permeable to all gases that have been investigated. These include H_2, N_2, O_2, Ar, CO, CO_2 (Walsby, 1971) and even C_4F_8 (with a diameter of 6.3 Å) (Walsby, 1982b). As a result of this permeability, the contents of gas vesicles equilibrate with the gases dissolved in the surrounding media, which in turn equilibrate with the overlying atmosphere. The question then arises as to why water molecules, which are considerably smaller than C_4F_8, either do not enter the vesicle or do not condense inside. It has been argued that it is the large surface tension of water that prevents liquid water from entering hydrophobic pores and causes the convex surface of any droplets that might form from water vapour to lose molecules to planar and concave water surfaces (Walsby, 1994). However, a macroscopic concept, such as surface tension, does not translate well to a molecular scale.

At the molecular level, water in the 3D liquid has the distinction of being able to accept two hydrogen bonds and donate two. Giving up any of these has a cost. On the other hand, water readily forms single file, rapidly diffusing chains in carbon nanotubes (Hummer *et al.*, 2001; Kyakuno *et al.*, 2011; Maniwa *et al.*, 2002). The stability of these 1D chains depends on dispersive attractive interactions with the smooth, electron dense nanotube walls. When such dispersive interactions are turned off in molecular simulations, the water molecules vacate the nanotubes in favour of 3D hydrogen bonding outside the nanotubes (Hummer *et al.*, 2001). Since dispersive interactions are very short range, one can imagine attenuating them in the real world by two means. One relies on hydrogenation of the surface to hold water molecules at a greater distance from the more electron rich second row atoms of the surface. Another frustration mechanism relies on nano-scale roughness to prevent the electron dense portions of the water molecules from approaching the electron dense portions of the wall molecules without having to break too many hydrogen bonds (Mittal and Hummer, 2010; Nakajima *et al.*, 2001; Wenzel, 1949). Simulations with a water–wall interaction energy minimum of 1.2 kcal mol^{-1}, have shown a transition from a hydrophilic surface to a hydrophobic surface at a roughness amplitude of about 3 Å and a wavelength of 11.6 Å (Mittal and Hummer, 2010).

While either of these mechanisms might retard access of water molecules to the interior of gas vesicles, it seems doubtful that they could prevent access altogether, given that water molecules are known to intercalate into hydrophobic pockets of proteins (Ernst *et al.*, 1995; Yu *et al.*, 1999). The question then is what prevents water vapour from condensing inside the vesicle. Condensation can occur either on a surface (heterogeneous nucleation) or in the gas phase (homogeneous nucleation). The heterogeneous pathway might be suppressed by either of the above mechanisms for reducing attractions to the surface: aliphatic residues present a hydrogen rich surface and the 'knob-and-holes' topology of an antiparallel β-sheet seems likely to offer the requisite roughness. The homogeneous pathway is intrinsically difficult [which is why aerosols in the atmosphere play an important role in the hydrological cycle (Andreae and Crutzen, 1997; Kaufman *et al.*, 2002) and cloud seeding is an option for stimulating rain (Bruintjes, 1999)]. In fact, for super-saturation levels of 2–4 (as would occur upon cooling by 10–20°C around 300 K), rates of homogeneous water nucleation have been estimated at ~1 event per cm^3 per second (Wolk *et al.*, 2002). Thus, for a vesicle like that in Fig. 10.1, with a volume of less than 10^{-14} cm^3, it would take more than 10^{14} s, or about a million years, for homogeneous nucleation to occur. Even for super-saturation levels of 6 (as would occur upon cooling by ~ 30°C around 300 K), rates of homogeneous water nucleation rise only to ~ 10^5 events per cm^3 per second (Wolk *et al.*, 2002), such that nucleation in a volume of less than 10^{-14} cm^3 would take more than 10^9 s, still on the order of decades. Clearly these are not biologically relevant time scales.

Taken together, the gaseous interior of gas vesicles is plausibly maintained by the combination of a shell sufficiently tight that water molecules can only pass through in vapour form, an interior surface that is not conducive to heterogeneous nucleation, and a cavity size that makes homogeneous nucleation highly unlikely during the lifetime of the cell. The question then is how nature builds this hollow object in the first place.

Since vesicles grow from small bicones, it can be imagined that several protein molecules initially come together at the tips, exposing their aliphatic surfaces to one another. If not rigidly held, the result will be a collapse of the aliphatic surfaces on one another. But if a structure forms that is strong enough to maintain a space < 5 nm between the aliphatic surfaces, then a vapour bubble is expected to form spontaneously between the surfaces (Ball, 2012; Chandler, 2005; Lum *et al.*, 1999; Patel *et al.*, 2012). The non-trivial remainder of the story involves growing the vesicles from the small bicones by adding more protein molecules. Since, theory estimates the critical cluster size for nucleating water condensation at 10–30 water molecules (Gao *et al.*, 1999), clusters of this size need to be excluded from the vesicle interior while the protein units are being inserted.

Molecular structure

The very features that make gas vesicles particularly interesting also make them poorly suited for standard structural studies. Since gas vesicle subunits cannot be dissolved without irreversible denaturation, they cannot be studied by solution NMR or crystallized for 3-D diffraction experiments. Even high-resolution cryo-electron microscopy is defeated by the multiple scattering that results from the large density difference between the inside and outside of the vesicle. Nonetheless, significant clues to the molecular structure of gas vesicles are now available.

Primary and secondary structure of GvpA

Consistent with the very stringent specifications required for gas vesicle function, the amino

```
AVEKTNSSSS LAEVIDRILD KGIVIDAWVR VSLVGIELLA IEARIVIASV ETYLKYAEAV LTQSAAVPA

CCC-----CC HHHHHHHHHH HCEEEEEE-- ---CCEEEEE EEEEEEEEEH HHHHHHHHCC CCCCCCCCC
```

Figure 10.2 Primary sequence of GvpA in *Anabaena flos-aquae* (top) and the secondary structures inferred from TALOS analysis of NMR chemical shifts (bottom) (Sivertsen *et al.*, 2010). C = coil, H = helix, E = sheet and dashes mark residues with insufficient spectroscopic information for TALOS analysis. Comparison with GvpA sequences of other organisms shows that the entire sequence is highly conserved among cyanobacteria and the sequence between the two terminal coils is highly conserved across all organisms (Sivertsen *et al.*, 2009). Underscored residues call attention to the charged residues within each of the β-strands and likely centres of a β-turn between the β-strands.

acid sequence of GvpA (Fig. 10.2) is highly homologous across prokaryotes from eubacteria to archaea (Sivertsen *et al.*, 2009; Walker *et al.*, 1984). At the same time, GvpA has no significant homology with other proteins (Ezzeldin *et al.*, 2012; Strunk *et al.*, 2011). Thus direct structural analogies cannot be drawn. Nevertheless all secondary structure prediction tools, which of course are based on structural data for other proteins, converge in assigning secondary elements of comparable length in the sequence

coil (c1) – helix (α1) – strand (β1) –
strand (β2) – helix (α2) – coil (c2)

An important feature of the putatively β-strand portions of the protein is that one set of alternating residues in each is exclusively aliphatic. Thus, each of the β-strands can present a purely aliphatic face. Facing inward, such surfaces could provide the hydrophobicity that has been attributed to the vesicle interior and would suppress heterogeneous nucleation of liquid water.

Experimental insights into secondary structure come primarily from spectroscopy. Both CD (Walsby, 1994) and FTIR (Strunk *et al.*, 2011) spectra are consistent with a mixture of α-helix and β-strand structures, and analysis of backbone chemical shifts assigned by solid state NMR spectroscopy (Sivertsen *et al.*, 2010) has provided site-specific details that accord well with the predictions (included in Fig. 10.2).

Tertiary and quaternary structure of GvpA

Because the GvpA monomer is small, it is not easy to distinguish between intramolecular and intermolecular interactions. Therefore, we consider tertiary and quaternary structure together.

Early X-ray diffraction studies not only measured the unit cell dimensions described above, but also detected an 0.47 nm periodicity typical of strand-strand spacing in β-sheets (Blaurock and Walsby, 1976). The sharpness of this reflection indicates that β-sheet extends over long distances in the vesicle wall. In recent years, such sheets have been recognized as the backbone of amyloid fibrils, which have in common with gas vesicles that they are insoluble in detergents.

Other experiments provide more details about the β-sheets. FTIR spectra show evidence of an anti-parallel arrangement (Strunk *et al.*, 2011) which is not common in amyloids (Shewmaker *et al.*, 2011), but which is thought to provide greater strength via more regular hydrogen bonds. Strength is also enhanced by the orientation of the hydrogen bonds close the 'magic angle' (54.7°) relative to the vesicle axis (Walsby, 1994): early X-ray studies of partially oriented vesicles (Blaurock and Walsby, 1976) found that the β-strands are tilted 36° from the vesicle axis which corresponds to an inclination of 54° for the hydrogen bonds. Subsequent atomic force microscopy (McMaster *et al.*, 1996) has detected features along the vesicle ribs that are consistent with the β-strand spacing and orientation originally deduced from the X-ray diffraction.

An intra-molecular anti-parallel β-sheet is readily achieved by a turn between the β1 and β2 strands. Significantly, the primary sequence of GvpA is such that, if the β-turn comprises an odd number of residues, then one side of the sheet will present only aliphatic residues. Interestingly, solid state NMR spectra of gas vesicles show duplicate resonances, of similar intensity, in the peptide backbone between the β1 and β2 strands, suggesting that GvpA occurs in two conformations that differ in the β-turn (Sivertsen *et al.*, 2009). This metamorphic behaviour is confirmed by solid state NMR correlation spectroscopy that finds four sets of anti-parallel β-β contacts, one β1–β1, one β2–β2, and two β1-β2 (Bayro *et al.*, 2012). Since the β1–β1 and β2–β2 contacts must be intermolecular, both sets of β1–β2 contacts must be intra-molecular. This evidence points to anti-symmetric GvpA dimers as the fundamental building block of an open ended anti-parallel β-sheet (about which more below).

Essentially nothing is known about the locations of the α-helices except that they seem not to obstruct access of the atomic force microscope probe to the β-sheet (McMaster *et al.*, 1996). One might therefore surmise that they are confined to, and comprise the major portion of, the protruding ridges between ribs and help to hold them together. In any case, MALDI-TOF mass spectrometry studies of proteolysed vesicles have shown that N- and C-termini are located at the outer surface of the vesicle (Belenky *et al.*, 2004). The protein assembly seems to otherwise be very compact as the other tested portions of the peptide backbone are inaccessible to proteases (Belenky *et al.*, 2004). A similar picture of amide accessibility to solvent has been drawn from solid state NMR studies of $^{2}H_{2}O$ exchange (Sivertsen *et al.*, 2010). Furthermore, the extent of the exposure of the N-terminus seems to be quite limited: antibodies raised against a peptide corresponding to 13 residues of the N-terminus bind only near the tips of the vesicles, indicating that the full epitope is not exposed in other parts of the vesicle (Powell *et al.*, 1991).

Structure of GvpC

Native GvpC in *Anabaena flos-aquae* incorporates five repeats of 33 amino acid residues, portions of each of which remain bound to vesicles after trypsin digestion (Fig. 10.3). Thus the 1:25 binding ratio of GvpC to GvpA corresponds to a 1:5 binding ratio of 33 residue repeats to GvpA (Buchholz *et al.*, 1993). In mutants of GvpC constructed with just three or four repeats, it is the 1:5 ratio of repeats to GvpA that is preserved (Kinsman *et al.*, 1995). Thus it appears that each repeat finds its own binding site on the vesicle surface and that binding to multiple sites is what enables GvpC to strengthen the vesicle.

The secondary structure of GvpC is predicted to be essentially entirely α-helical which would lend the molecule a rod-like shape (Dunton *et al.*, 2006). This may explain why vesicles grown in the presence of longer GvpC's develop larger diameters (and lower collapse pressures) than vesicles grown in the presence of shorter GvpC's (Dunton and Walsby, 2005). [This process is to be distinguished from adding GvpCs of various lengths back to the washed shells of mature vesicles which increases the collapse pressure (Kinsman *et al.*, 1995) even though it is not expected to alter the diameters.)]

The prevailing hypothesis has been that GvpC binds across the vesicle ribs, helping to hold them together. Given the relative dimensions of putatively helical, 33 residue GvpC repeats ($33 \times 0.15\,nm = 4.95\,nm$ long) and the vesicle ribs (4.57 nm spacing), this scenario can be accommodated with the GvpC tilted at 22.6° from the vesicle axis, as shown in Fig. 10.4. Furthermore, if the 36° tilt of the β-strands of GvpA is in the opposite direction, the GvpC repeats

```
                  MISLMAKIRQEHQSIAEK    N-terminus
VAELSLETREFLSVTTAKRQEQAEKQAQELQAF        repeat 1
YKDLQETSQQFLSETAQARIAQAEKQAQELLAF        repeat 2
HKELQETSQQFLSATAQARIAQAEKQAQELLAF        repeat 3
YQEVRETSQQFLSATAQARIAQAEKQAQELLAF        repeat 4
HKELQETSQQFLSATADARTAQAKEQKESLLKF        repeat 5
RQDLFVSIFG                              C-terminus
```

Figure 10.3 Primary sequence of GvpC in *Anabaena flos-aquae*, emphasizing the five 33 residue repeats. The underlined sequences are within fragments that adhere to gas vesicles after cleavage of R-X and K-X peptide bonds by digestion with trypsin. (Adapted from Dunton *et al.*, 2006.)

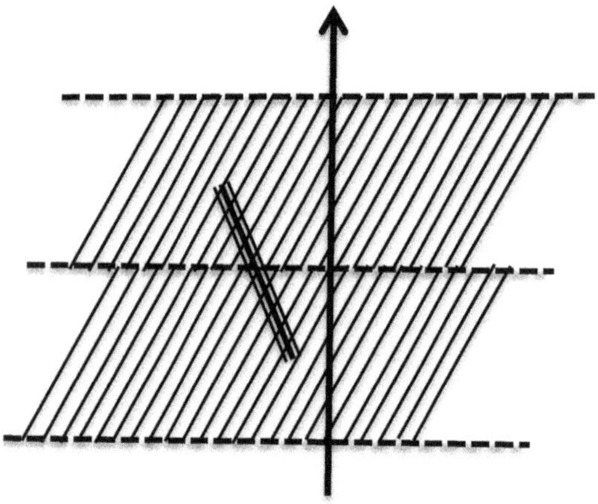

Figure 10.4 Plausible disposition of GvpC on the outer vesicle surface. The arrow indicates the direction of the vesicle axis, dashed lines represent interfaces between ribs, thin solid lines represent β-strands of GvpA, and the triple line represents one of the 33residue repeats of a GvpC molecule. The GvpA strands and the GvpC repeat are tilted, in opposite senses, at 36° and 23° respectively, relative to the vesicle axis. In this configuration, the GvpC repeat crosses nine GvpA strands. To give a sense of the potential for securing the interface between ribs, the GvpC repeat is arbitrarily situated to straddle the interface between two ribs, forming contacts with strands in both.

would cross them at an angle of 23° + 36° = 59°. At that angle, a 4.95 nm repeat of GvpC would project a distance of 4.24 nm perpendicular to the β-strands. Thus, given the 0.47 nm periodicity of β-strands in β-sheet, a single GvpC repeat would cross 9 β-strands or 4.5 GvpA monomers. This could roughly account for the 1:5 binding ratio of GvpC repeats to GvpA monomers (Walsby, 1994). However, there is no direct experimental information bearing on the orientation of GvpC molecules on the vesicle surface.

Molecular modelling

Molecular modelling strategies differ in the nature of the input, which can include constraints from experiment, molecular mechanics force fields, or some combination of the two. The three major efforts to-date illustrate these differences.

The first effort (Sivertsen *et al.*, 2009) sought to determine what GvpA arrangements could be qualitatively consistent with the experimental findings to that point, in particular an extended sheet of antiparallel β-strands, oriented at 36° to the vesicle axis and presenting a hydrophobic face to the vesicle interior. In this exercise, electrostatics came to play an important role (Fig. 10.5). The β1-strand carries an aspartate and an arginine four residues apart while the β2-strand carries a glutamate and an arginine two residues apart. Thus the only way for an extended β-sheet to stabilize all four charged residues in ion pairs, instead of just two of them, is to include antiparallel β1–β1 and antiparallel β2–β2 interfaces. In other words, the fundamental building block of the vesicle must be an anti-parallel dimer. However, in order to set the strands at the experimentally observed 36° and maintain the purely aliphatic character of one face of the sheet, it is necessary to also suppose that the required odd-numbered β-turn is shifted by one residue in alternating monomers. The resulting 'asymmetric dimer model' predicts that there will be two different chemical shifts for atoms at critical positions, such as in the β-turn and other transitions between secondary structure elements. It also predicts four different sets of inter-strand contacts, a β1–β2 set within each of the two types of monomer, and β1-β1 and β2-β2 sets between monomers. All of these predictions have been borne out in solid state NMR experiments (Bayro *et al.*, 2012; Sivertsen *et al.*, 2009). The model does not address the disposition of the helix or coil portions of the protein, for lack of relevant experimental information.

Relying on experimental constraints, the asymmetric dimer model does not make use of force fields within the protein (except in the qualitative sense of pairing oppositely charged residues) or between the protein and solvent. *De novo* modelling takes the opposite tack of asking how force fields alone would have GvpA fold. This is a non-trivial proposition because folding energies comprise small differences between large numbers. Thus far, there have been two such efforts (Ezzeldin *et al.*, 2012; Strunk *et al.*, 2011). Both:

- model haloarchaeal gas vesicles;
- start with the secondary structure predicted from the primary sequence;

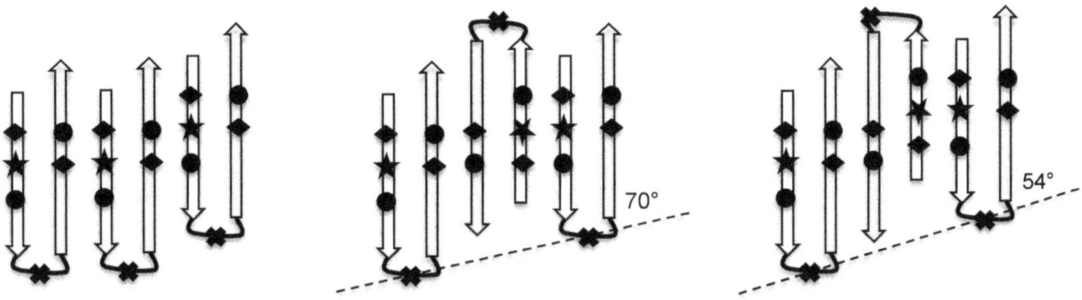

Figure 10.5 Schematic representation of three ways of forming an extended, anti-parallel β-sheet along the vesicle rib, noting placement of the carboxylic acids (♦), arginines (●), glycines (✖), and tryptophans (★) that occur within the sheet. All β-turns are odd number so that all the non-aliphatic residues are on the side of the sheet facing the reader. Left: with a monomeric unit cell, intermolecular interactions cannot pair all charged residues located within the sheet. Middle: with a symmetric dimer unit cell, all charged residues within the sheet are paired in inter-molecular salt bridges, but with an incorrect inclination of the strands to the rib (vesicle axis) of 70° (20°). Right: with an asymmetric dimer unit cell, in which β-turns in alternating molecules are centred on valine instead of glycine, all charged residues within the sheet are also paired in inter-molecular salt bridges, but with a correct inclination of the strands to the rib (vesicle axis) of 54° (36°). Note that the angles do not appear to be of the labelled magnitude because the distances between strands have been rendered on a larger scale than the distances along strands in order to enhance clarity.

- run numerous simulations of further monomer folding in continuum solvent;
- identify a single conformer as the best;
- find that this conformer features an odd-numbered β-turn, centred on the glycine, such that one face of the intramolecular β1–β2 sheet is purely aliphatic;
- dock rigid replicas of this one monomer in an antiparallel dimer with an inter-molecular β2–β2 sheet that forms salt bridges between the glutamate and arginine residues that are spaced two residues apart on each;
- dock rigid replicas of this dimer in a row to form a vesicle rib; and
- finally dock rigid replicas of this rib to form a vesicle wall.

However, the two models diverge in particulars. The first *de novo* model (Strunk *et al.*, 2011) predicts a structure for the GvpA of *Haloferax mediterranei* in which there is relatively little β-sheet because the β2 strands are abbreviated and the monomer conformation has α1 situated in a manner that interferes with the formation of any inter-molecular β1–β1 contacts. This seems difficult to reconcile with the strong inter-molecular β-sheet signal seen in FTIR spectra after H-D exchange in this species (Strunk *et al.*, 2011). In addition, α2 overlies the intra-molecular β-sheet, where it contributes nothing to stabilizing inter-rib contacts.

The second *de novo* model (Ezzeldin *et al.*, 2012) finds a monomer conformation for the GvpA of *Halobacterium* sp. *NRC-1* with full length β-strands and both helices on the hydrophilic side of the intra-molecular β-sheet. Snug docking of the monomers, and then the dimers, is achieved by enforcing the same ion pairing that was invoked in developing the asymmetric dimer model (Sivertsen *et al.*, 2009). As a result, an extended β-sheet is formed via substantial inter-molecular β1–β1 and β2–β2 contacts. However, because all the monomers have identical β-turns centred on the glycine, the dimer is symmetrical and the β-strands lie at an angle of ~20° relative to the vesicle axis. Furthermore, the α-helices overlie the β-strands, where they contribute nothing to stabilizing inter-rib contacts.

What lessons might be learned from these efforts? The most obvious is that routine *de novo* methods are not well suited to modelling a metamorphic protein. In the case of gas vesicles, the possibility of a second monomer fold, specifically with the β-turn displaced by one residue (Bayro *et al.*, 2012; Sivertsen *et al.*, 2009), needs to be considered.

In fact, it is remarkable that even one sensible monomer fold was found *de novo* in uniform solvent, given that folded GvpA monomers are not known to occur free in solution. *In vivo* they are most likely associated with chaperones when they are not embedded in vesicles. Since we do not know anything about the chaperones, the best way to model GvpA monomers might be at a gas–water interface rather than in uniform solvent.

A third issue is rigid docking. It seems unlikely that the monomer conformation does not adjust to its neighbours in the vesicle. In particular, the arrangement of the α-helices seems weakly constrained in the monomer and flexibility in the regions at both ends of each helix should allow them adopt to favourable interactions with neighbours at the edges of the β-sheets (i.e. between ribs). Such interactions would strengthen the vesicle, which would seem the best use of these fairly large appendages. Simulations of this process should be possible by applying suitable periodic boundary conditions.

Prospects

Although a great deal of progress has been made in understanding gas vesicles, they still hold intrigue. While we have some well developed ideas about how they maintain their gas filled interiors, it remains difficult to imagine how they build them in the first place: even with the benefit of presumed chaperones, it seems quite a feat to insert monomers (or dimers) into a growing vesicle without leaking clusters of water in the process. The reverse process, by which cells dismantle gas vesicles (to descend in the water column or to recycle protein from collapsed vesicles) is also non-trivial. Indeed, from an anthropocentric point of view, the latter is of particular interest given the role of amyloids in human disease and our current helplessness at dismantling them.

References

Ahlborn, F. (1895). Uber die Wasserblute *Byssus flos-aquae* und ihr Verhalten gegen Druck. Verh. Naturwiss. Ver. Hamburg III, 25.

Andreae, M.O., and Crutzen, P.J. (1997). Atmospheric aerosols: Biogeochemical sources and role in atmospheric chemistry. Science *276*, 1052–1058.

Ball, P. (2012). Nanobubbles are not a Superficial Matter. Chemphyschem *13*, 2173–2177.

Bayro, M.J., Daviso, E., Belenky, M., Griffin, R.G., and Herzfeld, J. (2012). An Amyloid Organelle, Solid-state NMR Evidence for Cross-beta Assembly of Gas Vesicles. J. Biol. Chem. *287*, 3479–3484.

Beard, S.J., Davis, P.A., Iglesias-Rodriguez, D., Skulberg, O.M., and Walsby, A.E. (2000). Gas Vesicle genes in *Planktothrix* spp. from Nordic lakes: strains with weak gas vesicles possess a longer variant of gvpC. Microbiology *146*, 2009–2018.

Beard, S.J., Handley, B.A., and Walsby, A.E. (2002). Spontaneous mutations in gas vesicle genes of *Planktothrix* spp. affect gas vesicle production and critical pressure. FEMS Microbiol. Lett. *215*, 189–195.

Belenky, M., Meyers, R., and Herzfeld, J. (2004). Subunit Structure of Gas Vesicles: A MALDI-TOF Mass Spectrometry Study. Biophys. J. *82*, 321A.

Blaurock, A.E., and Walsby, A.E. (1976). Crystalline Structure of Gas Vesicle Wall from Anabaena-Flos-Aquae. J. Mol. Biol. *105*, 183–199.

Bowen, C.C., and Jensen, T.E. (1965). Blue-green algae: fine structure of the gas vacuoles. Science *147*, 1460–1462.

Bruintjes, R.T. (1999). A review of cloud seeding experiments to enhance precipitation and some new prospects. Bull. Amer. Meteorol. Soc. *80*, 805–820.

Buchholz, B.E.E., Hayes, P.K., and Walsby, A.E. (1993). The Distribution of the Outer Gas Vesicle Protein, Gvpc, on the Anabaena Gas Vesicle, and Its Ratio to Gvpa. J. Gen. Microbiol. *139*, 2353–2363.

Carugo, O. (2008). Amino acid composition and protein dimension. Protein Sci. *17*, 2187–2191.

Chandler, D. (2005). Interfaces and the driving force of hydrophobic assembly. Nature *437*, 640–647.

Chu, L.J., Chen, M.C., Setter, J., Tsai, Y.S., Yang, H.Y., Fang, X.F., Ting, Y.S., Shaffer, S.A., Taylor, G.K., von Haller, P.D., *et al.* (2011). New structural proteins of *Halobacterium salinarum* gas vesicle revealed by comparative proteomics analysis. J. Proteome Res. *10*, 1170–1178.

Damerval, T., Houmard, J., Guglielmi, G., Csiszar, K., and Tandeau de Marsac, N. (1987). A developmentally regulated gvpABC operon is involved in the formation of gas vesicles in the cyanobacterium *Calothrix* 7601. Gene *54*, 83–92.

Damerval, T., Castets, A.-M., Houmard, J., and Tandeau de Marsac, N. (1991). Gas vesicle synthesis in the cyanobacterium *Pseudanabaena* sp.: occurrence of a single photoregulated gene. Mol. Microbiol. *5*, 657–664.

Dunton, P.G., and Walsby, A.E. (2005). The diameter and critical collapse pressure of gas vesicles in Microcystis are correlated with GvpCs of different length. FEMS Microbiol. Lett. *247*, 37–43.

Dunton, P.G., Mawby, W.J., Shaw, V.A., and Walsby, A.E. (2006). Analysis of tryptic digests indicates regions of GvpC that bind to gas vesicles of Anabaena flos-aquae. Microbiology-SGM *152*, 1661–1669.

Englert, C., Kruger, K., Offner, S., and Pfeifer, F. (1992a). Three Different but Related Gene Clusters Encoding Gas Vesicles in Halophilic Archaea. J. Mol. Biol. *227*, 586–592.

Englert, C., Wanner, G., and Pfeifer, F. (1992b). Functional-Analysis of the Gas Vesicle Gene-Cluster of the Halophilic Archaeon *Haloferax-Mediterranei* Defines the Vac-Region Boundary and Suggests a Regulatory Role for the Gvpd Gene or Its Product. Mol. Microbiol. *6*, 3543–3550.

Ernst, J.A., Clubb, R.T., Zhou, H.X., Gronenborn, A.M., and Clore, G.M. (1995). Demonstration of Positionally Disordered Water within a Protein Hydrophobic Cavity by NMR. Science *267*, 1813–1817.

Ezzeldin, H.M., Klauda, J.B., and Solares, S.D. (2012). Modeling of the major gas vesicle protein, GvpA: From protein sequence to vesicle wall structure. J. Struct. Biol. *179*, 18–28.

Gao, G.T., Oh, K.J., and Zeng, X.C. (1999). Effect of uniform electric field on homogeneous vapor-liquid nucleation and phase equilibria. II. Extended simple point charge model water. J. Chem. Phys. *110*, 2533–2538.

Hayes, P.K., and Powell, R.S. (1995). The GvpA/C Clustr of Anabaena flos-aquae has Multiple Copies of a Gene Encoding GvpA. Arch. Microbiol. *164*, 50–57.

Hayes, P.K., and Walsby, A.E. (1984). An investigation into the recycling of gas vesicle protein derived from collapsed gas vesicles. J. Gen. Microbiol. *130*, 1591–1596.

Hayes, P.K., and Walsby, A.E. (1986). The inverse correlation between width and strength of gas vesicles in cyanobacteria. Br. Phycol. J. *21*, 191–197.

Hayes, P.K., Walsby, A.E., and Walker, J.E. (1986). Complete amino-acid-sequence of cyanobacterial gas-vesicle protein indicates a 70 residue molecule that corresponds in size to the crystallographic unit-cell. Biochem. J. *236*, 31–36.

Hayes, P.K., Buchholz, B., and Walsby, A.E. (1992). Gas Vesicles are strengthened by the outer-surface protein, Gvpc. Arch. Microbiol. *157*, 229–234.

Hofacker, A., Schmitz, K.M., Cichonczyk, A., Sartorius-Neef, S., and Pfeifer, F. (2004). GvpE- and GvpD-mediated transcription regulation of the p-*gvp* genes encoding gas vesicles in *Halobacterium salinarum*. Microbiology-SGM *150*, 1829–1838.

Horne, M., Englert, C., Wimmer, C., and Pfeifer, F. (1991). A DNA region of 9 kbp contains all genes necessary for gas vesicle synthesis in halophilic Archaebacteria. Mol. Microbiol. *5*, 1159–1174.

Hummer, G., Rasaiah, J.C., and Noworyta, J.P. (2001). Water conduction through the hydrophobic channel of a carbon nanotube. Nature *414*, 188–190.

Jones, D.D., and Jost, M. (1970). Isolation and chemical characterization of gas vacuole membranes from *Microcystis aeruginosa* Kuetz. emend. Elenkin. Arch. Mikrobiol. *70*, 43–64.

Jost, M. (1965). Die Ultrastruktur von *Oscillatoria rubescens* D. C. Arch. Mikrobiol. *50*, 211–245.

Kaufman, Y.J., Tanre, D., and Boucher, O. (2002). A satellite view of aerosols in the climate system. Nature *419*, 215–223.

Kinsman, R., and Hayes, P.K. (1997). Genes encoding proteins homologous to halobacterial Gvps N, J, K, F, and L are located downstream of gvpC in the cyanobacterium *Anabaena flos-aquae*. DNA Sequ. *7*, 97–106.

Kinsman, R., Walsby, A.E., and Hayes, P.K. (1995). GvpCs with reduced numbers of repeating sequence elements bind to and strengthen cyanobacterial gas vesicles. Mol. Microbiol. *17*, 147–154.

Klebahn, H. (1895). Gasvakuolen. ein Bestandteil der Zellen der Wasserblutebildenden Phycochromaceen. Flore (Jena) *80*, 241.

Kyakuno, H., Matsuda, K., Yahiro, H., Inami, Y., Fukuoka, T., Miyata, Y., Yanagi, K., Maniwa, Y., Kataura, H., Saito, T., *et al.* (2011). Confined water inside single-walled carbon nanotubes: global phase diagram and effect of finite length. J. Chem. Phys. *134*.

Lehmann, H., and Jost, M. (1971). Kinetics of assembly of gas vacuoles in blue-green alga *Microcystis Aeruginosa* Kuetz Emend Elekin. Arch. Mikrobiol. *79*, 59–68.

Lehmann, H., and Jost, M. (1972). Assembly of gas vacuoles in a cell-free system of blue-green alga *Microcystis Aeruginosa* Kuetz Emend Elenkin. Arch. Mikrobiol. *81*, 100–102.

Lum, K., Chandler, D., and Weeks, J.D. (1999). Hydrophobicity at small and large length scales. J. Phys. Chem. B *103*, 4570–4577.

Maniwa, Y., Kataura, H., Abe, M., Suzuki, S., Achiba, Y., Kira, H., and Matsuda, K. (2002). Phase transition in confined water inside carbon nanotubes. J. Phys. Soc. Japan *71*, 2863–2866.

McMaster, T.J., Miles, M.J., and Walsby, A.E. (1996). Direct observation of protein secondary structure in gas vesicles by atomic force microscopy. Biophys. J. *70*, 2432–2436.

Mlouka, A., Comte, K., Castets, A.M., Bouchier, C., and Tandeau de Marsac, N. (2004). The gas vesicle gene cluster from *Microcystis aeruginosa* and DNA rearrangements that lead to loss of cell buoyancy. J. Bacteriol. *186*, 2355–2365.

Mittal, J., and Hummer, G. (2010). Interfacial thermodynamics of confined water near molecularly rough surfaces. Faraday Discus. *146*, 341–352.

Nakajima, A., Hashimoto, K., and Watanabe, T. (2001). Recent studies on super-hydrophobic films. Monatshefte Chemie *132*, 31–41.

Offner, S., Ziese, U., Wanner, G., Typke, D., and Pfeifer, F. (1998). Structural characteristics of halobacterial gas vesicles. Microbiology-SGM *144*, 1331–1342.

Offner, S., Hofacker, A., Wanner, G., and Pfeifer, F. (2000). Eight of fourteen *gvp* genes are sufficient for formation of gas vesicles in halophilic archaea. J. Bacteriol. *182*, 4328–4336.

Onsager, L. (1949). The effects of shape on the interaction of colloidal particles. Ann. New York Acad. Sci. *51*, 627–659.

Patel, A.J., Varilly, P., Jamadagni, S.N., Hagan, M.F., Chandler, D., and Garde, S. (2012). Sitting at the Edge: How Biomolecules use Hydrophobicity to Tune Their Interactions and Function. J. Phys. Chem. B *116*, 2498–2503.

Powell, R.S., Walsby, A.E., Hayes, P.K., and Porter, R. (1991). Antibodies to the N-terminal sequence of Gvpa bind to the ends of gas vesicles. J. Gen. Microbiol. *137*, 2395–2400.

Shewmaker, F., McGlinchey, R.P., and Wickner, R.B. (2011). Structural insights into functional and pathological amyloid. J. Biol. Chem. *286*, 16533–16540.

Shukla, H.D., and DasSarma, S. (2004). Complexity of gas vesicle biogenesis in *Halobacterium* sp. strain NRC-1: identification of five new proteins. J. Bacteriol. *186*, 3182–3186.

Sivertsen, A.C., Bayro, M.J., Belenky, M., Griffin, R.G., and Herzfeld, J. (2009). Solid-state NMR evidence for inequivalent GvpA subunits in gas vesicles. J. Mol. Biol. *387*, 1032–1039.

Sivertsen, A.C., Bayro, M.J., Belenky, M., Griffin, R.G., and Herzfeld, J. (2010). Solid-State NMR Characterization of Gas Vesicle Structure. Biophys. J. *99*, 1932–1939.

Smith, R.V., and Peat, A. (1967). Comparative structure of the gas-vacuoles of blue-green algae. Archiv fur Mikrobiologie *57*, 111–122.

Strodtmann, S. (1895). Die Ursache des Schwebvermogens bei den Cyanophyceen. Biol. Centralbl. *15*, 113–115.

Strunk, T., Hamacher, K., Hoffgaard, F., Engelhardt, H., Zillig, M.D., Faist, K., Wenzel, W., and Pfeifer, F. (2011). Structural model of the gas vesicle protein GvpA and analysis of GvpA mutants in vivo. Mol. Microbiol. *81*, 56–68.

Tandeau de Marsac, N., Mazel, D., Bryant, D.A., and Houmard, J. (1985). Molecular cloning and nucleotide sequence of a developmentally regulated gene from the cyanobacterium calothrix-PCC-7601 – a gas vesicle protein gene. Nucleic Acids Res. *13*, 7223–7236.

Waaland, J.R., and Branton, D. (1969). Gas vacuole development in a blue-green alga. Science *163*, 1339–1341.

Walker, J.E., and Walsby, A.E. (1983). Molecular-weight of gas-vesicle protein from the planktonic cyanobacterium Anabaena-Flos-Aquae and implications for structure of the vesicle. Biochem. J. *209*, 809–815.

Walker, J.E., Hayes, P.K., and Walsby, A.E. (1984). Homology of gas vesicle proteins in cyanobacteria and halobacteria. *J.* Gen. Microbiol. *130*, 2709–2715.

Walsby, A.E. (1969). Permeability of blue-green algal gas-vacuole membranes to gas. Proc. R. Soc. Lond. Series B *173*, 235–255.

Walsby, A.E. (1971). Pressure Relationships of Gas Vacuoles. Proc. R. Soc. London Series B *178*, 301–326.

Walsby, A.E. (1972). Structure and function of gas vacuoles. Bacteriol. Rev. *36*, 1–32.

Walsby, A.E. (1975). Gas vesicles. Annu. Rev. Plant Physiol. Plant Mol. Biol. *26*, 427–439.

Walsby, A.E. (1978). Properties and buoyancy-providing role of gas vacuoles in Trichodesmium Ehrenberg. Br. Phycol. J. *13*, 103–116.

Walsby, A.E. (1982a). The elastic compressibility of gas vesicles. Proc. R. Soc. Lond. Series B *216*, 355–368.

Walsby, A.E. (1982b). Permeability of Gas Vesicles to Perfluorocyclobutane. J. Gen. Microbiol. *128*, 1679–1684.

Walsby, A.E. (1994). Gas vesicles. Microbiol. Rev. *58*, 94–144.

Walsby, A.E., and Bleything, A. (1988). The dimensions of cyanobacterial gas vesicles in relation to their efficiency in providing buoyancy and withstanding pressure. J. Gen. Microbiol. *134*, 2635–2645.

Walsby, A.E., and Buckland, B. (1969). Isolation and purification of intact gas vesicles from a blue-green alga. Nature *224*, 716–717.

Walsby, A.E., and Hayes, P.K. (1988). The minor cyanobacterial gas vesicle protein, Gvpc, is attached to the outer surface of the gas vesicle. J. Gen. Microbiol. *134*, 2647–2657.

Walsby, A.E., Avery, A., and Schanz, F. (1998). The critical pressures of gas vesicles in *Planktothrix rubescens* in relation to the depth of winter mixing in Lake Zurich, Switzerland. J. Plankton Res. *20*, 1357–1375.

Wenzel, R.N. (1949). Surface roughness and contact angle. J. Phys. Colloid Chem. 53, 1466–1467.

Whitton, B.A., and Peat, A. (1969). On Oscillatoria Redekei van Goor. Archiv fur Mikrobiologie *68*, 362–376.

Wolk, J., Strey, R., Heath, C.H., and Wyslouzil, B.E. (2002). Empirical function for homogeneous water nucleation rates. J. Chem. Phys. *117*, 4954–4960.

Worcester, D.L. (1975). Neutron Diffraction Studies of Biological Membrane Components. Brookhaven Symp. Biol. *27*, 37–57.

Yu, B., Blaber, M., Gronenborn, A.M., Clore, G.M., and Caspar, D.L.D. (1999). Disordered water within a hydrophobic protein cavity visualized by x-ray crystallography. Proc. Natl. Acad. Sci. U.S.A. *96*, 103–108.

Motility in Unicellular and Filamentous Cyanobacteria

11

Bianca Brahamsha and Devaki Bhaya

Abstract

Both unicellular and filamentous cyanobacteria exhibit diverse types of motility, including the ability to move on surfaces and to swim in liquids. Motile behaviour can be regulated such that cyanobacteria can respond to important environmental signals such as light or chemical gradients. In some cases, components of the motility machinery, such as Type IV pili that allow certain unicellular species to move on surfaces have been characterized. In others, as is the case for swimming in certain marine unicellular species, and gliding in filamentous genera, the motility apparatus has yet to be identified. We describe here recent advances that have led to the identification of multiple photoreceptors and novel proteins involved in motility. The complex signal transduction pathways governing motile behaviour and attempts to model this behaviour are discussed. We conclude with some of the major challenges that remain in our understanding of the complexities of motile behaviour.

Introduction

Ever since Leuwenhoek's observations of 'animalcules' under the microscope, the motility of microorganisms has fascinated microbiologists. Many prokaryotes have been observed to move either in liquid or on moist surfaces, and this phenomenon was variously described as 'swimming', 'twitching', 'swarming' or 'gliding'. Both unicellular and filamentous microorganisms exhibit motility and cells can move as individuals or in large coordinated groups. Furthermore, motile behaviour can be regulated which allows microorganisms to respond to various environmental signals such as light or to chemical gradients. In some cases, the motility apparatus required for movement and the signal transduction cascade that allows for a response to environmental cues are well understood, while in others they remain to be discovered. It is also apparent that motility is a complex and regulated phenomenon where more than one mechanism for movement can be used depending on the lifestyle and environmental conditions (Jarrell and McBride, 2008). For instance, some bacteria encode both flagella and Type IV pili (TFP), while others can produce extracellular materials that aid or are required for motility.

Unicellular and filamentous cyanobacteria as well as many photosynthetic microalgae exhibit motile behaviours, and their responses to light are particularly well studied (Gomelsky and Hoff, 2011; Hader and Lebert, 2009; Hoff *et al.*, 2009; Jekely, 2009). The rich history of the study of their motility (initiated by Treviranus in 1817 and continued by Engelmann in Germany), stretches well over a hundred years and has been covered in many excellent and detailed reviews (see next section). Nevertheless, we feel it is important to keep this history in perspective, particularly for younger researchers, because many fundamental issues that were addressed are still relevant today. Thus, we start with a brief overview of the major mileposts that describe the various motility phenomena in unicellular and filamentous cyanobacteria. These careful experiments and resultant hypotheses set the stage for molecular and mechanistic approaches, although in this chapter, we primarily focus on recent advances. We conclude with a brief discussion of the challenges that lie ahead in advancing our understanding of the complexities of motile behaviour.

Mileposts in the history of motility in cyanobacteria

We have roughly divided the study of motility in cyanobacteria into three periods. Initially, the behaviour of motile cyanobacteria, bacteria and photosynthetic algae was observed using available microscopes. These careful studies and ingenious experiments by Treviranus, Engelmann and others, beginning from the early 1800s, documented the movement and light-responsiveness of several species of filamentous cyanobacteria and are summarized in excellent early reviews (Castenholz, 1982; Drews, 2005; Hader, 1987, 1988; Nultsch, 1975; Nultsch and Hader, 1979). In the second phase, starting in the early 1960s, several groups including those of Castenholz, Nultsch, Hader, and Clayton, among others, carried out numerous experiments that attempted to answer some of the fundamental questions regarding photoperception, signal transduction and mechanisms for motility, as well as the role of photosynthesis and metabolism in controlling motility responses. Most of these experiments were carried out with filamentous cyanobacteria (*Anabaena* sp., *Phormidium* sp. and *Oscillatoria* sp.), some of them isolated from hot springs (Castenholz, 1969), which exhibited rapid movements called 'gliding'. This motility was carefully described and it was recognized that gliding motility could look similar in different organisms yet the mechanisms that controlled these movements could be quite distinct (Castenholz, 1982; Diehn, 1976). The two models proposed to explain gliding in filamentous cyanobacteria, which are still being considered today, movement through mucilage secretion and movement through the production of surface waves, originated during this period. These phenomenological and environmentally relevant studies set the stage for the third or current phase in understanding motility, which relies strongly on molecular and biochemical tools to understand the mechanistic aspects of motility.

In the third phase, starting in the early 1990s, the availability of complete genome sequences, the development of molecular genetic methods, and the advent of new imaging techniques, have made possible a mechanistic and molecular understanding of motility. For instance, the use of mutants has revealed that surface proteins were required for the mysterious swimming behaviour of certain marine *Synechococcus* species. For freshwater unicellular cyanobacteria, several important advances have been made in describing the motor apparatus required for surface motility as well as the photo-perception responses that rely on multiple photoreceptors and complex signal transduction cascades. In addition, the modelling of motile cells in directional light has led to a better understanding of phototactic behaviour. In filamentous cyanobacteria, several models to explain gliding behaviour have been proposed and some are currently being tested. We anticipate that in the next few years these approaches will continue to advance our knowledge of surface-dependent and swimming motility in cyanobacteria. They should also provide insights into the ecological advantages that motility provides to photosynthetic organisms that cope with fluctuations in their environment, particularly light intensity and quality.

In this chapter, we focus on the advances made in the last two decades and identify the remaining challenges. Cyanobacteria have evolved a variety of means of movement, some of which share features with the types of motility exhibited by other bacteria. When sufficient similarities exist, we have used the accepted prokaryotic motility nomenclature (Jarrell and McBride, 2008). We will use **twitching motility** to refer to the surface-dependent motility of certain unicellular cyanobacteria that requires Type IV pili (or TFP); **swimming motility** to refer to the non-flagellar swimming exhibited by certain marine *Synechococcus* species; and **gliding motility** to refer to the surface-dependent movement of filamentous cyanobacteria for which no organelle has yet been identified.

Swimming motility in unicellular marine *Synechococcus* sp.

In 1985, Waterbury and colleagues reported that certain strains of marine *Synechococcus* were capable of swimming in liquids at speeds ranging from 5 to 25 µm/s (Waterbury *et al.*, 1985). This discovery was remarkable for two reasons: first, unlike other cyanobacteria which require contact with a surface in order to move, these strains of *Synechococcus* moved in liquids, and second, they did so in the absence of flagella or any other readily visible organelle. Flagellated bacteria are propelled in liquids by the rotation of flagella and this imparts certain features to their motile behaviour (Jarrell and McBride, 2008). Like flagellated bacteria, swimming cells of *Synechococcus* rotate about their longitudinal axis as they translocate, are slowed down and ultimately immobilized by increasing viscosity, and cells that become attached by one pole to a glass surface rotate about their point of attachment (Waterbury *et al.*, 1985; Willey, 1988). Hence, like flagellated bacteria, swimming *Synechococcus* generate both torque and thrust. However in spite of the use of a variety of approaches to detect and visualize flagella (Willey, 1988), reviewed by Brahamsha (Brahamsha, 1999), these were not seen, and the mechanism for swimming remains elusive.

Molecular approaches to swimming

The development of genetic tools for certain strains of marine *Synechococcus* has made it possible to address this problem at a molecular level (Brahamsha, 1996; McCarren and Brahamsha, 2005; McCarren *et al.*, 2005). Early work combining biochemistry and reverse genetics revealed that an abundant 130 kDa cell-surface glycoprotein, SwmA, was required for the generation of thrust in *Synechococcus* sp. strain WH 8102 (hereafter Syn WH 8102) (Brahamsha, 1996). A survey of other motile strains of marine *Synechococcus* revealed that, unlike non-motile isolates, they too produced an abundant polypeptide of similar molecular weight (Brahamsha, 1996) and encoded genes highly similar to *swmA* (Brahamsha, unpublished). SwmA encodes the S-layer that covers the surface of strain WH 8102 (McCarren and Brahamsha, 2005) but how it is involved in swimming is not understood. Interestingly, SwmA shows some similarity, particularly in repeated putative Ca^{2+}-binding motifs, to oscillin, a glycoprotein which forms fibrils on the surface of the gliding cyanobacterium *Phormidium uncinatum* (Hoiczyk, 1998) (discussed below) and to other S-layer and Repeats-in-toxin (RTX) proteins (Brahamsha, 1996), including the surface protein HlyA (sll1951) identified in the freshwater unicellular cyanobacterium, *Synechocystis* sp. PCC 6803 (hereafter Syn PCC 6803) (Bhaya *et al.*, 1999).

Transposon mutagenesis screens for motility mutants identified three loci in the genome of Syn WH 8102 that encode proteins required for swimming (McCarren and Brahamsha, 2005). Mutations in genes at two of these loci (chromosomal region 1, which includes *swmA* (SYNW0085), and chromosomal region 2), affect the production and the localization of SwmA (McCarren and Brahamsha, 2009). Some of these genes encode homologues of type I secretion systems, which have been shown to be involved in export of S-layer proteins in other organisms, and homologues of glycosyltransferases that might play a role in the glycosylation of SwmA. Chromosomal region 3 contains a very large gene (32 kb long, SYNW0953) which encodes the protein SwmB. SwmB (10,791 aa) is a 1.2 MDa cell-surface protein with several repetitive domains. In terms of its large size, repetitive architecture, and atypical amino acid usage (McCarren and Brahamsha, 2007), SwmB resembles a number of other giant cell-surface proteins (Reva and Tummler, 2008) such as adhesins, although none of these have been implicated in motility.

An examination of the genome of *Synechococcus* sp. strain WH 8102 corroborates that swimming in this species represents a unique type of motility (Palenik *et al.*, 2003). None of the proteins typically required for flagellar-dependent motility (flagellar motor, flagellar filament proteins, and chemotaxis proteins) were encoded on the genome. Six open reading frames (ORFs) associated with the biogenesis and function of TFP (homologues of *pilB, C, D, Q,* and *T*) were encoded on the genome but they do not encode the full complement of proteins required for pilus assembly and function, and pilin subunit homologues could not be identified. Furthermore, pili have not been observed in Syn WH 8102, nor does Syn WH 8102 exhibit surface-associated twitching (McCarren and Brahamsha, 2005). In addition, we insertionally inactivated the *pilT* and *pilC* homologues in Syn WH 8102, and neither of these inactivations had an effect on swimming (McCarren and Brahamsha, unpublished).

Cell surface proteins and the envelope structure of swimming *Synechococcus* sp.

The envelope structures of two strains of motile *Synechococcus*, Syn WH 8102 (McCarren *et al.*, 2005) and Syn WH 8113 (Samuel *et al.*, 2001), have been examined by a variety of electron microscopy techniques, including conventional methods as well as cryo-fixation and freeze-fracture and etching. This has revealed a multilayered Gram-negative architecture characteristic of cyanobacteria. In addition, these motile strains have an S-layer forming a paracrystalline array external to the outer membrane that can be observed using cryofixation and freeze substitution methods (McCarren *et al.*, 2005). In addition, in one of the strains (Syn WH8113), 5-nm-wide and 150-nm-long fibrils or spicules appear to cover the surface of the cell (Samuel *et al.*, 2001); however, such spicules were not observed in Syn WH 8102 and hence may not be characteristic of all motile strains. Studies with mutants (McCarren and Brahamsha, 2007; McCarren *et al.*, 2005) revealed that in Syn WH 8102, the S-layer is composed of the SwmA protein, while the giant SwmB protein is arranged in a punctate manner on the cell surface. Mutants lacking SwmA exhibit a fibrillar layer covering the outer membrane; the nature of this fibrillar layer is not currently known. While these studies point to the cell surface as being important for this type of motility, the role SwmA or SwmB plays is not yet clear. Both are abundant cell-surface proteins and may constitute part of the motility apparatus or their role in swimming may be indirect. It is possible that in their absence, outer membrane integrity is affected in such a way that as-yet unidentified proteins required for motility cannot be localized correctly or cannot interact properly.

Models for swimming in *Synechococcus* sp.

Models have been proposed for how marine *Synechococcus* might swim in the absence of flagella, and one of these, self-electrophoresis, in which cells establish and use an electrical field to propel themselves, has been ruled out experimentally (Pitta and Berg, 1995). Given the lack of any observable external organelles, models have focused on how the cell surface may generate thrust. In 1996, Ehlers *et al.* (1996) presented a model in which travelling surface waves generated by contractions and expansions of the outer membrane could create sufficient thrust to allow cells to swim at experimentally observed speeds. The rotation of the cell body during swimming could be explained by waves that are not axially symmetrical. Pitta and Berg showed that calcium was required for swimming and that inhibitors of voltage-gated calcium channels blocked motility, and proposed that calcium depolarization may serve to generate local swellings on the cell surface which could then be propagated along the surface to produce waves (Pitta *et al.*, 1997). Samuel *et al.* (2001) proposed that the spicules seen on the surface of Syn WH 8113 could be used to transmit the expansions and contractions to the surface and in effect function as oars to generate travelling waves in a manner analogous to that of ciliated eukaryotes. The absence of spicules in Syn WH 8102 implies that such structures are not required for motility in all swimming strains, and the Ehlers model (Ehlers *et al.*, 1996) does not require them.

Recently, another mechanism by which these surface waves could be generated has been proposed (Nan and Zusman, 2011; Ehlers and Oster, 2012). This model is based on one proposed for adventurous motility (A-motility), the gliding of individual cells of *Myxococcus xanthus*, in which motor proteins (MotA and MotB homologues) move along a continuous looped helical track composed of the cytoskeletal protein MreB (Nan and Zusman, 2011; Nan *et al.*, 2011). The motors carry cargo consisting of multiprotein complexes encoded by certain genes required for gliding in *M. xanthus*. These motor–cargo complexes distort the cell surface as they move along the track, generating surface waves that propel the cells. Ehlers and Oster (Ehlers and Oster, 2012) postulate the existence of such a helical track in *Synechococcus* and propose that rotation of the track itself, or of motor–cargo complexes moving along it could generate waves on the cell surface. In their model, the S-layer composed of SwmA plays a critical role. Because of its tilted geometry on the cell surface, it creates asymmetry and amplifies the travelling waves in a manner to provide sufficient mechanical coupling to the fluid in which the cells are swimming (Fig. 11.1). This model explains the rotational behaviour of cells during swimming and also that of SwmA mutants, which cannot swim but can still rotate when attached to a surface.

There is at present no experimental evidence for the occurrence of waves or undulations on the cell surface of swimming *Synechococcus*. The waves are predicted to be of an amplitude not resolvable by conventional light microscopy, but techniques such as total internal reflection fluorescence microscopy which were used to view deformations of the cell surface in *Myxococcus* (Nan *et al.*, 2011), as well as atomic force microscopy, should be applicable to *Synechococcus* and are currently being tested (Brahamsha, unpublished). While the presence of a helical cytoskeletal track has not been demonstrated in marine *Synechococcus*, the genome of Syn WH 8102 encodes an *mreB* homologue, and the small molecule A22, which depolymerizes the MreB cytoskeleton in other bacteria, rapidly inhibits swimming, implying

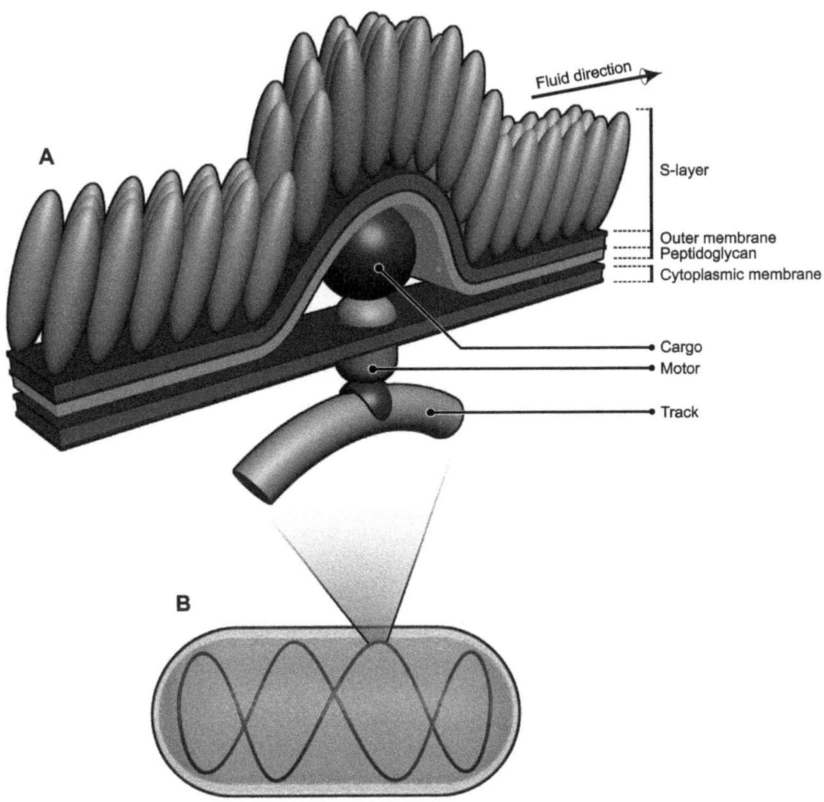

Figure 11.1 Surface wave model for non-flagellar swimming in marine *Synechococcus*. These drawings represent the model described in Nan *et al.* (2011) and Ehlers and Oster (2012). In this model, motor complexes carrying a cargo (yet-to-be identified motility proteins) move along a helical track and, as they do so, cause deformations of the cell surface, generating surface waves. In marine *Synechococcus* these waves are amplified by the tilt of the S-layer (≈60°) to provide sufficient mechanical coupling to the surrounding fluid. In (A) the motor–cargo complexes are moving in the same direction as the fluid, against the grain or tilt of the S-layer. Another version of the model proposes that such surface waves could also be generated by rotation of the helical track itself, driven by motors attached to the peptidoglycan layer. (B) A cell of marine *Synechococcus* showing the helical track which is proposed to either serve as the track over which motor–cargo complexes move, or to rotate itself. While the presence of such a track has not been demonstrated in marine *Synechococcus*, the genome of Syn WH 8102 encodes an *mreB* homologue and the small molecule A22 which de-polymerizes the MreB cytoskeleton rapidly inhibits swimming, implying that an intact MreB cytoskeleton is required for swimming in *Synechococcus*.

that an intact MreB cytoskeleton is required for swimming in *Synechococcus* (Brahamsha, in preparation) as it is for A-motility in *M. xanthus* (Mauriello *et al.*, 2010). These results suggest that aspects of the helical rotor model proposed for *M. xanthus* may indeed be applicable to *Synechococcus.* Like other forms of swimming or gliding in prokaryotes that are powered by an ion motive force, the energy source for swimming in *Synechococcus* is the sodium motive force (Willey *et al.*, 1987), implying that the motor converting chemical into mechanical energy is located in the cytoplasmic membrane.

Surface dependent motility in unicellular freshwater cyanobacteria

The phenomenon of surface dependent motility

Motility exhibited by certain unicellular freshwater cyanobacteria isolates was noted by groups at the Pasteur Institute (Rippka *et al.*, 1979; Stanier *et al.*, 1971) and elsewhere. It was observed that Syn PCC 6803 cells exhibited slow motility (typically 1–2 μm/min) on plates, but that this phenotype was often lost under standard laboratory conditions, possibly because it is not a characteristic that is under strong selection under normal maintenance conditions on plates (Choi *et al.*, 1999). Indeed, Syn PCC 6803 isolates currently used in different laboratories often have either subtle or significant differences in their motile behaviour, which has led to some confusion in the literature and problems with data interpretation and consistency (Bhaya *et al.*, 2000; Choi *et al.*, 1999; Wilde *et al.*, 2002; Yoshihara *et al.*, 2000). Whole genome sequencing is becoming relatively easy and inexpensive, and recently a few isolates of Syn PCC 6803 have been re-sequenced and correlated with certain motility phenotypes (Kaneko *et al.*, 1996; Kanesaki *et al.*, 2012; Trautmann *et al.*, 2012). In the future, it may be possible to identify exact changes at the gene/nucleotide level when isolates demonstrate phenotypic changes. For several years, the unicellular freshwater cyanobacterium Syn PCC 6803 has been extensively used for studies relating to photosynthesis and metabolism. Syn PCC 6803 has also become the preferred model system for the study of motility because of its consistent phototaxis phenotype, rapid doubling time, ease of genetic manipulation and ability to create targeted mutants.

Characteristics of motility and phototaxis in Syn PCC 6803

The simplest assay for surface-dependent motility is to place cells on a moist surface, such as low concentration agarose (0.4–0.8%); cells typically exhibit a slow surface dependent twitching behaviour (Bhaya *et al.*, 1999; Burriesci and Bhaya, 2008). If cells are placed in a directional light source, phototactic motility is easily visualized under a dissecting microscope as cells moving in large groups that typically make finger-like projections towards the light source (~1 mm in 6 h, or ~3 cell lengths per minute) (Fig. 11.2). A second, more quantitative method, is to use time-lapse video microscopy to track the behaviour of single cells or groups of cells (Burriesci and Bhaya, 2008; Choi *et al.*, 1999; Ng *et al.*, 2003). These time-lapse movies can be analysed using commercial or custom cell-tracking software to track various parameters (e.g. velocity, direction) of individual cells under various conditions and to build models of phototactic behaviour (see later).

Surface-dependent Syn PCC 6803 motility (Bhaya, 2004; Bhaya *et al.*, 1999; Yoshihara and Ikeuchi, 2004) shares characteristics with the behaviour of several pathogenic Gram-negative bacteria and the soil-dwelling bacterium *Myxococcus xanthus.* In these cases it has been associated with Type IV pili (TFP)-dependent motility, also known as twitching motility, because of the slow and 'jerky' movement of individual or groups of cells (Craig and Li, 2008; Mattick, 2002; Nan and Zusman, 2011; Pelicic, 2008). Choi *et al.* (1999) tracked the motion of individual Syn PCC 6803 cells suspended in media on glass slides and demonstrated that very low levels of white light (1 μmol/m^2/s) saturated the phototactic response, and that threshold fluence rates were as low as 0.002 μmol m^{-2} s^{-1}. They suggested that Syn PCC 6803 cells sensed light direction rather than light intensity and that ATP generated by oxidative phosphorylation was required for phototaxis, but that photosynthesis was not essential. A

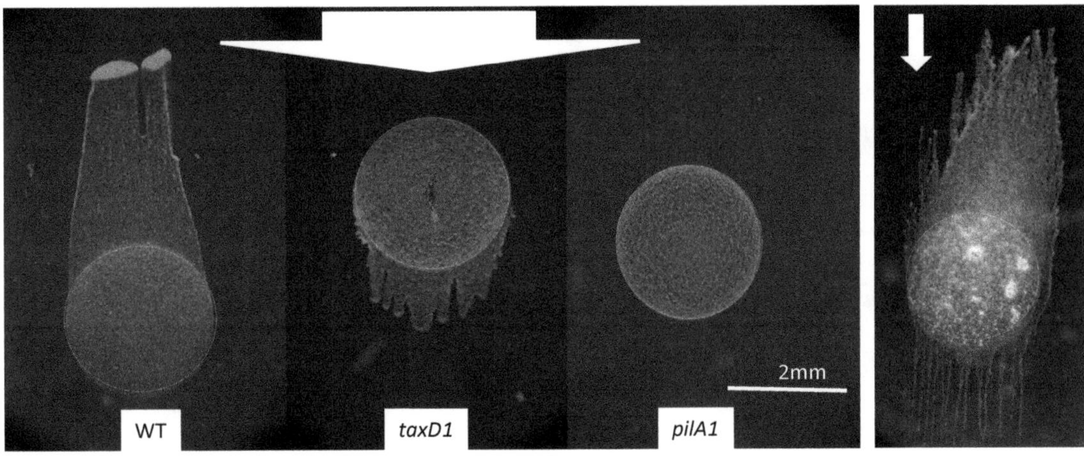

Figure 11.2 Phototactic behaviour of Syn PCC 6803, mutant cells and thermophilic *Synechococcus* sp. A directional phototaxis assay on 0.4% agarose plates was performed for 48 h. Wild-type (WT) cells (left) show positive phototaxis with groups of cells moving out from the original drop, making finger-like projections towards the light (white arrow); a *taxD1* mutant (middle) exhibits negative phototaxis; a *pilA1* mutant (right) is non-motile. Size of the drop is ~2–3 mm (marked with dotted white lines). Right panel: Thermophilic *Synechococcus* sp. isolated from microbial mats cells exhibiting both positive and negative phototaxis, in directional light. Note that most cells are positively phototactic with a smaller number of cells exhibiting motility away from light. A colour version of this figure is available in the plate section at the back of the book.

limited action spectrum indicated that phototaxis was maximal in red light and that UV-A (360 nm) caused negative phototaxis. In these experiments, phototaxis did not exhibit red/far red reversibility, which is expected of typical phytochrome-based behaviour. Ng *et al.* (2003) used a modified system to monitor motility of individual Syn PCC 6803 cells, which attached to, and moved along a glass surface. By using a number of different wavelengths and fluence rates, they established that positive phototaxis was a low fluence rate response, and that at higher fluence rates cells exhibited negative phototaxis, particularly in the blue light regions of the spectrum. This assay was also used to establish the light responses of mutants that lacked positive phototaxis (see later) (Ng *et al.*, 2003). More recently, these assays have been improved by the ability to use LEDs that have very specific wavelengths and bandwidths so that reproducibility and accuracy can be maintained (Burriesci and Bhaya, 2008).

Surface appendages in unicellular cyanobacteria

Several unicellular cyanobacterial species have been examined by electron microscopy revealing the presence of appendages around the cell, variously referred to as mucilage, filaments, fimbriae, sheaths, pili and spinae (Vaara, 1982). Although the role and biochemical nature of these diverse structures have yet to be completely described, it is likely that these extracellular layers play several important roles, including a role in motility (Pereira *et al.*, 2009). We focus on Syn PCC 6803 since molecular approaches to dissecting motility have led to a fairly detailed understanding of phototaxis and motility (see later).

Syn PCC 6803 cells are typically coccoidal in shape with a diameter of ~1 µm; cells examined by electron microscopy, using uranyl acetate based negative staining, indicated that there are at least two pilus morphotypes at the cell surface (Bhaya *et al.*, 1999, 2000). The cell is uniformly covered with a dense layer of very thin, brush-like pili, with an average diameter of 3–4 nm and a length of ~1 µm (Fig. 11.3). The biochemical characterization of these pili has not yet been attempted and it is possible that they are composed of polysaccharides or glycoproteins that constitute the extracellular matrix (Pereira *et al.*, 2009). Cells also have tufts of thick pili (average diameter 6–8 nm) that extend several cell

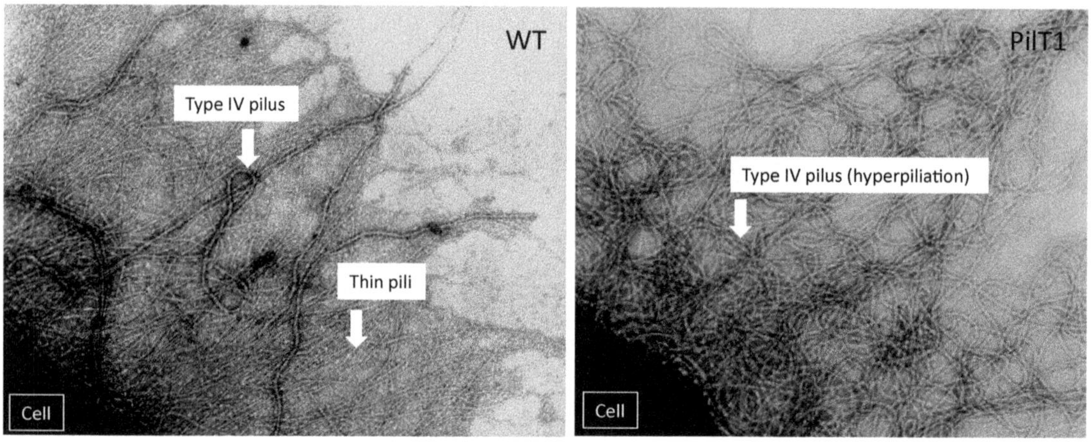

Figure 11.3 Transmission electron microscopy of Syn PCC 6803 WT and *pilT1* mutant cells. Cells were stained with 1% uranyl acetate. WT cells show both TFP and thin pili at the surface while the *pilT1* mutant is hyper-piliated with a large excess of TFP.

lengths (~4–5 μm) and which have been identified as TFP (see below). Cells grown on soft agar surfaces often appear to be connected by these pili, when visualized by electron microscopy (Bhaya *et al.*, 2000). The exact number and location of these thick pili (i.e. TFP) in a cell are not yet known although in rod-shaped bacteria, such as *M. xanthus*, TFP are typically found at the poles (Hartzell and Youderian, 1995; Nan and Zusman, 2011; Sogaard-Andersen, 2004).

Analysis of Type IV pilus-dependent motility in Syn PCC 6803

A reverse genetic approach was attempted to identify the motility apparatus in Syn PCC 6803. Because of the well-documented presence of surface pili, the twitching-like motility behaviour of cells, and the fact that flagella have never been identified in cyanobacteria, certain candidate genes were targeted for inactivation (Bhaya *et al.*, 1999). A *pilA1* (sll1694) gene encoding a protein with some similarity to a pilin, the repeating subunit that comprises a pilus (Craig and Li, 2008; Mattick, 2002; Pelicic, 2008) was identified and insertionally inactivated. The *pilA1* mutant was completely non-motile (Fig. 11.2). This experiment provided the first direct evidence that surface-dependent motility in Syn PCC 6803 was likely to be mediated by TFP. Consistent with its non-motile phenotype, the *pilA1* mutant completely lacked the thick pili (i.e. TFP), although the normal complement of thin pili were still present (Bhaya *et al.*, 1999).

Although the requirement for the structural subunit of pili (i.e. pilins encoded by *pilA1*) for TFP-dependent motility was clearly established (Bhaya *et al.*, 1999), inspection of the Syn PCC 6803 genome showed the existence of at least five other *pilA*-like genes. This similarity was confined mostly to the N-terminus. Inactivation of these genes (*pilA2–pilA6*) had no obvious effect on motility (Bhaya *et al.*, 1999; Yoshihara *et al.*, 2000, 2001) and their role in motility is not clear, although minor pilins and pseudopilins have been identified in other organisms. Type IV pilus biogenesis is a complex process and requires several steps, including proteolytic processing and N-methylation of pre-pilin polypeptides, which is carried out by a novel bi-functional signal peptidase, PilD. Proteins with weak similarity to Type IV pilins (and which often share a conserved cleavage site) have been shown to function in extracellular protein secretion and in transformation competence (recent reviews include Proft and Baker, 2009; Burrows, 2012).

The critical role of TFP in surface-dependent motility of Syn PCC 6803 was further demonstrated by creating mutants in genes encoding components required for pilus biogenesis and assembly. As expected, mutants in *pilQ* (encoding the secretin pore), *pilD*, (encoding the membrane-bound signal peptidase) and *pilC*, *pilM*, *pilN* and *pilO* (encoding components of pilus biogenesis and assembly apparatus) were all non-motile and also lacked TFP at the cell surface (Bhaya *et al.*, 2000; Yoshihara *et al.*, 2001). Interestingly, several, but not all, of these pilus biogenesis mutants were also found to be defective in natural transformation (Fusseneger *et al.*, 1997; Long *et al.*, 2003; Nakasugi *et al.*, 2006; Yoshihara *et al.*, 2000).

Surface-dependent motility occurs when TFP are extended, and attach to a surface (non-specifically to a surface such as agarose or specifically to polysaccharide fibrils of neighbouring cells). Next, the TFP retract and consequently the cell body is pulled along; it has been calculated that TFP may exert a substantial force of up to ~100 pN, to mediate cell movement (Maier *et al.*, 2002; Merz *et al.*, 2000, Merz and Forest, 2002). TFP dependent motility is an energy (ATP) requiring process and the sequential retraction and extension of TFP requires the activity of two ATPases. These are the conserved motor proteins, PilB (for extension) and PilT (for retraction), located at the inner membrane (Okamoto and Ohmori, 2002). In contrast to some other microorganisms, Syn PCC 6803 has two *pilB*-like genes, *pilB1* and *pilB2* (Bhaya *et al.*, 2000). As expected, inactivation of *pilB1* results in a non-motile phenotype. However, *pilB2* mutants retain motility so the role of PilB2 is still to be elucidated (Yoshihara *et al.*, 2001).

Syn PCC 6803 also has two copies of a *pilT*-like gene (*pilT1* and *pilT2*), which would encode the PilT motor required for pilus retraction. Both the *pilT1* and *pilT2* genes were individually inactivated (Bhaya *et al.*, 2000) and surprisingly, exhibited very different phenotypes. The PilT1 and PilT2 proteins are similar, although PilT2 has an N-terminus extension of ~80 amino acids, rich in proline residues. The *pilT1* mutant is unable to retract its pili and as a consequence is non-motile. Since the *pilT1* cells continue to synthesize pilins which are extended at the cell surface as TFP, but are not retracted, this would be predicted to result in a hyper-piliated cell and this was convincingly demonstrated in electron micrographs of *pilT1* mutants (Fig. 11.3) (Bhaya *et al.*, 2000). Interestingly, *pilA1* mRNA levels increased eight-fold in the *pilT1* mutant, suggesting a link between pilin expression levels and pilus activity, although the details of how this regulation is mediated awaits further study. The *pilT2* mutant exhibited an unexpected phenotype: mutants moved *away* from a directional light source, while wild-type cells exhibited positive phototaxis. One possible interpretation of the 'negative phototaxis' phenotype is that PilT2 is required for positive phototaxis, directly or indirectly, so that an inactivation of this machinery unveils a negative phototaxis which is otherwise masked by the dominant positive phototaxis (Fig. 11.4). These results provided important evidence that phototaxis can be either towards or away from a light source (referred to as positive and negative phototaxis, respectively). Under most light conditions and wavelengths tested so far, positive phototaxis is the dominant response, but there are exceptions, as described later. This phenotype allows one to screen for mutants with aberrant motility and identify other components involved in motility and photoperception (see later sections).

There are three reports that pilins in Syn PCC 6803 are glycosylated (containing xylose and fucose residues) or post-translationally modified (via acetylation or trimethylation of the C-terminal lysine of pilin) and that these have an effect on pilus biogenesis and/or function. Two proteins, a putative nucleotidyl transferase (encoded by sll0899) and a putative kinase (encoded by slr1443) may be required for this process (Kim *et al.*, 2004, 2009, 2011). The multiple and complex post-translational modifications and glycosylation of pilins in other organisms have also been reported to play an important role in infection, biofilm formation and adhesion (Aas *et al.*, 2006; Chamot-Rooke *et al.*, 2011; Marceau *et al.*, 1998).

Action spectra for phototaxis

The fact that Syn PCC 6803 cells exhibit both negative and positive phototaxis (Figs. 11.2 and 11.4) implies that the cell has (i) the ability to sense light direction, intensity and/or quality via photoreceptors, (ii) a signal transduction mechanism that allows the motility apparatus to respond to these light cues, and (iii) the capacity to make decisions about directional movement which may also depend on the metabolic or photosynthetic status of the cell (Mullineaux, 2001). Several of these features have been addressed successfully in the last few years using a variety of approaches (for

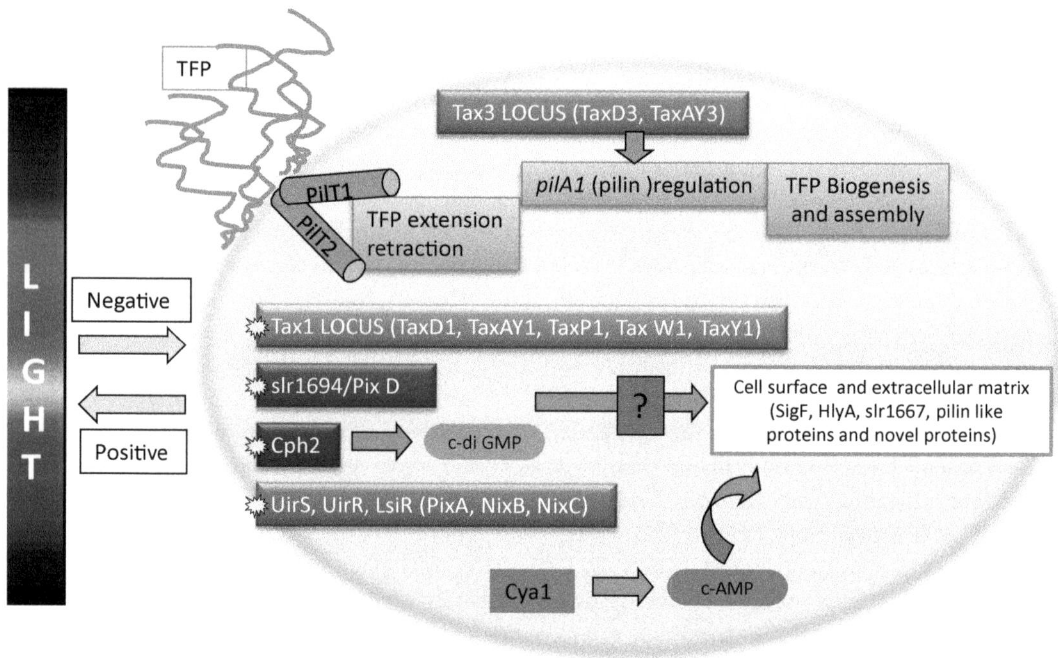

Figure 11.4 Model of positive and negative phototaxis in Syn PCC 6803. The major identified photoreceptors (shown with stars) are indicated (see Table 11.1 for more details). The Tax3 locus, which is required for TFP biogenesis and assembly, is also shown. PilT1 and PilT2 are required for TFP activity. The exact roles of cAMP and di-cGMP in regulating the cell surface and various other proteins are not known, although it is known that many of these proteins play a role on motility. A colour version of this figure is available in the plate section at the back of the book.

reviews see Bhaya, 2004; Yoshihara and Ikeuchi, 2004).

The action spectra for phototaxis have been measured for some cyanobacteria and can be quite complex and vary significantly between species. The action spectrum for phototaxis often overlaps with the absorption spectra of the dominant pigments in the cell, phycobilins and chlorophyll. However, inhibitors of photosynthesis do not have a major effect on phototactic orientation, or on gliding speed. Thus energy derived from photosynthesis may not be directly required for phototaxis, although addition of glucose increases motility (Bhaya *et al.*, 2006; Choi *et al.*, 1999; Nultsch, 1975). As mentioned earlier, Syn PCC 6803 cells have a very low threshold fluence rate (0.001 μmol/m^2/s) required to elicit phototaxis, and the response saturates at ~1 μmol/m^2/s (Choi *et al.*, 1999; Ng *et al.*, 2003).

When motility assays were carried out in a large spectrograph (in which both light wavelength and fluence rates can be precisely controlled), the thermophilic unicellular cyanobacterium *Thermosynechococcus elongatus* (or *Synechococcus elongatus*) exhibited markedly different action spectra that were dependent on fluence rates. At low fluence rates, red light was the only major elicitor of positive phototaxis, while at higher fluence rates (10 μmol/m^2/s^1) several action peaks (at 530, 570, 640 and 680 nm) were observed, and at even higher fluence rates, the red action peaks (at 640 and 680 nm) disappeared and far-red action peaks (at 720 and 740 nm) appeared. Consistent with this result, cells illuminated simultaneously with red and far-red wavelengths showed a drastic reduction in phototaxis, which the authors suggested might indicate the possible involvement of phytochrome (Kondou *et al.*, 2001). In Syn PCC 6803, a similar action spectrum was also reported although there were some striking differences such as the strong negative phototaxis in UV-A light (360 nm) (Choi *et al.*, 1999; Ng *et al.*, 2003).

Thermophilic *Synechococcus* isolates from microbial mats at Yellowstone National Park exhibited both negative and positive phototaxis. This complex dual behaviour might reflect strain-specific preferences for different light conditions (Ramsing *et al.*, 1997) (Fig. 11.2). Interestingly, *Synechococcus* isolates from these microbial mats which have recently been sequenced encode several components in common with Syn PCC 6803, including genes required for Type IV pilus biogenesis and photoreceptors that include TaxD1 and phytochrome-like photoreceptors (Bhaya *et al.*, 2007; Ulijasz *et al.*, 2009). In summary, these results lead to the conclusion that multiple photoreceptors with different spectral characteristics are likely to be involved in phototactic responses (see next section). Cyanobacteria encode a very large number of potential photoreceptors, including phytochromes, cryptochromes and blue-light photoreceptors, among others, making identification of their roles in phototaxis both revealing and challenging (Auldridge and Forest, 2011; Ikeuchi and Ishizuka, 2008; Rockwell and Lagarias, 2010).

Photoreceptors and signal transduction in phototaxis

One of the powerful ways to identify potential photoreceptors and other unknown components of the signal transduction cascade required for phototaxis is to screen for mutants that exhibit aberrant phototactic responses (Fig. 11.2). A transposon-tagged library of mutants was generated and almost 300 individual mutants with an aberration in motility were identified in the initial screen of approximately 10,000 individual colonies (Bhaya *et al.*, 2001a). Identification of the lesions in these mutants revealed that many novel proteins were involved in phototaxis. In addition, a number of chemotaxis-like operons encoding proteins involved in pilus biogenesis, signal transduction components and photoreceptors were also identified (see later). The majority (~90%) of mutants were either non-motile or demonstrated a reduced motility phenotype, but approximately ten per cent of the mutants had lost positive phototaxis (which is the phenotype of wild-type cells) and instead exhibited only negative phototaxis in white light. This particular class of mutants was predicted to have lesions in photoreceptors or in components of the signal transduction cascade (Bhaya *et al.*, 2001b) and were explored further.

One of the first putative photoreceptors to be identified via the transposon mutagenesis was TaxD1 (SyPixJ1, Sll0041; see Fig. 11.2) in a mutant that exhibited negative phototaxis in white light (Bhaya *et al.*, 2001a). This protein was also independently identified by Yoshihara *et al.* using a reverse genetic approach (PisJ1, later renamed PixJ1) (Bhaya *et al.*, 2001b; Yoshihara *et al.*, 2000, 2002). The TaxD1 protein is encoded by *taxD1*, which is in the *Tax1* locus that has at least six genes [*taxP1, taxY1, taxW1, taxD1, taxD1′* (a truncated version of *taxD1*) and *taxAY1*]. It was noted that genes in this locus had similarity to the chemotaxis genes of *E. coli*, and mutants in all of these genes produced a negatively phototactic phenotype, suggesting that this entire locus (perhaps an operon) is required for positive phototaxis.

The TaxD1 protein is a novel protein that has some similarity to methyl-accepting chemoreceptors of *E. coli*. It has a well-conserved C-terminal methyl-accepting (MA) domain, with an adjacent HAMP domain (a domain present in histidine kinases, adenyl cyclases, methyl-accepting proteins and phosphatases) and, in addition, two GAF domains at the N-terminus. GAF domains are part of a superfamily of proteins that includes the PAS/PAC and LOV domains that bind diverse small molecules such as tetrapyrroles, flavins, flavinoids and nucleotide cofactors (Crosson *et al.*, 2003; Herrou and Crosson, 2011; Vreede *et al.*, 2003). These typically occur as tandem repeats, with a bewildering combination of domains often in a single protein; this is especially common in some filamentous cyanobacteria (Auldridge and Forest, 2011; Ikeuchi and Ishizuka, 2008; Montgomery and Lagarias, 2002; Pepper, 1998; Rockwell and Lagarias, 2010) and in most cases their physiological role is unknown.

TaxD1 homologues as well as several other proteins that contain GAF domains have been identified in cyanobacterial genomes and represent a very large class of relatively simple photosensory proteins that are related to the well-studied plant phytochromes. This class of putative photosensing GAF-containing proteins have recently been termed 'cyanobacteriochromes' (or CBCRs) and some have been shown to bind a linear tetrapyrrole (either phycocyanobilin or phycoviolobilin)

(Ikeuchi and Ishizuka, 2008). Here we will focus on the biochemical and spectroscopic characterization of the chromophore that binds to the GAF domains in TaxD1/SyPixJ1. When His-tagged SyPixJ1 was expressed in Syn PCC 6803 and isolated from detergent-solubilized membranes it contained a covalently bound linear tetrapyrrole (Yoshihara *et al.*, 2004). Quite unexpectedly, this version of the protein showed a reversible photoconversion between a blue absorbing form (Pb, ~430–435 nm peak) and a green light absorbing form (Pg, ~535 nm) when irradiated with blue or green light, respectively. Surprisingly, it did not exhibit absorption in the red or far-red area of the spectrum, which is typical of phytochromes. It was also established, using site directed mutagenesis, that the chromophore was bound to the second GAF domain at a conserved Cys-His region (Yoshihara *et al.*, 2004, 2006).

The identification of this unusual photoconverting behaviour of GAF domains has been followed by several recent studies analysing the photoconversion of GAF domains from other cyanobacteria. These include the TePixJ_GAF (Ishizuka *et al.*, 2007) and the cyanobacteriochrome Tlr0924 from the thermophilic unicellular cyanobacterium *T. elongatus* (Rockwell *et al.*, 2008), as well as AnPixJ from *Anabaena* sp. PCC7120, which likely binds phycocyanobilin and exhibits a red/green photoconversion (Fukushima *et al.*, 2011; Narikawa *et al.*, 2008). The identification and properties of CBCRs is an area of active research and the reader is referred to recent reviews (Auldridge and Forest, 2011; Ikeuchi and Ishizuka, 2008; Rockwell and Lagarias, 2010).

It is worth noting that cyanobacteria synthesize a very large number of photo-pigments, including chlorophyll, haems and phycobilins, flavins and carotenoids, many of which play an important role in photosynthesis. Thus, there is great scope for binding of different chromophores in proteins with multiple GAF domains. Furthermore, it has been shown that based on the binding of these chromophores and the pocket to which they are attached, their absorption characteristics can be dramatically different. In the context of photomovements, this is a particularly critical feature, since it suggests that cyanobacteria may have evolved the capacity to exquisitely respond to light qualities as they vary from dawn to midday to dusk. In a broader context, it has been shown that cyanobacteria also use light signals and chromophore-binding proteins (e.g. RcaE) for complex adaptive processes such as chromatic acclimation (Kehoe, 2010; Kehoe and Grossman, 1996) and circadian rhythms (e.g. SyCikA) (Ivleva *et al.*, 2006; Schmitz *et al.*, 2000).

Much less is known about the physical location of the photoreceptors, signal transduction components and the TFP. However, it has been demonstrated that the TaxD1 homologue in the rod-shaped *T. elongatus* is localized at the poles (Kondou *et al.*, 2002). In *E. coli*, the majority of chemoreceptor complexes are clustered in large arrays at the poles of the cell, and the flagella are peritrichously arranged (Gestwicki *et al.*, 2000; Maddock and Shapiro, 1993; Webre *et al.*, 2003). Although we know that a chemotaxis-like operon is required for positive phototaxis in Syn PCC 6803, many fundamental questions remain about how the light signal is perceived by an individual cell and the signal transduction pathway progresses from the TaxD1 photoreceptor, via mobile regulators and histidine kinases, to the TFP apparatus.

Other chemotaxis-like loci required for motility and phototaxis

The *tax3* locus was also initially identified by transposon mutagenesis (Bhaya *et al.*, 2001a) and by reverse genetics (Chung *et al.*, 2001; Yoshihara *et al.*, 2002). Mutants in the Tax3 locus exhibit a non-motile phenotype. Furthermore, electron micrographs of mutants of the putative receptor protein (TaxD3) and the cognate hybrid histidine kinase (TaxAY3) indicated that they lacked TFP (Bhaya *et al.*, 2001a). The N-terminus of TaxD3, which might contain a sensing or interacting domain, has a tetratricopeptide (TPR) domain, which is often associated with protein–protein interactions. Further biochemical approaches will be required to establish why mutants at this locus block pilus biogenesis. In addition to the *tax1* locus and the *tax3* locus there is a *tax2* locus (Bhaya, 2004), which encodes all the components of a chemotaxis signal transduction cascade, but it does not appear to have a strong effect on motility (Yoshihara *et al.*, 2001).

Interestingly, all three histidine kinases in the Tax loci (TaxAY1, TaxAY2 and TaxAY3) are hybrid sensor histidine kinases, with a fused CheY-like regulator domain at the C-terminus (in fact TaxAY3 has *two* tandemly arranged CheY domains at the C-terminus) (Ashby and Houmard, 2006; Mizuno and Matsubara, 2003). TaxAY1 and TaxAY2 are classified as Class II kinases with canonical histidine phosphotransfer (HPt) domains at the N-terminus but TaxAY3 does not have a well-conserved HPt domain (Dutta *et al.*, 1999; Stock, 1999). Hybrid histidine kinases and Class II kinases have not been examined as carefully as other simple histidine kinases, but it has been suggested that the presence of multiple HPt domains and attached CheY-like domains might allow for complex phosphorelay signals whereby several inputs can be integrated into the signalling pathway (Jung *et al.*, 2012; Mizuno and Matsubara, 2003; Wuichet *et al.*, 2010). In fact, other proteins that carry a Hpt domain have been implicated in motility (Yoshihara *et al.*, 2000). In *E. coli*, where the signal transduction cascade for chemotaxis has been very well–studied, it appears that information from the multifunctional chemoreceptors is integrated through the single histidine kinase/regulator pair (CheA/CheY). So far, it has not been established whether the three Tax loci in Syn PCC 6803 act independently or if there is any cross-talk between them. Recent work from several research groups suggests that chemoreceptors are in large macromolecular complexes often at the poles of a cell and that this is important for signal cross-talk and amplification (Hazelbauer *et al.*, 2008).

Multiple photoreceptors are involved in phototaxis

Although TaxD1 (SypixJ1) is the best characterized of the photoreceptors in terms of the chromophores, other photoreceptors have also been implicated in regulating phototaxis. These include the blue light photoreceptor, BLUF (PixD or Slr1694), the UV-A sensing photoreceptor (UirS, PixA) and the phytochrome-like multi domain Cph2, which are described below (Fig. 11.4 and Table 11.1). The fact that multiple photoreceptors have now been shown to play a role in phototaxis suggests that cyanobacteria must be exquisitely responsive to their light environment and we have only just begun to appreciate the complexity and interplay between these signal transduction pathways.

BLUF domain photoreceptors

There are at least three classes of blue-light receptors that contain the flavin chromophore. These include the phototrophins (which contain a LOV domain), the cryptochromes and the recently discovered BLUF domain proteins (Gomelsky and Klug, 2002; Herrou and Crosson, 2011; Losi and Gartner, 2012). Cyanobacteria encode BLUF proteins and it has been demonstrated that the photocycle of BLUF proteins is similar to that of other BLUF proteins such as AppA, that show a UV-visible absorption spectrum, and a fast return from the excited to the ground state (Masuda *et al.*, 2004), strongly suggesting that it can act as a reversible photoreceptor. In the case of Syn PCC 6803, it was shown that a mutation in BLUF (PixD or Slr1694) blocked positive phototaxis in red light but surprisingly, neither the wild-type nor the mutant showed any motility towards blue light (460 nm) (Kita *et al.*, 2005; Okajima *et al.*, 2005). Furthermore, it has been shown that Slr1694 is present as an oligomer in the cell and that it likely forms a hetero-oligomer with a response regulator, Slr1963 (PixE), with a stoichiometry of PixD10-PixE5 (Yuan *et al.*, 2006, 2011). Very recently, it has been suggested that this protein–protein interaction is required for signal transduction and phototactic behaviour. The disassembly kinetics of this hetero-oligomer is light-intensity dependent, raising the intriguing possibility that this may be a mechanism for sensing light intensity. Although further follow-up research is required, the possibility that photo-movements will be sensitive to both light intensity and light quality is quite plausible. How changes in PixS oligomerization are transduced to changes in phototactic behaviours is still to be established.

Cph2 and the role of cyclic di-GMP in phototaxis

The Cph2 (Sll0821) photoreceptor was initially implicated as inhibiting motility of Syn PCC 6803 towards blue light, since WT cells do not move

Table 11.1 Photoreceptors in Syn PCC 6803 with their characteristics and possible roles in phototaxis

Locus tag (size in nt)	Synonyms	Photoreceptor class	Mutant phenotype	Function	Domains (positions)	Chromo phores	Domain organization	References
sll0041 (3003)	*taxD1*, *pixJ1*, *pisJ1*, *SypixJ1*, *cheD*, *tsr*	Cyanobacteriochrome (Phytochrome superfamily)	Negative phototaxis (WL, UV, BL, GL, RL)	Required for positive phototaxis	Phycocyanobilin	Blue/Green receptor Pb, λmax = 435 nm Pg λmax = 535 nm		Bhaya (2004), Yoshihara *et al*. (2004)
slr1694 (453)	*pixD*, *SyPixD*, *synbluf*, *bluf*	BLUF	Non-motile (BL)	Required for positive phototaxis	Flavin	Blue receptor		Okajima *et al*. (2005)
slr0473 (2247)	*cph1*, *hik35*, *phy*	Phytochrome	Positive phototaxis	Not known	Phycocyanobilin	Red/Far red receptor Pr, λmax = 668 nm Pfr, λmax = 718 nm		Hughes *et al*. (1997), Lamparter *et al*. (2001)
sll0821 (3831)	*cph2*	Phytochrome	Positive phototaxis (UV, BL, WL)	Inhibits phototaxis towards blue light (role in c-diGMP signaling)	Phycoerythrobilin Phycocyanobilin	N-terminal module Pf, λmax = 644 nm Pfr, λmax = 695 nm C-terminal module (GAF: 939-1090) Pb, λmax = 400 nm Pg, λmax = 525 nm		Fiedler *et al*. (2005), Anders *et al*. (2010), Savakis *et al*. (2012)
slr1212 (2535)	*uirS*, *ETR1*	Phytochrome	Positive phototaxis (UV, BL, GL, RL)	Promotes negative phototaxis	Phycocyanobilin	UV receptor Pv, λmax = 382 nm Pg, λmax = 534 nm	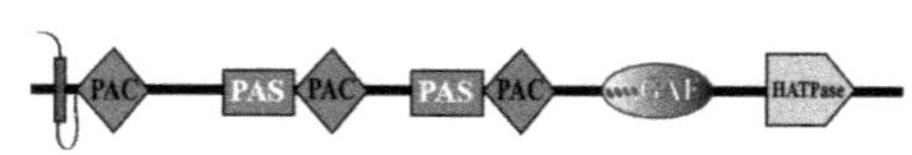	Song *et al*. (2011), Narikawa (2012)

towards blue light, but a *cph2* mutant exhibited motility towards blue light (Fiedler *et al.*, 2005) and towards UV-A (Moon *et al.*, 2011). This photoreceptor has a complex structure and contains six domains (GAF1-GAF2-GGDEF-EAL-GAF3-GGDEF) (Table 11.1). The first two GAF domains bind phycocyanobilin and behave as a P_r/P_{fr}-photosensory module (Anders *et al.*, 2011) while GAF3 binds a tetrapyrrole chromophore that can absorb blue light (Park *et al.*, 2000; Wu and Lagarias, 2000). Thus Cph2 has been called a 'composite phytochrome' with a Cph2-type phytochrome module at the N-terminus and a CBCR module at the C-terminus. In addition to these photoreceptor domains, Cph2 also contains GGDEF domains and EAL domains. GGDEF domains catalyse the synthesis of cyclic-di-GMP from two GTP molecules, while proteins with an EAL domain can catalyse cleavage of cyclic-di-GMP to linear 5′-pGpG. However, the N-terminal GGDEF1 domain is unlikely to have active diguanylate cyclase activity because it bears mutations in residues essential for catalytic activity (Savakis *et al.*, 2012). GGDEF and EAL domains are often found on the same polypeptide, and there are cases in which these domains contain mutations in their active site, although these mutants still show residual and/or regulatory activity (Cao *et al.*, 2010; Christen *et al.*, 2006; Hengge, 2009).

From studies using a number of specific mutants and complementing various background mutations under different light conditions, it was suggested that the terminal CBCR-GGDEF was critical for the blue light phototactic behaviour of Syn PCC 6803 cells (Savakis *et al.*, 2012). Furthermore, Cph2 may play an important role in sensing light across a broad range (in the green/blue/UVA regions of the spectrum). Changes in the light inputs are likely transduced to the GGDEF-EAL domains to finally result in a change in internal cyclic-di-GMP levels. Cyclic-di-GMP appears to be produced in a light-dependent manner by the C-terminal GGDEF2 domain and these high levels of cyclic-di-GMP inhibit motility in Syn PCC 6803. At this point it is not clear exactly how changes in cyclic-di-GMP levels are translated into behaviour of the motility apparatus (Savakis *et al.*, 2012). Notably, homologues of the hybrid photosensor Cph2 identified in Syn PCC 6803 are not universally present in cyanobacteria and so far have only been found in the unicellular *Cyanothece* sp. (PCC 7424 and PCC 7822) and in the filamentous *Lyngbya* sp. PCC 8106, raising the possibility that other novel photoreceptors may be present in other cyanobacteria.

A photoreceptor involved in UV-A/violet-regulated phototaxis

A locus containing four genes has recently been independently identified by two groups (Narikawa *et al.*, 2011; Song *et al.*, 2011) as being required for regulation of phototaxis. This region encodes a putative photoreceptor (UirS/PixA, Slr1212), and two putative response regulator proteins (UirR/NixB, Slr1213) and (LsiR/NixC, Slr1214) and the less well-characterized protein (Slr1215). Narikawa *et al.* introduced mutations into different background strains, either in cells that were constitutively positively phototactic (PCC-P) or negatively phototactic (PCC-N). They found that in the PCC-P background the *slr1212* mutant was negatively phototactic, while *slr1213* and *slr1214* mutants did not have an obvious altered phenotype. On the other hand, these same mutations when placed in the PCC-N background showed different effects, with *slr1212* exhibiting no obvious altered phenotype, while the *slr1213* and *slr1214* mutants appeared to reverse the negatively phototactic behaviour. Thus it was suggested that the opposing effects of PixA and NixBC imply that this pathway is involved in the switching between positive and negative phototaxis. Although the exact signal transduction pathway and mechanism by which this phototaxis switching behaviour is controlled remains to be determined, it is clear that multiple photoreceptors are involved in the phototactic behaviour of single-celled cyanobacteria, particularly in the UV/Blue/violet region of the spectrum.

Song *et al.* (2011) also used a reverse genetics approach that targeted this cluster of genes for inactivation and further study involving biochemical and expression studies. They reach similar conclusions to suggest that the UirR/UirS along with the PatA-like LsiR are part of a UV-A activated signalling system, which is required for negative phototaxis. Their model proposes that UirS is bound to membranes, where it responds to UV-A and thereby releases UirR. UirR, in turn, recognizes the LsiR promoter. Consistent with this model, constitutive expression of LsiR induces negative phototaxis under conditions where cells would exhibit positive phototaxis. Finally, because there is evidence that LsiR may be induced by other general stress responses, Song *et al.* (2011) speculate that it may be part of a system that allows the cell to integrate light signals from multiple photosensors. This allows the cell to respond to light signals in a dynamic way.

The role of cyclic AMP (cAMP) and surface architecture in phototaxis

The genome of Syn PCC 6803 encodes an adenylyl cyclase (Cya1) that synthesizes cAMP, and a cAMP receptor protein (Sycrp1) (Masuda and Ono, 2004; Yoshimura *et al.*, 2000). This suggests that cAMP may play a role as a second messenger, as occurs in many organisms such as *E. coli* and pathogenic bacteria such as *Mycobacterium* sp. (Kalia *et al.*, 2012; McDonough and Rodriguez, 2012). Ohmori and co-workers inactivated these genes in Syn PCC 6803 and observed that the corresponding mutant strains were non-motile and formed tight, dome-shaped colonies and that cAMP levels in the *cya1* mutant were 5% of normal levels. (Terauchi and Ohmori, 1999; Yoshimura *et al.*, 2002b). Blue light (450 nm) markedly increased cAMP levels within a few minutes, while red (630 nm) and far red (720 nm) did not (Ohmori and Okamoto, 2004; Terauchi and Ohmori, 1999). Cya1 was also independently identified in a transposon mutagenesis screen to identify motility mutants (Bhaya *et al.*, 2001b). Careful analysis of the motility of the *cya1* mutant demonstrated that they had an impaired motility on surfaces, which could be effectively rescued by externally added cAMP (Bhaya *et al.*, 2006) (Fig. 11.5).

The connection between blue light, cAMP levels and motility has been documented by Ohmori and co-workers in both unicellular and filamentous cyanobacteria (Ohmori and Okamoto, 2004). They found that the cellular cAMP level in the filamentous cyanobacterium, *Anabaena* sp. PCC7120, decreased in red light and increased in far red light (unlike the strong blue light-mediated effect on cAMP and motility in Syn PCC 6803). They also demonstrated that cAMP had multiple effects in the filamentous gliding cyanobacterium *Spirulina platensis*. It stimulated respiration and gliding motility and cells rapidly formed mats when cAMP (at concentrations of 1 μM) was

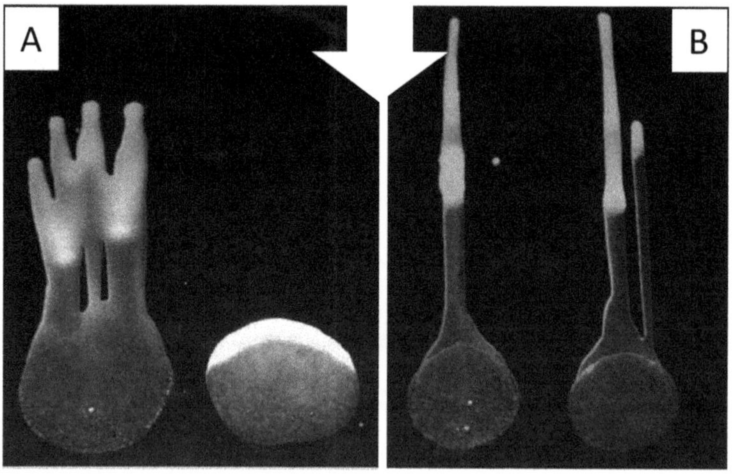

Figure 11.5 Effect of cAMP on phototaxis of Syn PCC 6803 and *cya1* mutant cells. WT (left), *cya1* mutant (right), were spotted on 0.4% agarose plates in the absence (A) or in the presence (B) of 0.1 mM cAMP. In the absence of added cAMP, *cya1* cells showed limited motility (A) but when cAMP is added to the plates, they exhibit typical finger-like projections, like WT cells. The arrow shows the direction of the light. Modified from Bhaya, 2006. A colour version of this figure is available in the plate section at the back of the book.

added to a cell suspension (Ohmori *et al.*, 1993). *S. platensis* cells have been reported to excrete cAMP into their surrounding medium and it is possible that cAMP-mediated motility changes may be important in the environment and possibly be part of a quorum sensing phenomenon, although these hypotheses have yet to be tested.

A transcriptomic analysis of the *sycrp1* mutant of *Synechocystis* identified a small set of down-regulated genes including *slr1667* (encoding CccS) and *slr1668* (CccP) and five other genes arranged in a putative operon (*slr2015* to *slr2019*) (Yoshimura *et al.*, 2002a). Sycrp1 binds to a region upstream of *slr1667* that contains a typical *E. coli* CRP-binding consensus sequence (5′-TGTGA-N_6-TCACA-3′) (Yoshimura *et al.*, 2002a). Transposon mutagenesis followed by identification of non-motile mutants also independently identified the gene cluster *slr2015*–*slr2019* (Bhaya *et al.*, 2001b). Recently it has been suggested that CccS and CccP are responsible for the architecture of the cell surface. Many of the pleiotropic effects that were observed, such as release of yellow pigments, lack of TFP and release of proteins into the medium strongly resemble the phenotype of the SigF mutant, which was the first mutant linking TFP and cell–surface architecture to motility (Bhaya *et al.*, 1999).

In summary, these results suggest that cAMP is an important player in the signal transduction for motility. It is not clear if cAMP controls motility through a pathway mediated by Sycrp1, and/or through regulation of genes such as *slr1667*/*slr1668* and the *slr2015* gene cluster. It is known that twitching motility in *P. aeruginosa* is controlled by the CRP homologue Vfr protein, which can bind both cAMP and cGMP (Beatson *et al.*, 2002), and photo-sensing in the protist *Euglena* is mediated by a photoactivable adenylyl cyclase (Ntefidou *et al.*, 2003). Furthermore, the connection between activation of cAMP-controlled pathways leading to photomovements, in response to either blue or red light, is further evidence that phototaxis is a complex pathway in which many light inputs working via secondary messengers and other pathways regulate motility.

Novel proteins in search of a role in phototaxis: tetratricopeptide repeat (TPR) domain proteins and pentapeptide repeat proteins

In addition to the gene products and pathways described above, a genetic approach has revealed that there are even more components of the phototaxis pathways in Syn PCC 6803 that need to be understood (Bhaya *et al.*, 2001b). A number of mutants were identified in which genes encoding coiled-coil domain- or TPR domain-containing proteins were disrupted (Bhaya *et al.*, 2001b). TPR domains (PFAM identifier PF00515) are ubiquitous in both eukaryotes and prokaryotes and are one of several repeat motifs that can act as scaffolds to facilitate the formation of multi–protein complexes involved in numerous cellular processes ranging from transcription to protein translocation and degradation (Dandrea and Regan, 2003; Lamb *et al.*, 1995; Sun *et al.*, 2011). Three non-motile mutants, created by transposon tagging, had inactivated genes (*sll0183*, *sll0414* and *sll0301*) encoding proteins that contained a tandem pentapeptide repeat domain (Bhaya *et al.*, 2001b). This domain (PFAM identifier PF00805) is found in many organisms but is most common in cyanobacteria (Bateman *et al.*, 1998; Liu and Wolk, 2011).

Two putative serine-threonine kinase mutants, inactivated in genes designated *spkA* and *spkB*, were also shown to lack motility (Bhaya *et al.*, 2001b; Kamei *et al.*, 2001, 2003). Very little is known about how these kinases function and it was originally assumed that serine-threonine kinases were functional exclusively in eukaryotes. However, they have now been discovered in many bacteria where they might function in complex signalling pathways. In addition, a chaperone-like protein (Sll0058), the SigF sigma factor and other proteins (HlyA, Sll1951) have also been implicated in phototaxis (Bhaya *et al.*, 2001b). These findings underscore that we still have a lot to learn about the complex pathways that cells use to sense light and move.

Modelling photoresponses

When wild-type Syn PCC 6803 cells are exposed to light, cells begin to move, although not

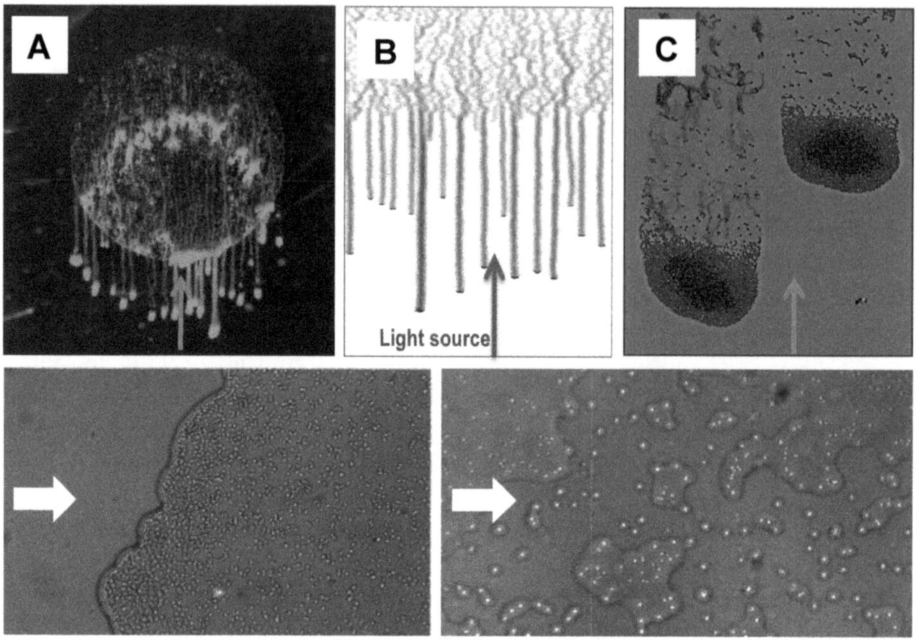

Figure 11.6 Model of phototactic and group behaviour. Top panel. (A) Phototactic behaviour of Syn PCC 6803 showing typical finger-like projections of groups of cells. (B) Simulation of phototactic behaviour using modelling. Note how this simulates the behaviour of cells as shown in (A). (C) Tracking behaviour of single cells. Each cell track is shown in a different colour. Lower panel. Left: A group of Syn PCC 6803 cells in an extracellular matrix as they move towards the light, Right: when small groups of cells are moving they also are surrounded by a distinct extracellular matrix, of unknown composition (modified from Galante *et al.*, 2012). A colour version of this figure is available in the plate section at the back of the book.

necessarily in the direction of light. They instead form small aggregations and eventually cells move towards the light. At the front of a spot of cells placed on an agarose surface, cells align along the boundary of the spot before forming the characteristic finger-like projections or swarms of cells (Fig. 11.6). However, when the motion of individual cells is tracked, it is clear that they display a quasi-random motion, although they exhibit a weak positive biased movement towards the light source (Burriesci and Bhaya, 2008; Galante *et al.*, 2012). We have also observed that as cells move along the surface they appear to release extracellular material (unpublished observations; Fig. 11.6), but the role and identity of this material is not yet known. It is possible to model these local interactions in Syn PCC 6803 and this may help us to predict the behaviour of mutant cells and understand how cells can sense light direction as well as the presence and behaviour of neighbouring cells. Some of the models that have been proposed imply that the typical behaviour of motile cells requires that these cells lay down extracellular material or extracellular polysaccharides as they move and this enhances motility (Burriesci and Bhaya, 2008; Galante *et al.*, 2012; Ursell *et al.*, 2013).

Certain filamentous cyanobacteria, such as *Phormidium uncinatum*, use photophobic responses to position themselves in areas of optimal light intensity (reviewed in Castenholz, 1982; Nultsch and Hader, 1988; Hader, 1988). In response to sudden spatial or temporal changes in the fluence rate, the filaments stop gliding and following a pause, reverse direction. In step-down responses (decrease in light intensity) the filaments thus move away from areas of darkness, while in step-up responses they avoid light that is too bright. Only a few cells at either end of a filament need to be exposed to the sudden change in light intensity in order for these responses to be elicited (Hader and Burkart, 1982). In step-down responses, the photosynthetic pigments are believed to be the photoreceptors and a model for signal transduction has been proposed in which changes in photosynthetic electron transport result in membrane depolarization. On the basis of ionophore and radiolabel uptake experiments, it has been proposed that this depolarization is amplified by Ca^{2+} fluxes, ultimately resulting in a difference in electrical potential between the ends of the filament, triggering a reversal (Hader, 1988). How the signal is transmitted to, and ultimately controls the motility machinery, remains to be determined. Less is known about the step up response where the action spectrum appears to be different than that for step down, showing a peak in the blue region of the spectrum (Hader, 1988). A recent computational model predicts that photophobic responses serve to optimize the distribution of a population with respect to a light field (Tamulonis, 2011) and hence are likely to play an important role in the environment. However, biochemical or *in silico* identification of potential photoreceptors and the regulation of signal transduction pathways in filamentous cyanobacteria still require further investigation.

Modelling photoresponses is a relatively new field and its importance lies in being able to predict motile behaviour of single cells or filaments or of motile communities and groups (Galante *et al.*, 2012; Tamulonis *et al.*, 2011; Ursell *et al.*, 2013). As more data about the behaviour of single cells and groups (both wild-type and mutant) from different species become available, it is possible that models of photoresponses will become more robust and predict behaviour patterns allowing us to understand the behaviour of motile organisms in the environment.

Gliding in filamentous cyanobacteria

Gliding motility is the active translocation of bacterial cells on a solid surface along their longitudinal axis. It is distinct from other types of surface locomotion such as twitching and swarming in that neither type IV pili, nor flagella are involved (McBride, 2001). Gliding is employed by several phylogenetically distinct groups of bacteria, including the *Chloroflexaceae*, the myxobacteria, members of the *Bacteroidetes*, and certain filamentous cyanobacteria. The mechanism is not well understood because neither structures nor organelles that could account for movement have been readily visible. Indeed it is probable that there is more than one mechanism underlying this type of movement (Jarrell and McBride, 2008; Mauriello *et al.*, 2010).

Characteristics of cyanobacterial gliding motility

Filamentous cyanobacteria of the order *Oscillatoriales* as well as some members of the *Nostocales* are capable of gliding and, unlike genera that are only transiently motile though the differentiation of hormogonia (discussed below), these generally glide at all times (Castenholz, 1982, 1989a,b). In all such gliders, movement requires contact with a surface and is associated with the production of extracellular polysaccharide mucilage or 'slime' that is left as a sheath or trail behind the gliding filament (Hoiczyk, 2000). Gliding speeds can range from 1 to 11 μm/s, with members of the *Nostocales* generally exhibiting slower gliding than that of the trichomes of the *Oscillatoriales* (Castenholz, 1989a). Members of the *Oscillatoriales* glide as straight filaments and rotate about their longitudinal axis as they move in a direction parallel to it (Hoiczyk, 2000). The direction of rotation can trace either a right-handed or left-handed helix and is a species-specific characteristic. Gliding filaments can also exhibit rapid reversals (Castenholz, 1982). Gliding members of the *Nostocales*, such as certain *Anabaena* species, do not rotate as they glide (Castenholz, 1989a), and some can move laterally as bent or U-shaped filaments (Hoiczyk and Baumeister, 1998; Nultsch *et al.*, 1979). These behaviours have intrigued researchers for decades and have led to several examinations of the ultrastructure of gliding cells with the intent of visualizing a motility apparatus. On the basis of these studies, which we describe below, two models have been proposed to account for this movement: surface waves and slime extrusion.

Surface wave model of gliding motility

Early work by Halfen and Castenholz (1970, 1971) with *Oscillatoria princeps* revealed the presence of an array of 6- to 9-nm-wide fibrils arranged in a helical fashion around the trichome and which appeared to be located between the peptidoglycan layer and the outer membrane. The right-handed helical arrangement of the fibrils corresponded with the direction of rotation of the trichomes, and its 60° pitch also matched that of a helix traced by a point on a trichome as it moved. They proposed a model whereby contraction of these fibrils would produce moving waves that could push against the sheath or substrate and propel cells in the opposite direction. The helical arrangement of the fibrils would account for the rotation of the trichome during gliding and indeed the orientation of the array in left- and right-hand rotating strains of *Oscillatoria* corresponded to the direction of rotation (Halfen and Castenholz, 1971).

More recently, using cryofixation techniques which preserve ultrastructure better than conventional chemical fixation, Hoiczyk and Baumeister described arrays of 10- to 12-nm-wide fibrils external to the outer membrane of *Oscillatoria, Lyngbya,* and *Phormidium* species (Hoiczyk and Baumeister, 1995). The orientation of the fibrillar arrays corresponded to the direction of rotation of the filament during gliding. Whether these fibrils are the ones described by Halfen and Castenholz (Halfen and Castenholz, 1971) is difficult to determine, as the strains of *Oscillatoria* used were not the same, and the fixation and staining techniques used for electron microscopy differed. The surface fibrils from *Phormidium* were purified and shown to be composed of oscillin, a single 646 amino acid Ca^{2+} binding glycoprotein. The oscillin is arranged in a helical fashion above an S-layer, and Hoiczyk and Baumeister (1997) have proposed that it plays a passive role in gliding, serving as a screw thread guiding the rotation of the filament during mucilage secretion (see below) (Fig. 11.7). Proton motive force is suggested to power gliding in *Phormidium* (Glagoleva *et al.*, 1980), which also supports a passive role for the oscillin array that is located above the S-layer, and is not close to the source of energy generation.

Fibrils in the cell envelope appear to be characteristic of the *Oscillatoriales*. David Adams' group has also shown the presence of a fibrillar array located in the cell envelopes of several *Oscillatoria* strains (Adams *et al.*, 1999; Adams, 2001; Read *et al.*, 2007). These fibrils are larger (25–30 nm wide) than the ones described by Halfen and Castenholz (1971) and Hoiczyk and Baumeister (1995, 1997) and, on the basis of the corrugated appearance of the outer membrane in thin sections, are proposed to be located between the outer membrane and the peptidoglycan. The authors propose a model similar to Halfen and Castenholz's in which they speculate that rhythmical undulations originating in the fibrillar array could then be transmitted through the outer membrane and serve to push exuded mucilage with sufficient force to propel the cell (Read *et al.*, 2007). A mathematical analysis of such model (Siddiqui *et al.*, 2001) predicts that sufficient force could be generated by such a mechanism to account for the observed gliding speeds of a number of gliding bacteria. The mechanism by which the undulations would be produced is unknown and at present there is no experimental evidence to support this idea. Physical deformation is difficult to detect in microorganisms and the amplitude of surface waves may be below the resolution limit of the light microscope, but even in *Oscillatoria princeps,* which has a trichome diameter of 35 μm, no surface deformations or undulations have been observed (Castenholz, 1982; Halfen and Castenholz, 1970). The role of cell envelope fibrils in gliding remains to be determined, and questions remain. For instance, are the fibrils seen in the various strains related by homology and/or do they serve the same function? Nevertheless, given the fact that some of these have been purified (Adams *et al.*, 1999; Hoiczyk and Baumeister, 1997) and the genes encoding one of them, oscillin, is known (Hoiczyk and Baumeister, 1997), it should be possible to localize them unambiguously using antibody staining techniques, as well as to generate knockouts using reverse genetics in tractable strains. In addition, it is possible that there may be other ways of generating surface waves in cyanobacteria. One way may be through the rotation of helical cytoskeletal elements as has been proposed for *M. xanthus* (Nan *et al.*, 2011).

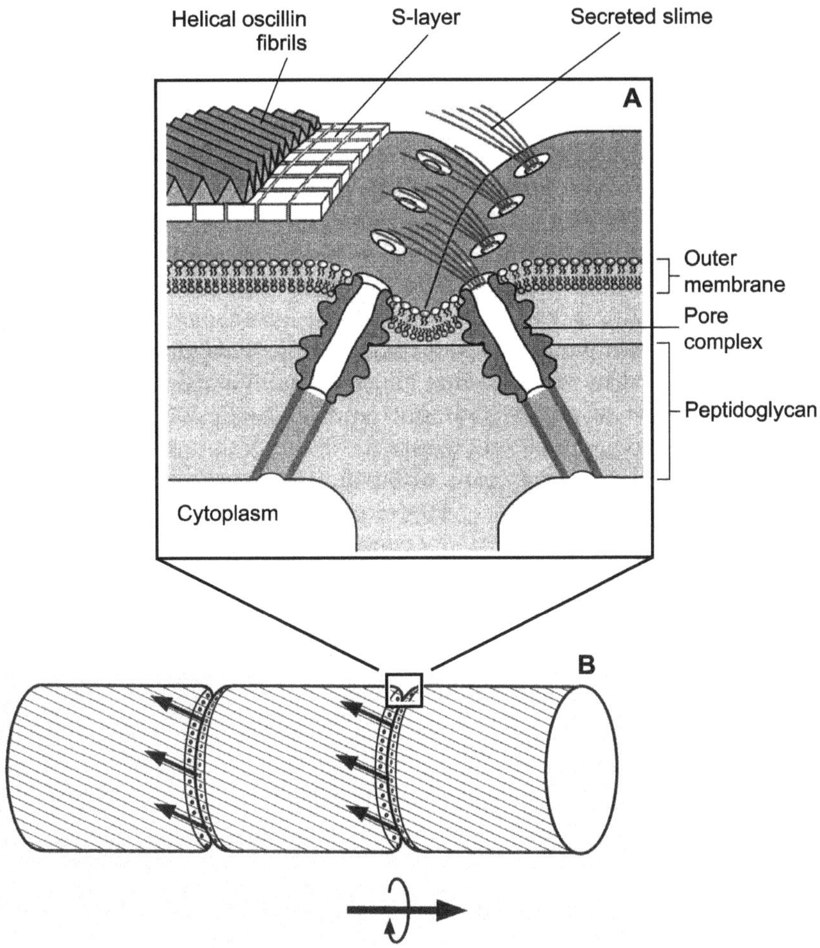

Figure 11.7 The slime extrusion model for gliding in filamentous cyanobacteria. (A) Model of the cell envelope of *P. uncinatum* showing the surface location of the oscillin fibrils, the pores located at the cell junction through which mucilage is secreted, and individual junctional pore complexes which traverse the outer membrane and peptidoglycan layers and extend into the cytoplasm. (B) Portion of a trichome of *P. uncinatum* showing mucilage being secreted at the junctional pores (small arrows) and being guided by the helically arranged oscillin fibrils. Secretion and flow of the slime that comes in contact with both the surface of the trichome and the substrate surface cause the cell to move in the opposite direction (large arrow). The cell body rotates as a consequence of the helical arrangement of the oscillin fibrils. (A) is reprinted from (Hoiczyk, 1998) with permission from Elsevier, and (B) is reprinted from (Hoiczyk and Baumeister, 1997) with permission from John Wiley and Sons.

Slime extrusion model of gliding motility

Gliding is always accompanied by the secretion of mucilage or slime. Staining of cells with India ink shows that the slime is translocated in bands or ribbons in a helical fashion, with a pitch of 60° (Halfen and Castenholz, 1971; Hoiczyk, 1998) and a handedness that corresponds to the orientation of the oscillin surface fibrils. The rate of elongation of the slime threads corresponds to that of movement and the threads emanate at cross walls where junctional pores are located (Hoiczyk and Baumeister, 1998) (Fig. 11.7). Such pore-like structures have been observed at cell junctions in a number of filamentous cyanobacteria and are arranged at an angle, forming a ring around the trichome at the septa between cells, and appear to extend from the cell membrane through the peptidoglycan layer (Halfen and Castenholz, 1971; Hoiczyk and Baumeister, 1995; Lamont, 1969;

Pankratz and Bowen, 1963). The junctional pores have been suggested to be the sites of mucilage secretion (Hoiczyk and Baumeister, 1998; Pankratz and Bowen, 1963). In the *Oscillatoriales*, the junctional pores appear to be arranged circumferentially. In contrast, in *Anabaena variabilis* strain B1403-4b, which does not rotate while gliding but moves laterally, the junctional pores are not circumferential but rather are arranged laterally. In addition, in this strain, slime secretion is only visualized on one side of the filament, the one away from the direction of movement (Hoiczyk and Baumeister, 1998).

Organelles consistent in diameter and shape with the pores seen in whole cells were isolated from *Phormidium uncinatum* and image reconstruction analysis revealed a 32-nm-long tube, 14 nm wide in the centre and 8 nm wide at each open end (Hoiczyk and Baumeister, 1998). In some instances, an additional 13-nm-long straight tube was found attached at one end and these complexes were proposed to span the entire cell envelope. These observations led Hoiczyk and Baumeister to propose that these structures represented junctional pore complexes and that these were secretion organelles composing the molecular motors responsible for gliding in cyanobacteria (Hoiczyk and Baumeister, 1998) (Fig. 11.7).

How then could slime secretion through pore complexes power motility? Pore-like complexes have also been seen in *Myxococcus xanthus*, another slime-producing gliding bacterium not closely related to cyanobacteria (Wolgemuth *et al.*, 2002). Wolgemuth and co-workers (Wolgemuth *et al.*, 2002) have proposed a model by which propulsive thrust is generated by the secretion of slime. They propose that the slime, a polyelectrolyte gel synthesized at the cytoplasmic membrane, enters the junctional pore, or 'nozzle', in a dehydrated form. As the slime is extruded through the nozzle, it is hydrated by water and it greatly expands. It is this expansion that forces the slime out of the nozzle and this produces thrust. A mathematical model, which takes into account the dimensions and geometry of the nozzle and assumes the slime behaves as a polyelectrolyte gel, predicts that the force exerted by the swelling is sufficient to propel the cells at the rates at which they have been observed to glide. This is in contrast to the model proposed by Siddiqui *et al.* (2001) in which the slime is propelled by the peristaltic waves produced by the fibrillar array.

This model is supported by several observations: the location of the pores corresponds to the sites of extrusion of slime threads; mutants that do not secrete polysaccharide (but which are pleiotropic as they also do not produce oscillin or the S-layer) are non-motile (Hoiczyk and Baumeister, 1997); the rate of extrusion of the slime threads matches that of gliding; and the direction of movement is opposite that of slime deposition. Of course, the latter two phenomena could be consequences of gliding, rather than being causative. Nevertheless, this is a testable model and future work should include establishing the chemical composition of the slime, identifying genes required for its biosynthesis, and characterization of the junctional pore complex at the molecular level. Since the junctional pore complex can be isolated from other envelope components (Hoiczyk and Baumeister, 1998) it should be possible, assuming it is composed of protein, to identify its component polypeptides and hence the genes encoding them.

Hormogonia and motility

Hormogonia are short filaments that differentiate in certain heterocyst-forming and non-heterocyst-forming filamentous cyanobacteria from larger trichomes (Tandeau de Marsac, 1994; Herdman and Rippka, 1988). In the *Nostocales* (taxonomic subsection IV) and *Stigonematales* (taxonomic subsection V (Boone and Castenholz, 2001)) certain members of the genera *Nostoc*, *Fremyella*, and *Fischerella* (also called *Mastigocladus*), have vegetative trichomes that are not motile. Rather, under certain conditions which can include changes in nutrient and salt concentrations, light quality, and symbiotic associations (Meeks *et al.*, 2002), the vegetative filaments undergo fragmentation and differentiation to produce hormogonia, which are shorter gliding filaments consisting of smaller cells arising from cell division without concurrent growth. Hormogonia are thought to function in dispersal and as the infective units in plant symbiotic associations and have been shown to be chemotactic towards plant exudates (Knight and Adams, 1996; Nilsson *et al.*, 2006). Motility is transient and following a period of time (48–72 h) (Meeks *et al.*, 2002), the hormogonia cease to glide, differentiate heterocysts and re-enter the vegetative cycle.

Hormogonia glide at speeds that range from 0.7 to 3 µm/s (Duggan *et al.*, 2007; Hernandez-Muniz and Stevens, 1987) and can also glide through viscous environments at speeds of 0.5 µm/s (Robinson *et al.*, 2007). Movement can be, but is not always, accompanied by rotation of the filament about its longitudinal axis (Hernandez-Muniz and Stevens, 1987). As hormogonia glide, they deposit 'slime' trails (Hernandez-Muniz and Stevens, 1987; Robinson *et al.*, 2007), and extrusion of material from junctional areas has been observed by atomic force microscopy in *Mastigocladus laminosus* (Robinson *et al.*, 2007), although the rate of extrusion has not been reported for hormogonia nor correlated to the speed of gliding as was done for *Phormidium* and *Anabaena cylindrica* (Hoiczyk and Baumeister, 1998; Walsby, 1968).

In addition to the reduction in cell size, differentiation of motile hormogonia is accompanied by the appearance of pores in the peptidoglycan layer at junctions between cells (Damerval *et al.*, 1991; Robinson *et al.*, 2007). Robinson *et al.* (2007) report that hormogonia of *M. laminosus* can have up to three concentric rings of pores or 'nozzles' near the septa, separating cells and they postulate that these are the sites of slime secretion. Also, unlike the vegetative filament, hormogonia are peritrichously piliated (Dick and Stewart, 1980; Damerval *et al.*, 1991; Duggan *et al.*, 2007). Hence hormogonial gliding shares morphological characteristics of both filamentous gliding and TFP-mediated motility.

Molecular approaches to hormogonia gliding

The ability to genetically manipulate *Nostoc punctiforme* has made it possible to address aspects of its physiology and behaviour at the molecular level (Cohen *et al.*, 1994). The genome of *N. punctiforme* encodes fifteen proteins with similarity to proteins required for TFP-dependent motility including *pilT, D, A, B, C, M, N, O, Q*. Insertional inactivation of the *pilT* and *pilD* homologues altered surface piliation and the *pilT* mutant was non-motile, implying that motility in hormogonia of this species may require type IV pili (Duggan *et al.*, 2007). Transcriptional profiling during hormogonia differentiation in *N. punctiforme* revealed that several genes showing similarity to those encoding known motility proteins are up-regulated (Campbell *et al.*, 2007). These include five genes encoding TFP components, as well the *hmp* locus, which contains genes encoding proteins with similarity to proteins comprising a chemotaxis-like signal transduction system. In addition, a cluster of genes that encode putative polysaccharide synthesis and type II secretion proteins, the *hps* locus, is up-regulated in hormogonia (Campbell *et al.*, 2007; Risser and Meeks, 2013). Mutations in the *hmp* locus genes result in mutants that differentiate non-motile hormogonia that also do not secrete polysaccharide (Risser and Meeks, 2013). The *hps* locus encodes glycosyl transferases, pseudopilins, and proteins containing prepilin peptidase domains, and mutations in these genes affect motility and polysaccharide secretion (Risser and Meeks, 2013). These authors propose that this locus may encode the components of the junctional pore complex. Homologues of these genes are present in the genomes of all the filamentous cyanobacteria sequenced to date and appear to be confined to these organisms suggesting a common mechanism for motility (Risser and Meeks, 2013). The availability of a genetic system and a cast of candidate genes encoding both proposed structures and polysaccharide biosynthetic pathways should permit rapid progress on these issues.

Conclusions

In this chapter we have focused on recent advances made in the field over the last 20 years. Many important discoveries have been made in our understanding of the mechanistic aspects and regulation of motility, such as the identification of novel proteins involved in motility and the identification of photoreceptors and complex signal transduction pathways. These results underscore the complexity of motile behaviours in unicellular and filamentous cyanobacteria and point the way for future research. Some of the challenging questions that remain are mentioned below.

For unicellular cyanobacteria such as Syn PCC

6803, the motility apparatus consisting of TFP has largely been uncovered, and work is now centred on understanding the complex signal transduction and regulatory cascades governing phototaxis and motile behaviour. For instance, how do cells sense light intensity and direction and how does phototactic behaviour of individual cells and those that move as coordinated groups differ? What is the role of novel proteins and small molecules such as cAMP and cyclic-di-GMP in controlling motility? How are signals from photosynthesis and other metabolic pathways involved in controlling the motile behaviour of cyanobacteria?

For swimming and gliding cyanobacteria, the motility machinery has yet to be uncovered. What is the motility apparatus for swimming? How do the surface proteins SwmA and SwmB function in swimming? Do they constitute part of the motility apparatus or do they play a secondary structural role? How is the sodium-motive force transduced to the motor? Can we detect surface waves? In gliding cyanobacteria, what are the components of the junctional pore complex? Is the complex required for gliding? Is secretion of mucilage required for gliding? How is the proton-motive force utilized by the motility apparatus in gliding? What is the role of TFP in hormogonia gliding?

Ultimately, the goal is to understand the ecological implications of motility. What is the importance and role of motility in the natural environment? Does phototactic behaviour impart an advantage to motile cells when they are in light that is potentially damaging e.g. UV or high light? Unlike other motile unicellular cyanobacteria such as *Synechocystis*, marine swimming *Synechococcus* are not phototactic (Willey and Waterbury, 1989). Rather, they are chemotactic towards nitrogenous compounds at environmentally relevant concentrations. Does this provide them with a competitive advantage in the oligotrophic environments in which they are found where N is a limiting nutrient (Willey and Waterbury, 1989)? What receptors and signal transduction pathways might they employ? What signal transduction pathways do chemotactic filamentous cyanobacteria and hormogonia employ? In this review, we have attempted to highlight important advances and raise other unanswered questions. Given the advances in the genetic tractability of a variety of cyanobacteria, we are confident that rapid progress will be made in answering at least some of these questions.

Acknowledgements

We would like to dedicate this chapter to our colleagues, Richard Castenholz and John Waterbury for their original contributions to the study of motile behaviour in cyanobacteria and their continued enthusiasm and support.

BB would like to thank Jack Meeks for many helpful comments and for sharing data prior to publication; Egbert Hoiczyk for kindly providing originals of figures and Anthony Daulo for figure preparation. DB would like to thank Alexandra Grote, Anchal Chandra and Rosanna Chau for their careful reading of the chapter and preparation of the figures and table. DB acknowledges the NSF and the Carnegie Institution of Science for research support.

References

Aas, F.E., Egge-Jacobsen, W., Winther-Larsen, H.C., Lovold, C., Hitchen, P.G., Dell, A., and Koomey, M. (2006). *Neisseria gonorrhoeae* type IV pili undergo multisite, hierarchical modifications with phosphoethanolamine and phosphocholine requiring an enzyme structurally related to lipopolysaccharide phosphoethanolamine transferases. J. Biol. Chem. *281*, 27712–27723.

Adams, D.G. (2001). How do cyanobacteria glide? Microbiol. Today *28*, 131–133.

Adams, D.G., Ashworth, D., and Nelmes, B. (1999). Fibrillar array in the cell wall of a gliding filamentous cyanobacterium. J. Bacteriol. *181*, 884–892.

Anders, K., von Stetten, D., Mailliet, J., Kiontke, S., Sineshchekov, V.A., Hildebrandt, P., Hughes, J., and Essen, L.O. (2011). Spectroscopic and photochemical characterization of the red-light sensitive photosensory module of Cph2 from *Synechocystis* PCC 6803. Photochem. Photobiol. *87*, 160–173.

Ashby, M.K., and Houmard, J. (2006). Cyanobacterial two-component proteins: structure, diversity, distribution, and evolution. Microbiol. Mol. Biol. Rev. *70*, 472–509.

Auldridge, M.E., and Forest, K.T. (2011). Bacterial phytochromes: more than meets the light. Crit. Rev. Biochem. Mol. Biol. *46*, 67–88.

Bateman, A., Murzin, A.G., and Teichmann, S.A. (1998). Structure and distribution of pentapeptide repeats in bacteria. Protein Sci. *7*, 1477–1480.

Beatson, S.A., Whitchurch, C.B., Sargent, J.L., Levesque, R.C., and Mattick, J.S. (2002). Differential regulation of twitching motility and elastase production by Vfr in *Pseudomonas aeruginosa*. J. Bacteriol. *184*, 3605–3613.

Bhaya, D. (2004). Light matters: phototaxis and signal transduction in unicellular cyanobacteria. Mol. Microbiol. *53*, 745–754.

Bhaya, D., Watanabe, N., Ogawa, T., and Grossman, A.R. (1999). The role of an alternative sigma factor in motility and pilus formation in the cyanobacterium *Synechocystis* sp. strain PCC6803. Proc. Natl. Acad. Sci. U.S.A. *96*, 3188–3193.

Bhaya, D., Bianco, N.R., Bryant, D., and Grossman, A. (2000). Type IV pilus biogenesis and motility in the cyanobacterium *Synechocystis* sp. PCC6803. Mol. Microbiol. *37*, 941–951.

Bhaya, D., Takahashi, A., and Grossman, A.R. (2001a). Light regulation of type IV pilus-dependent motility by chemosensor-like elements in *Synechocystis* PCC6803. Proc. Natl. Acad. Sci. U.S.A. *98*, 7540–7545.

Bhaya, D., Takahashi, A., Shahi, P., and Grossman, A.R. (2001b). Novel motility mutants of *Synechocystis* strain PCC 6803 generated by in vitro transposon mutagenesis. J. Bacteriol. *183*, 6140–6143.

Bhaya, D., Nakasugi, K., Fazeli, F., and Burriesci, M.S. (2006). Phototaxis and impaired motility in adenylyl cyclase and cyclase receptor protein mutants of *Synechocystis* sp. strain PCC 6803. J. Bacteriol. *188*, 7306–7310.

Bhaya, D., Grossman, A.R., Steunou, A.S., Khuri, N., Cohan, F.M., Hamamura, N., Melendrez, M.C., Bateson, M.M., Ward, D.M., and Heidelberg, J.F. (2007). Population level functional diversity in a microbial community revealed by comparative genomic and metagenomic analyses. ISME J. *1*(8), 703–713.

Boone, D.R., and Castenholz, R.W. (2001). Phylum BX Cyanobacteria. In Bergey's Manual of Systematic Bacteriology volume 1. Staley, J.T., ed. (Springer, New York), pp. 473–600.

Brahamsha, B. (1996). An abundant cell-surface polypeptide is required for swimming by the nonflagellated marine cyanobacterium *Synechococcus*. Proc. Natl. Acad. Sci. U.S.A. *93*, 6504–6509.

Brahamsha, B. (1999). Non-flagellar swimming in marine *Synechococcus*. J. Mol. Microbiol. Biotechnol. *1*, 59–62.

Burriesci, M., and Bhaya, D. (2008). Tracking phototactic responses and modeling motility of *Synechocystis* sp. strain PCC6803. J. Photochem. Photobiol. B *91*, 77–86.

Burrows L.L. (2012). *Pseudomonas aeruginosa* twitching motility:type IV pili in action. Annu. Rev. Microbiol. *66*, 493–520.

Campbell, E.L., Summers, M.L., Christman, H., Martin, M.E., and Meeks, J.C. (2007). Global gene expression patterns of *Nostoc punctiforme* in steady-state dinitrogen-grown heterocyst-containing cultures and at single time points during the differentiation of akinetes and hormogonia. J. Bacteriol. *189*, 5247–5256.

Cao, Z., Livoti, E., Losi, A., and Gartner, W. (2010). A blue light-inducible phosphodiesterase activity in the cyanobacterium *Synechococcus elongatus*. Photochem. Photobiol. *86*, 606–611.

Castenholz, R.W. (1969). Thermophilic blue-green algae and the thermal environment. Bacteriol. Rev. *33*, 476–504.

Castenholz, R.W. (1982). Motility and Taxes. In The Biology of Cyanobacteria, Carr, N.G., and Whitton, B.A., eds. (University of California Press, Berkeley), pp. 413–440.

Castenholz, R.W. (1989a). Order Nostocales. In Bergey's Manual of Systematic Bacteriology, Staley, J.T., ed. (Williams and Wilkins, Baltimore), pp. 1780–1782.

Castenholz, R.W. (1989b). Order Oscillatoriales. In Bergey's Manual of Systematic Bacteriology, Staley, J.T., ed. (Williams and Wilkins, Baltimore), pp. 1771–1780.

Chamot-Rooke, J., Mikaty, G., Malosse, C., Soyer, M., Dumont, A., Gault, J., Imhaus, A.F., Martin, P., Trellet, M., Clary, G., *et al.* (2011). Posttranslational modification of pili upon cell contact triggers *N. meningitidis* dissemination. Science *331*, 778–782.

Choi, J.S., Chung, Y.H., Moon, Y.J., Kim, C., Watanabe, M., Song, P.S., Joe, C.O., Bogorad, L., and Park, Y.M. (1999). Photomovement of the gliding cyanobacterium *Synechocystis* sp. PCC 6803. Photochem. Photobiol. *70*, 95–102.

Christen, B., Christen, M., Paul, R., Schmid, F., Folcher, M., Jenoe, P., Meuwly, M., and Jenal, U. (2006). Allosteric control of cyclic di-GMP signaling. J. Biol. Chem. *281*, 32015–32024.

Chung, Y.H., Cho, M.S., Moon, Y.J., Choi, J.S., Yoo, Y.C., Park, Y.I., Lee, K.M., Kang, K.W., and Park, Y.M. (2001). *ctr1*, a gene involved in a signal transduction pathway of the gliding motility in the cyanobacterium *Synechocystis* sp. PCC 6803. FEBS Lett. *492*, 33–38.

Cohen, M.F., Wallis, J.G., Campbell, E.L., and Meeks, J.C. (1994). Transposon mutagenesis of *Nostoc* sp. strain ATCC 29133, a filamentous cyanobacterium with multiple cellualr differentiation alternatives. Microbiology *140*, 3233–3240.

Craig, L., and Li, J. (2008). Type IV pili: paradoxes in form and function. Curr. Opin. Struct. Biol. *18*, 267–277.

Crosson, S., Rajagopal, S., and Moffat, K. (2003). The LOV domain family: photoresponsive signaling modules coupled to diverse output domains. Biochemistry *42*, 2–10.

Damerval, T., Guglielmi, G., Houmard, J., and De Marsac, N.T. (1991). Hormogonium differentiation in the cyanobacterium *Calothrix* – a photoregulated developmental process. Plant Cell 3, 191–201.

Dandrea, L.D., and Regan, L. (2003). TPR proteins: the versatile helix. Trends Biochem. Sci. *28*, 655–662.

Dick, H., and Stewart, W.D.P. (1980). The occurrence of fimbriae on a nitrogen-fixing cyanobacterium which occurs in lichen symbiosis. Arch. Microbiol. *109*, 107–109.

Diehn, B. (1976). Photomovement of microorganisms. Photochem. Photobiol. 23, 455–456.

Drews, G. (2005). Contributions of Theodor Wilhelm Engelmann on phototaxis, chemotaxis, and photosynthesis. Photosynth. Res. *83*, 25–34.

Duggan, P.S., Gottardello, P., and Adams, D.G. (2007). Molecular analysis of genes in *Nostoc punctiforme* involved in pilus biogenesis and plant infection. J. Bacteriol. *189*, 4547–4551.

Dutta, R., Qin, L., and Inouye, M. (1999). Histidine kinases: diversity of domain organization. Mol. Microbiol. *34*, 633–640.

Ehlers, K., and Oster, G. (2012). On the mysterious propulsion of *Synechococcus*. PLoS One *7*, e36081.

Ehlers, K.M., Samuel, A.D., Berg, H.C., and Montgomery, R. (1996). Do cyanobacteria swim using traveling surface waves? Proc. Natl. Acad. Sci. U.S.A. *93*, 8340–8343.

Fiedler, B., Borner, T., and Wilde, A. (2005). Phototaxis in the cyanobacterium *Synechocystis* sp. PCC 6803: role of different photoreceptors. Photochem. Photobiol. *81*, 1481–1488.

Fukushima, Y., Iwaki, M., Narikawa, R., Ikeuchi, M., Tomita, Y., and Itoh, S. (2011). Photoconversion mechanism of a green/red photosensory cyanobacteriochrome AnPixJ: time-resolved optical spectroscopy and FTIR analysis of the AnPixJ-GAF2 domain. Biochemistry *50*, 6328–6339.

Fusseneger, M., Rudel, T., Barten, R., Ryll, R., and Meyer, T.F. (1997). Transformation competence and type-4 pilus biogenesis in *Neisseria gonorrhoeae* – a review. Gene *192*, 125–134.

Galante, A., Wisen, S., Bhaya, D., and Levy, D. (2012). Modeling local interactions during the motion of cyanobacteria. J. Theor. Biol. *309*, 147–158.

Gestwicki, J.E., Lamanna, A.C., Harshey, R.M., McCarter, L.L., Kiessling, L.L., and Adler, J. (2000). Evolutionary conservation of methyl-accepting chemotaxis protein location in Bacteria and Archaea. J. Bacteriol. *182*, 6499–6502.

Glagoleva, T.N., Glagolev, A.N., Gusev, M.V., and Nikitina, K.A. (1980). Protonmotive force supports gliding in cyanobacteria. FEBS Lett. *117*, 49–53.

Gomelsky, M., and Hoff, W.D. (2011). Light helps bacteria make important lifestyle decisions. Trends Microbiol. *19*, 441–448.

Gomelsky, M., and Klug, G. (2002). BLUF: a novel FAD-binding domain involved in sensory transduction in microorganisms. Trends Biochem. Sci. *27*, 497–500.

Hader, D. (1987). Photomovement. In The Cyanobacteria, Fay, P., and Van Baalen, C., eds. (Elsevier, Amsterdam), pp. 325–345.

Hader, D.P. (1988). Signal perception and amplification in photoresponses of cyanobacteria. Biophys. Chem. *29*, 155–159.

Hader, D.P., and Burkart, U. (1982). Enhanced models for photophobic responses of the blue-green alga, *Phormidium uncinatum*. Plant Cell Physiol. *23*, 1391–1400.

Hader, D.P., and Lebert, M. (2009). Photoorientation in photosynthetic flagellates. Methods Mol. Biol. *571*, 51–65.

Halfen, L.N., and Castenholz, R.W. (1970). Gliding in a blue-green alga: a possible mechanism. Nature *225*, 1163–1165.

Halfen, L.N., and Castenholz, R.W. (1971). Gliding motility in the blue-green alga *Oscillatoria princeps*. J. Phycol. *7*, 133–145.

Hartzell, P.L., and Youderian, P. (1995). Genetics of gliding motility and development in *Myxococcus xanthus*. Arch. Microbiol. *164*, 309–323.

Hazelbauer, G.L., Falke, J.J., and Parkinson, J.S. (2008). Bacterial chemoreceptors: high-performance signaling in networked arrays. Trends Biochem. Sci. *33*, 9–19.

Hengge, R. (2009). Principles of c-di-GMP signalling in bacteria. Nat. Rev. Microbiol. *7*, 263–273.

Herdman, M., and Rippka, R. (1988). Cellular differentiation: hormogonia and baeocytes. Methods Enzymol. *167*, 232–242.

Hernandez-Muniz, W., and Stevens, S.E. (1987). Characterization of the motile hormogonia of *Mastigocladus laminosus*. J. Bacteriol. *169*, 218–223.

Herrou, J., and Crosson, S. (2011). Function, structure and mechanism of bacterial photosensory LOV proteins. Nat. Rev. Micro. *9*, 713–723.

Hoff, W.D., van der Horst, M.A., Nudel, C.B., and Hellingwerf, K.J. (2009). Prokaryotic phototaxis. Methods Mol. Biol. *571*, 25–49.

Hoiczyk, E. (1998). Structural and biochemical analysis of the sheath of *Phormidium uncinatum*. J. Bacteriol. *180*, 3923–3932.

Hoiczyk, E. (2000). Gliding motility in cyanobacteria: observations and possible explanations. Arch. Microbiol. *174*, 11–17.

Hoiczyk, E., and Baumeister, W. (1995). Envelope structure of four gliding filamentous cyanobacteria. J. Bacteriol. *177*, 2387–2395.

Hoiczyk, E., and Baumeister, W. (1997). Oscillin, an extracellular, Ca2+-binding glycoprotein essential for the gliding motility of cyanobacteria. Mol. Microbiol. *26*, 699–708.

Hoiczyk, E., and Baumeister, W. (1998). The junctional pore complex, a prokaryotic secretion organelle, is the molecular motor underlying gliding motility in cyanobacteria. Curr. Biol. *8*, 1161–1168.

Ikeuchi, M., and Ishizuka, T. (2008). Cyanobacteriochromes: a new superfamily of tetrapyrrole-binding photoreceptors in cyanobacteria. Photochem. Photobiol. Sci. *7*, 1159–1167.

Ishizuka, T., Narikawa, R., Kohchi, T., Katayama, M., and Ikeuchi, M. (2007). Cyanobacteriochrome TePixJ of *Thermosynechococcus elongatus* harbors phycoviolobilin as a chromophore. Plant Cell Physiol. *48*, 1385–1390.

Ivleva, N.B., Gao, T., LiWang, A.C., and Golden, S.S. (2006). Quinone sensing by the circadian input kinase of the cyanobacterial circadian clock. Proc. Natl. Acad. Sci. U.S.A. *103*, 17468–17473.

Jarrell, K.F., and McBride, M.J. (2008). The surprisingly diverse ways that prokaryotes move. Nat. Rev. Microbiol. *6*, 466–476.

Jekely, G. (2009). Evolution of phototaxis. Philos. Trans. R. Soc. Lond. B Biol. Sci. *364*, 2795–2808.

Jung, K., Fried, L., Behr, S., and Heermann, R. (2012). Histidine kinases and response regulators in networks. Curr. Opin. Microbiol. *15*, 118–124.

Kalia, D., Merey, G., Nakayama, S., Zheng, Y., Zhou, J., Luo, Y., Guo, M., Roembke, B.T., and Sintim, H.O. (2012). Nucleotide, c-di-GMP, c-di-AMP, cGMP, cAMP, (p)ppGpp signaling in bacteria and implications in pathogenesis. Chem. Soc. Rev. *42*, 305–341.

Kamei, A., Yuasa, T., Orikawa, K., Geng, X., and Ikeuchi, M. (2001). A eukaryotic-type protein kinase, SpkA, is required for normal motility of the unicellular cyanobacterium *Synechocystis* sp. strain PCC 6803. J. Bacteriol. *183*, 1505–1510.

Kamei, A., Yoshihara, S., Yuasa, T., Geng, X., and Ikeuchi, M. (2003). Biochemical and functional characterization of a eukaryotic-type protein kinase, SpkB, in the cyanobacterium, *Synechocystis* sp. PCC 6803. Curr. Microbiol. *46*, 296–301.

Kaneko, T., Sato, S., Kotani, H., Tanaka, A., Asamizu, E., Nakamura, T., Miyajima, N., Hirosawa, M., Sugiura, M., Sasamoto, S., *et al.* (1996). Sequence analysis of the genome of the unicellular cyanobacterium *Synechocystis* sp. strain PCC 6803. II. Sequence determination of the entire genome and assignment of potential protein-coding regions. DNA Res. *3*, 109–136.

Kanesaki, Y., Shiwa, Y., Tajima, N., Suzuki, M., Watanabe, S., Sato, N., Ikeuchi, M., and Yoshikawa, H. (2012). Identification of substrain-specific mutations by massively parallel whole-genome resequencing of *Synechocystis* sp. PCC 6803. DNA Res. *19*, 67–79.

Kehoe, D.M. (2010). Chromatic adaptation and the evolution of light color sensing in cyanobacteria. Proc. Natl. Acad. Sci. U.S.A. *107*, 9029–9030.

Kehoe, D.M., and Grossman, A.R. (1996). Similarity of a chromatic adaptation sensor to phytochrome and ethylene receptors. Science *273*, 1409–1412.

Kim, Y.H., Park, Y.M., Kim, S.J., Park, Y.I., Choi, J.S., and Chung, Y.H. (2004). The role of Slr1443 in pilus biogenesis in *Synechocystis* sp. PCC 6803: involvement in post-translational modification of pilins. Biochem. Biophys. Res. Commun. *315*, 179–186.

Kim, Y.H., Kim, J.Y., Kim, S.-Y., Lee, J.H., Lee, J.S., Chung, Y.-H., Yoo, J.S., and Park, Y.M. (2009). Alteration in the glycan pattern of pilin in a nonmotile mutant of *Synechocystis* sp. PCC 6803. Proteomics *9*, 1075–1086.

Kim, Y.H., Park, K.H., Kim, S.-Y., Ji, E.S., Kim, J.Y., Lee, S.K., Yoo, J.S., Kim, H.S., and Park, Y.M. (2011). Identification of trimethylation at C-terminal lysine of pilin in the cyanobacterium *Synechocystis* PCC 6803. Biochem. Biophys. Res. Commun. *404*, 587–592.

Kita, A., Okajima, K., Morimoto, Y., Ikeuchi, M., and Miki, K. (2005). Structure of a cyanobacterial BLUF protein, Tll0078, containing a novel FAD-binding blue light sensor domain. J. Mol. Biol. *349*, 1–9.

Knight, C.D., and Adams, D.G. (1996). A method for studying chemotaxis in nitrogen fixing cyanobacterium-plant symbioses. Physiol. Mol. Plant Path. *49*, 73–77.

Kondou, Y., Nakazawa, M., Higashi, S., Watanabe, M., and Manabe, K. (2001). Equal-quantum action spectra indicate fluence-rate-selective action of multiple photoreceptors for photomovement of the thermophilic cyanobacterium *Synechococcus elongatus*. Photochem. Photobiol. *73*, 90–95.

Kondou, Y., Mogami, N., Hoshi, F., Kutsuna, S., Nakazawa, M., Sakurai, T., Matsui, M., Kaneko, T., Tabata, S., Tanaka, I., *et al.* (2002). Bipolar localization of putative photoreceptor protein for phototaxis in thermophilic cyanobacterium *Synechococcus elongatus*. Plant Cell Physiol. *43*, 1585–1588.

Lamb, J.R., Tugendreich, S., and Hieter, P. (1995). Tetratrico peptide repeat interactions: tp TPR or not to TPR? Trends Biochem. Sci. *20*, 257–259.

Lamont, H.C. (1969). Shear-oriented microfibrils in mucilaginous investments of 2 motile oscillatoriacean blue-green algae. J. Bacteriol. *97*, 350–358.

Liu, J., and Wolk, C.P. (2011). Mutations in genes *patA* and *patL* of *Anabaena* sp. strain PCC 7120 result in similar phenotypes, and the proteins encoded by those genes may interact. J. Bacteriol. *193*, 6070–6074.

Long, C.D., Tobiason, D.M., Lazio, M.P., Kline, K.A., and Seifert, H.S. (2003). Low-level pilin expression allows for substantial DNA transformation competence in *Neisseria gonorrhoeae*. Infect. Immun. *71*, 6279–6291.

Losi, A., and Gartner, W. (2012). The evolution of flavin-binding photoreceptors: an ancient chromophore serving trendy blue-light sensors. Annu. Rev. Plant Biol. *63*, 49–72.

Maddock, J.R., and Shapiro, L. (1993). Polar location of the chemoreceptor complex in the *Escherichia coli* cell. Science *259*, 1717–1723.

Maier, B., Potter, L., So, M., Long, C.D., Seifert, H.S., and Sheetz, M.P. (2002). Single pilus motor forces exceed 100 pN. Proc. Natl. Acad. Sci. U.S.A. *99*, 16012–16017.

Marceau, M., Forest, K., Beretti, J.L., Tainer, J., and Nassif, X. (1998). Consequences of the loss of O-linked glycosylation of meningococcal type IV pilin on piliation and pilus-mediated adhesion. Mol. Microbiol. *27*, 705–715.

Masuda, S., and Ono, T.A. (2004). Biochemical characterization of the major adenylyl cyclase, Cya1, in the cyanobacterium *Synechocystis* sp. PCC 6803. FEBS Lett. *577*, 255–258.

Masuda, S., Hasegawa, K., Ishii, A., and Ono, T.A. (2004). Light-induced structural changes in a putative blue-light receptor with a novel FAD binding fold sensor of blue-light using FAD (BLUF); Slr1694 of *Synechocystis* sp. PCC6803. Biochemistry *43*, 5304–5313.

Mattick, J.S. (2002). Type IV pili and twitching motility. Annu. Rev. Microbiol. *56*, 289–314.

Mauriello, E.M., Mignot, T., Yang, Z., and Zusman, D.R. (2010). Gliding motility revisited: how do the myxobacteria move without flagella? Microbiol. Mol. Biol. Rev. *74*, 229–249.

McBride, M.J. (2001). Bacterial gliding motility: multiple mechanisms for cell movement over surfaces. Annu. Rev. Microbiol. *55*, 49–75.

McCarren, J., and Brahamsha, B. (2005). Transposon mutagenesis in a marine *Synechococcus* strain: isolation of swimming motility mutants. J. Bacteriol. *187*, 4457–4462.

McCarren, J., and Brahamsha, B. (2007). SwmB, a 1.12-megadalton protein that is required for nonflagellar swimming motility in *Synechococcus*. J. Bacteriol. *189*, 1158–1162.

McCarren, J., and Brahamsha, B. (2009). Swimming motility mutants of marine *Synechococcus* affected in production and localization of the S-layer protein SwmA. J. Bacteriol. *191*, 1111–1114.

McCarren, J., Heuser, J., Roth, R., Yamada, N., Martone, M., and Brahamsha, B. (2005). Inactivation of *swmA* results in the loss of an outer cell layer in a swimming *Synechococcus* strain. J. Bacteriol. *187*, 224–230.

McDonough, K.A., and Rodriguez, A. (2012). The myriad roles of cyclic AMP in microbial pathogens: from signal to sword. Nat. Rev. Microbiol. *10*, 27–38.

Meeks, J.C., Campbell, E.L., Summers, M.L., and Wong, F.C. (2002). Cellular differentiation in the cyanobacterium *Nostoc punctiforme*. Arch. Microbiol. *178*, 395–403.

Merz, A.J., and Forest, K.T. (2002). Bacterial surface motility: slime trails, grappling hooks and nozzles. Curr. Biol. *12*, R297–303.

Merz, A.J., So, M., and Sheetz, M.P. (2000). Pilus retraction powers bacterial twitching motility. Nature *407*, 98–102.

Mizuno, T., and Matsubara, M. (2003). Role of the histidine-containing phosphotransfer domain (HPt) in the multistep phosphorelay through the anaerobic hybrid sensor, ArcB. In Histidine Kinases in Signal Transduction, Inouye, M., and Dutta, R., eds. (Academic Press, San Diego), pp. 166–191.

Montgomery, B.L., and Lagarias, J.C. (2002). Phytochrome ancestry: sensors of bilins and light. Trends Plant Sci. *7*, 357–366.

Moon, Y.J., Kim, S.Y., Jung, K.H., Choi, J.S., Park, Y.M., and Chung, Y.H. (2011). Cyanobacterial phytochrome Cph2 is a negative regulator in phototaxis toward UV-A. FEBS Lett. *585*, 335–340.

Mullineaux, C.W. (2001). How do cyanobacteria sense and respond to light? Mol. Microbiol. *41*, 965–971.

Nakasugi, K., Svenson, C.J., and Neilan, B.A. (2006). The competence gene, *comF*, from *Synechocystis* sp. strain PCC 6803 is involved in natural transformation, phototactic motility and piliation. Microbiology *152*, 3623–3631.

Nan, B., and Zusman, D.R. (2011). Uncovering the mystery of gliding motility in the myxobacteria. Annu. Rev. Genet. *45*, 21–39.

Nan, B.Y., Chen, J., Neu, J.C., Berry, R.M., Oster, G., and Zusman, D.R. (2011). Myxobacteria gliding motility requires cytoskeleton rotation powered by proton motive force. Proc. Natl. Acad. Sci. U.S.A. *108*, 2498–2503.

Narikawa, R., Fukushima, Y., Ishizuka, T., Itoh, S., and Ikeuchi, M. (2008). A novel photoactive GAF domain of cyanobacteriochrome AnPixJ that shows reversible green/red photoconversion. J. Mol. Biol. *380*, 844–855.

Narikawa, R., Suzuki, F., Yoshihara, S., Higashi, S., Watanabe, M., and Ikeuchi, M. (2011). Novel photosensory two-component system (PixA-NixB-NixC) involved in the regulation of positive and negative phototaxis of cyanobacterium *Synechocystis* sp. PCC 6803. Plant Cell Physiol. *52*, 2214–2224.

Ng, W.O., Grossman, A.R., and Bhaya, D. (2003). Multiple light inputs control phototaxis in *Synechocystis* sp. strain PCC6803. J. Bacteriol. *185*, 1599–1607.

Nilsson, M., Rasmussen, U., and Bergman, B. (2006). Cyanobacterial chemotaxis to extracts of host and nonhost plants. FEMS Microbiol. Ecol. *55*, 382–390.

Ntefidou, M., Iseki, M., Watanabe, M., Lebert, M., and Hader, D.P. (2003). Photoactivated Adenylyl Cyclase Controls Phototaxis in the Flagellate *Euglena gracilis*. Plant Physiol. *133*, 1517–1521.

Nultsch, W. (1975). Phototaxis and photokinesis. In Primitive Sensory and communication systems, Carlile, M.J., ed. (Academic Press, London), pp. 29–90.

Nultsch, W., and Hader, D. (1979). Photomovement of motile microorganisms. Photochem. Photobiol. *29*, 423–437.

Nultsch, W., and Hader, D.P. (1988). Photomovement in motile microorganisms-II. Photochem. Photobiol. *47*, 837–869.

Nultsch, W., Schuchart, H., and Hohl, M. (1979). Investigations on the phototactic orientation of *Anabaena variabilis*. Arch. Microbiol. *122*, 85–91.

Ohmori, K., Hirose, M., and Ohmori, M. (1993). An increase in the intracellular concentration of cAMP triggers formation of an algal mat by the cyanobacterium *Spirulina platensis*. Plant Cell Environ. *34*, 169–171.

Ohmori, M., and Okamoto, S. (2004). Photoresponsive cAMP signal transduction in cyanobacteria. Photochem. Photobiol. Sci. *3*, 503–511.

Okajima, K., Yoshihara, S., Fukushima, Y., Geng, X., Katayama, M., Higashi, S., Watanabe, M., Sato, S., Tabata, S., Shibata, Y., *et al.* (2005). Biochemical and functional characterization of BLUF-type flavin-binding proteins of two species of cyanobacteria. J. Biochem. (Tokyo) *137*, 741–750.

Okamoto, S., and Ohmori, M. (2002). The cyanobacterial PilT protein responsible for cell motility and transformation hydrolyzes ATP. Plant Cell Physiol. *43*, 1127–1136.

Palenik, B., Brahamsha, B., Larimer, F.W., Land, M., Hauser, L., Chain, P., Lamerdin, J., Regala, W., Allen, E.E., McCarren, J., *et al.* (2003). The genome of a motile marine *Synechococcus*. Nature *424*, 1037–1042.

Pankratz, H.S., and Bowen, C.C. (1963). Cytology of blue-green algae. 1. Cells of *Symploca muscorum*. Am. J. Bot. *50*, 387–395.

Park, C.M., Kim, J.I., Yang, S.S., Kang, J.H., Shim, J.Y., Chung, Y.H., Park, Y.M., and Song, P.S. (2000). A second photochromic bacteriophytochrome from *Synechocystis* sp PCC 6803: Spectral analysis and down-regulation by light. Biochemistry 37, 10840–10847.

Pelicic, V. (2008). Type IV pili: e pluribus unum? Mol. Microbiol. *68*, 827–837.

Pepper, A.E. (1998). Molecular evolution: old branches on the phytochrome family tree. Curr. Biol. *8*, R117–120.

Pereira, S., Zille, A., Micheletti, E., Moradas-Ferreira, P., De Philippis, R., and Tamagnini, P. (2009). Complexity of cyanobacterial exopolysaccharides: composition, structures, inducing factors and putative genes involved in their biosynthesis and assembly. FEMS Microbiol. Rev. 33, 917–941.

Pitta, T.P., and Berg, H.C. (1995). Self-electrophoresis is not the mechanism for motility in swimming cyanobacteria. J. Bacteriol. *177*, 5701–5703.

Pitta, T.P., Sherwood, E.E., Kobel, A.M., and Berg, H.C. (1997). Calcium is required for swimming by the nonflagellated cyanobacterium *Synechococcus* strain WH8113. J. Bacteriol. *179*, 2524–2528.

Proft, T., and Baker, E.N. (2009). Pili in Gram-negative and Gram-positive bacteria – structure, assembly and their role in disease. Cell Mol. Life Sci. *66*, 613–635.

Ramsing, N.B., Ferris, M.J., and Ward, D.M. (1997). Light-induced motility of thermophilic *Synechococcus* isolates from Octopus Spring, Yellowstone National Park. App. Environ. Microbiol. *63*, 2347–2354.

Read, N., Connell, S., and Adams, D.G. (2007). Nanoscale visualization of a fibrillar array in the cell wall of filamentous cyanobacteria and its implications for gliding motility. J. Bacteriol. *189*, 7361–7366.

Reva, O., and Tummler, B. (2008). Think big–giant genes in bacteria. Environ. Microbiol. *10*, 768–777.

Rippka, R., Deruelles, J., Waterbury, J.B., Herdman, M., and Stanier, R.Y. (1979). Generic assignments, strain histories, and properties of pure cultures of cyanobacteria. J. Gen. Microbiol. *111*, 1–61.

Risser, D.D., and Meeks, J.C. (2013). Comparative transcriptomics with a motility deficient mutant leads to identification of a novel polysaccharide secretion system in *Nostoc punctiforme*. Mol. Microbiol. *87*, 884–893.

Robinson, W.B., Mealor, A.E., Stevens, S.E., and Ospeck, M. (2007). Measuring the force production of the hormogonia of *Mastigocladus laminosus*. Biophys. J. *93*, 699–703.

Rockwell, N.C., and Lagarias, J.C. (2010). A brief history of phytochromes. Chemphyschem *11*, 1172–1180.

Rockwell, N.C., Njuguna, S.L., Roberts, L., Castillo, E., Parson, V.L., Dwojak, S., Lagarias, J.C., and Spiller, S.C. (2008). A second conserved GAF domain cysteine is required for the blue/green photoreversibility of cyanobacteriochrome Tlr0924 from *Thermosynechococcus elongatus*. Biochemistry *47*, 7304–7316.

Samuel, A.D.T., Peterson, J., and Reese, T.S. (2001). Envelope structure of *Synechococcus* sp. WH8113, a nonflagellated swimming cyanobacterium. BMC Microbiol. *1*, 4.

Savakis, P., De Causmaecker, S., Angerer, V., Ruppert, U., Anders, K., Essen, L.O., and Wilde, A. (2012). Light-induced alteration of c-di-GMP level controls motility of *Synechocystis* sp. PCC 6803. Mol. Microbiol. *85*, 239–251.

Schmitz, O., Katayama, M., Williams, S.B., Kondo, T., and Golden, S.S. (2000). CikA, a bacteriophytochrome that resets the cyanobacterial circadian clock. Science *289*, 765–768.

Siddiqui, A.M., Burchard, R.P., and Schwarz, W.H. (2001). An undulating surface model for the motility of bacteria gliding on a layer of non-Newtonian slime. Non-Linear Mechan. *36*, 743–761.

Sogaard-Andersen, L. (2004). Cell polarity, intercellular signalling and morphogenetic cell movements in *Myxococcus xanthus*. Curr. Opin. Microbiol. *7*, 587–593.

Song, J.Y., Cho, H.S., Cho, J.I., Jeon, J.S., Lagarias, J.C., and Park, Y.I. (2011). Near-UV cyanobacteriochrome signaling system elicits negative phototaxis in the cyanobacterium *Synechocystis* sp. PCC 6803. Proc. Natl. Acad. Sci. U.S.A. *108*, 10780–10785.

Stanier, R.Y., Kunisawa, R., Mandel, M., and Cohen-Bazire, G. (1971). Purification and properties of unicellular blue-green algae (order *Chroococcales*). Bacteriol. Rev. *35*, 171–205.

Stock, J. (1999). Signal transduction: gyrating protein kinases. Curr. Biol. *9*, r364-r367.

Sun, S., Chen, J., Li, W., Altintas, I., Lin, A., Peltier, S., Stocks, K., Allen, E.E., Ellisman, M., Grethe, J., *et al.* (2011). Community cyberinfrastructure for Advanced Microbial Ecology Research and Analysis: the CAMERA resource. Nucleic Acids Res. *39*, D546–551.

Tamulonis, C., Postma, M., and Kaandorp, J. (2011). Modeling filamentous cyanobacteria reveals the advantages of long and fast trichomes for optimizing light exposure. PLoS ONE *6*, e22084.

Tandeau de Marsac, N. (1994). Differentiation of hormogonia and relationships with other biological processes. In The Molecular Biology of Cyanobacteria, Bryant, D.A., ed. (Kluwer Academic Publishers, Boston), pp. 825–842.

Terauchi, K., and Ohmori, M. (1999). An adenylate cyclase, Cya1, regulates cell motility in the cyanobacterium *Synechocystis* sp. PCC 6803. Plant Cell Physiol. *40*, 248–251.

Tonjum, T., Freitag, N.E., Namork, E., and Koomey, M. (1995). Identification and characterization of *pilG*, a highly conserved pilus-assembly gene in pathogenic *Neisseria*. Mol. Microbiol. *16*, 451–464.

Trautmann, D., Voss, B., Wilde, A., Al-Babili, S., and Hess, W.R. (2012). Microevolution in cyanobacteria: Re-sequencing a motile substrain of *Synechocystis* sp. PCC 6803. DNA Res. 19:435–48.

Ulijasz, A.T., Cornilescu, G., von Stetten, D., Cornilescu, C., Velazquez Escobar, F., Zhang, J., Stankey, R.J., Rivera, M., Hildebrandt, P., and Vierstra, R.D. (2009). Cyanochromes are blue/green light photoreversible photoreceptors defined by a stable double cysteine linkage to a phycoviolobilin-type chromophore. J. Biol. Chem. *284*, 29757–29772.

Ursell, T., Chau, R.M.W., Wisen, S., Bhaya, D., and Huang, K.C. (2013). Motility enhancement through surface modification is sufficient for cyanobacterial community organization during phototaxis. PLoS Comput. Biol. *9*, e1003205.

Vaara, T. (1982). The outermost surface structures in chroococcacean cyanobacteria. Can. J. Microbiol. *28*, 929–941.

Vreede, J., Van Der Horst, M.A., Hellingwerf, K.J., Crielaard, W., and Van Aalten, D.M. (2003). PAS domains – common structure, common flexibility? J. Biol. Chem. *278*, 18434–18439.

Walsby, A.E. (1968). Mucilage secretion and movements of blue-green algae. Protoplasma *65*, 223–230.

Waterbury, J.B., Willey, J.M., Franks, D.G., Valois, F.W., and Watson, S.W. (1985). A cyanobacterium capable of swimming motility. Science *230*, 74–75.

Webre, D.J., Wolanin, P.M., and Stock, J.B. (2003). Bacterial chemotaxis. Curr. Biol. *13*, R47–49.

Wilde, A., Fiedler, B., and Borner, T. (2002). The cyanobacterial phytochrome Cph2 inhibits phototaxis towards blue light. Mol. Microbiol. *44*, 981–988.

Willey, J.M. (1988). Characterization of swimming motility in a marine cyanobacterium. Woods Hole Oceanographic Institution and Massachussetts Institute of Technology, Cambridge, MA.

Willey, J.M., and Waterbury, J.B. (1989). Chemotaxis toward nitrogenous compounds by swimming strains of marine *Synechococcus* spp. Appl. Environ. Microbiol. *55*, 1888–1894.

Willey, J.M., Waterbury, J.B., and Greenberg, E.P. (1987). Sodium-coupled motility in a swimming cyanobacterium. J. Bacteriol. *169*, 3429–3434.

Wolgemuth, C., Hoiczyk, E., Kaiser, D., and Oster, G. (2002). How myxobacteria glide. Curr. Biol. *12*, 369–377.

Wu, S.H., and Lagarias, J.C. (2000). Defining the bilin lyase domain: lessons from the extended phytochrome superfamily. Biochemistry *39*, 13487–13495.

Wuichet, K., Cantwell, B.J., and Zhulin, I.B. (2010). Evolution and phyletic distribution of two-component signal transduction systems. Curr. Opin. Microbiol. *13*, 219–225.

Yoshihara, S., and Ikeuchi, M. (2004). Phototactic motility in the unicellular cyanobacterium *Synechocystis* sp. PCC 6803. Photochem. Photobiol. Sci. *3*, 512–518.

Yoshihara, S., Suzuki, F., Fujita, H., Geng, X., and Ikeuchi, M. (2000). Novel putative photoreceptor and regulatory genes required for the positive phototactic movement of the unicellular motile cyanobacterium *Synechocystis* sp. PCC 6803. Plant Cell Physiol. *41*, 1299–1304.

Yoshihara, S., Geng, X., Okamoto, S., Yura, K., Murata, N., Go, M., Ohmori, M., and Ikeuchi, M. (2001). Mutational analysis of genes involved in pilus structure, motility and transformation competency in the unicellular motile cyanobacterium *Synechocystis* sp. PCC 6803. Plant Cell Physiol. *42*, 63–73.

Yoshihara, S., Geng, X., and Ikeuchi, M. (2002). *pilG* gene cluster and split *pilL* genes involved in pilus biogenesis, motility and genetic transformation in the cyanobacterium *Synechocystis* sp. PCC 6803. Plant Cell Physiol. *43*, 513–521.

Yoshihara, S., Katayama, M., Geng, X., and Ikeuchi, M. (2004). Cyanobacterial phytochrome-like PixJ1 holoprotein shows novel reversible photoconversion between blue- and green-absorbing forms. Plant Cell Physiol. *45*, 1729–1737.

Yoshihara, S., Shimada, T., Matsuoka, D., Zikihara, K., Kohchi, T., and Tokutomi, S. (2006). Reconstitution of blue-green reversible photoconversion of a cyanobacterial photoreceptor, PixJ1, in phycocyanobilin-producing *Escherichia coli*. Biochemistry *45*, 3775–3784.

Yoshimura, H., Hisabori, T., Yanagisawa, S., and Ohmori, M. (2000). Identification and characterization of a novel cAMP receptor protein in the cyanobacterium *Synechocystis* sp. PCC 6803. J. Biol. Chem. *275*, 6241–6245.

Yoshimura, H., Yanagisawa, S., Kanehisa, M., and Ohmori, M. (2002a). Screening for the target gene of cyanobacterial cAMP receptor protein SYCRP1. Mol. Microbiol. *43*, 843–853.

Yoshimura, H., Yoshihara, S., Okamoto, S., Ikeuchi, M., and Ohmori, M. (2002b). A cAMP receptor protein, SYCRP1, is responsible for the cell motility of *Synechocystis* sp. PCC 6803. Plant Cell Physiol. *43*, 460–463.

Yuan, H., Anderson, S., Masuda, S., Dragnea, V., Moffat, K., and Bauer, C. (2006). Crystal structures of the *Synechocystis* photoreceptor slr1694 reveal distinct structural states related to signaling. Biochemistry *45*, 12687–12694.

Yuan, H., Dragnea, V., Wu, Q., Gardner, K.H., and Bauer, C.E. (2011). Mutational and structural studies of the PixD BLUF output signal that affects light-regulated interactions with PixE. Biochemistry *50*, 6365–6375.

Cellular Differentiation in Filamentous Cyanobacteria

12

Iris Maldener, Michael L. Summers and Assaf Sukenik

Abstract

The ability of vegetative cells that comprise the growing filament of cyanobacteria of the order Nostocales and Stigonematales to differentiate into distinct cell types is comprehensively discussed in this chapter. Three types of differentiated cells are covered: nitrogen fixing heterocysts, dormant akinetes and motile hormogonia. While the first type represents a terminal developmental process, the two latter cell types maintain the ability to resume the morphology and functions of vegetative cells including the renewal of cell division. Here we review the current knowledge and understanding of these differentiation processes that occur in cyanobacteria; from environmental signals that trigger the differentiation to the complex regulatory mechanisms that lead to morphological and functional alterations.

Introduction

Several strains of cyanobacteria grow as long branched or unbranched filaments, consisting of hundreds of so-called vegetative cells performing oxygenic photosynthesis. Species of the orders Nostocales and Stigonematales, (sections IV and V of the five taxonomic sections described by Rippka *et al.*, 1979) exhibit sophisticated processes of cell differentiation forming dormant akinetes, nitrogen fixing heterocysts and motile hormogonia (Fig. 12.1). Cell differentiation evolved as survival strategy and adaptation to changing environments. Probably, ancestors of filamentous cyanobacteria of sections IV and V were the first bacteria in the biosphere that developed multicellularity. While spore-like akinetes separate from each other and from their adjacent vegetative cells to form mature free cells, the heterocysts stay connected to the filament. There they fulfil their special task, staying in close contact to the vegetative cells with which they exchange metabolites and signalling molecules. The short filaments of motile hormogonia must also require a sophisticated intercellular signalling system to move coordinated in one direction. Cell–cell communication seems to be a requisite for the differentiation of all different cell forms, a process that we are just beginning to understand. This review focuses on the function and morphology of the three different cell types of cyanobacteria and touches some aspects of the regulation of their differentiation.

Heterocysts

Heterocystous cyanobacteria can be considered as true multicellular organisms in which two cell types exhibit a division of metabolic labour: vegetative cells fix CO_2 and heterocysts fix N_2. Differentiation and maintenance of the cellular pattern require a differential transcriptional program in heterocysts versus vegetative cells in addition to a highly sophisticated mechanism of intercellular communication. The cells of the filaments are connected; they depend on each other and they are able to cooperate by exchanging signalling molecules and metabolites. Thus, the establishment of growth under N_2-fixing conditions involves three major aspects: (1) morphological and functional differentiation of a vegetative cell to be able to fulfil a different metabolic task; (2) development and maintenance of

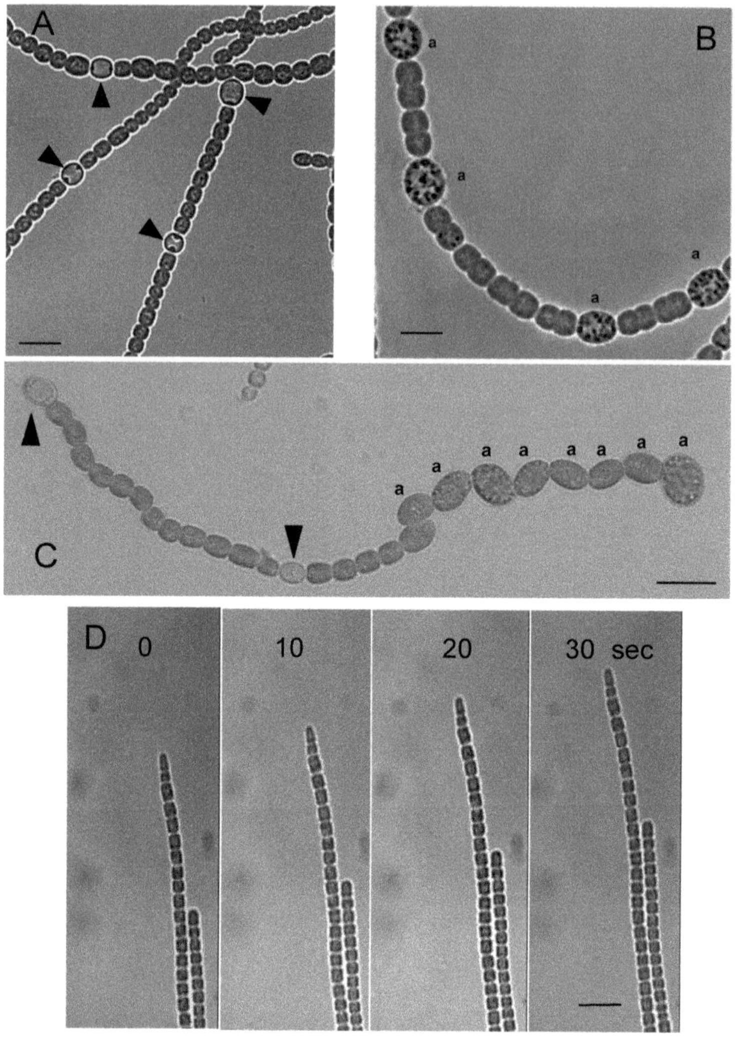

Figure 12.1 Differentiated cells in filamentous cyanobacteria. *Nostoc punctiforme* (A, C, D) and *Aphanizomenon ovalisporum* (B). Heterocysts indicated with arrowheads (t) and akinetes with the letter a. Time-lapse micrographs taken over 30 s show gliding hormogonia on an agar surface (D). Bar = 10 µm. (Micrographs courtesy of Josef Lehner, 2011 (A), and Susanne Berendt, 2012 (D); PhD theses, University of Tübingen, Germany.)

a one-dimensional pattern of cell distribution; and (3) communication between the cells of the multicellular organism. In the following sections we focus on the morphological changes leading to a functional mature heterocyst with reference to the communication structures between the cells of the filament (see also Chapter 13). The results described herein came from research done with two well-characterized and genetically accessible strains *Nostoc punctiforme* ATCC 29133 (Fig. 12.1) and *Anabaena* sp. PCC 7120. The reader is referred to the following excellent reviews for a more detailed description of different aspects of heterocyst differentiation in cyanobacteria: Adams and Duggan (1999), Flores and Herrero (2010), Golden and Yoon (2003), Kumar *et al.* (2010), Maldener and Muro-Pastor (2010), Meeks and Elhai (2002), Muro-Pastor and Hess (2012), Wolk *et al.* (1994), Xu *et al.* (2008), Zhang *et al.* (2006), and Zhao and Wolk (2008).

Heterocyst differentiation and function

The heterocyst is designed to protect nitrogenase from the atmospheric O_2 as well as from O_2 produced by photosynthetically active adjacent vegetative cells (Figs. 12.1, 12.2 and 12.3) (Fay, 1992; Gallon, 1992). The heterocysts provide a micro-oxic environment suitable for nitrogenase. Accelerated respiration, cessation of photosynthetic production of oxygen, and deposition of an envelope only slightly penetrable by O_2 are the principal means by which heterocysts create internal micro-oxic conditions. Heterocysts form only when the filaments are deprived of combined nitrogen and differentiate from a small portion of the vegetative cells located at different sites along the filament, depending on the strain. In some genera such as *Calothrix* or *Cylindrospermopsis* heterocysts are formed at the ends of the filament. In *Anabaena* spp. or *Nostoc* spp., intercalary heterocysts differentiate from vegetative cells at non-random intervals. During diazotrophic

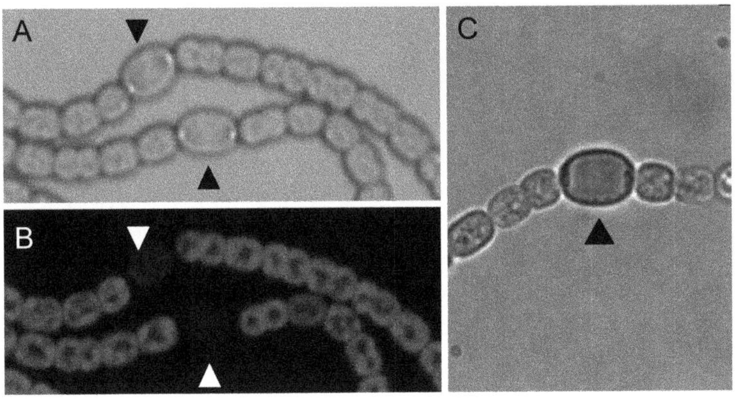

Figure 12.2 Filaments of *Anabaena* sp. PCC 7120 showing special features of heterocysts. The bright-field image (A) shows two heterocysts (arrowheads) attached to vegetative cells. Heterocysts lack red autofluorescence (B). Alcian blue stains the polysaccharides in the envelope of the heterocyst (C). Bar, 10 μm. Micrographs courtesy of Alicia M. Muro-Pastor, CSIC, Sevilla, Spain. A colour version of this figure is available in the plate section at the back of the book.

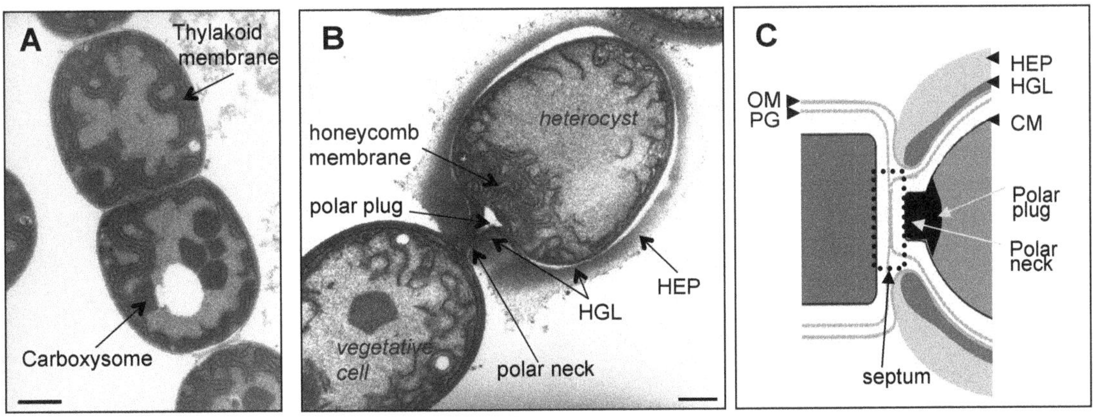

Figure 12.3 Electron micrographs and a scheme of filaments of *Anabaena* sp. PCC 7120 grown with nitrate (A) or with N_2 as the nitrogen source (B). Some cellular structures are indicated. Bars: 0.6 μm. (C) Scheme of the details of ultrastructure at the connection between a vegetative cell (left side of the image) and a heterocyst (right side of the image), based on (B). HEP: heterocyst envelope polysaccharide layer; HGL: heterocyst glycolipid layer; CM: cytoplasmic membrane; OM: outer membrane; PG: peptidoglycan layer. A colour version of this figure is available in the plate section at the back of the book.

growth new heterocysts arise approximately midway between two pre-existing ones. Once the differentiated heterocyst reaches a 'point of no return' it cannot resume the morphology and metabolic functions of a vegetative cell; cell division stops, differentiation progresses and the mature functional heterocyst is formed. Nearby vegetative cells are supplied with nitrogen compounds delivered from the mature N_2-fixing heterocyst and, in return, provide the heterocysts with fixed carbon used to generate reducing power required for nitrogen fixation.

Under laboratory conditions the differentiation process is induced by removing a combined nitrogen source from the growth medium (nitrogen step-down). Immature heterocysts (proheterocysts) become visible in the light microscope after staining the polysaccharide layer with Alcian Blue (Fig. 12.2C shows staining of a mature heterocyst). After maturation, heterocysts are morphologically distinct when viewed in the light microscope, nitrogenase activity increases and diazotrophic growth begins. In the first hours after nitrogen step-down, all cells of the filaments start to degrade their phycobiliproteins (Thiel, 1990; Wood and Haselkorn, 1980). After the onset of nitrogen fixation, the vegetative cells start to synthesize new phycobiliproteins, whilst in heterocysts their formation stays repressed, hence mature heterocysts are easily distinguishable from vegetative cells by the absence of autofluorescence (Fig. 12.2B) (Kumazaki *et al.*, 2013). RuBisCo as well as components of photosystem II (PSII) disappear, and newly synthesized enzymes of respiration, including cytochrome oxidases, provide the heterocyst with energy and remove remaining oxygen at the same time (Jones and Haselkorn, 2002; Valladares *et al.*, 2003, 2007; Wolk *et al.*, 1994). In heterocysts metabolism is changed from a photoautotrophic to a photoheterotrophic lifestyle, in which stored glycogen and organic metabolites from vegetative cells are used for production of reducing power via the oxidative pentose phosphate pathway (Summers *et al.*, 1995). In addition, ATP is produced by light-dependent cyclic electron flow around photosystem I (PSI) (Wolk *et al.*, 1994). Expression of genes encoding enzymes necessary for nitrogen fixation, such as those necessary for the formation of the intricate metal clusters of nitrogenase and hydrogenase, respectively, are induced during differentiation. The *nif* (nitrogen fixation) genes are expressed only after genome rearrangements occur, making possible heterocyst-specific nitrogenase expression (Golden *et al.*, 1985).

Filament growth relies on metabolite exchange between the two cell types of the filament (Wolk, 1968; Wolk *et al.*, 1974, 1976). The fixed nitrogen (NH_4^+) is incorporated into amino acids, through the GS/GOGAT (glutamine synthetase/glutamine-2-oxoglutarate-amido-transferase) pathway. Interestingly, glutamate, the substrate for GS is not made in heterocysts and has to be imported from vegetative cells. Fixed nitrogen is presumably transported to the vegetative cells of the filament as amino acids, including glutamine (Thomas *et al.*, 1977). Three amino acid ATP-binding-cassette (ABC)-transporter systems, accounting for 98% of the amino acid uptake, have been described in *Anabaena* sp. PCC 7120 that are involved in diazotrophic growth (Montesinos *et al.*, 1995; Pernil *et al.*, 2008; Picossi *et al.*, 2005). The disaccharide sucrose appears to be translocated to the carbon-starved heterocyst (Curatti *et al.*, 2002; Cumino *et al.*, 2007; López-Igual *et al.*, 2010; Vargas *et al.*, 2011). In the heterocyst the saccharide serves as a source of reducing equivalents in the form of NADPH and ferredoxin for respiration (helping to maintain the micro-oxic environment) and nitrogen fixation (Adams and Duggan, 1999; Wolk *et al.*, 1994).

The morphology of heterocysts

The ultra-structure of heterocysts was first investigated in the 1960s, when the function of this type of cell was still a matter of controversial discussion (Fay *et al.*, 1968). Later the structural differences between vegetative cells and heterocysts were comprehensively described (Adams and Duggan, 1999; Lang and Fay, 1971). One of the most obvious characteristics is the appearance of a conspicuous refractile envelope outside of the cell wall (Dunn and Wolk, 1970). The heterocyst-specific envelope consists of three distinct layers: an innermost, laminated layer, a homogeneous layer, and an outermost fibrous layer (Fig. 12.3). The chemical structure of the layers has been determined: the laminated layer consists of heterocyst-specific glycolipids (HGL) that are long-chain polyhydroxy alcohols glycosidically linked to hexose. Two different HGLs have been identified in *Anabaena* sp. PCC 7120 in which the hexose is glucose, and that vary by one OH-group in the glycone side chain (Gambacorta *et al.*, 1996). The homogeneous layer consists of heterocyst envelope polysaccharide (HEP), comprised of repeating units of branched oligosaccharides. The fibrous layer, which may be an un-compacted outer region of the homogeneous layer (Wolk *et al.*, 1994), will not be considered as a separate layer in the following text. Mutational analysis and physiological studies have indicated that the HGL layer is a barrier for the entrance of oxygen that is protected by the polysaccharide sheath, which by itself does not serve as a barrier for gases (Walsby, 1985; Wolk, 2000). Both layers, HEP and HGL, are essential for nitrogen fixation under oxic conditions, since mutants defective in any of the numerous genes involved in synthesis and deposition of the envelope are not able to grow on N_2 as sole nitrogen source (Ernst *et al.*, 1992). However, if nitrogen starvation is induced under anaerobic conditions those mutants develop nitrogenase activity after a while.

The formation of the specialized envelope of heterocysts involves, in addition to several regulatory proteins, a number of enzymes for synthesis and export of envelope material (Nicolaisen *et al.*, 2009) (Tables 12.1 and 12.2). For most of the genes involved in envelope formation in *Anabaena* sp. PCC 7120, it has been shown that they are induced in the medium phase of heterocyst differentiation (6–12 h), correlating with the ultra-structural analysis of morphological development. Synthesis of the HEP-layer polysaccharides is achieved by the action of numerous glycosyl transferases. Some of their genes are located in the genome on the so called HEP cluster (*alr2825–all2841*) (Huang *et al.*, 2005). Also glycosyl transferases HepB (*alr3698*) and Alr3699 are involved in HEP synthesis (Maldener *et al.*, 2003; Wang *et al.*, 2007). Mutations in genes

Table 12.1 Genes that are involved in the formation of the heterocyst homogenous layer and their putative function

Orfs	Gene name	Phenotype[a]	Function[b]	Reference
alr2825–alr2841	'HEP island'	HGL^+, HEP^-	Diverse proteins for HEP and LPS biosynthesis	Huang *et al.* (2005)
alr2834	hepC	HGL^+, HEP^-	UDP-galactose-lipid-carrier transferase	Zhu *et al.* (1998)
alr2835	hepA	HGL^+, HEP^-	ABC transporter	Holland and Wolk (1990)
alr0117	hepN	HGL^+, HEP^-	His kinase	Fan *et al.* (2006)
all2760	hepS	HGL^+, HEP^-	Ser/Thr kinase	Fan *et al.* (2006)
all4496	hepK	HGL^+, HEP^-	Two component system	Zhou and Wolk (2003)
alr0442	$devR_A$	HGL^+, HEP^-	Two component system	Zhou and Wolk (2003)
alr3698	hepB	HGL^+, HEP^-	Glycosyl transferase	Maldener *et al.* (2003), Wolk (2000)
alr3699	hepB2	HGL^+, HEP^-	Glycosyl transferase	Wang *et al.* (2007)
all4160		HGL^+, HEP^-	Glycosyl transferase Anti sigma-factor	Wang *et al.* (2007)
alr1086	henR	HGL^+, HEP^-	Response regulator	Fan *et al.* (2006)
alr5348		HGL^+, HEP^-	Spo0J-like domain	Fan *et al.* (2005)
all4388		HGL^+, HEP^-	Polysaccharide exporter	Maldener *et al.* (2003), Lechno-Yossef *et al.* (2011)
all1711	hepP	HGL^+, HEP^-	MSF protein, glycoside exporter	López-Igual *et al.* (2012)

[a]HEP: the polysaccharide (homogeneous) layer is not present, as shown by Alcian blue staining and/or electron microscopy.
[b]A putative function could be assigned from the mutant phenotype and sequence comparison analysis.

Table 12.2 Genes that are involved in the formation of the heterocyst laminated layer and their putative function

Orfs	Gene name	Phenotype[a]	Function[b]	Reference
all0813	hglK	TLC	Deposition of HGL (unknown protein)	Black *et al*. (1995)
alr2887	hgdD, tolC	TLC	HGL export (TolC-like outer-membrane channel)	Maldener *et al*. (2003), Moslavac *et al*. (2007), Staron *et al*. (2011)
alr3710	devB	TLC	HGL exporter, membrane fusion protein (Type I secretion system with TolC/HgdD)	Fiedler *et al*. (1998), Staron *et al*. (2011)
alr3711	devC	TLC	HGL exporter, inner membrane factor (Type I secretion system with TolC/HgdD)	Fiedler *et al*. (1998), Staron *et al*. (2011)
alr3712	devA	TLC	HGL exporter, ATPase (Type I secretion system with TolC/HgdD)	Fiedler *et al*. (1998), Staron *et al*. (2011)
all5341	hglT	TLC	Glycosylation of aglycone (Glycosyl transferase)	Awai and Wolk (2007)
all5343		TLC, delamination	Deposition of HGL as laminae, (Glycosyl transferase)	Fan *et al*. (2005)
all5345	hgdA	TLC, intracellular GL	Spatial deposition of HGL (Oxidoreductase)	Fan *et al*. (2005)
all5346	hgdC	TLC, intracellular GL	Spatial deposition of HGL devC-homologue (ABC transporter)	Fan *et al*. (2005)
all5347	hgdB	TLC, HGL-HEP-HGL, timing of deposition	Timing of deposition of HGL devB-homologue (ABC transporter)	Fan *et al*. (2005)
alr5348		TLC, HEP$^-$	Envelope formation (Spo0J, transcriptional regulator)	Fan *et al*. (2005)
alr5351	hglE$_A$	HGL$^-$	Biosynthesis of aglycones	Campbell *et al*. (1997), Fan *et al*. (2005)
alr5352	hglF	HGL$^-$	Biosynthesis of aglycones	Fan *et al*. (2005)
alr5353	hglG	HGL$^-$	Biosynthesis of aglycones	Fan *et al*. (2005)
alr5354	hglD	HGL$^-$	Biosynthesis of aglycones	Bauer *et al*. (1997)
alr5355	hglC	HGL$^-$	Biosynthesis of aglycones	Bauer *et al*. (1997)
alr5356	hglA	HGL$^-$	Biosynthesis of aglycones	Fan *et al*. (2005)
alr5357	hglB, hetM	HGL$^-$	Biosynthesis of aglycones	Black and Wolk (1994), Maldener *et al*. (2003)

[a]TLC: heterocyst glycolipids are detected in lipid extracts on thin layer chromatography (TLC) however the laminated layer is not detectable in electron microscopy. HGL$^-$: heterocyst glycolipids cannot be detected in lipid extracts on TLC and the laminated layer is not present. HEP-: the polysaccharide (homogeneous) layer is not present.
[b]A putative function could be assigned from the mutant phenotype and sequence comparison analysis.

for sugar/polysaccharide export, *all4388* and *all1711* (*hepP*), lead to loss of HEP layer formation (Ernst *et al.*, 1992; Lechno-Yossef *et al.*, 2011; López-Igual *et al.*, 2012; Maldener *et al.*, 2003). The *hepA* gene (*alr2835*) codes for an ABC transporter that could be involved in lipid A export and is essential for HEP-layer formation (Holland and Wolk, 1990; Huang *et al.*, 2005; Zhu *et al.*, 1998). HEP layer formation is an important early step in morphological differentiation and is obviously regulated by different systems including the two-component system HepK-DevR, the global transcription factor NtcA (see below) and others, such as HepN, HepS, and HenK (Fan *et al.*, 2006; Lechno-Yossef *et al.*, 2006; López-Igual *et al.*, 2012; Wang *et al.*, 2007; Zhao and Wolk,

2003, 2008). How these systems interplay is still unknown.

Numerous genes involved in synthesis and correct deposition of the heterocyst glycolipids (HGL) have been identified and the function of their products assigned (Awai *et al.*, 2009; Fan *et al.*, 2005). Many of those genes are clustered in the genome of *Anabaena* sp. PCC 7120 (*all5343–alr5358*). Some of these genes encode enzymes involved in lipid synthesis, while others encode homologues of the ABC transporter DevBCA (Alr3710/3711/3712). As with the HEP genes, many HGL genes are differentially expressed under the control of the heterocyst regulators NtcA and HetR (see below). Mutation of the DevBCA transporter leads to formation of immature heterocysts lacking the laminated layer (Fiedler *et al.*, 1998a,b; Maldener *et al.*, 1994; Staron *et al.*, 2011). A mutant in *alr2887*, encoding a TolC-like protein, shows the same phenotype (Ernst *et al.*, 1992; Maldener *et al.*, 2003; Moslavac *et al.*, 2007) and it has been shown that the DevBCA together with TolC forms a Type I secretion system to export the heterocyst-specific glycolipids (Staron *et al.*, 2011) (Fig. 12.4). A glycosyl-transferase, encoded by *all5341*, has been shown to be necessary for glycosylation of the glycolipid aglycone (Awai and Wolk, 2007). Table 12.2 summarizes HGL genes that are currently known to be involved in the formation of the laminated layer.

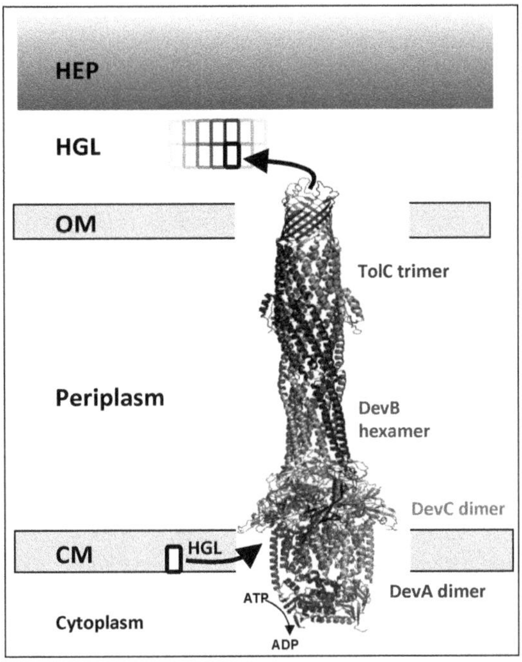

Figure 12.4 Model of HGL transport by the Type I secretion system DevBCA-TolC of *Anabaena* sp. PCC 7120. HGLs are built at the cytoplasmic membrane (CM) and transferred by the DevAC channel to the periplasmic DevB protein, which connects to the outer membrane (OM) protein TolC. After passage through the periplasm and the OM, the HGLs are deposited under the HEP layer to form the laminated layer. Model courtesy of Peter Staron, PhD thesis, University of Tübingen, Germany, 2012. A colour version of this figure is available in the plate section at the back of the book.

Deposition of envelope material beyond the outer membrane requires the transfer of the components through the cell wall. Therefore, flexibility in the cell wall structure seems to be important for heterocyst differentiation, perhaps by altering the cross linking inside the peptidoglycan allowing a penetration of the envelope material. Hence, it is likely that cell wall modifying enzymes play a role in heterocyst formation. The N-acetylmuramyl-L-alanine amidase HcwA (*all0093* gene product), a homologue to the AmiC-protein of *E. coli*, is involved in HEP-layer formation and is essential for heterocyst differentiation in *Anabaena* sp. PCC 7120 (Zhu *et al.*, 2001). Also, mutation in the upstream *amiC1* gene, *all0092*, inhibited heterocyst formation (Berendt *et al.*, 2012). In *N. punctiforme* ATCC 29133 mutation of the *amiC2* gene, Np_F1846, abolished differentiation of hormogonia, heterocysts, and akinetes and resulted in distorted filaments with aberrant cell division and cell size (Lehner *et al.*, 2011). The reason for the pleiotropic phenotype of *amiC* mutants in both strains might be related to the inability to exchange molecules, which has been observed by fluorescence recovery after photobleaching studies using the fluorophore calcein (Mullineaux *et al.*, 2008), implying a complete loss of cell communication capacity. Recently the structural basis for this phenotype was resolved: the amidase is necessary to create more than 150 nanopores, 20 nm wide, forming a nanopore array in the centre of the septal murein. These pores seem to constitute a framework for cell joining structures that traverse the murein sacculi (Lehner, 2012; Lehner *et al.*, 2013) (see also

Chapter 13). The *amiC1* mutant of *Anabaena* sp. PCC 7120, which cannot differentiate heterocysts, also lacks the nanopore array that is present in the wild-type (Lehner *et al.*, 2012). Mutations in other cell wall related genes, genes for lipopolysaccharide synthesis, *all4829* and *all4830* (Lázaro *et al.*, 2001; Xu *et al.*, 1997), and murein modification, *pbpB (alr5101)*, lead to a Fox⁻ phenotype (Lázaro *et al.*, 2001).

Most species of cyanobacteria have an inner thylakoid membrane system harbouring the components of respiratory and photosynthetic electron transport chains (Fig. 12.3A) (see also Chapter 4). The thylakoid membrane appears within the periphery of the protoplast of vegetative cells of *Anabaena* sp. PCC 7120 (Fig. 12.3A). It has been suggested that heterocysts contain two types of intra-cytoplasmic membrane systems: the thylakoid membranes harbouring elements of the photosynthetic apparatus and respiration and the so-called honeycomb membrane (Fig. 12.3B) (Lang and Fay, 1971; Murry *et al.*, 1981; Wolk *et al.*, 1994). The latter membrane is localized near the polar regions of the heterocysts, where it adjoins to the heterocyst neck. There, components of a highly active respiratory electron transport chain are localized (Murry *et al.*, 1981; Wolk *et al.*, 1994). In addition, the specific fluorescence of PSI could be localized in the polar region of the heterocysts, indicating that the polar membranes harbour both electron transport chains (Kumazaki *et al.*, 2013). Recently, protein FraH, encoded by *alr1603*, was discovered which could be involved in the formation of the honeycomb membrane (Merino-Puerto *et al.*, 2011).

In the electron microscope the polar structures of the heterocyst are visible: the protoplast is reduced to a neck, which ends in a narrow septum between heterocyst and adjacent vegetative cells (Fig. 12.3B). There, a thick glycolipid layer ensheathes the neck. During differentiation the envelope layers constrict the septum to this narrow area. Due to these structural alterations, the contact surface between vegetative and nitrogen-fixing cells is extremely reduced. This area probably has an influence on solute exchange between the two cell types and the communication along the filament (Walsby 2007; Wolk *et al.*, 1994). Oxygen that is liberated by the vegetative cells can diffuse through the septum. By entering the heterocyst at the neck and nearby areas, it gets captured and rapidly reduced to water by respiration in the polar honeycomb membrane. The fact that connections between vegetative cells and heterocysts are so narrow most likely reduces the danger of oxidative damage (Walsby, 2007). However, oxygen can also diffuse in at any of the lateral sides of the heterocyst. Therefore it is likely that respiration also takes place at lateral membranes. This is supported by the observation that inner membranes occur around the periphery of the heterocyst (Fig. 12.3B). At the heterocyst pole, the so-called polar plug is found. It consists, principally or solely, of the nitrogen-storage molecule cyanophycin (Lang and Fay, 1971; Lang *et al.*, 1972; Ziegler *et al.*, 2001) (see Chapter 9). The relevance of the accumulation of cyanophycin at the polar region is unclear (Ziegler *et al.*, 2001). Recent studies showed that intercellular exchange of small hydrophilic molecules occurs much faster in mutants defective in synthesis of cyanophycin and hence lacking the polar plug (Mullineaux *et al.*, 2008). Intracellular granules present in vegetative cells including (i) carboxysomes, which contain aggregates of RuBisCo; (ii) glycogen granules, a storage form of carbohydrate in cyanobacteria and (iii) structured granules of cyanophycin (except the polar plug) mostly disappear in mature heterocysts (Fig. 12.3).

The developmental stages of heterocyst differentiation on an ultra-structural level are shown in Fig. 12.5. About four hours after nitrogen step-down under our laboratory conditions, all cells of the filament still look alike. After six hours, alterations are first detected in distinctive immature heterocysts, termed proheterocysts. Their size looks larger than that of adjacent vegetative cells that, in contrast to proheterocysts, continue to divide. The adjacent vegetative cells do not show any sign of differentiation. The intercellular septa of proheterocysts constrict as soon as the HEP-layer appears. The thylakoid membranes then start to reorganize. Carboxysomes become smaller, but glycogen granules remain visible. After eight to twelve hours intracellular membranes begin to condense near the poles,

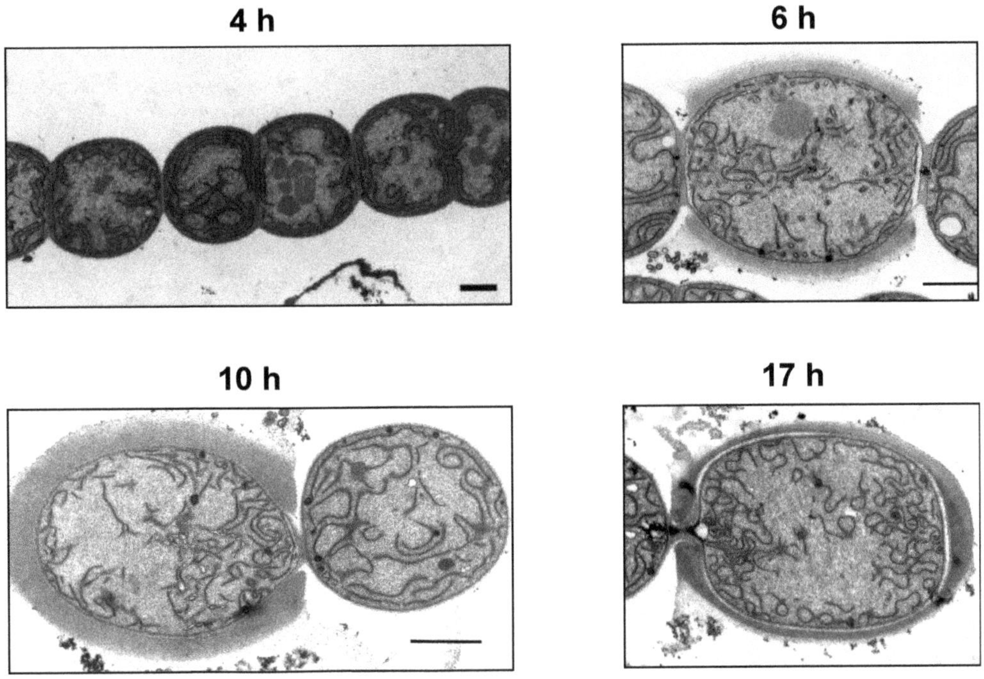

Figure 12.5 Electron micrographs of filaments of *Anabaena* sp. PCC 7120, taken from filaments incubated at different time points after nitrogen step-down. Bars, 0.6 μm.

glycogen granules and carboxysomes disappear, and the homogeneous layer becomes thickened. The septa between proheterocysts and vegetative cells further narrow. After about ten hours, reorganization of the intracellular membranes becomes more obvious. The most prominent of the two heterocyst-specific glycolipids can be detected as early as three hours after nitrogen starvation in lipid extracts (Moslavac *et al.*, 2007), however, the laminated layer is first observed much later at thirteen hours. This suggests that between synthesis and deposition of the glycolipids, a certain period of time is needed to assure that the HEP layer is ready. During this time the HGL-export system DevBCA-TolC is made and functions as schematically depicted in Fig. 12.4. HGLs are built at the cytoplasmic membrane and transferred by the DevAC channel to the periplasmic DevB protein, which connects to the outer membrane protein TolC. After passage through the periplasm and the outer membrane, the HGLs are deposited under the HEP layer to form the laminated layer.

Regulation of heterocyst differentiation

The dynamic changes in gene expression patterns during nitrogen starvation have been demonstrated by different methods. Micro-array analysis has been used to follow the differential expression of several hundreds genes during nitrogen starvation in both *Anabaena* sp. PCC 7120 and *N. punctiforme* (Campbell *et al.*, 2007, 2008; Ehira and Ohmori, 2006b; Ehira *et al.*, 2003; Lechno-Yossef *et al.*, 2006; Xu *et al.*, 2008). A time course of transcript abundance by RNA-seq has identified the differential transcriptome of *Anabaena* sp. PCC 7120 (Flaherty *et al.*, 2011). Based on comprehensive differential RNA-seq, a global map of transcriptional start sites has been defined, and genes regulated by HetR (see below) have been identified (Mitschke *et al.*, 2011).

The global nitrogen regulator NtcA and the heterocyst specific regulator HetR control initiation of heterocyst differentiation. NtcA is a transcriptional regulator of the CRP family and binds to consensus sequences in NtcA-regulated promoters (Herrero *et al.*, 2001, 2004). The binding of NtcA is influenced by the C-to-N balance and hence the 2-oxoglutarate level of the cell (Olmedo-Verd *et al.*, 2008; Tanigawa *et al.*, 2002; Vazquez-Bermudez *et al.*, 2003). *ntcA* and *hetR* mutants of *Anabaena* sp. PCC 7120 are not able to differentiate heterocysts and do not express many heterocyst-specific genes (Ehira and Ohmori, 2006a,b; Fiedler *et al.*, 2001; Frías *et al.*, 1994; Hernandez *et al.*, 2006; Herrero *et al.*, 2004; Muro-Pastor *et al.*, 1999, 2002; Wei *et al.*, 1994). *hetR* is one of the earliest genes required for development (Black *et al.*, 1993; Buikema and Haselkorn, 1991). Fig. 12.6B shows the expression of GFP under control of the *hetR* promoter in the heterocysts of filaments of *Anabaena* sp. PCC 7120. Unlike NtcA, HetR seems to be specifically involved in cell differentiation, including that of akinetes in the case of *N. ellipsosporum* (Leganés *et al.*, 1994), and is not required for growth in the presence of combined nitrogen. *hetR* encodes a unique protein with no known homologues, which shows rapid turn-over and which is able to bind DNA (Camargo *et al.*, 2012; Huang *et*

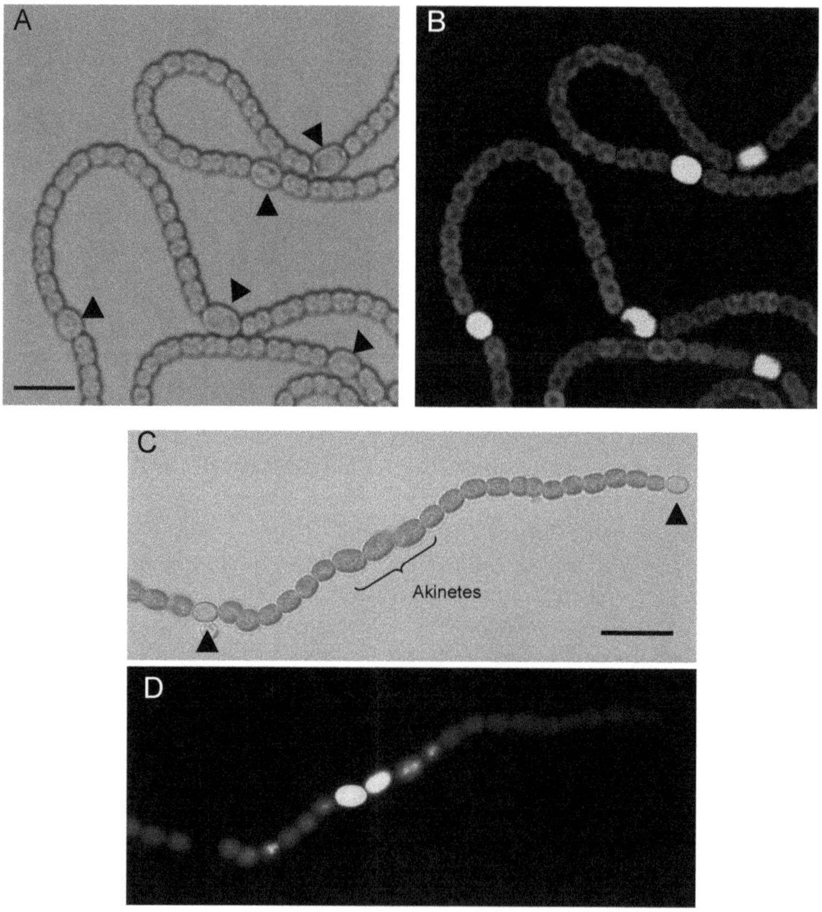

Figure 12.6 Differential gene expression in heterocysts and akinetes. Bright field (A and C) and overlay of autofluorescence and green fluorescence (B and D) of filaments of *Anabaena* sp. PCC 7120 (P_{hetR}-*gfp*) grown on N_2 (A and B) and of *N. punctiforme* ($P_{NpunF0784}$-*gfp*) 23 days following phosphate starvation (C and D). *NpunF0784* encodes a small heat shock protein in filaments of *Nostoc punctiforme*. The promoter region of each gene was cloned in front of a promoterless *gfp* gene encoding the green fluorescence protein and transferred on a self-replicating plasmid to each strain. Transcription of *hetR* occurs only in heterocysts (arrowheads) and that of *NpunF0784* primarily in akinetes. Bar, 10 µm. Micrographs courtesy of Alicia Muro-Pastor, CSIC, Sevilla, Spain (*hetR*) and Adrian Paz, CSU Northridge, Northridge, California (*NpunF0784*). A colour version of this figure is available in the plate section at the back of the book.

al., 2004; Kim *et al.*, 2011; Zhou *et al.*, 1998). A palindromic DNA motif has been suggested as the binding site of HetR (Higa and Callahan, 2010). Using differential RNA-seq more than 200 transcripts were identified to be under the control of HetR, 158 other transcriptional start sites were under control of NtcA (Mitschke *et al.*, 2011). In the same study a palindromic DNA motif (DIF$^+$) was identified in the promoter regions of heterocyst specific genes to be essential for differential expression (Mitschke *et al.*, 2011). NtcA and HetR exert a positive feedback on each other's expression (Muro-Pastor *et al.*, 2002). The NrrA protein is an NtcA-regulated response regulator (Ehira and Ohmori, 2006b; Muro-Pastor *et al.*, 2006) that is somehow involved in the mutual dependency of HetR and NtcA. Although *nrrA* mutants are able to produce heterocysts, they show altered expression of *hetR*, and NrrA has been shown to bind to the *hetR* promoter (Ehira and Ohmori, 2006a). Recently two protein phosphatases (PrpJ1 and PrpJ2) have been identified that are required for the mutual regulation of *ntcA* and *hetR* in *Anabaena* sp. PCC 7120 (Jang *et al.*, 2009).

Besides the early regulatory factors HetR and NtcA, several other gene products have been assigned a role at an early step during differentiation. HetC is a protein that shows homology to bacterial ABC exporters (Khudyakov and Wolk, 1997). *hetC* mutants show a pattern of non-fluorescent cells, an indication of differentiation, but there is no subsequent morphological differentiation and the non-fluorescent cells continue to divide (Xu and Wolk, 2001). *hetF* mutants also lack heterocysts (Wong and Meeks, 2001) and the HetF protein seems to be involved in HetR turnover (Risser and Callahan, 2008). Overexpression of HetL, a pentapeptide repeat protein, stimulates heterocyst formation but *hetL* mutants show normal heterocyst differentiation and diazotrophic growth (Liu and Golden, 2002). *patA* encodes a protein with CheY-like phospho-acceptor, helix–turn–helix and PATAN domains and also seems to be involved in HetR turnover (Liang *et al.*, 1992; Makarova *et al.*, 2006; Risser and Callahan, 2008). *patA* mutants produce heterocysts that are located mostly at the ends of the filaments. Products of a gene cluster comprising open reading frames *hetZ*, *patU5* and *patU3* are also involved in the regulation of heterocyst differentiation and pattern formation (Zhang *et al.*, 2007). For further detailed descriptions of regulatory aspects and pattern formation, the reader is referred to recently published reviews covering this topic (Aldea *et al.*, 2008; Flores and Herrero, 2010; Herrero *et al.*, 2013; Kumar *et al.*, 2010; Zhang *et al.*, 2006; Zhao and Wolk., 2008).

Signals other than nitrogen deficiency might also induce heterocyst differentiation (Adams and Carr, 1981; Campbell *et al.*, 1993; Sato and Wada, 1996). Proteins involved in the initial signalling are likely to integrate signals of light, temperature, progression of the cell cycle and, perhaps, circadian information. Because cyanobacteria lack 2-oxoglutarate dehydrogenase, 2-oxoglutarate accumulates under conditions in which the GS/GOGAT pathway is not operative, thus reflecting the carbon/nitrogen status (for reviews, see Luque and Forchhammer, 2008; Zhang *et al.*, 2006). Signal transduction might also occur via calcium and Ca^{2+}-binding proteins, including CcbP from *Anabaena* sp. PCC 7120 (Torrecilla *et al.*, 2000; Zhao *et al.*, 2005) and cyclic-di-GMP, a second messenger involved in the regulation of cellular differentiation processes in other bacteria (Neunuebel and Golden, 2008).

Determination of the pattern

How the individual cells of the filament sense the nitrogen status, start to develop and to form an inhibitor to prevent neighbouring cells from differentiation, thereby establishing a pattern of semi-regularly spaced heterocysts, is an exciting focus of investigation (Meeks *et al.*, 2002; Xu *et al.*, 2008; Zhang *et al.*, 2006; Zhao and Wolk., 2008). Currently it is not clear whether all cells of the filament, strings of cells, perhaps those that are in the same state of the cell cycle, or a single cell initiates the first steps of differentiation (Toyoshima *et al.*, 2010). From expression studies of early regulatory genes, it is known that a cluster of several cells starts the program of differentiation. In any case, after two to four hours the cluster gets resolved into single cells (Meeks and Elhai, 2002; Wilcox *et al.*, 1973; Zhao and Wolk., 2008). A regulatory system of activators and inhibitors of differentiation regulate the patterning of heterocysts. Probably, the non-random spacing pattern relies on interactions of autoregulatory HetR (see above) and the diffusible inhibitors PatS and HetN, both of which contain the pentapeptide RGSGR (Callahan and Buikema, 2001; Huang *et al.*, 2004; Yoon and Golden, 1998). The pentapeptide alone is sufficient to inhibit heterocyst differentiation *in vivo*, when added to a culture, and to inhibit binding of HetR to DNA *in vitro*. Expression of *patS* and *hetN* occurs in developing heterocysts and heterocysts, respectively, and depends on HetR. It is thought that the inhibitor rapidly diffuses to the neighbouring vegetative cells resulting in an inhibitor gradient along the filament (Callahan and Buikema, 2001; Yoon and Golden, 1998). Recently it has been shown that the inhibitors may influence posttranscriptional decay of the activator HetR (Risser and Callahan, 2009). Another inhibitor of heterocyst differentiation has been identified in *N. punctiforme*. A mutant in the membrane protein PatN shows an increased frequency of heterocysts, resulting in reduced growth rate and nitrogen fixation (Risser *et al.*, 2012). The protein shows biased inheritance in the vegetative cells, which results from localization to one half of a cell. After cell division it is inherited by only one daughter cell. As a result of PatN inheritance, PatN is unevenly distributed in the filament. PatN appears to inhibit PatA, which counteracts the negative effects of PatS and HetN on heterocyst differentiation. Hence, inactivation of PatN has a positive effect on heterocyst frequency as stated above.

Akinetes

Akinetes (from the Greek 'akinetos' – motionless) are resting cells of species members of the Nostocales and Stigonematales orders. These are spore-like, non-motile cells that differentiate from vegetative cells and serve a perennating role. Akinetes are larger and thicker than vegetative cells and contain large amounts of reserve materials and essential cellular metabolic pools. The akinete shape differs among species from sphere to oblate spheroid and their distribution and position within a filament (trichome) is frequently used as a taxonomic feature. The reader is referred to the following reviews for a more detailed description of different aspects of akinete differentiation: Adams and Duggan (1999); Herdman (1988); Herdman and Rippka (1988); and Kaplan-Levy *et al.* (2010).

Structure, composition and localization

Akinetes differ from vegetative cells by structure and morphological features. They are larger than

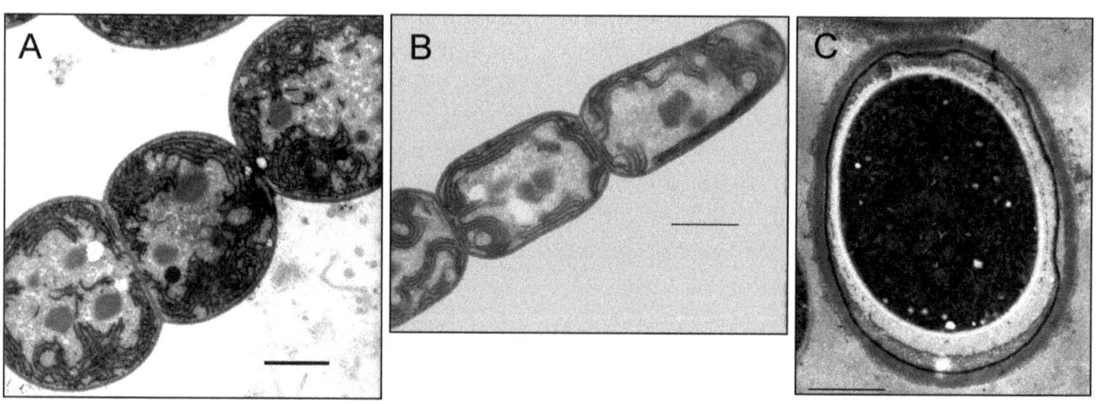

Figure 12.7 Electron micrographs of ultra thin sections through filaments of *N. punctiforme* and *A. variabilis* showing different cell types. A culture of *N. punctiforme* grown in the presence of ammonia and nitrate, having vegetative cells (A), was induced for hormogonia differentiation by N-step-down and shifting to red light (Berendt *et al.*, 2012) (B). A culture of *A. variabilis* was depleted for phosphate for 2 weeks resulting in unicellular akinetes (C). Bar, 1 μm. Micrograph in (C) courtesy of Rebeca Perez, University of Tübingen, Germany.

vegetative cells or heterocysts (Figs. 12.1B and C, 12.6 and 12.7) and, in some strains, can be up to 10-fold larger than vegetative cells. Akinetes are surrounded by a thickened cell wall and a multilayered extracellular envelope (Herdman, 1987, 1988; Nichols and Adams, 1982), they contain glycogen and distinctive granules of cyanophycin, but lack gas vesicles (Cardemil and Wolk, 1976, 1979; Simon, 1987; Wildman *et al.*, 1975). The thylakoid system of akinetes is substantially reduced (Fig. 12.7C) but never completely lost (Miller and Lang, 1968). In akinetes of *Nostoc* sp. PCC 7524, the mean cellular content of cyanophycin was eightfold higher than in vegetative cells (Sutherland *et al.*, 1979). However, accumulation of cyanophycin was not specific for akinete development; vegetative cells also accumulated glycogen and cyanophycin when entering the stationary phase (Herdman, 1987). Incubation of *Anabaena cylindrica* with the arginine analogue, canavanine (Nichols and Adams, 1982), and mutation of the arginine biosynthesis gene, *argL*, in *Nostoc ellipsosporum* (Leganés *et al.*, 1998), resulted in the production of akinetes lacking cyanophycin, suggesting that cyanophycin accumulation is not essential for the formation of akinetes, thus leaving the role of cyanophycin in the germinating akinetes unresolved.

The mean cellular content of RNA, DNA and protein was similar in vegetative cells and akinetes of *Nostoc* sp. PCC 7524 (Sutherland *et al.*, 1979), whereas the akinetes of *A. cylindrica* contained the same amount of RNA, but more than twice as much DNA, and ten times as much protein as vegetative cells (Simon, 1977). In *Aphanizomenon ovalisporum*, DAPI staining demonstrated the accumulation of nucleic acids in developed akinetes (Sukenik *et al.*, 2009). Normally, cyanobacteria are polyploids with several chromosome copies in vegetative cells. The chromosomal content in akinetes of *A. ovalisporum* increased 15-fold concomitant with accumulation of ribosomes. This massive accumulation of nucleic acids in akinetes was likely supported by phosphate supplied from inorganic polyphosphate bodies that were abundantly present in vegetative cells, but notably absent from akinetes (Sukenik *et al.*, 2012). Polyphosphate bodies were also rare in mature akinetes of *Nostoc* sp. PCC 7524, although they were commonly present in vegetative cells during akinete differentiation (Sukenik *et al.*, 2013).

The position of akinetes along the trichome varies among cyanobacterial species and strains. In some cases heterocysts are reported to influence their location (Wolk, 1966). For example, akinetes develop immediately adjacent to heterocysts in *A. cylindrica* but several cells away from the heterocyst in *Anabaena circinalis* (Fay *et al.*, 1984; Li *et al.*, 1997). In other cases such as in *N. punctiforme*, akinetes develop initially midway between heterocysts with subsequent formation in strings, showing a gradient of decreasing maturity away from the first to develop (Fig. 12.1C and 12.6C, 12.6D). In *A. ovalisporum*, akinetes at early stages of development are either separated from each other by several vegetative cells (Fig. 12.1B) or form separated strings of two or more akinetes along the trichome. In many species, akinetes of various sizes and differentiation stages are found in the same trichome, indicating the unsynchronized mode of akinete development (Sukenik *et al.*, 2013). While the placement of akinetes in relation to heterocysts was attributed to the production of cyanophycin by heterocysts and its accumulation in akinetes (Adams and Duggan, 1999), the physiological and biochemical features that determine which of the vegetative cells along a trichome will differentiate into an akinete are rather complex and presumably associate with the availability of resources other than nitrogen. These could be the cellular phosphate status, efficient DNA duplication capabilities and an appropriate phase of the cell cycle. The role of these features has yet to be verified as well as the communication and signalling between the differentiating akinetes and the adjacent vegetative cells or heterocysts.

Metabolic activities

The metabolic activity of mature akinetes is rather reduced whereas enhanced metabolism is expected during the differentiation and maturation processes. CO_2 fixation showed reduced rates in *A. cylindrica* and *Nostoc* sp. PCC 7524 (Fay, 1969b; Rai *et al.*, 1985, 1987; Sutherland *et al.*, 1985), the respiration rate was often elevated but lost in older akinetes, presumably in relation to the maturation process (Chauvat *et al.*, 1982; Herdman, 1987; Yamamoto, 1976). Isolated akinetes of *Nostoc spongiforme* respired in the dark, evolved oxygen in the light and retained residual capability to synthesize proteins and lipids (Thiel and Wolk, 1983). Developing akinetes of *A. cylindrica* were metabolically active but the activity significantly decreased as they mature (Fay, 1969b). Mature akinetes of *A. cylindrica* were reported to have little chlorophyll and no functional PSI (Fay, 1969b). Akinetes of *Anabaena doliolum* lost both chlorophyll and phycocyanin following incubation in the dark for several weeks (Singh and Sunita, 1974). *In vivo* fluorescence measurements of *Anabaena variabilis* akinetes suggested that the akinetes lacked a functional PSII, although the reaction centre chlorophyll was present (Bjorn *et al.*, 1983). PsbA protein (D1 polypeptide) of PSII was detected in akinetes (and other cell types) of the cyanobionts within leaf cavities of *Azolla caroliniana* Willd (Braun-Howland and Nierzwicki-Bauer, 1990). The stoichiometric ratio of PSI to PSII in *A. ovalisporum* akinetes remained more or less the same as in vegetative cells and the cellular content of PsbA (D1) and PsaC, a PSI subunit, proteins per cell volume remained fairly stable (Sukenik *et al.*, 2007). Furthermore, preliminary immunoblotting experiments indicated the presence of the RuBisCo large subunit in *A. ovalisporum* akinetes (Kaplan-Levy *et al.*, 2010). Changes in the photosynthetic activities of individual vegetative cells and akinetes were demonstrated in trichomes of *A. ovalisporum* during akinete formation (Sukenik *et al.*, 2007). It was further shown that mature isolated akinetes retained only residual photosynthetic capacity as the photosynthetic machinery was modified. In mature akinetes of *A. ovalisporum* the phycobilisome antenna was reduced in size and apparently detached from the reaction centres (Sukenik *et al.*, 2009). Similarly, the disappearance of phycocyanin from *A. cylindrica* akinetes was reported (Fay, 1969a) in accordance with observations on diminished photosynthetic activity in isolated akinetes (Fay, 1969b). The reduction of the phycobilisome pool in *A. ovalisporum* akinetes was linked with an increase in the expression of *nblA* encoding a protein that facilitates the degradation of the phycobiliproteins (Kaplan-Levy *et al.*, 2010). Enhanced expression level of *nblA* was also demonstrated for akinetes of the *zwf* mutant of *N. punctiforme* (Campbell *et al.*, 2007). Thus, it was postulated that in *A. ovalisporum* and *N. punctiforme* the reduction in the phycobilisome pool in mature akinetes is aimed at minimizing absorption of light energy to diminish potential damage to reaction centres during dormancy. The presence of reaction centre complexes in mature akinetes ensures a prompt recovery of photosynthesis upon germination, to energize essential cellular processes (Kaplan-Levy *et al.*, 2010).

Genome multiplication and the high polyploidy found in akinetes of *A. ovalisporum* require DNA replication. This DNA duplication process was inhibited by low concentrations of nalidixic acid, an inhibitor of DNA replication in prokaryotes. Consequently akinetes formation was ceased (Sukenik *et al.*, 2013).

Signals and signalling towards differentiation

No single environmental factor has been identified to trigger akinete differentiation. A major, although not the only, trigger for akinete development is light (Adams and Duggan, 1999). For example, in *Nostoc* sp. PCC 7524 cultivated in the presence of excess inorganic nutrients, akinetes differentiated as light availability was reduced due to the culture self-shading (Sutherland *et al.*, 1979). However, high light intensities triggered the formation of akinetes in *Cylindrospermopsis raciborskii* (Moore et al., 2005). In *A. ovalisporum* high light intensity synergistically affected akinete formation in conjunction with potassium deficiency (Sukenik *et al.*, 2007, 2013). In various *Anabaena* isolates, temperature was important for triggering akinete differentiation (Li *et al.*, 1997), whereas in *C. raciborskii* the formation of akinetes was triggered by an initial temperature shock, by the frequency of temperature fluctuations and by high light intensity (Moore *et al.*, 2005). Light quality also plays a role in the control of akinete formation. Green rather than white light stimulated akinete differentiation in *Gloeotrichia* (Wyman and Fay, 1986). In the toxic cyanobacterium *A. circinalis*, red or green irradiance was much more effective for akinete production than blue light. For cells grown under a predominantly red, white or green irradiance, even short exposures to blue light substantially reduced the number of akinetes, suggesting that blue light inhibits akinete formation (Thompson *et al.*, 2009).

Nutrient status and other abiotic factors have been identified as potential triggers for akinete differentiation. In *A. circinalis,* phosphate limitation appeared to be the major trigger, whereas limitations for N, inorganic C, iron, trace elements or light had no effect on the development of akinetes (van Dok and Hart, 1996). In *N. punctiforme,* akinetes were induced within 2 weeks of starvation for phosphate (Meeks and Elhai, 2002). However, phosphorus was required to allow full development of akinetes in *C. raciborskii* (Moore *et al.,* 2003, 2005) and in *A. circinalis* (Fay *et al.,* 1984). The significance of phosphate in the differentiation of akinetes was demonstrated for *A. ovalisporum* (Sukenik *et al.,* 2013). As mentioned above, it has been postulated that the observed accumulation of nucleic acids during akinete formation is supported by phosphate supplied from inorganic polyphosphate bodies that were abundantly present in vegetative cells, but notably absent from akinetes. Recruited phosphate from internal and external sources during akinete differentiation is deposited in DNA (Sukenik *et al.,* 2012).

In *Anabaena doliolum* (Rao *et al.,* 1987) and *Anabaena torulosa* (Sarma and Khattar, 1993) a critical C:N ratio appeared to be important for akinete differentiation. In *A. ovalisporum* deprivation of potassium ions (K^+) triggered the formation of akinetes. A burst of akinete formation was observed within 1–2 weeks after the induction (K^+ depletion) was imposed (Sukenik *et al.,* 2007). However, other environmental conditions such as high light and optimal temperature (18–24°C) were needed to guarantee akinete formation in *A. ovalisporum* under potassium deficiency (Sukenik *et al.,* 2013). A potential link to carbon balance is exhibited by a *N. punctiforme zwf* mutant that formed functional akinetes during dark incubation in the presence of fructose (Argueta and Summers, 2005). The glucose 6-phosphate dehydrogenase (*zwf*) mutation stops entry of sugars into the oxidative pentose phosphate pathway, the major route of carbon catabolism in cyanobacteria, which also blocks dark heterotrophic growth using fructose (Summers *et al.,* 1995).

It has been postulated that diverse stimuli, reported to affect akinete formation, induce a common physiological trigger – perhaps decreased cell division or low energy – which results in akinete development (Adams and Duggan, 1999). The collective observations on the environmental stimuli that trigger akinete formation in different cyanobacterial species and strains are mostly consistent with cellular energy limitation and cessation of cell division as primary signals for akinete differentiation. The metabolic imbalance triggering akinete formation in the *zwf* mutant of *N. punctiforme* however did not decrease ATP for 6 days following dark incubation with fructose (concurrent with akinete formation), supporting the role of a cessation of cell division as a signal for akinete induction. Diverse external cues, such as light, temperature or nutritional status, are translated into a secondary internal signal in a specific vegetative cell, within a trichome. This internal signal eventually triggers the cell differentiation into an akinete. The observation that only few cells on a filament differentiate into akinetes suggests that the signal transduction pathways activated by the environmental cues need to intercalate with cellular factors that are present only during a certain phase of the cell cycle. Hence, we suggest that the stage in the cell cycle dictates if the external signals will induce akinete differentiation of a given vegetative cell. However the question of how an external signal promotes differentiation is still unclear.

Genes associated with akinete differentiation

The elucidation of the molecular mechanism that regulates the akinete differentiation process lagged behind heterocyst studies until recently. The data accumulated on heterocyst formation suggests that these cells and akinetes share some commonalities in the molecular pathway of cell differentiation (Zhang *et al.,* 2006). Four genes were found to be involved in both differentiation processes as mentioned above:

1. *hepA* encodes an ABC transporter required for the deposition of polysaccharides in the envelope of both cell types in *A. variabilis* (Leganés *et al.,* 1994);
2. *devR* encodes a response regulator of a two-component system. (Campbell *et al.,* 1996);

3 *hetR*, which is a regulator of many heterocyst genes (see above), is expressed in akinetes and its mutation in *Nostoc ellipsosporum* resulted in a failure to form either heterocysts or akinetes (Leganés *et al.*, 1994), but was not required for akinete formation in *N. punctiforme* (Wong and Meeks, 2001);

4 *argL* encodes an *N*-acetylglutamate semialdehyde dehydrogenase, an enzyme involved in L-arginine biosynthesis. Mutation of *argL* in *N. ellipsosporum* resulted in smaller than wild-type akinetes that lacked cyanophycin granules and failed to germinate (Leganés *et al.*, 1998).

A common regulatory pathway for differentiation of heterocysts and akinetes has been suggested (Campbell *et al.*, 2007), including the possibility that the HetR regulatory pathway is involved in the akinete differentiation process. However unlike in heterocyst differentiation, this pathway might be activated at later stages of akinete differentiation, leading to akinete maturation (Kaplan-Levy *et al.*, 2010). In heterocysts the HetR pathway leads to activation of several processes: (1) the synthesis of a polysaccharide envelope a process in which DevR is implicated; (2) deposition of a glycolipid layer (possibly via expression of *aet*; see below); (3) cessation of cell division; and (4) cessation of oxygenic photosynthetic activity (Zhao and Wolk, 2008). The first three processes are also essential for the formation and maturation of akinetes.

The establishment of a reliable system that gives synchronous akinete differentiation in the *zwf* mutant of *N. punctiforme* (Argueta and Summers, 2005) allowed the application of a high-throughput technology to study akinete formation in cyanobacteria. Three novel genes involved in akinete differentiation were detected (Argueta *et al.*, 2006). The genes were designated: *aet* (NpF0062), which is an akinete-expressed gene encoding an ABC transporter with high similarity to the *E. coli* MsbA lipopolysaccharide transporter; *aapN* (NpF5999), which encodes an akinete aminopeptidase belonging to the M28 peptidase family; and *hap* (NpR4070) a hormogonium/akinete-expressed protease homologous to the α-subunit group of the M16 zinc-dependent proteases.

Global gene expression in cultures of the *N. punctiforme zwf* mutant three days into the akinete differentiation process revealed the up-regulation of 255 genes, 41% of which encoded known proteins (Campbell *et al.*, 2007). An increase in transcript levels of four transcriptional regulators or transcription-related proteins was recorded: two transcription factors of the Crp family and two sigma factors. In addition, increase in the transcript of *amiC1 (NpF1845)* encoding an enzyme that degrades peptidoglycan linker bonds was reported. The expression of the *nblA* gene increased as well (Campbell *et al.*, 2007). In *A. ovalisporum* genes orthologous to *avaK*, *aet* and *nblA* were highly expressed in isolated akinetes as compared with their expression level in vegetative cells of an exponentially grown culture (Kaplan-Levy *et al.*, 2010). The 3-day *zwf* mutant akinete microarray experiment identified a small heat shock protein (sHSP, Npun_F0784) that was up-regulated ~7-fold. An Npun_F0784-GFP transcriptional reporter strain was used to confirm high levels of transcription in akinetes, with expression occurring as they become morphologically discernible based on cell size (Fig. 12.6D). Whether this sHSP is required as a chaperone for proteins newly synthesized during akinete differentiation or as a protective mechanism to survive harsh environmental conditions during extended dormancy remains to be determined.

Germination

Germination of akinetes is a complex coordinated process triggered by various ambient conditions (Karlsson-Elfgren *et al.*, 2004; Reynolds, 1972). In many cases the environmental stimuli that trigger akinete germination correspond to those supporting growth of vegetative cultures. In addition, sediment mixing and resuspension play an important role in germination as they relocate the akinetes from the bottom sediment into the water column and the photic zone. In some cases germination was reported as a photoperiodic (day-length) dependent process occurring only after a maturation period was completed (Karlsson-Elfgren and Brunberg, 2004). Germination was shown to be dependent on light intensity and did not take place in the dark or in the presence of DCMU, a photosynthesis inhibitor (Yamamoto, 1976), suggesting the direct involvement of the photosynthetic apparatus in germination. The most active spectral range for germination was between 620 and 630 nm, coinciding with the maximum light absorption by C-phycocyanin (Nichols and Adams, 1982). Akinetes of *A. ovalisporum* germinated within a temperature range of 18–25°C, but germination yield was low and unsynchronized. Bottom sediments collected from Lake Kinneret and incubated under laboratory conditions in N-free medium yielded many filaments of *A. ovalisporum*, pointing to the role of akinetes in the establishment of a new population (Hadas *et al.*, 1999).

The germination of akinetes was described in detail for several Nostocales species (Baker and Bellifemine, 2000; Braune and Doehler, 1996; Hori *et al.*, 2003; Kaplan-Levy *et al.*, 2010; Moore *et al.*, 2004). Germination of akinetes begins with reorganization of cellular material followed by elongation and cell division that occurs inside the akinete. Expansion of the cells results in an increase in turgor pressure, which consequently leads to the opening of the akinete envelope on one of the terminal side poles and emergence of the germling from the akinete envelope. The open envelope may remain associated with the developing filament for some time. It is unclear whether the envelope disruption is assisted by enzymatic activity or if it occurs solely due to the increased internal pressure resulting from the expanded cells. In some cyanobacteria such as *N. punctiforme*, during germination the entire akinete wall may dissolve and, hence, not be microscopically visible (Adams and Duggan, 1999; Meeks *et al.*, 2002). Morphological and biochemical alterations in the germling lead to a fully developed young trichome (Moore *et al.*, 2004). Germination of *Cyanospira* akinetes was accompanied by *de novo* protein synthesis that took place prior to the first cell division (Sili *et al.*, 1994). In *A. circinalis* photosynthetic activity provided the energy for akinete germination (Kezhi *et al.*, 1985), but the rate of germination was determined by the respiratory oxygen uptake of the akinetes, in a temperature dependent manner (Fay, 1988). An additional stage (although not an obligatory one) in the germination process is the development of gas vacuoles that support successful flotation of germlings and trichomes in the water column. In *Gloeotrichia echinulata* the newly formed filaments float for 2–4 days after germination (Karlsson-Elfgren and Brunberg, 2004).

Hormogonia

Some filamentous cyanobacteria are capable of forming hormogonia, broadly defined as shorter filaments of a filamentous cyanobacteria released from a longer parent trichome that are in some way distinguishable from the parent (Rippka *et al.*, 1979). Hormogonia can form in both heterocystous and non-heterocyst forming cyanobacteria and are typically distinguished from the parent filament by differences in cell size, shape or motility (Fig. 12.1D and 12.7C). Gas vesicles are characteristic of hormogonia in some but not all species. Hormogonia of heterocystous cyanobacteria always lack heterocysts, even if combined nitrogen is absent. When present during hormogonia induction, heterocysts break loose from the filament and the intervening trichomes become the hormogonia. In non-heterocyst forming strains, shorter filaments are formed by necridial cells that die and lyse, fragmenting the filament into shorter hormogonial fragments (Lamont, 1969). Hormogonia are a transient stage in the life-cycle of these cyanobacteria. Following their formation, hormogonia de-differentiate back to the vegetative state within a few days and resume growth. Strains of *Nostoc* and *Calothrix* require a period of vegetative growth before they can again form hormogonia (Meeks and Elhai, 2002).

The reader is directed to previous excellent reviews for more information about hormogonia and their similarities with baeocytes regarding dispersal (Herdman and Rippka, 1988), their occurrence and variations in morphology, induction conditions, and properties among cyanobacteria (Tandeau de Marsac, 1994), and an overview of a well-studied model system in *N. punctiforme* (Meeks *et al.*, 2002).

Function

The primary function attributed to hormogonia in nature is short-distance dispersal. Cyanobacteria that grow successfully as filaments are eventually confined to the location they occupy, since movement is limited by their extreme length. Growth only magnifies the problem as filaments become entangled with themselves, each other, the underlying substrate, or other members of the community. Hormogonia provide the release from such containment by allowing shorter fragments to escape potentially stressful conditions by bulk flow of the surrounding medium, buoyancy (by formation of gas vesicles in planktonic strains), or self-directed escape using motility (typically gliding motility for benthic strains, as illustrated by time-lapse photography in Fig. 12.1D). Positive phototaxis may be used to keep the cells in the proper light environment for growth (Lazaroff, 1973). The ability of filamentous cyanobacteria to form hormogonia is therefore a selective advantage that allows habitation of ideal niches in the environment and enhances species survival. (For further discussion on motility, see Chapter 11.)

Some heterocystous cyanobacteria form symbioses with a plant host and provide fixed nitrogen for the plant (Adams, 2000; Adams and Duggan, 2012; Meeks, 1998; Rai *et al.*, 2000). Motile hormogonia function as the infective agent in forming such symbioses by entering into pre-existing plant cavities, or in the case of the *Gunnera–Nostoc* symbiosis, intracellularly (Bergman *et al.*, 2008). Positive chemotaxis towards symbiotic plant tissues (Knight and Adams, 1996) likely aides in the establishment of colonization. Subsequent de-differentiation into vegetative filaments inside the plant and formation of filaments containing large numbers of heterocysts establish the infection. Hormogonia are induced and attracted to specific areas of the plant by chemical signals termed hormogonia inducing factors (HIF), and once inside the plant, their escape is prevented by inhibiting hormogonia formation by putative chemical signals termed hormogonia repressing factors (HRF) (Meeks and Elhai, 2002).

Another function attributed to hormogonia in the genus *Scytonema* is attachment (Hernandez-Marine and Roldan, 2005). Biofilm samples from walls of illuminated caves and catacombs exhibited hourglass-shaped necridial cells that assume this shape due to encroachment of rounded neighbouring cells extended into the dying cell. Staining indicated that DNA and thylakoids were degraded in necridia, followed by the formation of a sticky substance that exhibited staining characteristics of polysaccharides. Upon lysis and release of the hormogonia, the sticky substance remained attached to one or both ends of the hormogonia. The authors postulated these sticky ends played a role in adherence, such as for the establishment of biofilms in the environment. This role could explain their observations of individual trichomes attached to the basal surface of these biofilms by their ends (Hernandez-Marine and Roldan, 2005).

Morphology and behaviour

Hormogonial cells of heterocyst-forming genera such as *Calothrix*, *Fischerilla*, *Nostoc* and *Tolypothrix* are typically smaller than vegetative cells (Tandeau de Marsac, 1994). This is due to repeated division without cell elongation or increase in trichome biomass. DNA replication halts (Herdman and Rippka, 1988), as does the net synthesis of protein and chlorophyll during the periods of hormogonium differentiation and gliding motility (Campbell and Meeks, 1989). Because cyanobacterial cells are thought to contain many copies of their chromosome, such division without DNA replication would diminish, but not eliminate, the genome copy number (Tandeau de Marsac, 1994).

A distinguishing feature of some hormogonia is the altered morphology of cells found at the ends of the trichome. The end-cells of *N. punctiforme* typically become tapered and cone-shaped (Figs. 12.1D and 12.7C). Upon de-differentiation under nitrogen-fixing conditions, the end cells are the first to become heterocysts (Campbell and Meeks, 1989; Meeks *et al.*, 2002). Aside from an obvious streamlining effect for gliding, the reason for this cone-shaped transformation remains unknown. The presence of apical pads at the tips of hormogonia in some *Scytonema* and *Leptolyngbya* species composed of polysaccharides is also distinctive. The apical pads arise from the remnants of lysed necridia and have a hypothesized role in attachment leading to biofilm formation (Hernandez-Marine and Roldan, 2005).

Motility is another feature that characterizes hormogonia in some strains (see also Chapter 11). Filamentous cyanobacteria move by a gliding mechanism that may involve pili (Duggan *et al.*, 2007) or directional secretion of polysaccharide slime (Hoiczyk and Baumeister, 1998). Since motile hormogonia are the infective unit that enters plant hosts for nitrogen-fixing symbioses, most of the laboratory work on hormogonia has focused on nitrogen-fixing *Nostoc* strains with motile hormogonia. The surface of motile hormogonia in the *Nostoc* genus is covered by type IV pili, whereas the non-motile vegetative filaments are not (Adams and Duggan, 2012). This makes pili, rather than the unorganized fibrillar arrays located between the peptidoglycan and outer membrane layers that are present in both cell types (Read *et al.*, 2007), a likely candidate for hormogonial motility in this strain.

Photo- and chemotaxis of a multicellular filament require coordination between the cells. Recently, it was shown that the cell wall amidase AmiC2 (NpF1846), described above, is abundant in the cross walls of the hormogonia (Lehner *et al.*, 2013). Since this enzyme creates a nanopore array in the peptidoglycan of the septa, and the *amiC2* mutant is not able to form hormogonia, the array might be a key feature in intercellular communication that is required in filamentous cyanobacteria and especially in hormogonia. The AmiC2 protein disappears in maturing akinetes, which do not require communication any more (Lehner *et al.*, 2013).

An easy to discern feature exhibited by hormogonia in some strains is the presence of refractile gas vesicles. Gas vesicles make filaments more buoyant in the water column and may enhance survival of pelagic filamentous cyanobacteria that may be lost from the photosphere by settling. The ability to differentiate into hormogonia has been a distinguishing feature separating the genus *Nostoc* from the non-hormogonia forming *Anabaena* genus (Rippka *et al.*, 1979). However, hormogonia production can be quickly lost upon continuous culture under laboratory conditions, and lead to erroneous conclusions of phylogeny, as illustrated by past *Nostoc*/*Anabaena* nomenclature disputes among cyanobacteriologists. Through the application of molecular genetics, the presence of the gene encoding the major gas vesicle protein *gvpA* has been used to reveal genus assignment of these closely related genera (Damerval *et al.*, 1989). The presence of *gvpA*, even if not expressed or functional, can be regarded as evidence of a *Nostoc* sp. (Tandeau de Marsac, 1994).

Differentiation signals and signalling

There appears to be no single environmental factor triggering hormogonia formation among the strains with this ability (Meeks and Elhai, 2002; Tandeau de Marsac, 1994). Various environmental changes in the laboratory, including those predicted to enhance as well as limit growth seem to elicit hormogonia formation. These include dilution of a stationary phase culture into fresh media (Rippka *et al.*, 1979), removal of nitrate (Armstrong *et al.*, 1983; Herdman and Rippka, 1988), addition of phosphate (Wood *et al.*, 1986) or iron (Douglas *et al.*, 1986), lowering osmolarity (Knight and Adams, 1996), lowering ionic strength (Armstrong *et al.*, 1983; Gantar *et al.*, 1993), or placement of liquid-grown cultures onto solid agar media (Hernendez-Muniz and Stevens, 1987). Light quality has also been used to trigger differentiation of *Calothrix* hormogonia, with red-light stimulation and green light repression. The action spectrum of light-regulated hormogonia formation appears similar to that of complementary chromatic acclimation (Damerval *et al.*, 1991b). (For a detailed description of complementary chromatic acclimation, see Chapter 6.)

Plant exudates of cyanobacterial symbionts have the ability to signal both chemotaxis and differentiation of vegetative cells into hormogonia, as well as to inhibit hormogonium formation. Extracts of nitrogen starved liverwort *Blasia pusilla*, a symbiotic partner of *Nostoc* sp., provide a chemoattractant for motile hormogonia (Knight and Adams, 1996). Other bryophytes such as *Anthoceros punctatus* produce an unidentified compound that induces hormogonia formation in many *Nostoc* species, but only when the plant has experienced nitrogen starvation (Campbell and Meeks, 1989). Factors stimulating hormogonium formation have also been reported for exudates of the angiosperm symbiotic partner *Gunnera* (Rasmussen *et al.*, 1994) and for wheat (Gantar *et al.*, 1993). Induction of hormogonia formation upon nitrogen starvation of the plant would increase infection frequency and therefore be a selective advantage in forming beneficial nitrogen-fixing symbiotic infections.

In *N. punctiforme*, reducing sugars such as sucrose, glucose, and fructose were shown to inhibit hormogonia formation, whereas arabinose or galactose did not (Khamar *et al.*, 2010). This has spatial relation to the sugar locations in and around the stem gland of its plant host *Gunnera manicata*, which has external arabinose but high internal concentrations of glucose and fructose. It was proposed that this arrangement would facilitate hormogonia entry into the glands, but once inside, hinder their escape by inhibiting further hormogonia formation (Khamar *et al.*, 2010).

Genes associated with hormogonia differentiation

Early work identified several genes that were transcriptionally up-regulated during differentiation of hormogonia. Evidence for a HRF first came from the study of genes identified using reporter transposon mutagenesis of *N. punctiforme*. Mutagenesis of two genes (*hrmA* and *hrmU*) caused an increase in hormogonia formation in the presence of its symbiotic partner beyond that normally observed in a wild-type strain (Cohen and Meeks, 1997). Subsequent exploration and analysis of the *hrm* locus (Campbell *et al.*, 2003) revealed a gene encoding a transcriptional repressor (*hrmR*) and an additional intervening gene (*hrmI*) that are likely transcribed as an operon (*hrmRIUA*), as well as additional genes *hrmE* and *hrmK* in this locus. HrmR was shown to bind regions upstream from its own gene and *hrmE* containing a 14 bp palindrome bound by HrmR. HrmR did not bind the promoters of these genes in the presence of galacturonate or lysate from *N. punctiforme* cells that had been exposed to HRF, indicating the presence of a small inducer molecule. Genes in the *hrm* locus encode proteins similar to enzymes for hexuronic acid catabolism (*hrmIUA*), a carbohydrate kinase (*hrmK*), and an aldehyde dehydrogenase (*hrmE*). Together, the sequence homology, mutant and biochemical analysis led to the hypothesis that the locus was induced by a HRF, most likely a sugar molecule, that when bound to HrmR caused de-repression of genes required for production of a metabolite inhibiting hormogonium formation (Campbell *et al.*, 2003). These results were further supported by a subsequent whole-genome transcriptomic analysis (Campbell *et al.*, 2008) that identified nearby genes (Npun_R5323–25; encoding an ABC transport system and a permease for sugars) induced in parallel with *hrmUA* and hypothesized to catalyse entry of the HRF molecule. This locus seems to be specific for plant symbiotic associations and was not induced upon hormogonia formation by dilution/nitrogen starvation (Campbell *et al.*, 2008). Recent mutagenesis of genes related to sugar transport in this locus and phenotypic characterization of the resulting mutants has shown that the glucose permease (GlcP; Npun_R5323) but not the ABC fructose transporter (Npun_R5327–24) is essential for establishing a successful symbiosis of *N. punctiforme* with its plant partner *A. punctatus* (Ekman *et al.*, 2013). It is interesting to note that unlike all previous mutants lacking symbiotic competence, the *glcP* mutant is capable of producing normal functioning heterocysts and motile hormognia. In addition to the hypothesis that sugars are involved in confinement of cyanobacteria to symbiotic associations through inhibition of the formation of motile hormogonia (Adams and Duggan, 2012; Khamar *et al.*, 2010), we must now include that sugar uptake and metabolism are also involved in establishing or maintaining symbiotic relationships.

The *sigH* gene encodes a group 2 sigma factor of RNA polymerase in *N. punctiforme* (Npun_R1771) that was induced in as little as 1.5 h following exposure to nitrogen starved *A. punctatus* (Campbell *et al.*, 1998). The only phenotype attributed to a mutant of this gene was a 6-fold increase in infection frequency of the plant without a similar increase in nitrogen-fixing activity. SigH was not required for induction of the previously discovered *hrm* locus described above, indicating multiple pathways for controlling symbiotic associations with the plant host. Non-regulatory structural genes such as those encoding gas vesicle proteins (Damerval *et al.*, 1991a) and pili (Duggan *et al.*, 2007) are also induced in hormogonia.

Large-scale genome-wide transcriptional profiling of hormogonia formation in *N. punctiforme* has identified a large number of genes (1827) that change expression in cultures 1 day following hormogonia induction by dilution/nitrogen starvation (Campbell *et al.*, 2007). A subsequent more detailed time course study comparing this induction condition with that produced with added plant HIF identified 1,328 genes common to both induction conditions (Campbell *et al.*, 2008). In both reports, approximately equal numbers of genes were up- and down-regulated, and large numbers of genes involved in transcriptional regulation implied that multiple regulatory events occur during development that could help to explain the massive number of regulated genes found during hormogonia differentiation.

Aside from the plethora of genes discovered, perhaps the most striking feature of the time-course data is a transient down-regulation of induced genes, and up-regulation of repressed genes at 12 h following hormogonia induction by dilution/nitrogen starvation. This transient change back towards expression typical of vegetative growth was termed a 'check point', and hypothesized to reflect a point at which a decision is made to either return to the vegetative state, or continue with the process of hormogonia formation. The HIF-dependent induction time-course array showed no such checkpoint, and was therefore inferred to be a stronger signal for hormogonia induction since there was no 'choice' available to these cultures (Campbell *et al.*, 2008). Regardless of the induction signal, the 1328 core genes common to both experiments will almost certainly contain those that define and describe the unique features exhibited by hormogonia.

Several recent single-gene studies have been published that address hormogonia formation. An *N. punctiforme* insertional mutant in Npun_F3173, a member of a polyketide synthase gene cluster, exhibited an accumulation of short filaments containing a single heterocyst in the absence of combined nitrogen, and continuous hormogonia formation in ~30% of the filaments of a culture throughout the entire growth curve after induction by dilution into fresh medium (Liaimer *et al.*, 2011). Hormogonia levels could be reduced in liquid by adding conditioned media or HPLC fractions of this media that had previously supported wild-type growth, and on solid media by close proximity of a wild-type colony. These results support the hypothesis that a diffusible molecule produced by this gene cluster is involved in normal cell differentiation control. In the mutant, hormogonia were not repressed when in symbiosis with a liverwort host plant, indicating that plant hormogonia repression signals occur downstream of the cyanobacterial signals. These results also indicate that the method for hormogonia induction by dilution of a nitrogen-replete culture into nitrogen-free fresh media may be due to the removal/imbalance of small molecule signals. It is interesting to note that Npun_F3173 was induced in a DNA microarray experiment following nitrogen starvation (Campbell *et al.*, 2007), identical to the conditions used for hormogonia induction by these researchers (Liaimer *et al.*, 2011), but not upon hormogonia induction by either a plant factor or dilution into fresh nitrogen-free media (Campbell *et al.*, 2008). These results suggest that the ketide may be more connected to heterocyst than to hormogonium differentiation. It will be interesting to study whether the presence of this proposed ketide is linked to repression of hormogonia formation.

N. punctiforme insertional mutants in Npun_R0896, a multi-domain adenyl cyclase (*cyaC*) exhibited three- to fourfold reduction in cAMP levels yet retained the ability to form hormogonia with wild-type like surface pili (Chapman *et al.*, 2008). The specific location of mutagenic insertions had opposite effects on the ability to form effective symbioses with a plant host. Mutations in the N-terminal end produced reduced symbiotic competence whereas insertion at the C-terminal adenyl cyclase domain produced increased infection frequencies relative to the wild-type strain. A luciferase reporter strain did not increase expression during hormogonium differentiation, although increased expression was observed in early heterocysts. Together these results indicate that normal levels of cAMP are not required for hormogonium differentiation, yet sensory GAF domains present in CyaC may be required for successful colonization of a host, perhaps via chemotaxis.

Conclusions

Cellular differentiation in filamentous cyanobacteria represents advanced properties among multicellular prokaryotes. While cellular differentiation in prokaryotes leads to the formation of spores and other dispersal cellular units (formation of spores in *Bacillus*, fruiting bodies of *Myxococcus*, and the motile swarmer cells of *Caulobacter*), the vegetative cells of filamentous cyanobacteria can differentiate into three different cell types, each of which has a different function: N_2-fixing heterocysts, spore-like akinetes, and motile hormogonia. Heterocysts attracted extended and intense research efforts and their function and differentiation process are relatively well understood. Beyond their unique structure, the prime motivation to study heterocysts was their clearly visual structure, their role in N_2 fixation, the simple and ubiquitous environmental trigger that induces their formation and their relatively rapid differentiation process. The diverse and advanced experimental approach that led to the current understanding of heterocyst function, structure and regulation paved the way to elucidate the formation of the two other differentiated cellular forms, akinetes and hormogonia.

Global approaches to the identification of elements involved in cell differentiation in cyanobacteria are of the greatest interest because of the wideness and number of elements being identified. However, a great deal of work is needed to understand the relationships between those elements and identify common elements that are involved in the differentiation processes. Nevertheless, it is important to keep in mind that interpretations based on transcription levels do not necessarily correspond to protein or activity levels. Posttranscriptional regulation orchestrated by non-coding RNAs, as found in many other bacterial responses to stress have been shown to be operative in these organisms (Mitschke *et al.*, 2011) and might be involved in differentiation processes. The increasing use of methods for cellular localization of proteins inside the filament, combined with high resolution microscopy, as well as transcriptomic/proteomic studies should lead to a comprehensive understanding of the pathways and relationships of components involved in sensing, functioning and controlling yet unexplored aspects of cyanobacterial differentiation such as ageing of the heterocyst, dedifferentiation of hormogonia and germination of akinetes.

Acknowledgements

The authors wish to thank various foundations and organizations that supported their research on cyanobacterial differentiation processes. A.S. was supported by the Joint German–Israeli Project (FKZ 02WT0985, WR803) funded by the German Ministry of Research and Technology (BMBF) and by Israel Ministry of Science and Technology (MOST) and the Israel Science Foundation (Grant No. 319/12). IM was funded by Deutsche Forschungsgemeinschaft (DFG; SBF766 and GRK1708). M.L.S. was supported by the United States National Institutes of Health (5SC1GM093998).

References

Adams, D.G. (2000). Symbiotic interactions. In The ecology of cyanobacteria Their diversity in time and space, Whitton, B.A., and Potts, M., eds. (Klewer Academic Publishers, Boston), pp. 523–561.

Adams, D.G., and Carr, N.G. (1981). The developmental biology of heterocyst and akinete formation in cyanobacteria. Crit. Rev. Microbiol. *9*, 45–100.

Adams, D.G., and Duggan, P.S. (1999). Heterocyst and akinete differentiation in cyanobacteria. New Phytol. *144*, 3–33.

Adams, D.G., and Duggan, P.S. (2012). Signalling in cyanobacteria-plant symbioses. In Signaling and Communication in Plant Symbiosis, Perotto, S., and Baluška, F., eds. (Springer-Verlag, Berlin), pp. 93–121.

Aldea, M.R., Kumar, K., and Golden, L.W. (2008). Heterocyst development and pattern formation. In Chemical Communication among Bacteria, Winans, S.C., and Bassler, B.L., eds. (ASM Press, Washington, DC), pp. 75–90.

Argueta, C., and Summers, M.L. (2005). Characterization of a model system for the study of *Nostoc punctiforme* akinetes. Arch. Microbiol. *183*, 338–346.

Argueta, C., Yuksek, K., Patel, R., and Summers, M.L. (2006). Identification of *Nostoc punctiforme* akinete-expressed genes using differential display. Mol. Microbiol. *61*, 748–757.

Armstrong, R.E., Hayes, P.K., and Walsby, A.E. (1983). Gass vacuole formation in hormogonia of *Nostoc muscorum*. J. Gen. Microbiol. *128*, 263–270.

Awai, K., and Wolk, C.P. (2007). Identification of the glycosyl transferase required for synthesis of the principal glycolipid characteristic of heterocysts of *Anabaena* sp. strain PCC 7120. FEMS Microbiol. Lett. *266*, 98–102.

Awai, K., Lechno-Yossef, S., and Wolk, C.P. (2009). Heterocyst Envelope Glycolipids. In Lipids in

Photosynthesis: Essential and Regulatory Functions, Wada, H., and Murata, N., eds. (Springer Science, Dordrecht, Netherlands), pp. 179–202.

Baker, P.D., and Bellifemine, D. (2000). Environmental influences on akinete germination of *Anabaena circinalis* and implications for management of cyanobacterial blooms. Hydrobiologia *427*, 65–73.

Berendt, S., Lehner, J., Zhang, Y.V., Rasse, T.M., Forchhammer, K., and Maldener, I. (2012). Cell wall amidase AmiC1 is required for cellular communication and heterocyst development in the cyanobacterium *Anabaena* PCC 7120 but not for filament integrity. J. Bacteriol. *194*, 5218–5227.

Bergman, B., Ran, L., and Adams, D.G. (2008). Cyanobacterial-plant symbioses: signaling and development. In The Cyanobactera: Molecular Biology, Genomics and Evolution, Herrero, A., and Flores, E., eds. (Caister Adademic Press, Norfolk, UK), pp. 447–473.

Bjorn, G.S., Braune, W., and Bjorn, L.O. (1983). Photochromic pigments in akinetes and pigment characteristics of akinetes in comparison with vegetative cells of *Anabaena-variabilis*. Physiol. Plant. *59*, 493–500.

Black, T.A., Cai, Y., and Wolk, C.P. (1993). Spatial expression and autoregulation of *hetR*, a gene involved in the control of heterocyst development in *Anabaena*. Mol. Microbiol. *9*, 77–84.

Braun-Howland, E.B., and Nierzwicki-Bauer, S.A. (1990). Occurrence of the 32 kDa Q_B-binding protein of photosystem II in vegetative cells, heterocysts and akinetes of *Azolla caroliniana* cyanobionts. Planta *180*, 361–371.

Braune, W., and Doehler, G. (1996). ^{15}N-uptake, influenced by UV-B radiation, and pattern of amino-acid pools during akinete germination in *Anabaena variabilis* (Cyanobacteria). J. Basic Microbiol. *36*, 219–227.

Buikema, W.J., and Haselkorn, R. (1991). Characterization of a gene controlling heterocyst differentiation in the cyanobacterium *Anabaena* 7120. Genes Dev. *5*, 321–330.

Callahan, S.M., and Buikema, W.J. (2001). The role of HetN in maintenance of the heterocyst pattern in *Anabaena* sp. PCC 7120. Mol. Microbiol. *40*, 941–950.

Camargo, S., Valladares, A., Flores, E., and Herrero, A. (2012). Transcription activation by NtcA in the absence of consensus NtcA-binding sites in an *Anabaena* heterocyst differentiation gene promoter. J. Bacteriol. *194*, 2939–2948.

Campbell, D., Houmard, J., and De Marsac, N.T. (1993). Electron transport regulates cellular differentiation in the filamentous cyanobacterium *Calothrix*. Plant Cell *5*, 451–463.

Campbell, E.L., and Meeks, J.C. (1989). Characteristics of hormogonia formation by symbiotic *Nostoc* spp. in response to the presence of *Anthoceros punctatus* or its extracellular products. Appl. Environ. Microbiol. *55*, 125–131.

Campbell, E.L., Hagen, K.D., Cohen, M.F., Summers, M.L., and Meeks, J.C. (1996). The *devR* gene product is characteristic of receivers of two-component regulatory systems and is essential for heterocyst development in the filamentous cyanobacterium *Nostoc* sp. strain ATCC 29133. J. Bacteriol. *178*, 2037–2043.

Campbell, E.L., Brahamsha, B., and Meeks, J.C. (1998). Mutation of an alternative sigma factor in the cyanobacterium *Nostoc punctiforme* results in increased infection of its symbiotic plant parner, *Anthoceros punctatus*. J. Bacteriol. *180*, 4938–4941.

Campbell, E.L., Wong, F.C.Y., and Meeks, J.C. (2003). DNA binding properties of the HrmR protein of *Nostoc punctiforme* responsible for transcriptional regulation of genes involved in the differentiation of hormogonia. Mol. Microbiol. *47*, 573–582.

Campbell, E.L., Summers, M.L., Christman, H., Martin, M.E., and Meeks, J.C. (2007). Global gene expression patterns of *Nostoc punctiforme* in steady-state dinitrogen-grown heterocyst-containing cultures and at single time points during the differentiation of akinetes and hormogonia. J. Bacteriol. *189*, 5247–5256.

Campbell, E.L., Christman, H., and Meeks, J.C. (2008). DNA microarray comparisons of plant factor and nitrogen-deprivation induced hormogonia reveal decision making transcriptional regulation patterns in *Nostoc punctiforme*. J. Bacteriol. *190*, 7382–7391.

Cardemil, L., and Wolk, C.P. (1976). The polysaccharides from heterocyst and spore envelopes of a blue-green alga. Methylation analysis and structure of the backbones. J. Biol. Chem. *251*, 2967–2975.

Cardemil, L., and Wolk, C.P. (1979). The polysaccharides from heterocyst and spore envelopes of a blue-green alga. Structure of the basic repeating unit. J. Biol. Chem. *254*, 736–741.

Chapman, K.E., Duggan, P.S., Billington, N.A., and Adams, D.G. (2008). Mutation at different sites in the *Nostoc punctiforme cyaC* gene, endoding the multiple-domain enzyme andenylate cyclase, results in different levels of infection of the host plant *Blasia pusilla*. J. Bacteriol. *190*, 1841–1847.

Chauvat, F., Corre, B., Herdman, M., and Joset-Espardellier, F. (1982). Energetic and metabolic requirements for the germination of akinetes of the cyanobacterium *Nostoc* PCC7524. Arch. Microbiol. *133*, 44–49.

Cohen, M.F., and Meeks, J.C. (1997). A hormogonium regulating locus, *hrmUA*, of the cyanobacterium *Nostoc punctiforme* strain ATCC 29133 and its response to an extract of a symbiotic plant partner *Anthoceros punctatus*. Mol. Plant Microbe Interact. *10*, 280–289.

Cumino, A.C., Marcozzi, C., Barreiro, R., and Salerno, G.L. (2007). Carbon cycling in *Anabaena* sp. PCC 7120. Sucrose synthesis in the heterocysts and possible role in nitrogen fixation. Plant Physiol. *143*, 1385–1397.

Curatti, L., Flores, E., and Salerno, G. (2002). Sucrose is involved in the diazotrophic metabolism of the heterocyst-forming cyanobacterium *Anabaena* sp. FEBS Lett. *513*, 175–178.

Damerval, T., Castets, A.-M., Guglielmi, G., Houmard, J., and Tanduau de Marsac, N. (1989). Occurrence and distribution of gas vesicle genes among cyanobacteria. J. Bacteriol. *171*, 1445–1452.

Damerval, T., Castets, A., Houmard, J., and Tandeau de Marsac, N. (1991a). Gas vesicle synthesis in the cyanobacterium *Pseudoanabaena* sp.: Occurence of a single photoregulated gene. Mol. Microbiol. *5*, 657–664.

Damerval, T., Guglielmi, G., Houmard, J., and Tandeau de Marsac, N. (1991b). Hormogonium differentiation in the cyanobacterium *Calothrix* sp.: A Photoregulated Developemental Process. Plant Cell 3, 191–201.

van Dok, W., and Hart, B.T. (1996). Akinete differentiation in *Anabaena circinalis* (cyanophyta). J. Phycol. *32*, 557–565.

Douglas, D., Peat, A., Whitton, B.A., and Wood, P. (1986). Influence of iron status on structure of the cyanobacterium *Calothrix* 7601. Gene *54*, 83–92.

Duggan, P.S., Gottardello, P., and Adams, D.G. (2007). Molecular analysis of genes in *Nostoc punctiforme* involved in pilus biogenesis and plant infection. J. Bacteriol. *189*, 4547–4551.

Dunn, J.H., and Wolk, C.P. (1970). Composition of the cellular envelopes of *Anabaena cylindrica*. J. Bacteriol. *103*, 153–158.

Ehira, S., and Ohmori, M. (2006a). NrrA directly regulates expression of *hetR* during heterocyst differentiation in the cyanobacterium *Anabaena* sp. strain PCC 7120. J. Bacteriol. *188*, 8520–8525.

Ehira, S., and Ohmori, M. (2006b). NrrA, a nitrogen-responsive response regulator facilitates heterocyst development in the cyanobacterium *Anabaena* sp. strain PCC 7120. Mol. Microbiol. *59*, 1692–1703.

Ehira, S., Ohmori, M., and Sato, N. (2003). Genome-wide expression analysis of the responses to nitrogen deprivation in the heterocyst-forming cyanobacterium *Anabaena* sp. strain PCC 7120. DNA Res. *10*, 97–113.

Ekman, M., Picossi, S., Campbell, E.L., Meeks, J.C., and Flores, E. (2013). *Nostoc* sugar transporter necessary to establish a cyanobacterium-plant sympiosis. Plant Phys. *161*, 1984–1992.

Ernst, A., Black, T., Cai, Y., Panoff, J.M., Tiwari, D.N., and Wolk, C.P. (1992). Synthesis of nitrogenase in mutants of the cyanobacterium *Anabaena* sp. strain PCC 7120 affected in heterocyst development or metabolism. J. Bacteriol. *174*, 6025–6032.

Fan, Q., Huang, G., Lechno-Yossef, S., Wolk, C.P., Kaneko, T., and Tabata, S. (2005). Clustered genes required for synthesis and deposition of envelope glycolipids in *Anabaena* sp. strain PCC 7120. Mol. Microbiol. *58*, 227–243.

Fan, Q., Lechno-Yossef, S., Ehira, S., Kaneko, T., Ohmori, M., Sato, N., Tabata, S., and Wolk, C.P. (2006). Signal transduction genes required for heterocyst maturation in *Anabaena* sp. strain PCC 7120. J. Bacteriol. *188*, 6688–6693.

Fay, P. (1969a). Cell differentiation and pigment composition in *Anabaena cylindrica*. Arch. Mikrobiol. *67*, 62–70.

Fay, P. (1969b). Metabolic activities of isolated spores of *Anabaena cylindrica*. J. Exp. Bot. *20*, 100–109.

Fay, P. (1988). Viability of akinetes of the planktonic cyanobacterium *Anabaena circinalis*. Proc. R. Soc. Lond. B *234*, 283–301.

Fay, P. (1992). Oxygen relations of nitrogen fixation in cyanobacteria. Microbiol. Rev. *56*, 340–373.

Fay, P., Stewart, W.D., Walsby, A.E., and Fogg, G.E. (1968). Is the heterocyst the site of nitrogen fixation in blue-green algae? Nature *220*, 810–812.

Fay, P., Lynn, J.A., and Majer, S.C. (1984). Akinete development in the planktonic blue-green alga *Anabaena circinalis*. Brit. Phycol. J. *19*, 163–174.

Fiedler, G., Arnold, M., Hannus, S., and Maldener, I. (1998a). The DevBCA exporter is essential for envelope formation in heterocysts of the cyanobacterium *Anabaena* sp. strain PCC 7120. Mol. Microbiol. *27*, 1193–1202.

Fiedler, G., Arnold, M., and Maldener, I. (1998b). Sequence and mutational analysis of the *devBCA* gene cluster encoding a putative ABC transporter in the cyanobacterium *Anabaena variabilis* ATCC 29413. Biochim. Biophys. Acta. *1375*, 140–143.

Fiedler, G., Muro-Pastor, A.M., Flores, E., and Maldener, I. (2001). NtcA-dependent expression of the *devBCA* operon, encoding a heterocyst-specific ATP-binding cassette transporter in *Anabaena* spp. J. Bacteriol. *183*, 3795–3799.

Flaherty, B.L., Van Nieuwerburgh, F., Head, S.R., and Golden, J.W. (2011). Directional RNA deep sequencing sheds new light on the transcriptional response of *Anabaena* sp. strain PCC 7120 to combined-nitrogen deprivation. BMC Genom. *12*, 332.

Flores, E., and Herrero, A. (2010). Compartmentalized function through cell differentiation in filamentous cyanobacteria. Nature Rev. Microbiol. *8*, 39–50.

Frías, J.E., Flores, E., and Herrero, A. (1994). Requirement of the regulatory protein NtcA for the expression of nitrogen assimilation and heterocyst development genes in the cyanobacterium *Anabaena* sp. PCC 7120. Mol. Microbiol. *14*, 823–832.

Gallon, J.R. (1992). Reconciling the incompatible: N_2 fixation and O_2. New Phytol. *122*, 571–609.

Gambacorta, A., Romano, I., Trincone, A., Soriente, A., Giordano, M., and Sodano, G. (1996). The heterocyst glycolipids of cyanobacteria. 5. Heterocyst glycolipids from five nitrogen-fixing cyanobacteria. Gazz. Chim. Ital. *126*, 653–656.

Gantar, M., Kerby, N.W., and Rowell, P. (1993). Colonization of wheat (*Triticum vulgare* L.) by N2-fixing cyanobacteria. III. The role of a hormogonia-promoting factor. New Phytol. *124*, 505–513.

Golden, J.W., and Yoon, H.S. (2003). Heterocyst development in *Anabaena*. Curr. Opin. Microbiol. *6*, 557–563.

Golden, J.W., Robinson, S.J., and Haselkorn, R. (1985). Rearrangement of nitrogen fixation genes during heterocyst differentiation in the cyanobacterium *Anabaena*. Nature *314*, 419–423.

Hadas, O., Pinkas, R., Delphine, E., Vardi, A., Kaplan, A., and Sukenik, A. (1999). Limnological and ecophysiological aspects of Aphanizomenon ovalisporum bloom in Lake Kinneret, Israel. J. Plankt. Res. *21*, 1439–1453.

Herdman, M. (1987). Akinetes: structure and function. In The cyanobacteria, Fay, P., and Van Baalen, C., eds. (Elsevier, Amsterdam), pp. 227–250.

Herdman, M. (1988). Cellular differentiation: Akinetes. Methods Enzymol. *167*, 222–232.

Herdman, M., and Rippka, R. (1988). Cellular Differentiation: Hormogonia and Baeocytes. Meth. Enzymol. *167*, 232–242.

Hernandez, J.A., Muro-Pastor, A.M., Flores, E., Bes, M.T., Peleato, M.L., and Fillat, M.F. (2006). Identification of a *furA cis* antisense RNA in the cyanobacterium *Anabaena* sp. PCC 7120. J. Mol. Biol. *355*, 325–334.

Hernandez-Marine, M., and Roldan, M. (2005). Adherance of hormogonia to substrata is mediated by polysaccharides produced by necridic cells. Algological Stud. *117*, 239–249.

Hernendez-Muniz, W., and Stevens, J., S.E. (1987). Characterization of the motile hormogonia of *Mastigocladus laminosus*. J. Bacteriol. *169*, 218–223.

Herrero, A., Muro-Pastor, A.M., and Flores, E. (2001). Nitrogen control in cyanobacteria. J. Bacteriol. *183*, 411–425.

Herrero, A., Muro-Pastor, A.M., Valladares, A., and Flores, E. (2004). Cellular differentiation and the NtcA transcription factor in filamentous cyanobacteria. FEMS Microbiol. Rev. *28*, 469–487.

Herrero, A., Picossi, S., and Flores, E. (2013). Gene expression during heterocyst differentiation. Adv. Bot. Res. *65*, 281–329.

Higa, K.C., and Callahan, S.M. (2010). Ectopic expression of *hetP* can partially bypass the need for *hetR* in heterocyst differentiation by *Anabaena* sp. strain PCC 7120. Mol. Microbiol. *77*, 562–574.

Hoiczyk, E., and Baumeister, W. (1998). The junctional pore complex, a prokaryotic secretion organelle, is the molecular motor underlying gliding motility in cyanobacteria. Curr. Biol. *8*, 1161–1168.

Holland, D., and Wolk, C.P. (1990). Identification and characterization of *hetA*, a gene that acts early in the process of morphological differentiation of heterocysts. J. Bacteriol. *172*, 3131–3137.

Hori, K., Okamoto, J.i., Tanji, Y., and Unno, H. (2003). Formation, sedimentation and germination properties of *Anabaena* akinetes. Biochem. Engineer. J. *14*, 67–73.

Huang, G., Fan, Q., Lechno-Yossef, S., Wojciuch, E., Wolk, C.P., Kaneko, T., and Tabata, S. (2005). Clustered genes required for the synthesis of heterocyst envelope polysaccharide in *Anabaena* sp. strain PCC 7120. J. Bacteriol. *187*, 1114–1123.

Huang, X., Dong, Y., and Zhao, J. (2004). HetR homodimer is a DNA-binding protein required for heterocyst differentiation, and the DNA-binding activity is inhibited by PatS. Proc. Natl. Acad. Sci. U.S.A. *101*, 4848–4853.

Jang, J., Shi, L., Tan, H., Janicki, A., and Zhang, C.-C. (2009). Mutual regulation of *ntcA* and *hetR* during heterocyst differentiation requires two similar PP2C-type protein phosphatases, PrpJ1 and PrpJ2, in *Anabaena* sp. strain PCC 7120. J. Bacteriol. *191*, 6059–6066.

Jones, K.M., and Haselkorn, R. (2002). Newly identified cytochrome c oxidase operon in the nitrogen-fixing cyanobacterium *Anabaena* sp. strain PCC 7120 specifically induced in heterocysts. J. Bacteriol. *184*, 2491–2499.

Kaplan-Levy, R.N., Hadas, O., Summers, M.L., Rücker, J., and Sukenik, A. (2010). Akinetes – dormant cells of cyanobacteria. In Dormancy of Cells and Organisms, Cerdà, J., Clark, M., and Lubzens, E., eds. (Springer, Berlin), pp. 269–300.

Karlsson-Elfgren, I., and Brunberg, A.-K. (2004). The importance of shallow sediments in the recruitment of *Anabaena* and *Aphanizomenon* (Cyanophyceae). J. Phycol. *40*, 831–836.

Karlsson-Elfgren, I., Rengefors, K., and Gustafsson, S. (2004). Factors regulating recruitment from the sediment to the water column in the bloom-forming cyanobacterium *Gleotrichia echinulata*. Freshwater Biol. *49*, 265–273.

Kezhi, B., Guoliang, W., and Cheng, C. (1985). Studies on the mechanism of light-dependent germination of akinetes of blue-green algae. Hydrobiologia *123*, 89–91.

Khamar, H.J., Breathwaite, E.K., Prasse, C.E., Fraley, E.R., Secor, C.R., Chibane, F.L., Elhai, J., and Chiu, W.-L. (2010). Multiple roles of soluble sugars in the establishment of *Gunnera-Nostoc* endosymbiosis. Plant Physiol. *154*, 1381–1389.

Khudyakov, I., and Wolk, C.P. (1997). *hetC*, a gene coding for a protein similar to bacterial ABC protein exporters, is involved in early regulation of heterocyst differentiation in *Anabaena* sp. strain PCC 7120. J. Bacteriol. *179*, 6971–6978.

Kim, Y., Joachimiak, G., Ye, Z., Binkowski, T.A., Zhang, R., Gornicki, P., Callahan, S.M., Hess, W.R., Haselkorn, R., and Joachimiak, A. (2011). Structure of transcription factor HetR required for heterocyst differentiation in cyanobacteria. Proc. Natl. Acad. Sci. *108*, 10109–10114.

Knight, C.D., and Adams, D.G. (1996). A method for studying chemotaxis in nitrogen fixing cyanobacterium-plant symbioses. Physiol. Mol. Plant Pathol. *49*, 73–77.

Kumar, K., Mella-Herrera, R.A., and Golden, J.W. (2010). Cyanobacterial heterocysts. Cold Spring Harb. Perspect. Biol. *2*, a000315.

Kumazaki, S., Akari, M., and Hasegawa, M. (2013). Transformation of thylakoid membranes during differentiation from vegetative cell into heterocyst visualized by microscopic spectral imaging. Plant Physiol. *161*, 1321–1333.

Lamont, H.C. (1969). Sacraficial dell death and trichome breakage in Oscillatorian blue-green alga: The role of Murine. Arch. Microbiol. *69*, 237–259.

Lang, N.J., and Fay, P. (1971). The heterocysts of blue-green algae II. Details of ultrastructure. Proc. Roy. Soc. Lond. B *178*, 193–203.

Lang, N.J., Simon, R.D., and Wolk, C.W. (1972). Correspondence of cyanophyicin granules with structured granules in *Anabaena cylindrica*. Arch. Microbiol. *83*, 313–320.

Lázaro, S., Fernández-Piñas, F., Fernández-Valiente, E., Blanco-Rivero, A., and Leganés, F. (2001). *pbpB*, a gene coding for a putative penicillin-binding protein, is required for aerobic nitrogen fixation in the cyanobacterium *Anabaena* sp. strain PCC7120. J. Bacteriol. *183*, 628–636.

Lazaroff, N. (1973). Photomorphogenesis and Nostocacean development. In The biology of blue-green algae, Carr, N.G., and Whitton, B.A., eds. (Blackwell Scientific Publications, Oxford), pp. 279–319.

Lechno-Yossef, S., Fan, Q., Ehira, S., Sato, N., and Wolk, C.P. (2006). Mutations in four regulatory genes have interrelated effects on heterocyst maturation in *Anabaena* sp. strain PCC 7120. J. Bacteriol. *188*, 7387–7395.

Lechno-Yossef, S., Fan, Q., Wojciuch, E., and Wolk, C.P. (2011). Identification of ten *Anabaena* sp. genes that under aerobic conditions are required for growth on dinitrogen but not for growth on fixed nitrogen. J. Bacteriol. *193*, 3482–3489.

Leganés, F., Fernández-Piñas, F., and Wolk, C.P. (1994). Two mutations that block heterocyst differentiation have different effects on akinete differentiation in *Nostoc ellipsosporum*. Mol. Microbiol. *12*, 679–684.

Leganés, F., Fernández-Piñas, F., and Wolk, C.P. (1998). A transposition-induced mutant of *Nostoc ellipsosporum* implicates an arginine-biosynthetic gene in the formation of cyanophycin granules and of functional heterocysts and akinetes. Microbiol. (Reading) *144*, 1799–1805.

Lehner, J. (2012). Funktionelle Charkterisierung der Zellwandproteine AmiC2 und SepJN und ihre Bedeutung für die Vielzelligkeit des filamentösen Cyanobakteriums *Nostoc punctiforme* ATCC 29133. PhD thesis (Faculty of Science, Eberhard Karls University of Tübingen).

Lehner, J., Zhang, Y., Berendt, S., Rasse, T.M., Forchhammer, K., and Maldener, I. (2011). The morphogene AmiC2 is pivotal for multicellular development in the cyanobacterium *Nostoc punctiforme*. Mol. Microbiol. 79, 1655–1669.

Lehner, J., Berendt, S., Forchhammer, K., and Maldener, I. (2012). Perforation of the septal murein by cell wall amidase AmiC is essential for cell–cell communication and cell differentiation in filamentous cyanobacteria of the order *Nostocales*. Paper presented at: ISPP 2012 (Porto, Portugal).

Lehner, J., Berendt, S., Dörsam, B., Perez, R., Forchhammer, K., and Maldener, I. (2013). Prokaryotic multicellularity: a nanopore array for bacterial cell communication. FASEB J. *27*, 2293–2300.

Li, R., Watanabe, M., and Watanabe, M.M. (1997). Akinete formation in planktonic *Anabaena* spp (Cyanobacteria) by treatment with low temperature. J. Phycol. *33*, 576–584.

Liaimer, A., Jenke-Kodama, H., Ishida, K., Hinrichs, K., Stangeland, J., Hertweck, C., and Dittmann, E. (2011). A polyketide interferes with cellular differentiation in the symbiotic cyanobacterium *Nostoc punctiforme*. Environ. Microbiol. Reports 3, 550–558.

Liang, J., Scappino, L., and Haselkorn, R. (1992). The *patA* gene product, which contains a region similar to CheY of *Escherichia coli*, controls heterocyst pattern formation in the cyanobacterium *Anabaena* 7120. Proc. Natl. Acad. Sci. U.S.A. *89*, 5655–5659.

Liu, D., and Golden, J.W. (2002). *hetL* overexpression stimulates heterocyst formation in *Anabaena* sp. strain PCC 7120. J. Bacteriol. *184*, 6873–6881.

López-Igual, R., Flores, E., and Herrero, A. (2010). Inactivation of a heterocyst-specific invertase indicates a principal role of sucrose catabolism in heterocysts of *Anabaena* sp. J. Bacteriol. *192*, 5526–5533.

López-Igual, R., Lechno-Yossef, S., Fan, Q., Herrero, A., Flores, E., and Wolk, C.P. (2012). A major facilitator superfamily protein, HepP, is involved in formation of the heterocyst envelope polysaccharide in the cyanobacterium *Anabaena* sp. strain PCC 7120. J. Bacteriol. *194*, 4677–4687.

Luque, I., and Forchhammer, K. (2008). Nitrogen assimilation and C/N balance sensing. In The Cyanobacteria Molecular Biology, Genomics and Evolution, Herrero, A., and Flores, E., eds. (Caister Academic Press, Norfolk, UK), pp. 335–382.

Makarova, K.S., Koonin, E.V., Haselkorn, R., and Galperin, M.Y. (2006). Cyanobacterial response regulator PatA contains a conserved N-terminal domain (PATAN) with an alpha-helical insertion. Bioinformatics *22*, 1297–1301.

Maldener, I., and Muro-Pastor, A.M. (2010). Cyanobacterial heterocysts. In Encyclopedia of Life Sciences (ELS) (John Wiley and Sons). doi:10.1002/9780470015902.a0000306.pub2

Maldener, I., Fiedler, G., Ernst, A., Fernández-Piñas, F., and Wolk, C.P. (1994). Characterization of *devA*, a gene required for the maturation of proheterocysts in the cyanobacterium *Anabaena* sp. strain PCC 7120. J. Bacteriol. *176*, 7543–7549.

Maldener, I., Hannus, S., and Kammerer, M. (2003). Description of five mutants of the cyanobacterium *Anabaena* sp strain PCC 7120 affected in heterocyst differentiation and identification of the transposon-tagged genes. FEMS Microbiol. Lett. *224*, 205–213.

Meeks, J.C. (1998). Symbiosis between nitrogen-fixing cyanobacteria and plants. BioScience *48*, 266–276.

Meeks, J.C., and Elhai, J. (2002). Regulation of cellular differentiation in filamentous cyanobacteria in free-living and plant-associated symbiotic growth states. Microbiol. Mol. Biol. Rev. *66*, 94–121; table of contents.

Meeks, J.C., Campbell, E.L., Summers, M.L., and Wong, F.C. (2002). Cellular differentiation in the cyanobacterium *Nostoc punctiforme*. Arch. Microbiol. *178*, 395–403.

Miller, M.M., and Lang, N.J. (1968). The fine structure of akinete formation and germination in *Cylindrospermum*. Arch. Microbiol. *60*, 303–313.

Mitschke, J., Vioque, A., Haas, F., Hess, W.R., and Muro-Pastor, A.M. (2011). Dynamics of transcriptional start site selection during nitrogen stress-induced cell differentiation in *Anabaena* sp. PCC7120. Proc. Natl. Acad. Sci. U.S.A. *108*, 20130–20135.

Montesinos, M.L., Herrero, A., and Flores, E. (1995). Amino acid transport systems required for diazotrophic growth in the cyanobacterium *Anabaena* sp. strain PCC 7120. J. Bacteriol. *177*, 3150–3157.

Moore, D., O'Donohue, M., Shaw, G., and Critchley, C. (2003). Potential triggers for akinete differentiation in an Australian strain of the cyanobacterium *Cylindrospermopsis raciborskii* (AWT 205/1). Hydrobiologia *506–509*, 175–180.

Moore, D., McGregor, G., and Shaw, G. (2004). Morphological changes during akinete germination in *Cylindrospermopsis raciborskii* (Nostocales, cyanobacteria). J. Phycol. *40*, 1098–1105.

Moore, D., O'Donohue, M., Garnett, C., Critchley, C., and Shaw, G. (2005). Factors affecting akinete differentiation in *Cylindrospermopsis raciborskii* (Nostocales, Cyanobacteria). Freshwater Biol. *50*, 345–352.

Moslavac, S., Nicolaisen, K., Mirus, O., Al Dehni, F., Pernil, R., Flores, E., Maldener, I., and Schleiff, E. (2007). A TolC-like protein is required for heterocyst development in *Anabaena* sp. strain PCC 7120. J. Bacteriol. *189*, 7887–7895.

Mullineaux, C.W., Mariscal, V., Nenninger, A., Khanum, H., Herrero, A., Flores, E., and Adams, D.G. (2008). Mechanism of intercellular molecular exchange in heterocyst-forming cyanobacteria. EMBO J. *27*, 1299–1308.

Muro-Pastor, A.M., and Hess, W.R. (2012). Heterocyst differentiation: from single mutants to global approaches. Trends Microbiol. *20*, 548–557.

Muro-Pastor, A.M., Valladares, A., Flores, E., and Herrero, A. (1999). The *hetC* gene is a direct target of the NtcA transcriptional regulator in cyanobacterial heterocyst development. J. Bacteriol. *181*, 6664–6669.

Muro-Pastor, A.M., Valladares, A., Flores, E., and Herrero, A. (2002). Mutual dependence of the expression of the cell differentiation regulatory protein HetR and the global nitrogen regulator NtcA during heterocyst development. Mol. Microbiol. *44*, 1377–1385.

Muro-Pastor, A.M., Olmedo-Verd, E., and Flores, E. (2006). All4312, an NtcA-regulated two-component response regulator in *Anabaena* sp. strain PCC 7120. FEMS Microbiol. Lett. *256*, 171–177.

Murry, M.A., Olafsen, A.G., and Benemann, J.R. (1981). Oxidation of diaminobenzidine in the heterocysts of *Anabaena cylindrica*. Curr. Microbiol. *6*, 201–206.

Neunuebel, M.R., and Golden, J.W. (2008). The *Anabaena* sp. strain PCC 7120 gene *all2874* encodes a diguanylate cyclase and is required for normal heterocyst development under high-light growth conditions. J. Bacteriol. *190*, 6829–6836.

Nichols, J.M., and Adams, D.G. (1982). Akinetes. In The biology of cyanobacteria, Carr, N.G., and Whitton, B.A., eds. (Blackwell Scientific Publications, Oxford), pp. 387–412.

Nicolaisen, K., Hahn, A., and Schleiff, E. (2009). The cell wall in heterocyst formation by *Anabaena* sp. PCC 7120. J. Basic Microbiol. *49*, 5–24.

Olmedo-Verd, E., Valladares, A., Flores, E., Herrero, A., and Muro-Pastor, A.M. (2008). Role of two NtcA-binding sites in the complex *ntcA* gene promoter of the heterocyst-forming cyanobacterium *Anabaena* sp. strain PCC 7120. J. Bacteriol. *190*, 7584–7590.

Pernil, R., Picossi, S., Mariscal, V., Herrero, A., and Flores, E. (2008). ABC-type amino acid uptake transporters Bgt and N-II of *Anabaena* sp. strain PCC 7120 share an ATPase subunit and are expressed in vegetative cells and heterocysts. Mol. Microbiol. *67*, 1067–1080.

Picossi, S., Montesinos, M.L., Pernil, R., Lichtle, C., Herrero, A., and Flores, E. (2005). ABC-type neutral amino acid permease N-I is required for optimal diazotrophic growth and is repressed in the heterocysts of *Anabaena* sp. strain PCC 7120. Mol. Microbiol. *57*, 1582–1592.

Rai, A.N., Rao, V.V., and Singh, H.V. (1985). The biology of cyanobacterial akinetes. J. Plant Sci. Res. *1*, 1–20.

Rai, A.N., Söderbäck, E., and Bergman, B. (2000). Cyanobacterium-plant symbioses. New Phytol. *147*, 449–481.

Rao, V.V., Ghosh, R., and Singh, H.N. (1987). Diazotrophic regulation of akinete development in the cyanobacterium *Anabaena doliolum*. New Phytol. *106*, 161–168.

Rasmussen, U., Johansson, C., and Bergman, B. (1994). Early communication in the *Gunnera–Nostoc* symbiosis: Plant-induced cell differentiation and protein synthesis in the cyanobacterium. Mol. Plant Microbe Interact. *6*, 696–702.

Read, N., Connell, S., and Adams, D.G. (2007). Nanoscale visualization of a fibrillar array in the cell wall of filamentous cyanobacteria and Its implications for gliding motility. J. Bacteriol. *189*, 7361–7366.

Reynolds, C.S. (1972). Growth, gas vacuolation and buoyancy in a natural population of a planktonic blue-green alga. Freshwater Biol. *2*, 87–106.

Rippka, R., Deruelles, J., Waterbury, J.B., Herdman, M., and Stanier, R.Y. (1979). Generic assignments, strain histories and properties of pure cultures of cyanobacteria. J. Gen. Microbiol. *111*, 1–61.

Risser, D.D., and Callahan, S.M. (2008). HetF and PatA control levels of HetR in *Anabaena* sp. strain PCC 7120. J. Bacteriol. *190*, 7645–7654.

Risser, D.D., and Callahan, S.M. (2009). Genetic and cytological evidence that heterocyst patterning is regulated by inhibitor gradients that promote activator decay. Proc. Natl. Acad. Sci. U.S.A. *106*, 19884–19888.

Risser, D.D., Wong, F.C., and Meeks, J.C. (2012). Biased inheritance of the protein PatN frees vegetative cells to initiate patterned heterocyst differentiation. Proc. Natl. Acad. Sci. U.S.A. *109*, 15342–15347.

Sarma, T.A., and Khattar, J.I.S. (1993). Akinete differentiation in phototrophic, photoheterotrophic and chemoheterotrophic conditions in *Anabaena torulosa*. Folia Microbiol. *38*, 335–340.

Sato, N., and Wada, A. (1996). Disruption analysis of the gene for a cold-regulated RNA-binding protein, *rbpA1*, in *Anabaena*: cold-induced initiation of the heterocyst differentiation pathway. Plant Cell Physiol. *37*, 1150–1160.

Sili, C., Ena, A., Materassi, R., and Vincenzini, M. (1994). Germination of desiccated aged akinetes of alkaliphilic cyanobacteria. Arch. Microbiol. *162*, 20–25.

Simon, R.D. (1977). Macromolecular composition of spores from the filamentous cyanobacterium *Anabaena cylindrica*. J. Bacteriol. *129*, 1154–1155.

Simon, R.D. (1987). Inclusion bodies in the cyanobacteria: cyanophycin, polyphosphate, and polyhedral bodies. In The Cyanobacteria: Current Research, Fay, P., and van Baalen, C., eds. (Elsevier/North Holland Biomedical Press, Amsterdam), pp. 199–226.

Singh, H.N., and Sunita, K. (1974). Biochemical study of spore germination in blue-green alga *Anabaena doliolum*. J. Exp. Bot. *25*, 837–845.

Staron, P., Forchhammer, K., and Maldener, I. (2011). Novel ATP-driven pathway of glycolipid export involving TolC protein. J. Biol. Chem. *286*, 38202–38210.

Sukenik, A., Beardall, J., and Hadas, O. (2007). Photosynthetic characterization of developing and mature akinetes of *Aphanizomenon ovalisporum* (cyanoprokaryota). J. Phycol. *43*, 780–788.

Sukenik, A., Stojkovic, S., Malinsky-Rushansky, N., Viner-Motzini, Y., and Beardall, J. (2009). Fluorescence approaches reveal variations in cellular composition during formation of akinetes in the cyanobacterium *Aphanizomenon ovalisporum*. Eur. J. Phycol. *44*, 309–317.

Sukenik, A., Kaplan-Levy, R.N., Mark Welch, J., and Post, A.F. (2012). Massive multiplication of genome and ribosomes in dormant cells (akinetes) of *Aphanizomenon ovalisporum* (Cyanobacteria). ISME J. *6*, 670–679.

Sukenik, A., Kaplan-Levy, R.N., Viner-Motzini, Y., Quesada, A., and Hadas, O. (2013). Potassium deficiency triggers the development of dormant cells (akinetes) in *Aphanizomenon ovalisporum* (Nostocales, Cyanoprokaryota). J. Phycol. *49*, 580–587.

Summers, M.L., Wallis, J.G., Campbell, E.L., and Meeks, J.C. (1995). Genetic evidence of a major role for glucose-6-phosphate dehydrogenase in nitrogen fixation and dark growth of the cyanobacterium *Nostoc* sp. strain ATCC 29133. J. Bacteriol. *177*, 6184–6194.

Sutherland, J.M., Herdman, M., and Stewart, W.D.P. (1979). Akinetes of the cyanobacterium *Nostoc* PCC 7524: macromolecular composition, structure and control of differentiation. J. Gen. Microbiol. *115*, 273–287.

Sutherland, J.M., Reaston, J., Stewart, W.D.P., and Herdman, M. (1985). Akinetes of cyanobacterium *Nostoc* pcc-7524 macromolecular and biochemical changes during synchronous germination. J. Gen. Microbiol. *131*, 2855–2864.

Tandeau de Marsac, N. (1994). Differentiation of hormogonia and relationships with other biological processes. In The Molecular Biology of Cyanobacteria, Bryant, D.A., ed. (Dordrecht: Kluwer Publishers), pp. 825–842.

Tanigawa, R., Shirokane, M., Maeda Si, S., Omata, T., Tanaka, K., and Takahashi, H. (2002). Transcriptional activation of NtcA-dependent promoters of *Synechococcus* sp. PCC 7942 by 2-oxoglutarate in vitro. Proc. Natl. Acad. Sci. U.S.A. *99*, 4251–4255.

Thiel, T. (1990). Protein turnover and heterocyst differentiation in the cyanobacterium *Anabaena variabilis*. J. Phycol. *26*, 50–54.

Thiel, T., and Wolk, C.P. (1983). Metabolic activities of isolated akinetes of the cyanobacterium *Nostoc spongiaeforme*. J. Bacteriol. *156*, 369–374.

Thomas, J., Meeks, J.C., Wolk, C.P., Shaffer, P.W., Austin, S.M., and Chien, W.-S. (1977). Formation of Glutamine from [^{13}N]ammonia, [^{13}N]dinitrogen, and [^{14}C]glutamate by Heterocysts Isolated from *Anabaena cylindrica*. J. Bacteriol. *129*, 1545–1555.

Toyoshima, M., Sasaki, N.V., Fujiwara, M., Ehira, S., Ohmori, M., and Sato, N. (2010). Early candidacy for differentiation into heterocysts in the filamentous cyanobacterium Anabaena sp. PCC 7120. Arch. Microbiol. *192*, 23–31.

Thompson, P.A., Jameson, I., and Blackburn, S.I. (2009). The influence of light quality on akinete formation and germination in the toxic cyanobacterium *Anabaena circinalis*. Harmful Algae *8*, 504–512.

Torrecilla, I., Leganés, F., Bonilla, I., and Fernández-Piñas, F. (2000). Use of recombinant aequorin to study calcium homeostasis and monitor calcium transients in response to heat and cold shock in cyanobacteria. Plant Physiol. *123*, 161–176.

Ungerer, J.L., Pratte, B.S., and Thiel, T. (2008). Regulation of fructose transport and its effect on fructose toxicity in *Anabaena* spp. J. Bacteriol. *190*, 8115–8125.

Valladares, A., Herrero, A., Pils, D., Schmetterer, G., and Flores, E. (2003). Cytochrome c oxidase genes required for nitrogenase activity and diazotrophic growth in *Anabaena* sp. PCC 7120. Mol. Microbiol. *47*, 1239–1249.

Valladares, A., Maldener, I., Muro-Pastor, A.M., Flores, E., and Herrero, A. (2007). Heterocyst development and diazotrophic metabolism in terminal respiratory oxidase mutants of the cyanobacterium *Anabaena* sp. strain PCC 7120. J. Bacteriol. *189*, 4425–4430.

Vargas, W.A., Nishi, C.N., Giarrocco, L.E., and Salerno, G.L. (2011). Differential roles of alkaline/neutral invertases in *Nostoc* sp. PCC 7120: Inv-B isoform is essential for diazotrophic growth. Planta *233*, 153–162.

Vazquez-Bermudez, M.F., Herrero, A., and Flores, E. (2003). Carbon supply and 2-oxoglutarate effects on expression of nitrate reductase and nitrogen-regulated genes in *Synechococcus* sp. strain PCC 7942. FEMS Microbiol. Lett. *221*, 155–159.

Walsby, A.E. (1985). The permeability of heterocysts to the gases nitrogen and oxygen. Proc. R. Soc. Lond. B *226*, 345–366.

Walsby, A.E. (2007). Cyanobacterial heterocysts: terminal pores proposed as sites of gas exchange. Trends Microbiol. *15*, 340–349.

Wang, Y., Lechno-Yossef, S., Gong, Y., Fan, Q., Wolk, C.P., and Xu, X. (2007). Predicted glycosyl transferase genes located outside the HEP island are required for formation of heterocyst envelope polysaccharide in *Anabaena* sp. strain PCC 7120. J. Bacteriol. *189*, 5372–5378.

Wei, T.F., Ramasubramanian, T.S., and Golden, J.W. (1994). *Anabaena* sp. strain PCC 7120 *ntcA* gene required for growth on nitrate and heterocyst development. J. Bacteriol. *176*, 4473–4482.

Wilcox, M., Mitchison, G.J., and Smith, R.J. (1973). Pattern formation in the blue-green alga *Anabaena*. II. Controlled proheterocyst regression. J. Cell. Sci. *13*, 637–649.

Wildman, R.B., Loescher, J.H., and Carol, L.W. (1975). Development and germination of akinetes of *Aphanizomenon flos-aquae*. J. Phycol. *11*, 96–104.

Wolk, C.P. (1966). Evidence of a role of heterocysts in the sporulation of a blue-green alga. Am. J. Bot. *53*, 260–262.

Wolk, C.P. (1968). Movement of carbon from vegetative cells to heterocysts in *Anabaena cylindrica*. J. Bacteriol. *96*, 2138–2143.

Wolk, C.P. (2000). Heterocyst formation in Anabaena. ASM Press, Washington, DC.

Wolk, C.P., Austin, S.M., Bortins, J., and Galonsky, A. (1974). Autoradiographic localization of ^{13}N after fixation of ^{13}N-labeled nitrogen gas by a heterocyst-forming blue-green alga. J. Cell Biol. *61*, 440–453.

Wolk, C.P., Ernst, A., and Elhai, J. (1994). Heterocyst metabolism and development. In The Molecular Biology of cyanobacteria, Bryant, D.A., ed. (Dordrecht, The Netherlands: Kluwer Academic Publishers), pp. 769–823.

Wolk, C.P., Thomas, J., Shaffer, P.W., Austin, S.M., and Galonsky, A. (1976). Pathway of nitrogen metabolism after fixation of ^{13}N-labeled nitrogen gas by the cyanobacterium, *Anabaena cylindrica*. J. Biol. Chem. *251*, 5027–5034.

Wong, F.C., and Meeks, J.C. (2001). The *hetF* gene product is essential to heterocyst differentiation and affects HetR function in the cyanobacterium *Nostoc punctiforme*. J. Bacteriol. *183*, 2654–2661.

Wood, N.B., and Haselkorn, R. (1980). Control of phycobiliprotein proteolysis and heterocyst differentiation in *Anabaena*. J. Bacteriol. *141*, 1375–1385.

Wood, P., Peat, A., and Whitton, B.A. (1986). Influence of phosphorous status on the fine structure of the cyanobacterium (blue-green algae) *Calothrix parietina*. Cytobiosis *47*, 89–99.

Wyman, M., and Fay, P. (1986). Interaction between light quality and nitrogen availability in the differentiation of akinetes in the planktonic cyanobacterium *Gloeotrichia echinulata*. Brit. Phycol. J. *21*, 147–154.

Xu, X., and Wolk, C.P. (2001). Role for *hetC* in the transition to a nondividing state during heterocyst differentiation in *Anabaena* sp. J. Bacteriol. *183*, 393–396.

Xu, X., Khudyakov, I., and Wolk, C.P. (1997). Lipopolysaccharide dependence of cyanophage sensitivity and aerobic nitrogen fixation in *Anabaena* sp. strain PCC 7120. J. Bacteriol. *179*, 2884–2891.

Xu, X., Elhai, J., and Wolk, C.P. (2008). Transcriptional and developmental responses by *Anabaena* to deprivation of fixed nitrogen. In The Cyanobacteria Molecular Biology, Genomics and Evolution, Herrero, A., and Flores, E., eds. (Caister Academic Press, Norfolk, UK), pp. 382–422.

Yamamoto, Y. (1976). Effect of some physical and chemical factors on the germination of akinetes of *Anabaena cylindrica*. J. Gen. Appl. Microbiol. *22*, 311–323.

Yoon, H.S., and Golden, J.W. (1998). Heterocyst pattern formation controlled by a diffusible peptide. Science *282*, 935–938.

Zhang, C.-C., Laurent, S., Sakr, S., Peng, L., and Bedú, S. (2006). Heterocyst differentiation and pattern formation in cyanobacteria: a chorus of signals. Mol. Microbiol. *59*, 367–375.

Zhang, W., Du, Y., Khudyakov, I., Fan, Q., Gao, H., Ning, D., Wolk, C.P., and Xu, X. (2007). A gene cluster that regulates both heterocyst differentiation and pattern formation in *Anabaena* sp. strain PCC 7120. Mol. Microbiol. *66*, 1429–1443.

Zhao, J., and Wolk., C.P. (2008). Developmental biology of heterocysts, 2006. In Myxobacteria: Multicellularity and Differentiation, Whitworth, D.E., ed. (ASM Press, Washington, DC), pp. 397–418.

Zhao, Y., Shi, Y., Zhao, W., Huang, X., Wang, D., Brown, N., Brand, J., and Zhao, J. (2005). CcbP, a calcium-binding protein from *Anabaena* sp. PCC 7120, provides evidence that calcium ions regulate heterocyst differentiation. Proc. Natl. Acad. Sci. U.S.A. *102*, 5744–5748.

Zhou, R., and Wolk, C.P. (2003). A two-component system mediates developmental regulation of biosynthesis of a heterocyst polysaccharide. J. Biol. Chem. *278*, 19939–19946.

Zhou, R., Wei, X., Jiang, N., Li, H., Dong, Y., Hsi, K.L., and Zhao, J. (1998). Evidence that HetR protein is an unusual serine-type protease. Proc. Natl. Acad. Sci. U.S.A. *95*, 4959–4963.

Zhu, J., Kong, R., and Wolk, C.P. (1998). Regulation of *hepA* of *Anabaena* sp. strain PCC 7120 by elements 5′ from the gene and by *hepK*. J. Bacteriol. *180*, 4233–4242.

Zhu, J., Jäger, K., Black, T., Zarka, K., Koksharova, O., and Wolk, C.P. (2001). HcwA, an autolysin, is required for heterocyst maturation in *Anabaena* sp. strain PCC 7120. J. Bacteriol. *183*, 6841–6851.

Ziegler, K., Stephan, D.P., Pistorius, E.K., Ruppel, H.G., and Lockau, W. (2001). A mutant of the cyanobacterium *Anabaena variabilis* ATCC 29413 lacking cyanophycin synthetase: growth properties and ultrastructural aspects. FEMS Microbiol. Lett. *196*, 13–18.

Cell–Cell Joining Proteins in Heterocyst-forming Cyanobacteria

13

Vicente Mariscal

Abstract

The filamentous, heterocyst-forming cyanobacteria have been studied for decades as prototypes of multicellular prokaryotic organisms. They grow as rows of hundreds of cells and have been described as reproducing by random trichome breakage, implying that the unit of growth is the filament. In the heterocyst-forming cyanobacteria, nitrogen fixation and photosynthesis are spatially separated in different cell types. It is well known that filament growth under diazotrophic conditions depends on the intercellular exchange of substances between the different cell types including regulatory signal molecules and metabolites, but the possible routes through which cells communicate have only recently been subjected to study. At least two different paths could be used for cell–cell communication: the continuous periplasm that is shared by all the cells in the filament and some proteinaceous structures present at the septa between adjacent cells that can mediate, operating as channels, direct intercellular transfer of small water-soluble molecules. This chapter focuses on intercellular communication through septal channels and describes the proteins that could compose such channels.

Electron microscopy of the cyanobacterial intercellular septa

The occurrence of intercellular connecting structures perpendicular to the septum between cells in filamentous cyanobacteria has called the attention of cyanobacteriologists since electron microscopy studies were first carried out in these microorganisms. Wildon and Mercer (1963) reported the analysis by transmission electron microscopy of the ultra structure of the heterocyst in *Anabaena* sp. They used samples stained with potassium permanganate and described the presence of thin, lightly stained structures that were present between heterocysts and vegetative cells and involved only the plasma membranes of the adjacent cells without any apparent direct contact between the protoplasts. These structures were afterwards noticed in the septum between consecutive vegetative cells in the filaments (Bisalputra *et al.*, 1975; Bultler and Allsopp, 1972). Lang and Fray (1971) introduced the term 'microplasmodesmata' to name these structures, emphasizing that, like plasmodesmata from plants, they may represent sites of more intimate contact between cells, establishing some type of intercellular connexion. However, although nothing similar to plant plasmodesmata (cytoplasmic bridges delimited by cytoplasmic membranes) has ever been demonstrated to occur in cyanobacteria (Flores *et al.*, 2006), the term 'microplasmodesmata' has been extensively used in the cyanobacterial literature to denote the cell–cell connecting structures seen by electron microscopy. Moreover, to the best of my knowledge, a relationship between the presence of these structures and intercellular transfer of metabolites has never been confirmed in filamentous cyanobacteria.

By means of freeze-fracture electron microscopy, cell–cell connecting structures have been observed as pits in the exterior leaflet of the plasma membrane and corresponding protrusions in the protoplasmic leaflet (Giddings and Staeheling, 1978, 1981). These structures occur at a circular area in the centre of the septum, with a variable number of structures and with a constant size of at least 40 nm in length and less than 5 nm in diameter (Haselkorn, 1978). Heterocyst-forming cyanobacteria have about 200–300 of these structures per septum between vegetative cells, and about 50 structures in the septa between heterocysts and vegetative cells. On the other hand, in filamentous strains that do not form heterocysts the number of cell–cell connecting structures counted in the septa between vegetative cells is only about 30–40 (Giddings and Staehelin, 1981).

In the model heterocyst-forming cyanobacterium *Anabaena* sp. PCC 7120 the nature and distribution within the septum of the cell–cell connecting structures have been recently addressed (Wilk *et al.*, 2011). In this work, the authors combined two complementary staining methods to prepare samples for electron tomography studies, concluding that the structures likely contain protein and do not involve plasma membrane continuity. The first staining method employed potassium permanganate. In this case, cell–cell connecting structures appear as thin stained strands, as previously reported. Tomographic series of the septum between cells clearly show that these structures have a length of 27 nm and a diameter of 5.5 nm and that they occur in a structured manner in the centre of the septum (Wilk *et al.*, 2011).

The second staining method used by Wilk *et al.* (2011) employs a combination of ruthenium red, osmium tetroxide and uranyl acetate that highlight surface-associated carbohydrates and polysaccharides. Because, when applying this method, peptidoglycan layers in the septal region were stained while the junctions were not, perforations that traverse the peptidoglycan layers but not the plasma membrane were clearly seen, having a length of about 26 nm and a diameter of about 14 nm (Wilk *et al.*, 2011). Similar perforations have been found in *Nostoc* spp. (Gorelova and Baulina, 2008; Lehner *et al.*, 2013). Gorelova and Baulina (2008) describe the presence of peptidoglycan perforations perpendicular to the septa between vegetative cells and heterocysts and between vegetative cells. In *Nostoc punctiforme*, high-resolution imaging of isolated peptidoglycan sacculi, void of proteins, have recently revealed the presence of a rigid disc-shaped septal structure containing an array of about 150 'holes', which have been named 'nanopores' (Lehner *et al.*, 2013).

The cell–cell connecting structures found in filamentous cyanobacteria show similar staining properties as the proteinaceous gap junctions of metazoans (Alberts *et al.*, 2002). Gap junctions are intercellular channels made of proteins that join the plasma membranes of adjacent cells and permit the free passage of ions and small molecules between cells. They are cylinders made of 6 units of transmembrane proteins (each bearing four transmembrane segments) called connexins, which assemble to form a connexon. The connexons form a channel that connect the cytoplasm of adjacent cells and hold the interacting plasma membranes at a fixed distance (Alberts *et al.*, 2002). Despite that open reading frames that encode proteins homologous to gap junction proteins from metazoans are not found in the *Anabaena* genome (Kaneko *et al.*, 2001), given that the *Anabaena* cell–cell connecting structures are found in the septum between cells and show size and staining properties similar to gap junctions, septal cell–cell connecting structures of filamentous cyanobacteria will hereafter be referred as 'septal junctions'.

All these observations suggest that the septum between cells in filamentous cyanobacteria is a structured section containing proteinaceous complexes that link the plasma membranes of the adjacent cells and perhaps allow intercellular molecular exchange between these cells. Septal peptidoglycan layers bear a nanopore array that may serve as a framework for the septal junction structures to traverse the rigid peptidoglycan. A model showing this interpretation is presented in Fig. 13.1.

Peptidoglycan remodelling proteins in the intercellular septa

In unicellular bacteria, cell division concludes with daughter cell separation by the action of LytM-type amidases, which specifically hydrolyse the septal cell wall peptidoglycan, and the Pal–Tol complex, which is involved in outer membrane

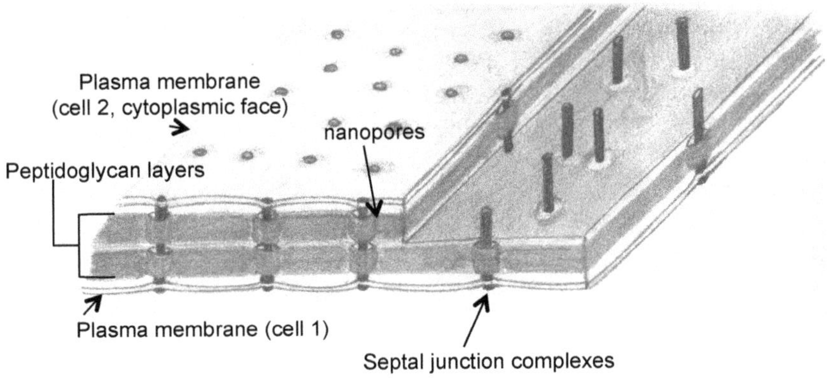

Figure 13.1 Model showing a portion of a septum between adjacent cells in filamentous cyanobacteria illustrating the septal junctions and the characteristic features described by electron microscopy. Septal junctions (black bars), which join the plasma membranes from the neighbouring vegetative cells, traverse the septal peptidoglycan through 'nanopores'.

invagination during septation (Vollmer *et al.*, 2008; Typas *et al.*, 2012). The scenario changes in filamentous cyanobacteria, where peptidoglycan hydrolysis and outer membrane invagination does not take place, which results in the presence of a common envelope that surround all the cells in the filament (Flores *et al.*, 2006). In cyanobacteria of the order *Nostocales* several proteins that are involved in cell wall synthesis and structure have been recently identified. These proteins are ConR in *Anabaena* sp. PCC 7120, and AmiC1 and AmiC2 in *Nostoc punctiforme* and *Anabaena* sp. PCC 7120.

ConR from *Anabaena* sp. PCC 7120 (Mella-Herrera *et al.*, 2011) forms part of a family of proteins that contains a LytR-CpsA-Psr domain, which has been associated with septum formation and cell wall maintenance in *Streptococcus pneumoniae* (Johnsborg and Havarstein, 2009). Mutation of *conR* causes a septum-formation defect that is most evident when the mutant is subjected to nitrogen deprivation. Under these conditions, abnormally long heterocysts that appear open at their ends are formed (Fan *et al.*, 2006; Mella-Herrera, 2011). Consequently, more oxygen can enter the heterocysts accounting for their low nitrogenase activity and impaired diazotrophic growth (Fan *et al.*, 2006). Although ConR function remains unknown, it has been postulated that, like other proteins containing LytR motifs, it could regulate proteins involved in the synthesis and maintenance of the cell wall, in particular those involved in heterocyst septum remodelling during heterocyst differentiation (Mella-Herrera *et al.*, 2011).

The AmiC-type amidases have been also found to contribute to septum formation in filamentous cyanobacteria (Lehner *et al.*, 2011; Berendt *et al.*, 2012). In *E. coli*, the AmiC enzymes are periplasmic peptidoglycan hydrolases that during cell division cleave covalent bonds in the murein sacculus, thus allowing daughter cell separation (Vollmer *et al.*, 2008). AmiC orthologues are found in the filamentous cyanobacteria *Anabaena* sp. PCC 7120 and *Nostoc punctiforme*, and some *amiC* mutants have been studied (Zu *et al.*, 2001; Lehner *et al.*, 2011, 2013; Berendt *et al.*, 2012). Inactivation of *amiC* amidase genes in *Anabaena* sp. PCC 7120 does not alter filament morphology, although a mutant of *amiC1* (open reading frame *alr0092* of the *Anabaena* genome) has been described to be deficient in heterocyst differentiation and diazotrophic growth (Berendt *et al.*, 2012) and a mutant of *amiC2* (*alr0093*, also known as *hcwA*) has been described to be deficient in aerobic diazotrophic growth (Zhu *et al.*, 2001). In contrast, the *amiC2* mutant of *Nostoc punctiforme* shows filament alterations caused by the use of aberrant cell division planes and, as in the case of strain PCC 7120, is impaired in heterocyst differentiation and diazotrophic growth (Lehner *et al.*, 2011). An AmiC2 protein modified to bear a Tat signal sequence (instead of its natural Sec signal sequence) and a C-terminal GFP fusion

(tat_{ss}-AmiC2-GFP) has been shown to preferentially localize to the septa of recently divided cells and to the septa of newly formed heterocysts. Cell wall lytic activity of AmiC2 from *Nostoc punctiforme* has been demonstrated *in vitro*, and heterologous expression of the AmiC2 catalytic domain restores the ability to complete daughter cell separation in a filamentous *E. coli amiA/B/C* triple mutant (Lehner *et al.*, 2013). Based on the observation that the septal discs of the *amiC* mutant from *Nostoc punctiforme* do not have nanopores, Lehner *et al.* (2013) have proposed a novel function for this enzyme in cyanobacteria, namely to drill holes in a highly organized manner into the septal peptidoglycan layers serving as a framework for allocation of septal junctions.

In conclusion, these studies point out the importance of peptidoglycan hydrolases in the maturation of the newly synthetized septum during cell division and in the remodelling of the septum during heterocyst differentiation. It appears that AmiC amidases are also involved in the remodelling of the peptidoglycan that surrounds the septal junction complexes, producing perforations (nanopores) through which such complexes can traverse the peptidoglycan.

Proteins of the intercellular septa in heterocyst-forming cyanobacteria

In several experiments of random transposon mutagenesis carried out in *Anabaena* sp. PCC 7120 searching for mutants unable to fix N_2 under oxic conditions (Fox$^-$ mutants), filament fragmentation mutants frequently appeared (Buikema and Haselkorn, 1991; Ernst *et al.*, 1992, Fan *et al.*, 2006, Nayar *et al.*, 2006). Characterization of some of such mutants led to the identification of three proteins involved in maintaining septum architecture and function: SepJ, FraC and FraD. Although these proteins do not show homology to connexins of gap junctions, they could be the components of the septal junctions because they have been localized to the intercellular septa and have been shown to provide cell–cell binding and intercellular transfer functions in *Anabaena* sp. PCC 7120.

The SepJ protein

Open reading frame *alr2338* of the genome of *Anabaena* sp. PCC 7120 (Kaneko *et al.*, 2001) was targeted in several transposon-induced Fox$^-$ mutants, including four of those generated by Nayar *et al.* (2007) and 21 out of the 1,076 generated by Fan *et al.* (2006). The phenotype of some of these mutants was studied in detail and the corresponding gene was named *fraG* (fragmentation, glycolipid; Nayar *et al.*, 2007) or *sepJ* (septal protein; Flores *et al.*, 2007).

Mutants of *sepJ* make short filaments of about eight cells in media with combined nitrogen (Flores *et al.*, 2007; Nayar *et al.*, 2007; Mariscal *et al.*, 2011). However, filament fragmentation in the *sepJ* mutants becomes most evident upon incubation in the absence of combined nitrogen, resulting in very short filaments of two or three cells. In these filaments, heterocyst differentiation is aborted at an early stage, just after the deposition of the polysaccharide layer of the heterocyst envelope, before the cells are committed to differentiate. The *sepJ* mutants produce aberrant pro-heterocysts containing two or three cells encapsulated by a polysaccharide sheath and are incapable of diazotrophic growth (Flores *et al.*, 2007; Nayar *et al.*, 2007; Mariscal *et al.*, 2011).

Expression of *sepJ* is regulated by nitrogen availability; *sepJ* transcript levels increase about three times when the filaments are subjected to nitrogen deprivation. However, the induction of *sepJ* does not require the heterocyst differentiation regulators NtcA and HetR (Flores *et al.*, 2007). The *sepJ* gene encodes a 751 amino acid protein that comprises three main domains: a 200 residue N-terminal coiled-coil region that could be involved in protein–protein interactions (Newman *et al.*, 2000; Lupas and Gruber, 2005) and is required for filament integrity and diazotrophic growth (Mariscal *et al.*, 2011); a linker domain of 211 amino acids rich in Pro and Ser residues that seems to be dispensable for SepJ function; and a 340 residue C-terminal permease domain, which is similar to proteins of the Drug and Metabolite Exporter (DME) family and provides intercellular molecular exchange functions required for diazotrophy (Flores *et al.*, 2007; Mariscal *et al.*, 2011; Fig. 13.2A).

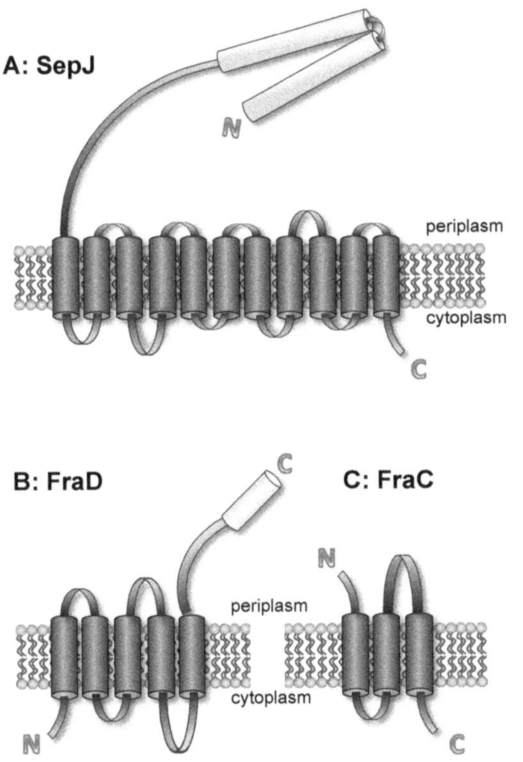

Figure 13.2 Topological models of the septal proteins from *Anabaena* sp. PCC 7120 (not to scale). The grey and white cylinders represent transmembrane segments and coiled-coil domains respectively.

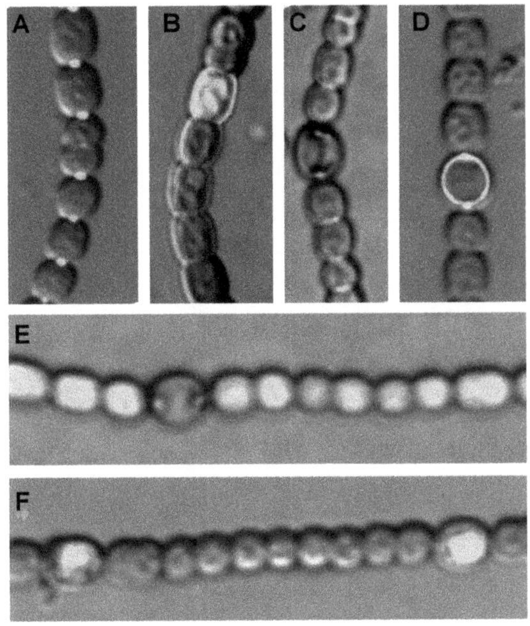

Figure 13.3 Uses of GFP in *Anabaena* sp. PCC 7120 as reporter of subcellular localization of proteins and cellular specificity of promoter activity. Figures show the localization of GFP in the septa between cells (A, SepJ–GFP fusion; Flores *et al.*, 2007), soluble in the periplasm (B, P_{patS}-GFP bearing a Tat signal peptide and a signal peptidase I cutting site [P_{patS}-tat$_{ss}$-spI$_{cs}$-GFP]; Mariscal *et al.*, 2007), integrated in the plasma membrane of vegetative cells (C, Amt1–GFP fusion; Merino-Puerto *et al.*, 2010), anchored to the plasma membrane of heterocysts (D, P_{patS}-GFP bearing a Tat signal peptide and signal peptidase II cutting site [P_{patS}-tat$_{ss}$-spII$_{cs}$-GFP]; Mariscal *et al.*, 2007), soluble in the cytoplasm of vegetative cells (E, P_{rbcL}-GFP; Yoon and Golden, 1998) and soluble in the cytoplasm of heterocysts (F, P_{patS}-GFP; Mariscal *et al.*, 2007). A colour version of this figure is available in the plate section at the back of the book.

In recent years, green fluorescent protein (GFP) fusions have become widely used for subcellular localization of proteins and for the study of cell-specific gene or protein expression in cyanobacteria. For this, the use of enhanced-fluorescence variants of the GFP and the separation of GFP fluorescence emission from autofluorescence of the cyanobacterial photosynthetic pigments has been crucial. Using these procedures, cytoplasmic, periplasmic and membrane-associated proteins have been localized in the model organism *Anabaena* sp. PCC 7120 (Mariscal *et al.*, 2008; Olmedo-Verd *et al.*, 2006; Merino-Puerto *et al.*, 2011b; Yoon and Golden, 1998; Fig. 13.3). One example relevant for this review chapter is a SepJ–GFP fusion, which is located at the cell poles in the intercellular septa between vegetative cells and between vegetative cells and heterocysts (Flores *et al.*, 2007; Fig. 13.3A). It has been shown that septal localization of SepJ-GFP depends on the presence of the coiled-coil domain but not of the linker domain (Mariscal *et al.*, 2011). Additionally, SepJ-GFP is seen localized to a ring, similar to a Z ring, in the middle of the cells when cell division starts, and it progressively focuses to the middle of the septum during cell division (Flores *et al.*, 2007; Mariscal and Flores, 2010). The ring-like distribution of SepJ suggessts that it might be recruited to the early division ring and interact with proteins of the divisome.

The predicted topology of SepJ suggests that the coiled coil and linker domains are placed in the periplasm, whereas the C-terminus of the protein would be located in the cytoplasm (Fig. 13.2A). This topology is supported by the fact that the GFP, which only folds properly in the cytoplasm, is readily detected when fused to the C-end of SepJ (Flores *et al.*, 2007). Additionally, when heterologously expressed in *E. coli*, the coiled-coil domain of SepJ has been found to be exported to and concentrated in the periplasm, indicating that it has the targeting signals required for export to this bacterial compartment (V. Mariscal, unpublished). The predicted periplasmic localization of the SepJ coiled-coil domain allows to proposse a role for it in binding adjacent cells by means of interactions between the coiled-coil domains of SepJ from each cell. Thus, SepJ could be one of the components of the septal junction complexes that have been observed by electron microscopy.

In order to establish a relationship between septal junctions and SepJ in *Anabaena* sp. PCC 7120, the intercellular septa of a *sepJ* deletion mutant were studied by means of electron tomography (Wilk *et al.*, 2011). Remarkably, septal junctions comparable of those in the wild-type cells were observed in the *sepJ* mutant. However, a more detailed analysis revealed that the mean distance between the cytoplasmic membranes of the adjacent cells was shorter in the *sepJ* mutant than in the wild-type (18 nm versus 27 nm). This is indicative of an alteration of the septa in the mutant, leading the authors to propose the existence of two types of septal junction complexes, each one with a different size (Wilk *et al.*, 2011). In conclusion, although SepJ is required for maintaining the normal cell–cell connections at the septum, additional components appear to contribute to septal junctions.

The Fra proteins

Other than SepJ, several proteins including FraC, FraD and FraE influence filament integrity in *Anabaena* sp. PCC 7120 (Bauer *et al.*, 1995; Merino-Puerto *et al.*, 2010). It is likely that these proteins are also components of septal junction complexes. As in the case of *sepJ* mutants, a *fraC* mutant was obtained by transposon mutagenesis and identified as a Fox$^-$ mutant that showed fragmentation under nitrogen deprivation (Buikema and Haselkorn, 1991).

The *fraC* gene is part of a cluster of genes containing *fraD* and *fraE*, which are co-transcribed from a promoter upstream of *fraC* (Merino-Puerto *et al.*, 2010). The *fraC* operon is expressed at relatively low, rather constant levels both in the presence and absence of combined nitrogen (Merino-Puerto *et al.*, 2010). In order to know the specific function of each of the *fra* genes, *fraC* and *fraD* were inactivated by introducing in-frame deletions (thus avoiding possible polar effects of the mutations) whereas *fraE*, which is the last gene in the *fraC* operon, was inactivated by insertion of a plasmid (Merino-Puerto *et al.*, 2010). The three *fra* mutants are able to grow at rates comparable to that of the wild-type in the presence of combined nitrogen. However, they are impaired in diazotrophic growth, and upon incubation in the absence of combined nitrogen, they show extensive filament fragmentation. Even though all three mutants apparently show morphologically mature heterocysts, electron microscopy reveals that the heterocyst neck is altered in the mutants: the cytoplasmic membrane close to the septum of the heterocyst is retreated towards the interior of the heterocyst, leading to a wider septum (Merino-Puerto *et al.*, 2011a). On the other hand, in the *fraE* mutant the entire heterocyst neck is missing. It is also worth noting that heterocysts of the *fraE* mutant appear less mature than heterocysts of the *fraC* and *fraD* mutants and sometimes the heterocyst polysaccharide layer is missing (Merino-Puerto *et al.*, 2011a). The fragmentation phenotype observed in all *fra* mutants under diazotrophic conditions could result from the detachment of the altered heterocysts from the filament. Moreover, in the *fra* mutants the malformed septum between heterocysts and vegetative cells could impair the transfer of metabolites required for sustaining diazotrophic growth, if transfer were to take place through septal junctions.

FraC, FraD and FraE are integral membrane proteins. FraC is a 179 residue protein that contains 3 transmembrane segments with a predicted periplasmic N-terminus and a cytoplasmic C-terminus and no other recognizable structural or functional domain (Merino-Puerto *et al.*, 2010; Fig. 13.2C). FraD is a 343 residue protein that contains 5 transmembrane segments with predicted cytoplasmic N-terminus and periplasmic

C-terminus (Fig. 13.2B). The 167 residue-long C-terminal portion of FraD contains two strongly predicted, albeit short, coiled-coil motifs that may be involved in protein–protein interactions. Although the amino-acid sequence of FraD does not show homology to SepJ, the domain architecture of both proteins is similar: an extracellular coiled-coil domain linked to a membrane-spanning domain (Fig. 13.2), and they could play similar functions. FraE is a 267 residue membrane protein bearing 5 transmembrane segments that shows homology to permeases of the ABC_2 type transporter family (IPR 013525) involved in the export of drugs and carbohydrates. This family includes proteins of the capsular polysaccharide export system of Gram-negative bacteria (Reizer *et al.*, 1992). The lack of the heterocyst neck together with the irregular polysaccharide layer deposition in the heterocysts of the *fraE* mutant suggests that FraE could be involved in the export of the polysaccharide layer that surrounds the heterocyst neck.

GFP fusions have been used to study FraC and FraD subcellular localization in *Anabaena* sp. PCC 7120. A GFP–FraD fusion localizes at the intercellular septa similar to, but not as focused as, SepJ-GFP. On the other hand, FraC-GFP localizes to a ring, similar to a Z ring, when cell division starts and mainly to the intercellular septa when cell division is completed, but localization is less defined than that observed for SepJ-GFP or GFP-FraD (Merino-Puerto *et al.*, 2010). Under diazotrophic conditions, the GFP signal is also detected in both strains in the poles of the heterocysts (Merino-Puerto *et al.*, 2010).

The septal localization of the native FraD protein in *Anabaena* sp. PCC 7120 has been confirmed by immunogold labelling with antibodies raised against its C-terminal (predicted periplasmic) segment (Merino-Puerto *et al.*, 2011a). In the wild-type strain, the labelling appeared along all the septum between vegetative cells, more precisely in the zone between the cytoplasmic membranes of the adjacent cells indicating that, as predicted, the C-terminal part of FraD is periplasmic. It is also worth noting that FraD, like SepJ-GFP, is detected in the nascent septa of cells that have started division (Merino-Puerto *et al.*, 2011a). Therefore, as is the case for SepJ, also FraC (see the previous paragraph) and FraD might interact with the divisome.

Several studies suggest that FraC, FraD and SepJ might interact influencing septal localization of each other (Merino-Puerto *et al.*, 2010, 2011a). On one hand, FraD is hardly detected by immunolocalization in a *fraC* mutant (Merino-Puerto *et al.*, 2011a). The facts that *fraC* and *fraD* are co-transcribed, that the *fraC* and *fraD* mutants show very similar (if not identical) phenotypes, and that FraC is required for localization of FraD suggest that both proteins act together, perhaps as a complex. On the other hand, SepJ-GFP localization is altered in the *fraC* and *fraD* mutants. In these mutants, SepJ remains in the septum, but not as focussed at the septum centre as it is in the wild-type strain, being the SepJ–GFP fusion detected along the entire length of the septum between cells (Merino-Puerto *et al.*, 2010). However, the initial localization of SepJ in a position similar to a cell division ring and in the nascent septa is not altered in the *fraC* or *fraD* mutants (Merino-Puerto *et al.*, 2010). Consequently, FraC and FraD proteins could be needed to retain SepJ in the middle of the septum but not for its putative initial positioning in the Z ring during cell division. In contrast, FraD localization is not altered in a *sepJ* mutant (Merino-Puerto *et al.*, 2011a).

Intercellular molecular exchange in filamentous cyanobacteria

The ability to visualize, track, and quantify molecules and events in living cells with high spatial and temporal resolution is essential for understanding physiology (Lippincott-Schwartz and Patterson, 2003). Thus, the use of tracers to study the functionality and properties of the structures that allow the transfer of substances between cells, together with the use of mutants affected in that transfer, has represented an important step forward in the knowledge of the physiology of filamentous cyanobacteria (Mullineaux *et al.*, 2008; Mariscal *et al.*, 2011). Intercellular molecular exchange has been shown to occur in several strains of *Anabaena* sp. and *Nostoc* sp. (Thomas *et al.*, 1977; Mullineaux *et al.*, 2008; Mariscal and Flores, 2010). Although the mechanism of intercellular transfer of compounds in the cyanobacterial filament remains unknown, two different conduits have been considered: a continuous periplasm (Flores *et al.*, 2006; Mariscal *et al.*, 2007; Nicolaisen *et al.*, 2009) and the septal junctions (Mullineaux *et al.*, 2008; Merino-Puerto *et al.*, 2011a). For the identification of this second way of communication between cells and of the proteins involved, the use of small fluorescent probes in different strains of *Anabaena* sp. has been crucial.

Fluorescent tracers and their use in cyanobacteria

Among the different non-toxic fluorescent dyes that have been used to study cell–cell communication, the most fruitful ones are calcein and carboxyfluorescein (Abbaci *et al.*, 2007, 2008). Both compounds are commercially available as non-fluorescent acetoxymethyl esters (Calcein AM and CFDA), which are sufficiently hydrophobic to traverse cell membranes, so that they can be loaded into the cells. In the cytoplasm, the ester groups of the molecules are hydrolysed by esterases to produce the fluorescent, hydrophilic and negatively charged calcein (622 Da) or carboxyfluorescein (375 Da), which no longer can traverse the cytoplasmic membrane.

Intercellular exchange of calcein and carboxyfluorescein has been shown to take place from cell to cell in different biological systems including cyanobacteria (Abbaci *et al.*, 2007; Mullineaux *et al.*, 2008; Mariscal *et al.*, 2011). This intercellular exchange has been studied by means of fluorescence recovery after photobleaching (FRAP) experiments. This technique consists in the irreversible photochemical bleaching of fluorescent molecules, caused by exposition to a light beam of appropriate wavelength and sufficient intensity in a small region of the cell, and recording by time-lapse microscopy (measuring fluorescence within the emission spectrum of the fluorochrome) the subsequent movement into the photobleached area of surrounding non-bleached fluorescent molecules (Lippincott-Schwartz and Patterson, 2003).

By means of FRAP, it has been observed that in filamentous, heterocyst-forming cyanobacteria of the genera *Anabaena* and *Nostoc* calcein and carboxyfluorescein diffuse between cells (Mullineaux *et al.*, 2008; Merino-Puerto *et al.*, 2011a; Lehner *et al.*, 2011). Calcein is exchanged between vegetative cells and between vegetative cells and heterocysts, although the exchange coefficient of the latter is significantly lower than that of the exchange between vegetative cells, perhaps due to the minor surface of contact between heterocysts and vegetative cells (Mullineaux *et al.*, 2008). Carboxyfluorescein is also exchanged between vegetative cells (Mariscal *et al.*, 2011) but, because heterocysts are not labelled, the exchange of this dye between vegetative cells and heterocysts cannot be studied. Because of their characteristic pK_a values (carboxyfluorescein $pK_a = 6.4$; calcein $pK_a = 5.0$), carboxyfluorescein fluorescence is much more sensitive to pH than calcein fluorescence. The pH of the heterocyst cytoplasm could be different from that of the vegetative cell cytoplasm, which could explain the differences in carboxyfluorescein fluorescence levels in the two cell types.

The fact that both small anionic tracers, calcein and carboxyfluorescein, can move from cell to cell by diffusion whereas the larger fluorescent probe GFP (27 kDa) does not (Yoon and Golden, 1998; Mariscal *et al.*, 2007) could provide information about the specificity of the channels in terms of size and, perhaps, charge discrimination.

Two different channels communicate cells in the *Anabaena* filament

The use of *Anabaena* mutants altered in calcein or carboxyfluorescein transfer has been essential for the identification of the proteins putatively involved in channel activity and the specificity of the channels that mediate intercellular molecular exchange. Mutants of *Anabaena* sp. PCC 7120 that lack the septal proteins SepJ, FraC or FraD are affected in calcein and carboxyfluorescein transfer, but to different extents (Mullineaux *et al.*, 2008; Mariscal *et al.*, 2011; Merino-Puerto *et al.*, 2010, 2011a). A null *sepJ* mutant is strongly affected in intercellular transfer of calcein, but only partially impaired in transfer of carboxyfluorescein (Fig. 13.4). Thus, SepJ has been proposed to be a component of a septal protein channel that

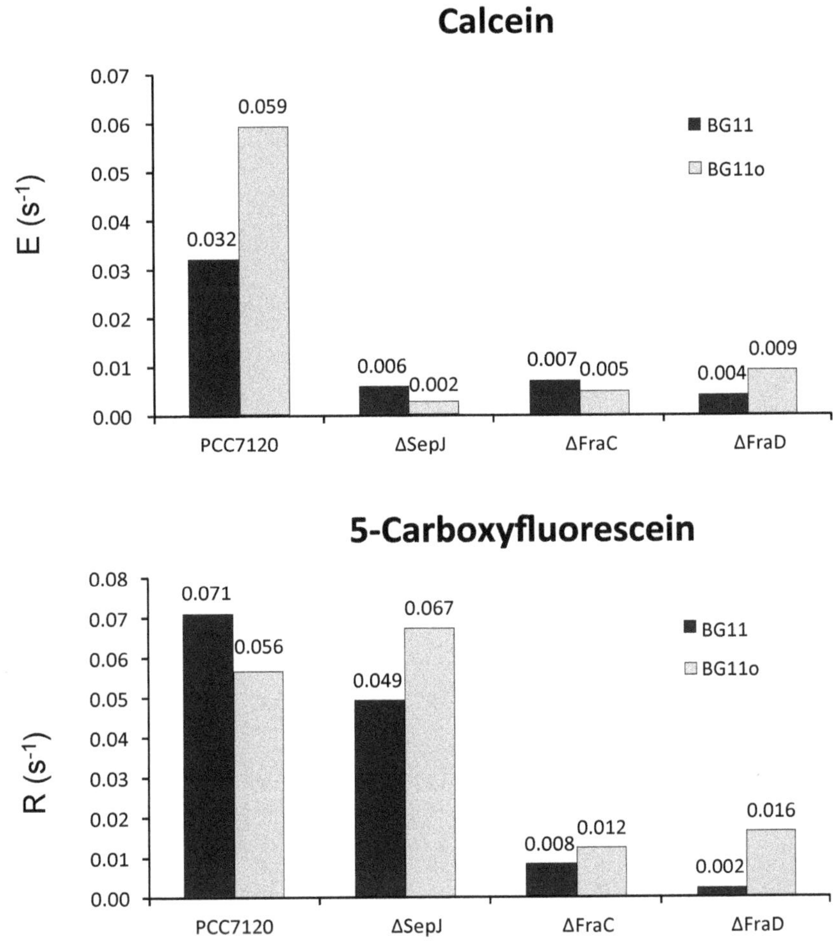

Figure 13.4 Exchange coefficients for calcein (*E*, s^{-1}) and recovery coefficients for carboxyfluorescein (*R*, s^{-1}) for exchange of probes between vegetative cells in different strains of *Anabaena*. Figures are the average of data previously reported (Mullineaux *et al.*, 2008; Merino-Puerto *et al.*, 2010, 2011a; Mariscal *et al.*, 2011).

allows intercellular transfer of calcein (Mullineaux *et al.*, 2008).

On the other hand, *fraC*, *fraD* and double *fraC–fraD* mutants show strong decreases in the transfer of both calcein and carboxyfluorescein (Fig. 13.4; Merino-Puerto *et al.*, 2010, 2011a). Because *fraC* and *fraD* mutants show an alteration in SepJ localization and SepJ is required for calcein transfer, the impairment of calcein transfer in the *fra* mutants could respond to an indirect effect. Thus, FraC and FraD could together participate in a separate pathway for the transfer of carboxyfluorescein that does not require SepJ (Merino-Puerto *et al.*, 2011a).

Other mutants that are affected in calcein transfer between cells are the *amiC1* mutant of *Anabaena* sp. PCC 7120 and the *amiC2* mutant of *Nostoc punctiforme* (Lehner *et al.*, 2011; Berendt *et al.*, 2012). As described above, cyanobacterial AmiC amidases are necessary for the maturation of the new septum during cell division, providing a framework for septal junctions to traverse septal peptidoglycan layers. It is possible that their mutation influence proper localization and activity of the channels involved in intercellular exchange.

In summary, two different protein complexes involving the septal proteins SepJ, on one hand, and FraC and FraD, on the other, have been implicated in intercellular transfer of compounds in the filament of *Anabaena* sp. PCC 7120 (Mullineaux *et al.*, 2008; Mariscal *et al.*, 2011; Merino-Puerto *et al.*, 2010, 2011a). SepJ channels would mediate calcein transfer whereas FraC/FraD channels would mediate carboxyfluorescein transfer. These two types of channels could operate concurrently in the septum between cells perhaps bearing different substrate specificities. Although the identity of the physiological substances transferred through these channels remains unknown, sucrose, glutamine and glutamate, largely agreed to be exchanged between heterocysts and vegetative cells and distributed between vegetative cells in the filament (Thomas *et al.*, 1977; López-Igual *et al.*, 2010; Vargas *et al.*, 2011), might be among the naturally transferred substrates. Additionally, alanine has been recently identified as a putatively exchanged substrate (Pernil *et al.*, 2010).

Concluding remarks

The analysis of the microscopic structure of filamentous cyanobacteria has revealed that the septum between cells is a section containing proteinaceous structures, here called 'septal junctions', which are allocated inside 'nanopores' that traverse the septal peptidoglycan layers. Several septal proteins, SepJ on one hand, and FraC/FraD on the other, that are essential for maintaining filament integrity and diazotrophic growth in *Anabaena* sp. PCC 7120 appear to be fundamental components of the *Anabaena* septal junctions. These proteins form part of two different types of channels that likely have specific selectivity for intercellular molecular exchange. Of particular importance will be to unravel the physiological substrates transferred through each of these types of channels.

Acknowledgements

Experimental work by V.M. was supported by grant numbers BFU2008-03811 and BFU2011-22762 from the Spanish government, co-financed by FEDER.

References

Abbaci, M., Barberi-Heyob, M., Stines, J.R., Blondel, W., Dumas, D., Guillemin, F., and Dileon, J. (2007). Gap junctional intercellular communication capacity by gap-FRAP technique: A comparative study. Biotechnol. J. *2*, 50–61.

Abbaci, M., Barberi-Heyob, M., Blondel, W., Guillemin, F., and Dileon, J. (2008). Advantages and limitations of commonly used methods to assay the molecular permeability of gap junctional intercellular communication. Biotechniques *45*, 33–62.

Alberts, B., Johnson, A., Lewis, J., Raff, M., Roberts, K., and Walter, P. (2002). Molecular Biology of the Cell, 4th edition (Garland Science, New York).

Bauer, C.C., Buikema, W.J., Black, K., and Haselkorn, R. (1995). A short-filament mutant of *Anabaena* sp. strain PCC 7120 that fragments in nitrogen-deficient medium. J. Bacteriol. *177*, 1520–1526.

Berendt, S., Lehner, J., Zhang, Y.V., Rasse, T.M., Forchhammer, K., and Maldener, I. (2012). Cell wall amidase AmiC1 is required for cellular communication and heterocyst development in the cyanobacterium *Anabaena* PCC 7120 but not for filament integrity. J. Bacteriol. *194*, 5218–5227.

Bisalputra, T., Oakley, B.R., Walker, D.C., and Shields, C.M., (1975). Microtubular complexes in blue-green algae. Protoplasma *86*, 19–28.

Buikema, W.J., and Haselkorn, R. (1991). Isolation and complementation of nitrogen fixation mutants of the cyanobacterium *Anabaena* sp. strain PCC 7120. J. Bacteriol. *173*, 1879–1885.

Butler, R.D., and Allsopp, A., (1972). Ultrastructural investigation in the Stigonemataceae (Cyanophyta). Arch. Mikrobiol. *82*, 283–299.

Ernst, A., Black, T., Cai, Y.P., Panoff, J.M., Tiwari, D.N., and Wolk, C.P. (1992). Synthesis of nitrogenase in mutants of the cyanobacterium *Anabaena* sp. strain PCC 7120 affected in heterocyst development or metabolism. J. Bacteriol. *174*, 6025–6032.

Fan, Q., Lechno-Yossef, S., Ehira, S., Kaneko, T., Ohmori, M., Sato, N., Tabata, S., and Wolk, C.P. (2006). Signal transduction genes required for heterocyst maturation in *Anabaena* sp. strain PCC 7120. J. Bacteriol. *188*, 6688–6693.

Flores, E., and Herrero, A. (2010). Compartmentalized function through cell differentiation in filamentous cyanobacteria. Nat. Rev. Microbiol. *8*, 39–50.

Flores, E., Herrero, A., Wolk, C.P., and Maldener, I. (2006). Is the periplasm continuous in filamentous multicellular cyanobacteria? Trends Microbiol. *14*, 439–443.

Flores, E., Pernil, R., Muro-Pastor, A.M., Mariscal, V., Maldener, I., Lechno-Yossef, S., Fan, Q., Wolk, C.P., and Herrero, A. (2007). Septum-localized protein required for filament integrity and diazotrophy in the heterocyst-forming cyanobacterium *Anabaena* sp. strain PCC 7120. J. Bacteriol. *189*, 3884–3890.

Gorelova, O.A., and Baulina, O.I. (2008). Ultrastructure of cyanobacterium *Nostoc* sp. f. *Blasia* cell forms in persisting populations. Microbiology (Moscow) *78*, 609–617.

Giddings, T.H., and Staeheling, L.A. (1978). Plasma membrane architecture of *Anabaena cylindrica*: occurrence of microplasmodesmata and changes associated with heterocyst development and the cell cycle. Cytobiologie *16*, 235–249.

Giddings, T.H., and Staeheling, L.A. (1981). Observation of microplasmodesmata in both heterocyst-forming and non-heterocyst forming filamentous cyanobacteria by freeze-fracture electron microscopy. Arch. Microbiol. *129*, 295–298.

Haselkorn, R. (1978). Heterocysts. Ann. Rev. Plant Physiol. *29*, 319–344.

Johnsborg, O., and Havarstein, L.S. (2009). Pneumococcal LytR, a protein from the LytR-CpsA-Psr family, is essential for normal septum formation in *Streptococcus pneumoniae*. J. Bacteriol. *191*, 5859–5864.

Kaneko, T., Nakamura, Y., Wolk, C.P., Kuritz, T., Sasamoto, S., Watanabe, A., Iriguchi, M., Ishikawa, A., Kawashima, K., Kimura, T., *et al.* (2001). Complete genomic sequence of the filamentous nitrogen-fixing cyanobacterium *Anabaena* sp. strain PCC 7120. DNA Res. *8*, 205–213.

Lang, N.J., and Fay, P. (1971). The heterocysts of blue-green algae II. Details of ultrastructure. Proc. R. Soc. Lond. B *178*, 193–203.

Lehner, J., Zhang, Y., Berendt, S., Rasse, T.M., Forchhammer, K., and Maldener, I. (2011). The morphogene AmiC2 is pivotal for multicellular development in the cyanobacterium *Nostoc punctiforme*. Mol. Microbiol. *79*, 1655–1669.

Lehner, J., Berendt, S., Dörsam, B., Pérez, R., Forchhammer, K., and Maldener, I. (2013). Prokaryotic multicellularity: a nanopore array for bacterial cell communication. FASEB J. doi: 10.1096/fj.12-225854.

Lippincott-Schwartz, J., and Patterson, G.H. (2003). Development and use of fluorescent protein markers in living cells. Science *300*, 87–91.

Lopez-Igual, R., Flores, E., and Herrero, A. (2010). Inactivation of a heterocyst-specific invertase indicates a principal role of sucrose catabolism in heterocysts of *Anabaena* sp. J. Bacteriol. *192*, 5526–5533.

Lupas, A.N., and Gruber, M. (2005). The structure of alpha-helical coiled coils. Adv. Protein. Chem. *70*, 37–78.

Mariscal, V., and Flores, E. (2010). Multicellularity in a heterocyst-forming cyanobacterium: pathways for intercellular communication. Adv. Exp. Med. Biol. *675*, 123–135.

Mariscal, V., Herrero, A., and Flores, E. (2007). Continuous periplasm in a filamentous, heterocyst-forming cyanobacterium. Mol. Microbiol. *65*, 1139–1145.

Mariscal, V., Herrero, A., Nenninger, A., Mullineaux, C., and Flores, E. (2011). Functional dissection of the three-domain SepJ protein joining the cells in cyanobacterial trichomes. Mol. Microbiol. *79*, 1077–1088.

Mella-Herrera, R.A., Neunuebel, M.R., and Golden, J.W. (2011). *Anabaena* sp. strain PCC 7120 *conR* contains a LytR-CpsA-Psr domain, is developmentally regulated, and is essential for diazotrophic growth and heterocyst morphogenesis. Microbiology *157*, 617–626.

Merino-Puerto, V., Mariscal, V., Mullineaux, C., Herrero, A., and Flores, E. (2010). Fra proteins influencing filament integrity, diazotrophy and localization of septal protein SepJ in the heterocyst-forming cyanobacterium *Anabaena* sp. Mol. Microbiol. *75*, 1159–1170.

Merino-Puerto, V., Schwarz, H., Maldener, I., Mariscal, V., Mullineaux, C., Herrero, A., and Flores, E. (2011a). FraC/FraD-dependent intercellular molecular exchange in the filaments of a heterocyst-forming cyanobacterium, *Anabaena* sp. Mol. Microbiol. *82*, 87–98.

Merino-Puerto, V., Mariscal, V., Schwarz, H., Maldener, I., Mullineaux, C., Herrero, A., and Flores, E. (2011b). FraH is required for reorganization of intracellular membranes during heterocyst differentiation in *Anabaena* sp. strain PCC 7120. J. Bacteriol. *193*, 6815–6823.

Mullineaux, C.W., Mariscal, V., Nenninger, A., Khanum, H., Herrero, A., Flores, E., and Adams, D.G. (2008). Mechanism of intercellular molecular exchange in heterocyst-forming cyanobacteria. EMBO J. *27*, 1299–1308.

Nayar, A.S., Yamaura, H., Rajagopalan, R., Risser, D.D., and Callahan, S.M. (2007). FraG is necessary for filament integrity and heterocyst maturation in the cyanobacterium Anabaena sp. strain PCC 7120. Microbiology *153*, 601–603.

Newman, J.R., Ethan, W., and Kim, P. (2000). A computationally directed screen identifying interacting coiled coils from *Saccharomyces cerevisiae*. Proc. Natl. Acad. Sci. U.S.A. *97*, 13203–13208.

Nicolaisen, K., Mariscal, V., Bredemeier, R., Pernil, R., Moslavac, S., López-Igual, R., Maldener, I., Herrero, A., Schleiff, E., and Flores, E. (2009). The outer membrane of a heterocyst-forming cyanobacterium is a permeability barrier for uptake of metabolites that are exchanged between cells. Mol. Microbiol. *74*, 58–70.

Olmedo-Verd, E., Muro-Pastor, A.M., Flores, E., and Herrero, A. (2006). Localized induction of the *ntcA* regulatory gene in developing heterocysts of *Anabaena* sp. strain PCC 7120. J. Bacteriol. *188*, 6694–6699.

Pernil, R., Herrero, A., and Flores, E. (2010). Catabolic function of compartmentalized alanine dehydrogenase in the heterocyst-forming cyanobacterium *Anabaena* sp. strain PCC7120. J. Bacteriol. *192*, 5165–5172.

Reizer, J., Reizer, A., and Saier, M.H. (1992). A new subfamily of bacterial ABC-type transport-systems catalyzing export of drugs and carbohydrates. Prot. Sci. *1*, 1326–1332.

Rippka, R., Deruelles, J., Waterbury, J.B., Herdman, M., and Stanier, R.Y. (1979). Generic assignments, strain stories and properties of pure cultures of cyanobacteria. J. Gen. Microbiol. *111*, 1–61.

Thomas, J., Meeks, J.C., Wolk, C.P., Shaffer, P.W., Austin, S.M., and Chien, W.S. (1977). Formation of glutamine from [^{13}N]ammonia, [^{13}N]dinitrogen, and [^{14}C] glutamate by heterocysts isolated from *Anabaena cylindrica*. J. Bacteriol. *129*, 1545–1555.

Typas, A., Banzhaf, M., Gross, C., and Vollmer, W. (2012). From regulation of peptidoglycan synthesis to bacterial growth and morphology. Nat. Rev. Microbiol. *10*, 123–135.

Vargas, W., Nishi. C., Giarrocco, L., and Salerno, G. (2011). Differential roles of alkaline/neutral invertases in *Nostoc* sp. PCC 7120: Inv-B isoform is essential for diazotrophic growth. Planta 233, 153–162.

Vollmer, W., Joris, B., Charlier, P., and Foster, S. (2008). Bacterial peptidoglycan (murein) hydrolases. FEMS Microbiol. Rev. *32*, 259–286.

Wade, M.H., Trosko, J.E., and Schlinder, M. (1986). A fluorescence photobleaching assay of gap junction-mediated communication between human cells. Science *232*, 525–528.

Wilk, L., Strauss, M., Rudolf, M., Nicolaisen, K., Flores, E., Kuhlbrandt, W., and Schleiff, E. (2011). Outer membrane continuity and septosome formation between vegetative cells in the filaments of *Anabaena* sp. PCC 7120. Cell. Microbiol. *13*, 1744–1754.

Wildon D.C., and Mercer, F.V. (1963). The ultrastructure of the heterocyst and akinete of blue-green algae. Archiv. Mikrobiol. *47*, 19–31.

Yoon, H.S., and Golden, J.W. (1998). Heterocyst pattern formation controlled by a diffusible peptide. Science *282*, 935–938.

Zhu, J., Jäger, K., Black, T., Zarka, K., Koksharova, O., and Wolk, C.P. (2001). HcwA, an autolysin, is required for heterocyst maturation in *Anabaena* sp. Strain PCC 7120. J. Bacteriol. *183*, 6841–6851.

Index

A

B

C

D

T

U

V

Z

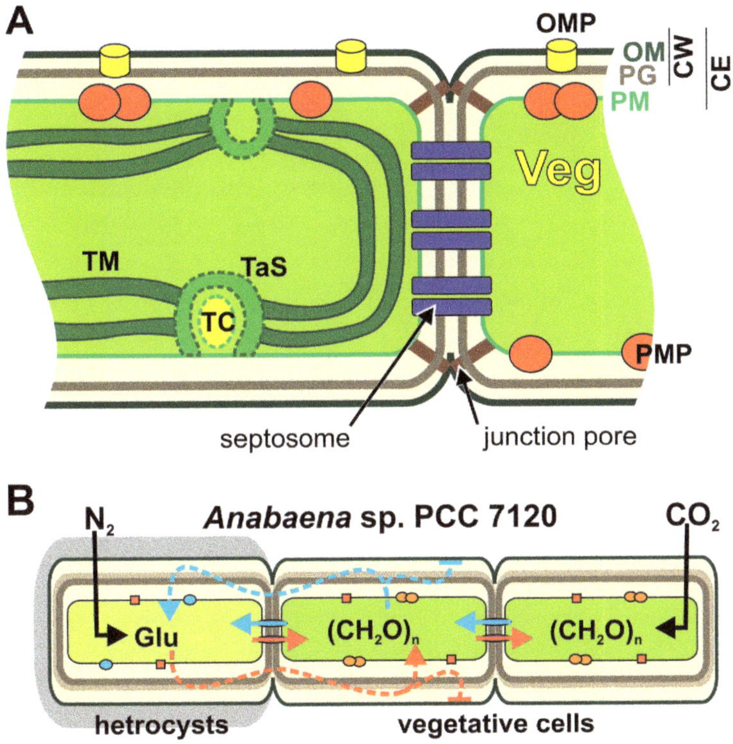

Figure 3.1 The membrane organization of cyanobacteria. (A) The structural composition of the cell envelope of cyanobacteria is shown as scheme. The outermost barrier is attached to the outer membrane (dark green) and the peptidoglycan layer (brown) and the plasma membrane (light green). At the septum border the periplasm is crossed by junction pores (dark brown), and in the septum by septosomes (blue). Outer membrane proteins (OMP, yellow) and plasma membrane proteins (red) are shown to indicate their existence. The inner-cellular structures formed by the thylakoid membranes (TM) are likely associated with the plasma membrane by the thylakoid associated semi-circular structures (TaS), which in some cases are associated with the thylakoid centres (TC). The figure is adapted from Wilk *et al.* (2011). (B) The N_2-fixing heterocysts and the CO_2-fixing vegetative cells co-exist in a single filament. Amino acids such as glutamine are nitrogen carriers that are transported to vegetative cells. In turn, sugars are carbon carriers produced by vegetative cells, which need to be transferred to heterocysts. Two possible transport routes are currently discussed. One considers the periplasm as a possible transport path for metabolites (dashed arrows), while another considers the septosomes as major structures for the transport of metabolites (solid arrows). The figure is adapted from Flores *et al.* (2006).

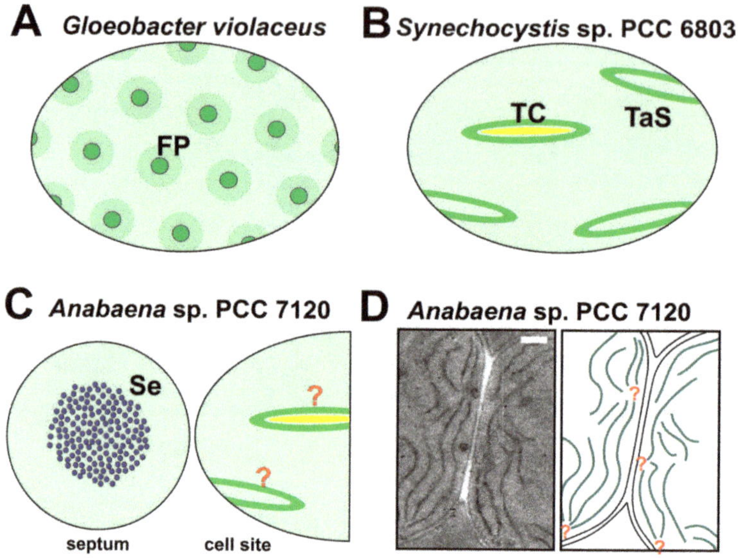

Figure 3.3 The plasma membrane of cyanobacteria contains specialized regions. (A) Shown is the distribution of the fluorescent patches (FP) observed in the plasma membrane of the unicellular cyanobacterium *Gloeobacter violaceus*, which contains the photosynthetic and respiratory machines. Per cell, 46 fluorescent patches have been found. The illustration is adapted from Rexroth *et al.* (2011). (B) The thylakoid membrane sheets of *Synechocystis* sp. PCC 6803 converge at sites adjacent to the plasma membrane (N=3 to 10 per cell; van de Meene *et al.*, 2006), which are likely coincident with the thylakoid associated semcircular structures (TaS, Stengel *et al.*, 2012). In addition, these sites are in N=0 to 4 cases associated with the described thylakoid centres (van de Meene *et al.*, 2006). (C) The septum of the filamentous cyanobacterium *Anabaena* sp. PCC 7120 contains up to 500 septosomes (Se). In this cyanobacterium, the thylakoid membranes converge at the plasma membrane in some cases. Although thylakoid-associated semicircular structures have not been described for this cyanobacterium, they might exist as well. (D) The image of the septum between two cells of *Anabaena* sp. PCC 7120 is shown to illustrate the convergence of the thylakoid membranes. On the right side the thylakoid membranes are drawn to scale, and question marks indicate possible positions of convergence close to the plasma membrane. The image was taken by Laura Wilk, MPI of Biophysics, Frankfurt, Germany. The cells were prepared as described in Wilk *et al.* (2011) by the method employing potassium permanganate. Scale bar, 200 nm.

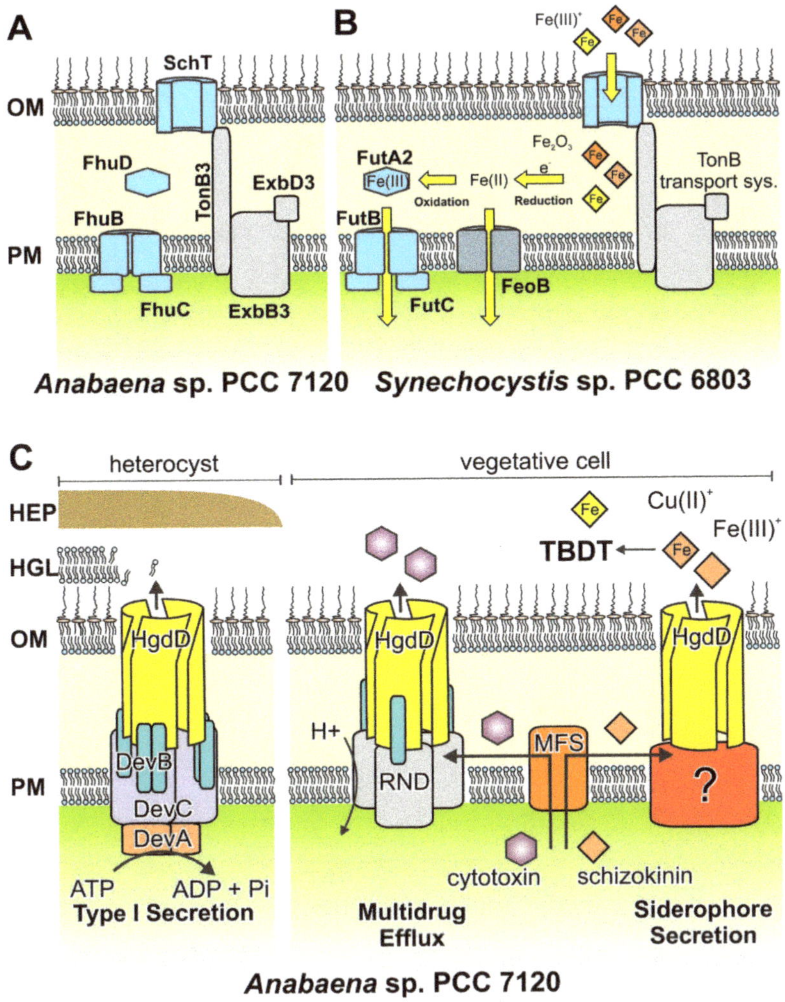

Figure 3.5 Iron uptake and metabolite export strategies. (A) Siderophore uptake: recognition and transport through the outer membrane of the siderophore schizokinen is mediated by the TonB-dependent-transporter SchT energized by TonB3 in complex with ExbB3 and ExbD3. The final uptake is regulated by a Fhu system in the periplasm and the plasma membrane. (B) Reductive uptake: In dilute heterogeneous environments iron can be free, bound to a variety of organic ligands (squares) or form ferric oxide colloids. In *Synechocystis* sp. PCC 6803 the mode of transport through the OM is not known. Once inside the periplasm, all iron species are reduced and unchelated Fe ions are transported through the PM. Free iron in the Fe(II) form can be taken up by transporters such as FeoB or can be re-oxidized to Fe(III) and taken up by transporters like FutABC. (C) The HgdD-dependent secretion systems. The homotrimeric TolC-like HgdD (yellow) is the outer membrane factor for many secretion processes in *Anabaena* sp. PCC 7120. HgdD is expressed in both vegative cells and heterocysts but shows a transient enhanced expression with a peak at nine hours after nitrogen deprivation. Together with the DevA, DevB and DevC (left), HgdD is involved in the secretion of heterocyst specific glycolipids to form the heterocyst specific glycolipid (HGL)-layer that is needed to maintain the micro-oxic environment for nitrogen fixation. DevB is the membrane fusion protein (MFP), DevC the permease and DevA the ATP-hydrolysing subunit of an ABC-transporter. Together they form a classical Type I secretion system. HgdD is additionally involved in the secretion of small molecular weight substances such as antibiotics, dyes (middle) and siderophores (right). The secretion of antibiotics and dyes depends on the membrane potential of the plasma membrane and likely functions in cooperation with an RND-family efflux pump protein such as *E. coli* AcrAB. For siderophore secretion both HgdD and SchE, a protein of the major facilitator superfamily (MFS), are required. Unlike the ABC transporter and RND-like proteins, MFS proteins are not encoded next to a membrane MFP and are unlikely to interact directly with HgdD. Therefore, a third component might be involved to charge HgdD in siderophore secretion.

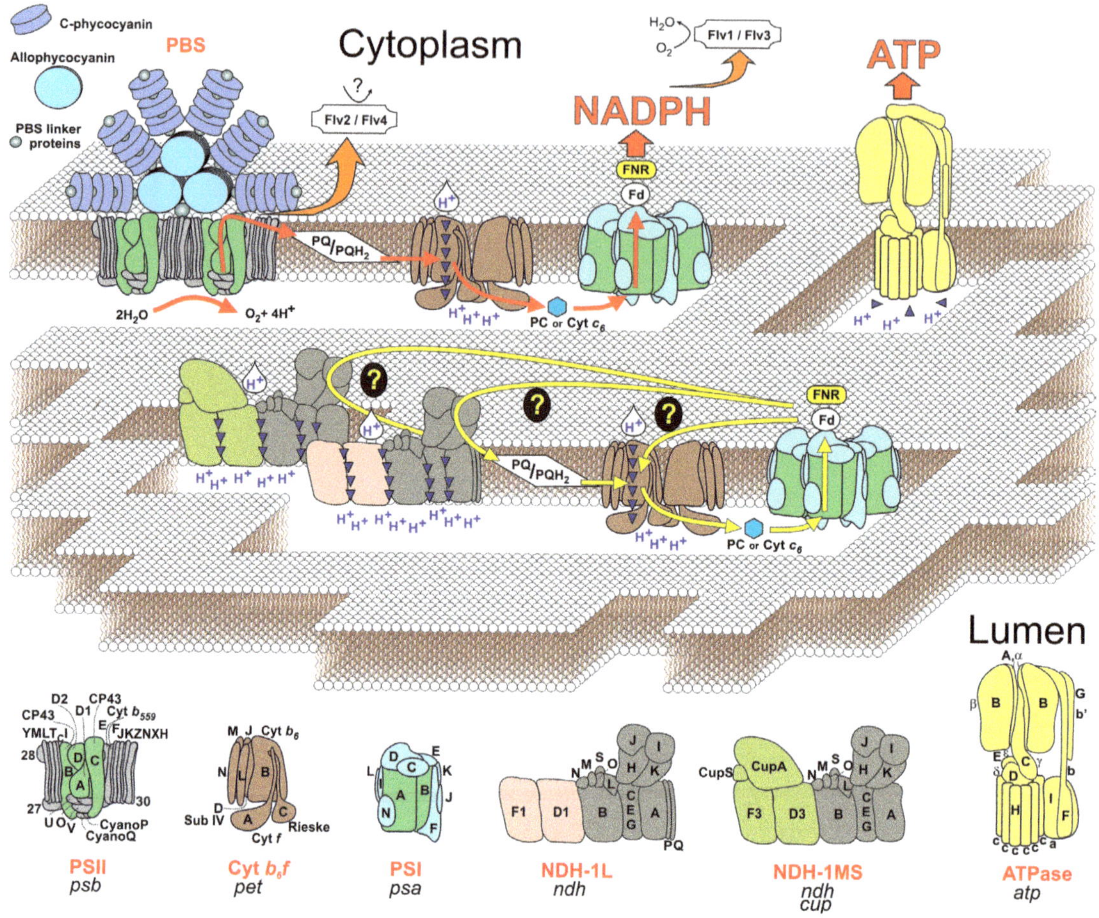

Figure 4.1 Scheme of photosynthetic protein complexes in the thylakoid membrane of *Synechocystis* 6803 and some associated proteins involved in the light reactions. Linear electron transport is shown by red arrows. Cyclic electron transport pathways shown by yellow arrows are still under debate and indicated with question marks. Subunit composition of protein complexes (in monomeric forms) is presented at the bottom of the picture. Subunits are designated by capital letters of the corresponding genes; a commonly used protein name is also indicated for some subunits. Blue triangles represent proton gradient channels.

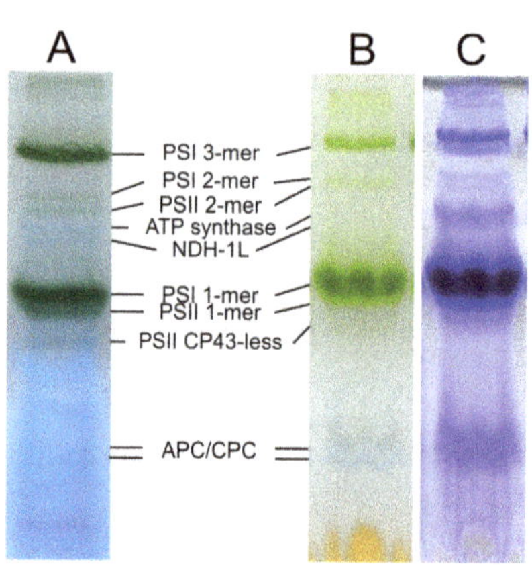

Figure 4.2 1-D native PAGE of thylakoid protein complexes. The crude thylakoid membrane fraction of *Synechocystis* 6803 was obtained as described in Zhang *et al.* (2004). Protein complexes were solubilized with 1.5% DM and separated by Blue native (BN-) PAGE (A) or Clear Native (CN-) PAGE (B) as described in Herranen *et al.* (2004) and Wittig *et al.* (2007). Gradient 4–12% PAGE was used in both (A) and (B). (C) CN-PAGE-separated complexes were stained with Coomassie Blue.

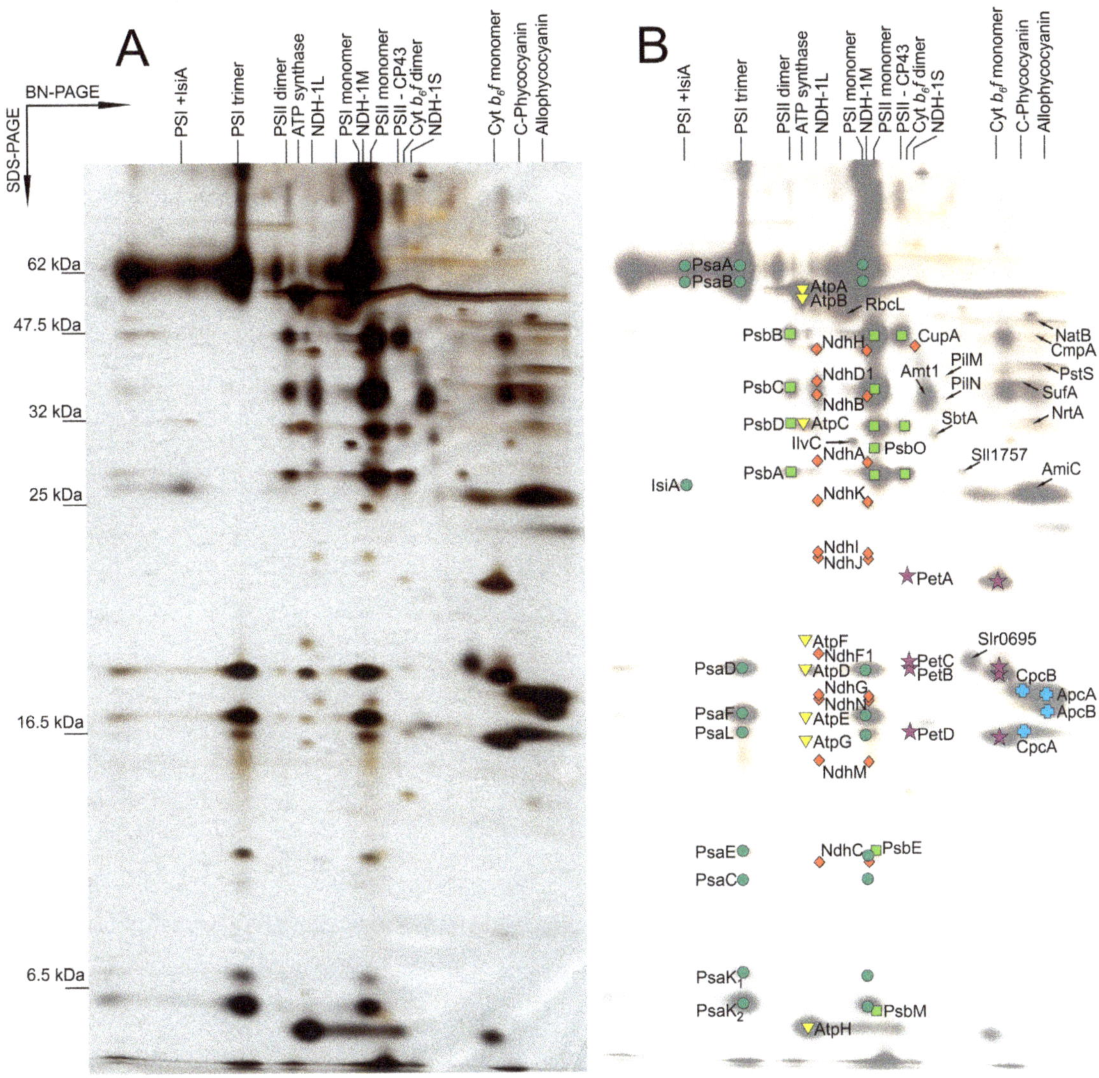

Figure 4.3 2-D BN/SDS-PAGE of thylakoid protein complexes from *Synechocystis* 6803 grown in high CO_2 conditions. (A) A silver-stained gel. (B) Schematic representation of (A), with subunits of complexes shown by following coloured symbols: phycobilisome components, blue crosses; PSII, green squares; Cyt b_6f, magenta stars; PSI, turquois circles; ATP synthase, yellow triangles; NDH-1, red diamonds. Identification of proteins is based on mass spectrometry. Proteins from cytosol or other membranes present as contaminants in the crude thylakoid membrane preparation are shown by arrows.

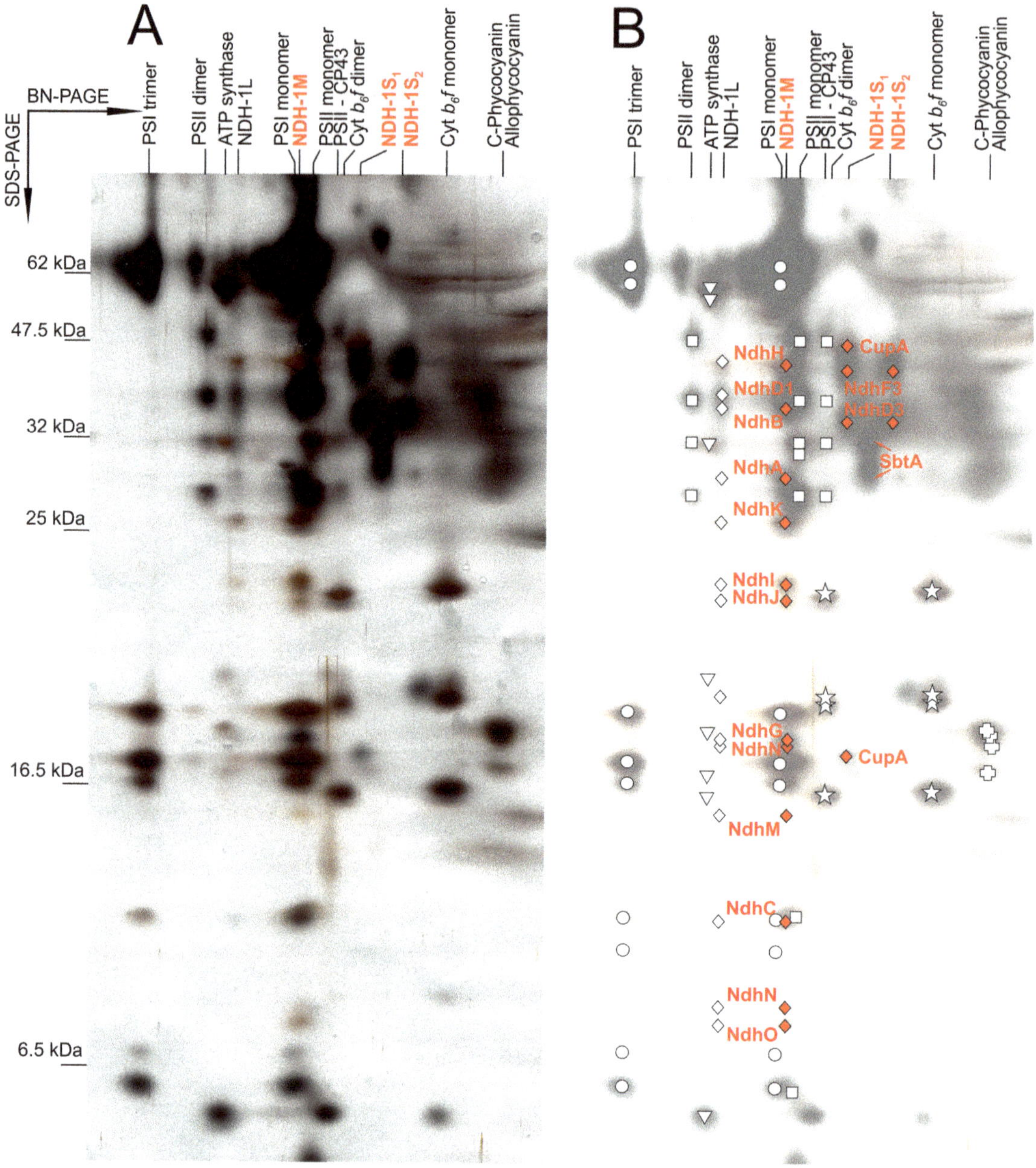

Figure 4.4 2-D BN/SDS-PAGE of thylakoid protein complexes from *Synechocystis* 6803 grown in low CO_2 conditions. (A) A silver-stained gel. (B) Schematic representation of (A), with subunits of complexes depicted by the same symbols as in Figure 4.3 but uncoloured. Identification of proteins is based on mass spectrometry. Proteins and complexes up-regulated in low CO_2 conditions (compared to high CO_2) are marked with red.

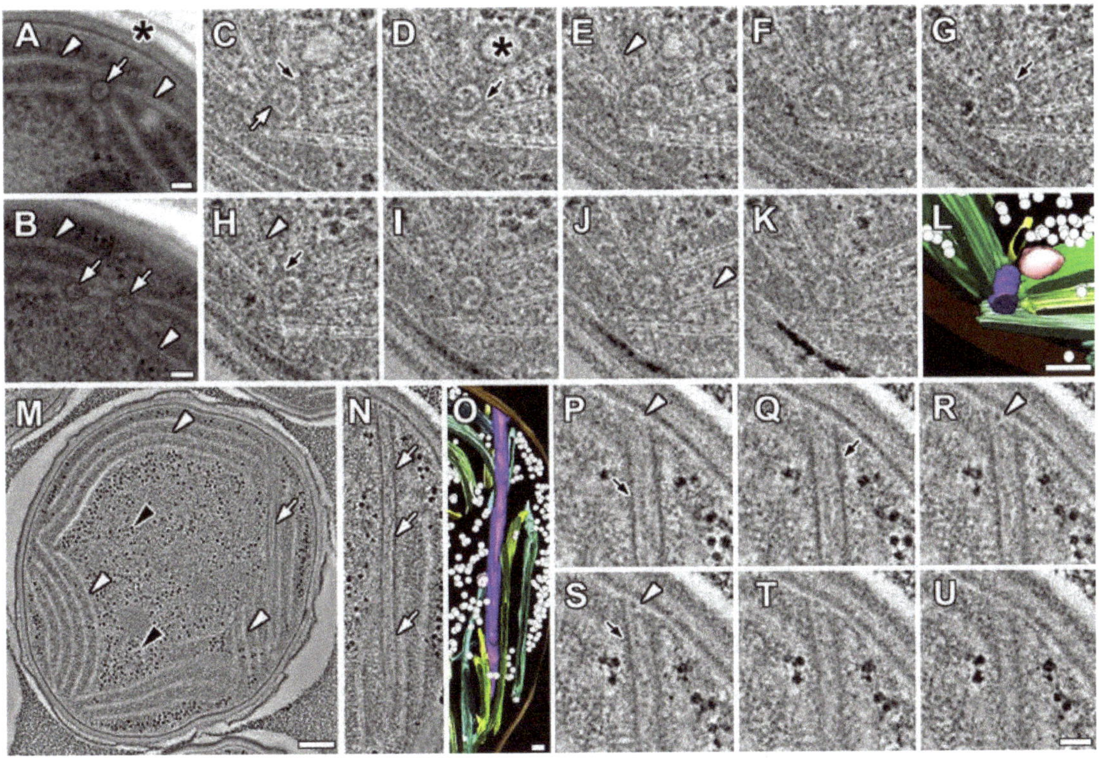

Figure 5.4 The thylakoid centre of *Synechocystis*. (A, B) Cross-sections of thylakoid membranes (white arrowheads) and thylakoid centres (white arrows). Asterisk: cell wall. Bars = 50 nm. (C–K) 1.4 nm slices taken through a selected cross-section of a thylakoid centre (white arrow), surrounding thylakoid membranes (white arrowhead) and the cytoplasm. Dense material (black arrows) coats the thylakoid centre, which appears to be associated with thylakoid membranes. Black asterisk: lipid body. Bar = 50 nm. (L) Model of a thylakoid centre reconstructed from the images shown in (C-K). Thylakoid membranes (green) surround the thylakoid centre (blue). Pink: lipid body; white: ribosomes; brown cytoplasmic membrane. Bar = 50 nm. (M) Electron micrograph of a *Synechocystis* cell with the thylakoid centre (white arrow) extending from one side to the other of the cell. White arrowheads: thylakoid membranes; black arrowheads: ribosomes. Bar = 200 nm. (N) Longitudinal view of a thylakoid centre (arrows) and a derived model (O) with associated thylakoid membrane pairs (green), ribosomes (white) and the cytoplasmic membrane (brown). Bar = 50 nm. (P-U) 1 nm sections of the thylakoid centre (black arrow) shown in (M) and (N) and its association (white arrowhead) with the cytoplasmic membrane and periplasmic space. Bar = 50 nm. From (van de Meene *et al.*, 2006), with kind permission of Springer Science and Business Media.

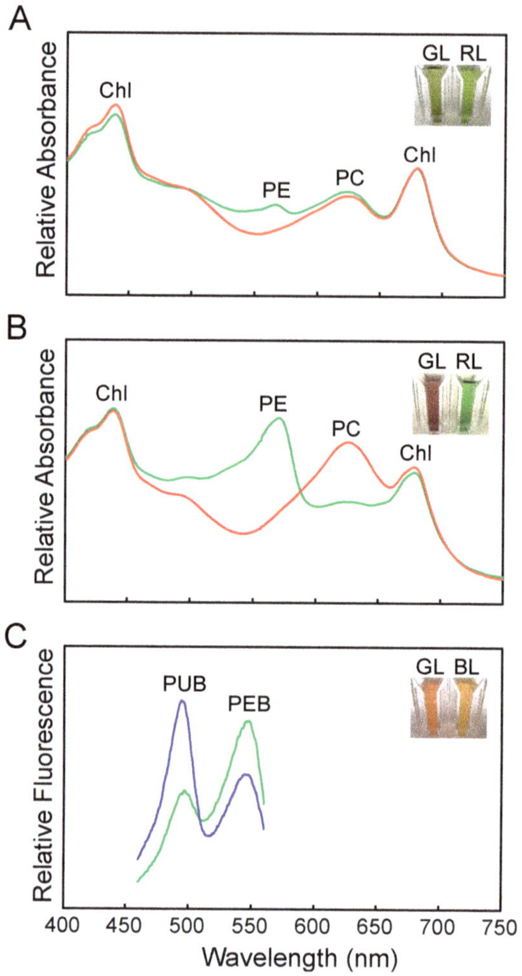

Figure 6.1 Chromatic acclimation responses and colour phenotypes. (A) Whole-cell absorption spectra of the CA2 strain *Nostoc punctiforme* after acclimation to GL and RL. (B) Whole-cell absorption spectra of the CA3 strain *F. diplosiphon* after acclimation to GL and RL. (C) Whole-cell fluorescence spectra of the CA4 strain *Synechococcus* sp. RS9916 after acclimation to GL and BL. The colour of each line denotes the light colour in which the cells were grown. Maxima for Chl, chlorophyll; PE, phycoerythrin; PC, phycocyanin; PUB, phycourobilin; and PEB, phycoerythrobilin are denoted. Insets show the colour phenotype of cells after acclimation to the specific light colour treatment indicated above each cuvette.

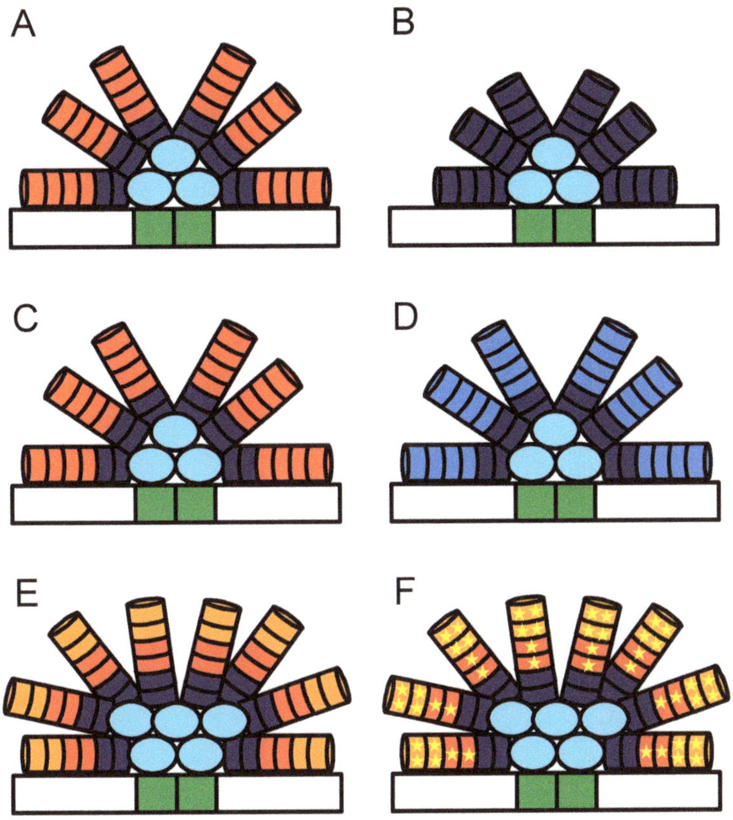

Figure 6.2 Phycobilisome structures among cyanobacterial species capable of chromatic acclimation. PBS of (A) CA2 species adapted to green light, (B) CA2 species adapted to red light, (C) CA3 species adapted to green light, (D) CA3 species adapted to red light, (E) CA4 species adapted to green light, and (F) CA4 species adapted to blue light. Yellow stars indicate the location and number of bilins changing from PEB to PUB upon a switch from green light to blue light. In the PBS models above, the thylakoid membranes are white, photosynthetic reaction centres are green, AP cores are light blue, PC1 discs are dark blue, PE discs are red, PC2 discs are medium blue, and PEII discs are orange.

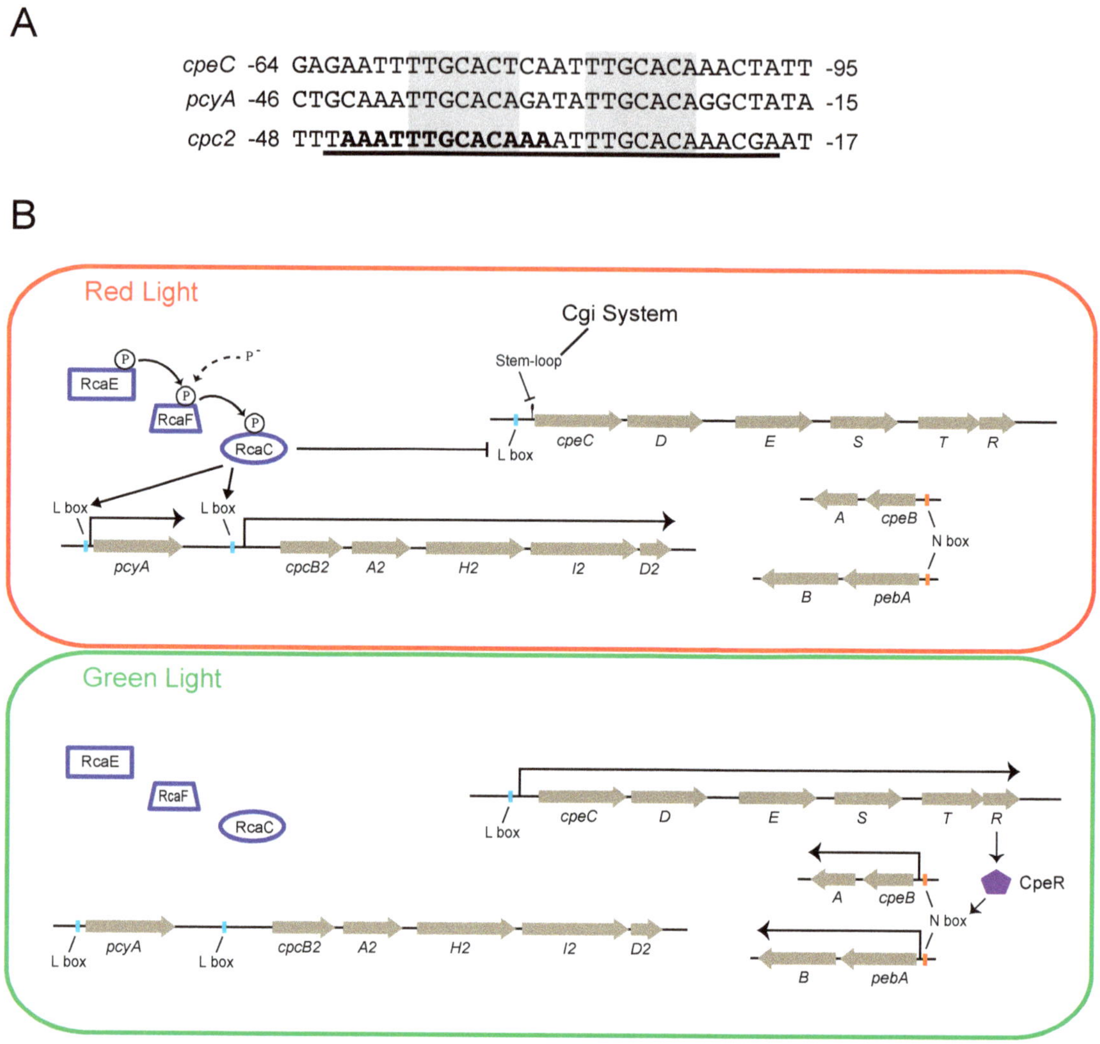

Figure 6.3 Model for CA3 regulation in *F. diplosiphon*. (A) Alignment of L box sequences upstream of *cpeC*, *pcyA*, and *cpc2* in *F. diplosiphon*. Note that the *cpeC* L box sequence is on the opposite strand, relative to transcription start, compared to *pcyA* and *cpc2*. Shaded letters indicate shared sequence identity with the consensus L box repeat. Bold letters indicate the direct repeat sequence, over lapping by two base pairs (Casey and Grossman, 1994). Numbers are for known transcription start sites in *F. diplosiphon*; for *cpc2*, numbering is based on the most recent determination (Alvey *et al.*, 2007). The line demarks the sequence that makes up the L box repeats and flanking sequence, as most recently defined (Bezy *et al.*, 2011). (B) The Rca system appears to be more phosphorylated in RL than GL, leading to the activation of *pcyA* and *cpc2* and the repression of *cpeC* through binding of RcaC to L boxes (indicated). In GL, the Rca system effect is lost and *cpeC* expression increases, leading to expression of CpeR and activation of other GL-expressed genes, possibly through N boxes (indicated). In RL, the Cgi system represses *cpeC* expression post-transcriptionally, requiring a stem–loop within the 5′ leader for this regulation, and this repression is relieved in GL.

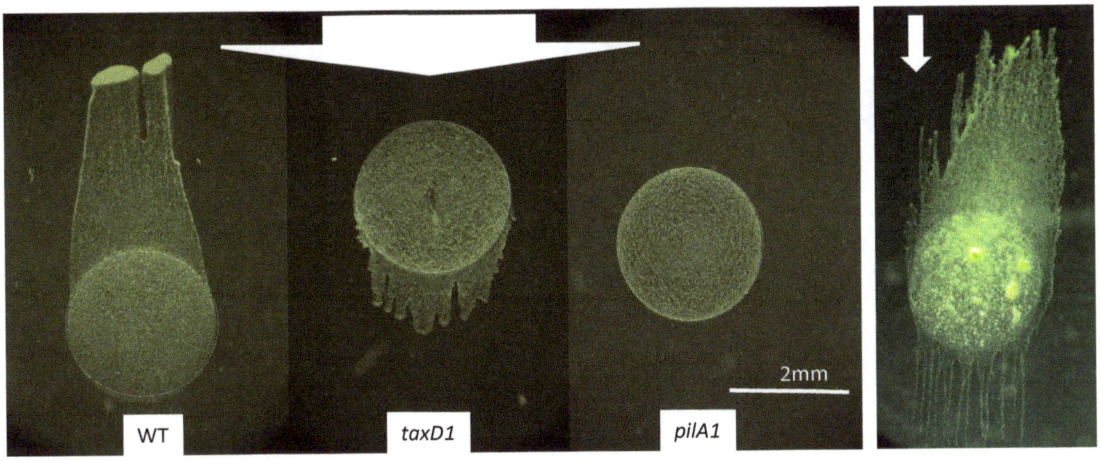

Figure 11.2 Phototactic behaviour of Syn PCC 6803, mutant cells and thermophilic *Synechococcus* sp. A directional phototaxis assay on 0.4% agarose plates was performed for 48 h. Wild-type (WT) cells (left) show positive phototaxis with groups of cells moving out from the original drop, making finger-like projections towards the light (white arrow); a *taxD1* mutant (middle) exhibits negative phototaxis; a *pilA1* mutant (right) is non-motile. Size of the drop is ~2–3 mm (marked with dotted white lines). Right panel: Thermophilic *Synechococcus* sp. isolated from microbial mats cells exhibiting both positive and negative phototaxis, in directional light. Note that most cells are positively phototactic with a smaller number of cells exhibiting motility away from light.

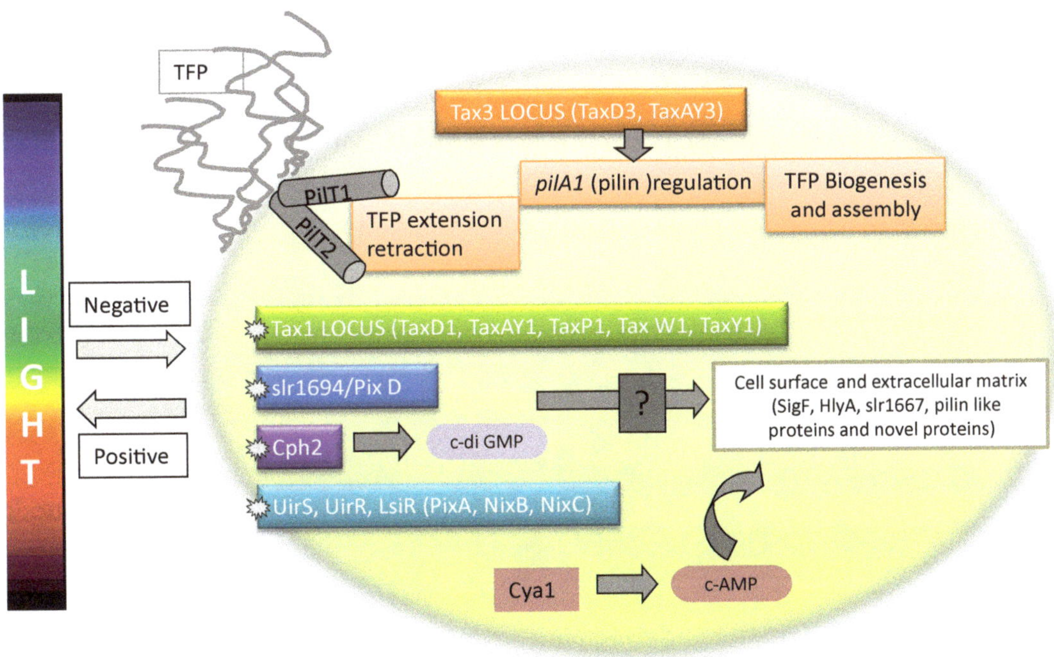

Figure 11.4 Model of positive and negative phototaxis in Syn PCC 6803. The major identified photoreceptors (shown with stars) are indicated (see Table 11.1 for more details). The Tax3 locus, which is required for TFP biogenesis and assembly, is also shown. PilT1 and PilT2 are required for TFP activity. The exact roles of cAMP and di-cGMP in regulating the cell surface and various other proteins are not known, although it is known that many of these proteins play a role on motility.

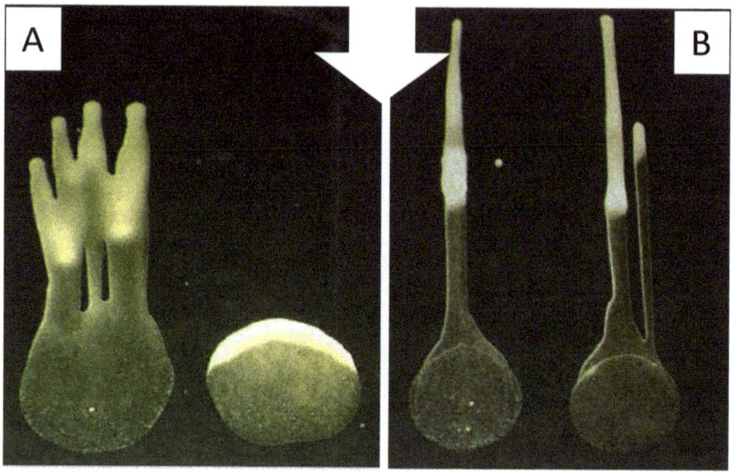

Figure 11.5 Effect of cAMP on phototaxis of Syn PCC 6803 and *cya1* mutant cells. WT (left), *cya1* mutant (right), were spotted on 0.4% agarose plates in the absence (A) or in the presence (B) of 0.1 mM cAMP. In the absence of added cAMP, *cya1* cells showed limited motility (A) but when cAMP is added to the plates, they exhibit typical finger-like projections, like WT cells. The arrow shows the direction of the light. Modified from Bhaya, 2006.

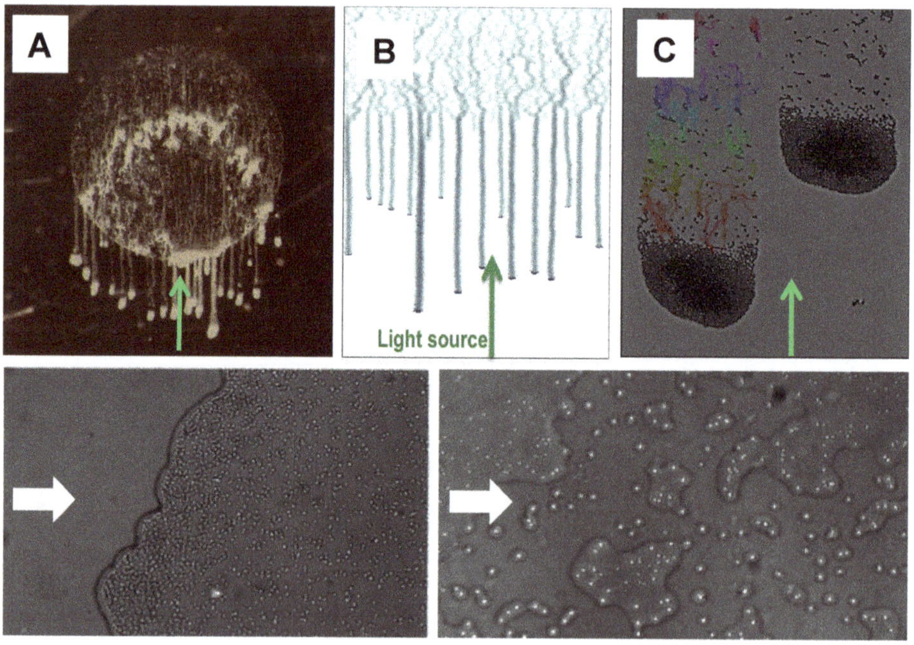

Figure 11.6 Model of phototactic and group behaviour. Top panel. (A) Phototactic behaviour of Syn PCC 6803 showing typical finger-like projections of groups of cells. (B) Simulation of phototactic behaviour using modelling. Note how this simulates the behaviour of cells as shown in (A). (C) Tracking behaviour of single cells. Each cell track is shown in a different colour. Lower panel. Left: A group of Syn PCC 6803 cells in an extracellular matrix as they move towards the light, Right: when small groups of cells are moving they also are surrounded by a distinct extracellular matrix, of unknown composition (modified from Galante *et al.*, 2012).

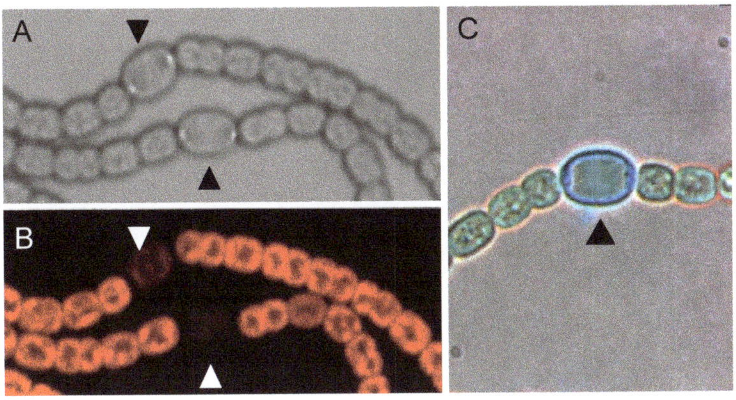

Figure 12.2 Filaments of *Anabaena* sp. PCC 7120 showing special features of heterocysts. The bright-field image (A) shows two heterocysts (triangles) attached to vegetative cells. Heterocysts lack red autofluorescence (B). Alcian blue stains the polysaccharides in the envelope of the heterocyst (C). Bar, 10 μm. Micrographs courtesy of Alicia M. Muro-Pastor, CSIC, Sevilla, Spain.

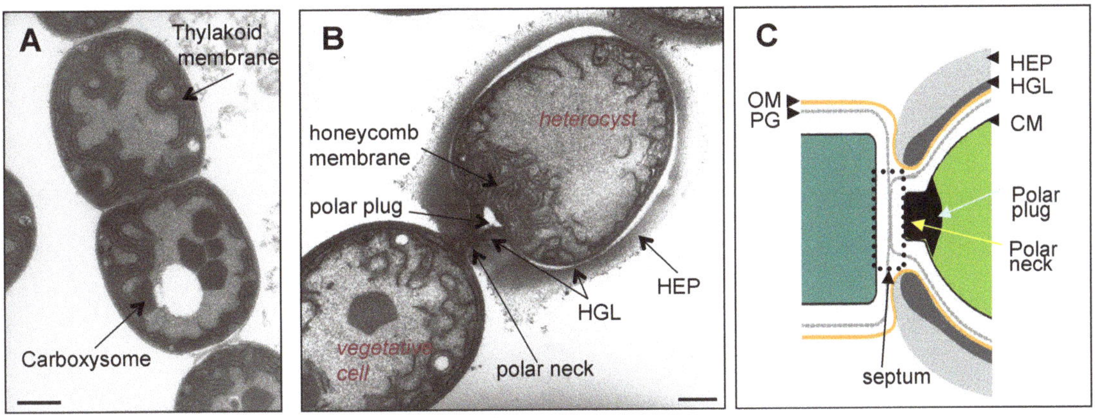

Figure 12.3 Electron micrographs and a scheme of filaments of *Anabaena* sp. PCC 7120 grown with nitrate (A) or with N_2 as the nitrogen source (B). Some cellular structures are indicated. Bars: 0.6 μm. (C) Scheme of the details of ultrastructure at the connection between a vegetative cell (left side of the image) and a heterocyst (right side of the image), based on (B). HEP: heterocyst envelope polysaccharide layer; HGL: heterocyst glycolipid layer; CM: cytoplasmic membrane; OM: outer membrane; PG: peptidoglycan layer.

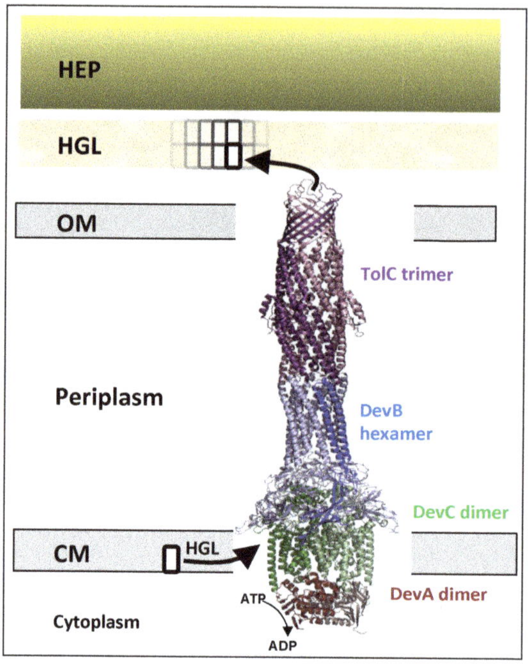

Figure 12.4 Model of HGL transport by the Type I secretion system DevBCA-TolC of *Anabaena* sp. PCC 7120. HGLs are built at the cytoplasmic membrane (CM) and transferred by the DevAC channel to the periplasmic DevB protein, which connects to the outer membrane (OM) protein TolC. After passage through the periplasm and the OM, the HGLs are deposited under the HEP layer to form the laminated layer. Model courtesy of Peter Staron, PhD thesis, University of Tübingen, Germany, 2012.

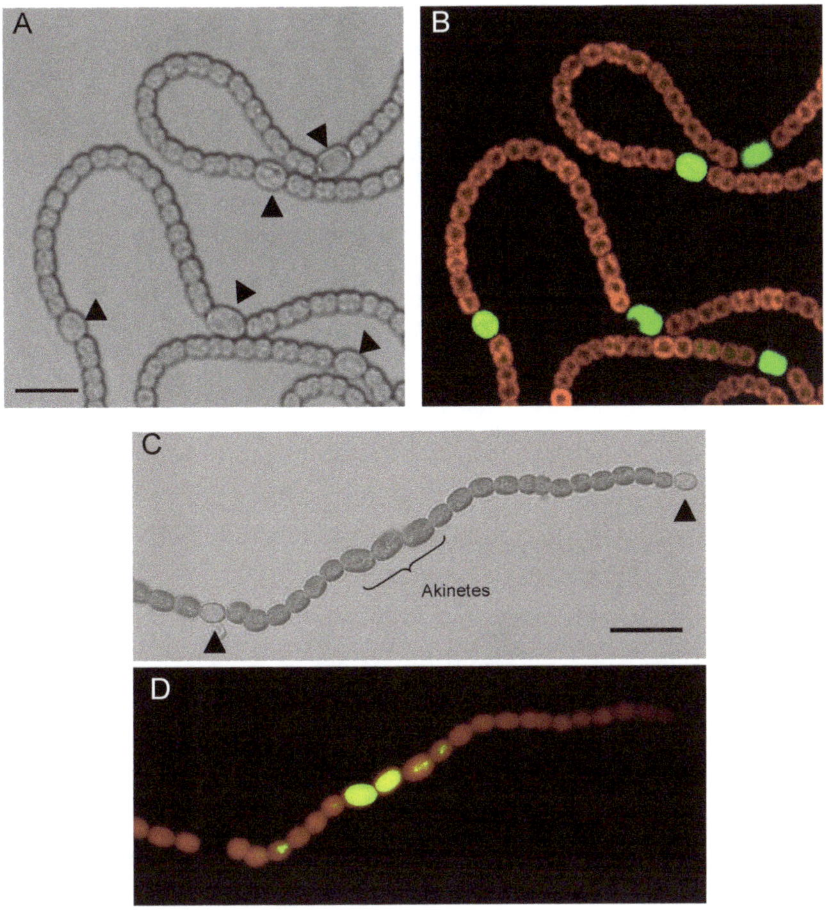

Figure 12.6 Differential gene expression in heterocysts and akinetes. Bright field (A and C) and overlay of autofluorescence and green fluorescence (B and D) of filaments of *Anabaena* sp. PCC 7120 (P_{hetR}-*gfp*) grown on N_2 (A and B) and of *N. punctiforme* ($P_{NpunF0784}$-*gfp*) 23 days following phosphate starvation (C and D). *NpunF0784* encodes a small heat shock protein in filaments of *Nostoc punctiforme*. The promoter region of each gene was cloned in front of a promoterless *gfp* gene encoding the green fluorescence protein and transferred on a self-replicating plasmid to each strain. Transcription of *hetR* occurs only in heterocysts (arrowheads) and that of *NpunF0784* primarily in akinetes. Bar, 10 μm. Micrographs courtesy of Alicia Muro-Pastor, CSIC, Sevilla, Spain (*hetR*) and Adrian Paz, CSU Northridge, Northridge, California (*NpunF0784*).

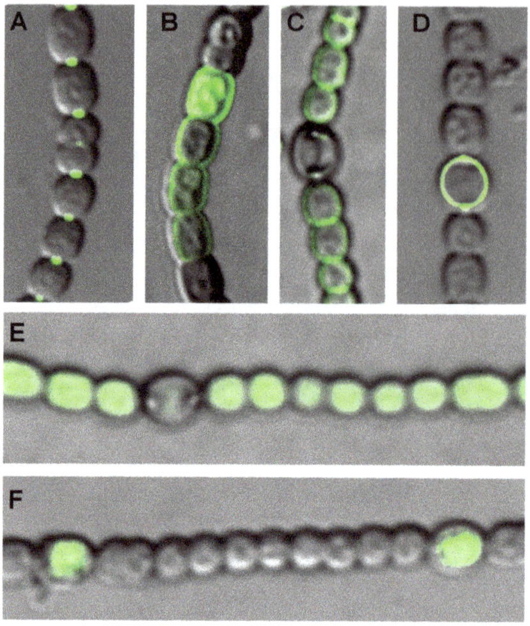

Figure 13.3 Uses of GFP in *Anabaena* sp. PCC 7120 as reporter of subcellular localization of proteins and cellular specificity of promoter activity. Figures show the localization of GFP in the septa between cells (A, SepJ–GFP fusion; Flores *et al.*, 2007), soluble in the periplasm (B, P_{patS}-GFP bearing a Tat signal peptide and a signal peptidase I cutting site [P_{patS}-tat_{ss}-spI_{cs}-GFP]; Mariscal *et al.*, 2007), integrated in the plasma membrane of vegetative cells (C, Amt1–GFP fusion; Merino-Puerto *et al.*, 2010), anchored to the plasma membrane of heterocysts (D, P_{patS}-GFP bearing a Tat signal peptide and signal peptidase II cutting site [P_{patS}-tat_{ss}-$spII_{cs}$-GFP]; Mariscal *et al.*, 2007), soluble in the cytoplasm of vegetative cells (E, P_{rbcL}-GFP; Yoon and Golden, 1998) and soluble in the cytoplasm of heterocysts (F, P_{patS}-GFP; Mariscal *et al.*, 2007).

www.ingramcontent.com/pod-product-compliance
Lightning Source LLC
LaVergne TN
LVHW071340210926
848486LV00003B/197
9781908230386